Mathematical Models in the Biosciences 1

Mathematical Models in the Biosciences 1

Michael Frame

Yale UNIVERSITY PRESS/NEW HAVEN & LONDON

Set in Adobe Garamond Pro and The Sans Pro type by Newgen North America

Printed in the United States of America,

Library of Congress Control Number: 2020940888
ISBN 978-0-300-22831-1 (paperback : alk.paper)

A catalogue record for this book is available from the British Library.

This paper meets the requirements of ANSI/NISO Z39.48-1992 (Permanence of Paper).

10 9 8 7 6 5 4 3 2 1

Contents

Preface

So far as we can tell, mathematics is universal. The substitution rule for integration works the same in physics, biology, chemistry, economics, and literature. See page 524 of *Gravity's Rainbow* [131] for the last example. This universality is a deep feature of mathematics, in some part responsible for Eugene Wigner's [174] writing on "The unreasonable effectiveness of mathematics in the natural sciences." It leads to obvious economies: one calculus course can, and often does, serve students who intend to major in fields other than math.

This book and its companion volume began as notes for two courses, second and third semester calculus–integration, infinite series, and vector calculus–for biosciences students. I was tricked into developing these courses by Steve Stearns from Yale's Department of Ecology and Evolutionary Biology. But when I began to look, the material fascinated me, and I loved the students. Clever, curious, serious, tolerant of my lame jokes, many already had thought carefully about empathy as a guide to the practice of medicine. Steve's one simple comment led to the book in your hands now, and to the companion volume.

But this project did have some rough patches. When I was a kid I wanted to know everything. By the time I started elementary school, my goal had focused to all of science. My father and I built an electric motor from a few nails, some wire, and a bit of copper pipe. He gave up a corner of his woodworking shop to become my lab. I made a thermocouple by twisting together copper and steel wire, and a galvanometer with a coil of wire, two magnetized needles, and a bit of cardboard and thread. I had a rock collection, chemistry set, telescope, microscope, electronics lab, optics lab, and Petri dishes for culturing soil bacteria, and did summer volunteer work in an archaeological dig along the Kanawha River where the St. Albans point arrowhead was discovered four blocks from our house. Every birthday and Christmas gift, everything I earned by mowing lawns, went to equip my lab, and to buy books. Great fun, until I turned twelve.

That year something happened. It should have been trivial, shouldn't have bothered me. It was a supplementary algebra problem to find the general term of a series. I figured out a solution, but it was clunky, mechanical. I didn't like it, but I couldn't do better. A few days later my teacher showed me an elegant, simple solution that broke my heart. I knew, *knew*, that I wasn't bright enough or imaginative enough to do what I wanted. A few really bad days followed. Then recalibration. I couldn't do everything, but I could do something. Eventually I picked the things that I understood best, math and physics, and there I stayed, until my late 50s. Then Steve Stearns made his remark about mathematical biology courses and I needed to learn biology. But biology is not like math; biology is complicated. Math is simple: a few basic ideas that you then rearrange in clever ways. Biology requires that you know so very many facts. These were my rough patches. (Despite help from many people, mentioned in the Acknowledgments, all errors are my responsibility.) So I learned as best I could, but my approach always was modeled on math. My infinitely patient med tech wife tried to adjust my efforts, and to some degree she succeeded. But likely I've not always found the elegant, simple approach. Nevertheless, far too late in life I've fallen in love with biology, too. This book is my attempt to explain how that happened.

That's part of the story of why I wrote this book. Some of the other reasons are cultural. In order to appreciate the remarkable universality of mathematics, you need to be familiar with a fair bit of mathematics and several fields of science; it is something you will appreciate near the end of your undergraduate days, but maybe not so much near their beginning. Consequently, this universality, highly valued by mathematics teachers, is not so obvious to our students.

Note that even children can understand an aspect of this universality: some of the ways mathematics is the language of nature. I recall evenings reading my high school geometry book. After really understanding a proof, looking out the window at the silhouettes of the trees, branches dark against the purpling sky, a few brilliant stars peeking between them, I knew the same geometry works all across the world and determines much of our understanding of the shape of trees, of snowflakes, of seashells. If other worlds are homes to creatures capable of contemplating their environments, they will know the same geometry. How is it possible to think of these things and not be amazed by our ability to understand–if only dimly– the structure of space and time, the workings of the things that fill them? To share this amazement with you is a reason I wrote the book you hold now.

Second, many people are better able to understand complicated concepts when they are presented in the context of a topic of natural interest. Improper integrals may seem a useless abstract technique, but are a bit clearer when viewed as a way to test the feasibility of a model of the growth rate of long-lived species. Abstraction, also much beloved by teachers, does not yet have intrinsic value for some students. By developing some ideas of mathematics in the context of biological systems, I hope to achieve two goals: to overcome this abstraction barrier at the introduction of these ideas, and to make them easier to remember because they are part of a world more obviously relevant to you.

Nevertheless, unadorned mathematics, especially geometry, is drop-dead beautiful. I've been thinking about geometry, seeing it in the wide world, exploring the twist and turn of shapes in my head, for over sixty years. Still, after all this time, watching the elegant dance of geometric figures behind my eyes takes my breath, stops my heart for a moment. Through the vagaries of situation, and the life, sickness, and death of beloved family, geometry has remained a constant. Not a friend–it cares not at all for me, and in fact, these brief anthromorphological thoughts are just annoying–but indifferent, aloof, inhabiting another world altogether. Glimpses of this world have given me a few moments of pleasure subtle beyond all common measure.

This brings me to the true purpose of writing this book. I want to help you find what you love to do. The real reason you're in school is to look around at people who are spending their lives doing what they love most in the world. Not all of us display this feeling openly; for some, you'll need to look closely. Then ask, "Can I see myself doing this for the rest of my life?" For most people you see, the answer will be "No," but if you're fortunate, for a few the answer will be "Maybe." I was very fortunate, I found early what I love to do. So now I want to help with your own search, in part by explicitly stating this goal, unnoticed by many. But mostly by posing a problem, asking a question.

> How does evolution discover and use the laws of mathematics?

Thinking about this problem could absorb a whole life, a quiet, deep, wonderful life. Wish I could start again and do this. Sadly not me, I'm far too old, but maybe you. This book is an attempt to give you some tools to help you decide if this is how you want to spend your life. For most, the answer will be "No thanks." But if even one or two answer "Yes," my oh my, what a life awaits you.

Ways to use this book

This book grew out of the first of two math for biosciences courses I developed at Yale. Roughly they cover the content of second and third semester calculus, with emphasis on concepts and examples for bioscience students. Along the way we develop some other mathematics useful in biosciences but not often seen in calculus classes. Many of these phenomena can be modeled, at least in broad strokes, with differential equations. Some of these, called pure-time equations, can be solved by integration. This provides a setting for learning some techniques of integration in Chapter 4.

Most interesting biological systems involve at least two variables interacting in nonlinear ways. Some examples are presented in Chapter 7. Before we try to analyze these systems, in Chapter 8 we study linear systems. For these the origin is the only fixed point, and the long-term behavior is determined by the type of this fixed point, which in turn is determined by the eigenvalues of the coefficient matrix. Linear algebra courses involve much more than the computation of eigenvalues, and because the eigenvalues of a 2×2 matrix are the solution of a quadratic equation, the stability of fixed points involves only a tiny, self-contained portion of a linear algebra course.

From this we move on to nonlinear differential equations in the plane. Exact solutions can be a challenge to find, but from the viewpoint of biological systems, this is a challenge we need not take too seriously. Often system parameters are estimated from experimental data, so effectively can be specified only in some range. Consequently, we abandon most attempts to find exact solutions and instead focus on geometric analyses of asymptotic properties of the solutions. As with linear systems, some forms of asymptotic behavior are determined by fixed points and their stability. For nonlinear systems this involves the computation of partial derivatives and eigenvalues of the derivative matrices. The computation of partial derivatives is straightforward. The next step is a bit more work. In addition to fixed points and trajectories connecting them, planar differential equations can exhibit two other asymptotic behaviors: closed trajectories (loops of some sort) and limit cycles. These are the topics of Chapter 9. That nothing else can occur is the content of the Poincaré-Bendixson theorem. In Sect. A.10 we sketch the main steps of a proof of the Poincaré-Bendixson theorem.

Differential equations with time-varying parameters lead to power series solutions, the topic of Chapter 10. (These problems also can be approached by the method of integrating factors, presented in Sect. A.1.) This is an opportunity to present a very quick trip through infinite series, one of the main topics of most standard calculus 2 courses. Although this topic involves some very pretty mathematics, because our main interest is in bioscience applications, we take a fast, direct route to understanding series solutions of differential equations.

Often population structures are modeled by probability distributions, what fraction of the population is in state 1, what fraction in state 2, and so on. This requires some background in probability, the subject of Chapter 11. In addition to combinatorial rules and distributions discrete and continuous, here we discuss Simpson's paradox and the causality calculus of Judea Pearl, and some examples of hypothesis tests and of Bayesian inference.

The most personal part of this project is Chapter 12, an account of my brother's experiences with clinical trials for chronic lymphocytic leukemia (CLL) at the James Cancer Center. Many people have similar stories for themselves or a family member. Similar in broad outline, but different in detail. And detail is where hope and heartbreak live. This story continues. Early in 2019, my brother's CLL evolved a way to ignore the effects of ibrutinib, which had controlled his CLL for several years. Several months later he began a clinical trial of ARQ 531, which has shown some promise with patients whose CLL has grown to tolerate ibrutinib. If you think you may become a physician, I hope you will read this chapter. It is my view of where and how empathy can fit into the life and work of a doctor.

I'll list the chapters I used for the first course, and suggest some alternates.

I began the course with Chapter 2. Likely this material is unfamiliar to most students and so is an effective view of the power of mathematical models. More common applications of calculus to biology are through differential equations, introduced in Chapters 3 and 4, the first to illustrate medical applications, the second to introduce techniques of integration, a staple of calculus 2, to solve some differential equations. A more standard course would continue with the topics of Chapter 5; I replaced this with Chapter 6, because length, area, and volume computations are not so central to medicine, but scaling relations are. Then on to Chapter 7 for differential equations in the plane, and then in Chapter 8 we cover the linear algebra, eigenvalues and eigenvectors, to solve systems of linear differential equations. Systems of nonlinear equations are the subject of Chapter 9, though I omitted Sect. 9.8. Then power series in Chapter 10 to solve non-autonomous equations. The course ended with a bit of probability (Sects. 11.1–11.4), the implications of Simpson's paradox a sobering end to the semester.

The course was fast paced, homework was essential, office hours were crowded. The level of biomedical examples attracted students who were not averse to extra work. Some substitutions are Chapter 5 for Chapter 6 and more of Chapter 11 or Chapter 9. Many other variations are possible, but all should be built around substantial biomedical examples.

Volume 2 contains these chapters.

Chapter 13 *Higher dimensions* Here we apply eigenvalue analysis to biological systems in 3 or more dimensions. Examples include virus dynamics and immune response dynamics. Also in this chapter we introduce chaos in differential equations and explore examples.

Chapter 14 *Stochastic models* Examples include age structured populations and Leslie matrices, and sensitivity analysis of population growth. The main theoretical result is the Perron-Frobenius theorem.

Chapter 15 *A tiny bit of genetics* Here we introduce the selection equation, Price's equation, the quasispecies equation, fitness landscapes, and bioinformatics.

Chapter 16 *Markov chains in biology* In this chapter we introduce matrix differential equations and apply them to ion channel dynamics, to the Clancy-Rudy model which identifies relations between gene mutations and cardiac arrhythmia, and to tumor suppressor genes.

Chapter 17 *Some vector calculus* Here we present the main topics of a vector calculus course: divergence, curl, and gradient; line and surface integrals, and the theorems of Gauss, Green, and Stokes. Applications include Bendixson's criterion for closed trajectories and the index of a vector field at a fixed point.

Chapter 18 *A glimpse of systems biology* This involves methods to study networks if biological agents. Transcription networks are our main example.

Chapter 19 *What's next?* We'll sample some evolutionary medicine, translational bioinformatics, and a topological method to analyze sequences of biological data.

Acknowledgments

Some books owe much to earlier books. In writing this text, I was informed and sometimes inspired by these: [3, 4, 5, 34, 50, 64, 72, 110, 111, 155, 176, 177]. If you are interested in, or become interested in, the conversation between biology and math, I encourage you to read all of them. Treasures await in each.

Many people helped me write this book. Garrett Odell, who taught my undergraduate topology course at RPI in 1972, introduced me to the power of mathematical biology, specifically, just how much of invagination in early gastrulation is due to mechanical forces on neighboring cells. This idea stayed in my head for forty-five years. I intended to thank Gary when I finished this book, a quiet voice from the end of my career to recall the start of his. But he died in May of 2018. I waited too long.

My colleague Ted Bick at Union College talked me into teaching a mathematical biology course in 1989. This was my first exposure to the breadth of the subject, immense possibilities then seen by me only as shadows. I intended to thank Ted when I finished this book, but he died in August 2016. You might think this loss of a dear friend would have made me contact Gary, but no. Sometimes I am just staggeringly blind.

So, if you feel you need to thank someone, don't wait. Be better than I've been. Expressed appreciation always improves the world.

The chapters on differential equations developed from a Union College course I taught, aided by colleagues Arnold Seiken and William Fairchild, enthusiastic and thoughtful members of the audience. Much of the material in Chapters 2 and 6 grew out of discussions, some of the most interesting I've ever had, with David Peak, my coauthor of [115]. Conversations with Yale colleagues Sandy Chang, John Hall, Miki Havlickova, Douglas Kankel, David Pollard, Richard Prum, William Segraves, Stephen Stearns, Andy Szymkowiak, Günter Wagner, and Robert Wyman were instructive and enjoyable. The Howard Hughes Medical Institute grant number 52006963 and the Yale College Dean's Office provided generous support. In addition, I've benefitted from the comments and suggestions of my students. Serious, interested students are a delight for any teacher, and I've been especially fortunate with the students I've gotten. In particular, thanks to Monique Arnold, Christopher Coyne, Mariana Do Carmo, Rafael Fernandez, Candice Gurbatri, Megan Jenkins, Misun Jung, Miriam Lauter, Regina Lief, Jonathan Marquez, Aala Mohamed, Susie Park, Miriam Rock, Ashley Schwarzer, Bijan Stephen, Paschalis Toskas, and Zaina Zayyad. And special thanks to Divyansh Agarwal, Aiyana Bobrownicki, Colleen Clancy, Noelle Driver, Liz Hagan, Fran Harris, Christina Stankey, and Taylor Thomas, who contributed substantial amounts of material–ideas, corrections and clarifications, some sections and many many exercises. This is a far better book because of their efforts.

My experiences on earlier projects with Yale University Press made me expect that working with my editor, Joseph Calamia, would be a pleasure. And it was. Thanks, Joe. When Joe left Yale University Press, I found the same high standards working with Erica Hanson, Jeffrey Schier and Elizabeth Sylvia.

Anonymous reviewers made suggestions that led to considerable improvements. These include the addition of Sect. 13.3 of volume 2 on Michelis-Menten enzyme kinetics, Chapter 17 of volume 2 on vector calculus, and Chapters B of this volume and B of volume 2 to replace Java programs, a moving target in these days of web vulnerabilities, with Mathematica code.

I've taught many premed students near the beginning of their careers. They anchored my grasp of the starting point of medical education. My appreciation of the goal of medical education comes from knowing thoughtful doctors who understood when it was necessary to take a big gamble and when it wasn't. In particular, I thank Dr. Steven Artz, who kept my father alive for many years, Dr. John Bubinak, who has kept my wife alive, Dr. John Byrd, who has kept my brother alive (more on this is recounted in Chapter 12), Dr. Daniel Geisser, who has kept me alive, and Dr. Richard Magliula for taking such good care of our large collection of formerly stray cats. All of these doctors have been very good at explaining disease and treatment, and when talking with my wife and me, all assumed we are intelligent and interested. With some other doctors we have had less satisfying interactions. If you become a physician, when you talk with your patients please remember they are not just ciphers.

The artist T. E. Breitenbach asked a wonderful question which led to Chapter 12. His insights always take our conversations in directions unexpected and enjoyable.

Amy Chang, the education director of the American Society for Microbiology, made me aware of John Jungck's interesting work on the place of bioinformatics in the undergraduate biology curriculum. Thanks, Amy.

Mike Donnally, next-door neighbor in my childhood and a dear friend all these years later, has been generous with his botanical knowledge. So has Mary Laine, also a dear friend for many years. That two such serious people become so animated when they talk about plants is a delight, a contagious delight. Evolution has found intricate, beautiful structures in the plant world, too. The effort to understand these constructs gives a different perspective on the dance of evolution that produced our DNA, our cells, our organs, our selves, and our diseases.

Curiosity is the most important force of the mind. The desire to know what's around the corner, what's over the horizon, was my faithful companion–or maybe I was its companion–through my long exploration of how biology and math dance together. But intellect alone is not enough to carry out a project that is so personally important. For that an emotional pole star is needed. My navigation is provided by my family: my brother Steve and his wife Kim, my sister Linda and her husband David, my nephew Scott and his wife Maureen, my wife Jean, and my late parents, Mary and Walter. Thoughts of them have helped me to explore the mechanism of empathy, to sculpt how I see its importance. For this, and for giving more understanding and affection than he deserved to a goofy little kid who "always had his nose buried in some durn book," I do not have words adequate for the thanks owed.

Special thanks to Jean Maatta, my med tech wife, for her limitless patience in explaining complicated, contingent biology to her geometer husband. More than any other experience, learning the biology needed for this book emphasized that

If biology always worked the same way, it would be a subset of math.

(This formulation occurred in a conversation with Dr. Magliula.) But it doesn't and it isn't. We are so far from an axiomatic basis for biology. It is complicated, many aspects interrelated, probably we still are asking questions that don't lead us in useful directions. This is unfamiliar ground for someone (me, for example) who has spent over half a century thinking very hard about geometry. I am grateful for my wife's patience and good humor in explaining again and again that my attempts to fit biology into mathematical categories were wrong-headed.

Still, biology and math can talk with one another. This book is a part of that conversation.

Chapter 1 Review

This book covers a selection of some topics included in any standard calculus 2 text, concrete instances of some methods of linear algebra, bits of the geometric theory of differential equations, and some probability, all presented in terms of biological and biomedical situations. In this chapter we sketch the calculus 1 material you need to know, and give a brief overview of topics we'll cover.

1.1 RULES FOR DIFFERENTIATION

Suppose f and g are differentiable functions and A and B are constants. Differentiation follows these rules:

Linearity rule: $(A \cdot f \pm B \cdot g)' = A \cdot f' \pm B \cdot g'$

Product rule: $(f \cdot g)' = f' \cdot g + f \cdot g'$

Quotient rule: $(f/g)' = (g \cdot f' - f \cdot g')/g^2$

Chain rule: $(f(g))' = f'(g) \cdot g'$

Here are a few familiar calculations

$(x^A)' = Ax^{A-1}$ $(\sin(x))' = \cos(x)$ $(\cos(x))' = -\sin(x)$

$(e^x)' = e^x$ $(A^x)' = \ln(A)A^x$ $(\ln(x))' = 1/x$ $(\log_A(x))' = 1/(x\ln(A))$

and examples of combinations of these

$$\tan(x)' = \left(\frac{\sin(x)}{\cos(x)}\right)' = \frac{\cos(x)\sin(x)' - \sin(x)\cos(x)'}{\cos^2(x)} = \frac{1}{\cos^2(x)} = \sec^2(x)$$

$$\left(e^{\sin(x)}\right)' = e^{\sin(x)}\sin(x)' = e^{\sin(x)}\cos(x)$$

$$\left(e^{\sin(x^3)}\right)' = e^{\sin(x^3)}\sin(x^3)' = e^{\sin(x^3)}\cos(x^3)3x^2$$

1.2 INTERPRETATIONS OF THE DERIVATIVE

Recall that the derivative of a (differentiable) function is the slope of the tangent line of the graph of the function. Also, the slope of the tangent line at the point $(x, f(x))$ is approximated by the slope of the secant line through the points $(x, f(x))$ and $(x+h, f(x+h))$, as shown in the left side of Fig. 1.1. This gives the familiar definition of the derivative

$$f'(x) = \lim_{h \to 0} \frac{f(x+h) - f(x)}{h}$$

the source of most of the results in Sect. 1.1.

The derivative has a dynamical interpretation, in addition to the geometric one just given. Exchange x for t and suppose $f(t)$ denotes position at time t. Then the definition of the derivative shows $f'(t)$ is the rate of change of the position, that is, the speed. Similarly, $f''(t)$ is the acceleration, the rate of change of the speed. The right side of Fig. 1.1 shows the graph of a cubic function (solid curve), its (quadratic) derivative (wide dashes), and its (linear) second

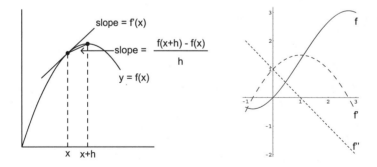

Figure 1.1. Left: the derivative as slope of the tangent line. Right: the derivative as rate of change.

derivative (narrow dashes). If the solid curve represents position, the wide dash curve represents speed and the narrow dash curve acceleration.

In class I asked, "If the first derivative is speed and the second derivative is acceleration, what's the third derivative?" Several students replied, "Jerk." Then from me, "And that's the only time you can say that to me without getting into trouble." General laughter. Comfortable people are more likely to ask questions, and laughter helps make people comfortable.

1.3 INTERMEDIATE AND MEAN VALUE THEOREMS

The intermediate value theorem and the mean value theorem are two basic tools from calculus 1. Because we'll use them later, we'll review them now.

On the left side of Fig. 1.2 we see an illustration of the intermediate value theorem: for any continuous function f, if $y = d$ lies between $y = f(a)$ and $y = f(b)$, then $d = f(c)$ for some $x = c$ between $x = a$ and $x = b$. The only circumstance for which such an $x = c$ would not exist is one where the graph of f has a jump around $y = d$, making f discontinuous.

We will use discontinuous functions, in Chapter 2 for example. Here we just mention that some of the clear properties of continuous functions need not be exhibited by discontinuous functions. Much of nature is discontinuous, so we must be careful with our models.

On the right side of Fig. 1.2 we see an illustration of the mean value theorem: for any continuously differentiable function f, for some $x = c$ between $x = a$ and $x = b$ the tangent line to the graph of $y = f(x)$ at $x = c$ is parallel to the secant line between $(a, f(a))$ and $(b, f(b))$. Recalling the interpretation of the

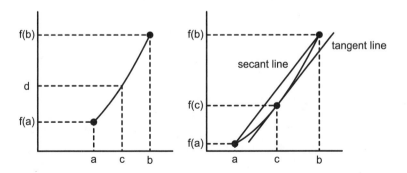

Figure 1.2. Illustrations of the intermediate value theorem (left) and mean value theorem (right).

derivative as the slope of the tangent line, we can restate the conclusion of the mean value theorem as

$$f'(c) = \frac{f(b) - f(a)}{b - a}$$

This is easiest to see if the secant line is horizontal. For example, if f is increasing at $x = a$, the tangent line has positive slope at $x = a$. In order to have $f(b) = f(a)$, somewhere between $x = a$ and $x = b$, f must be decreasing and so the tangent line must have negative slope. Somewhere between a tangent line of positive slope and a tangent line of negative slope, we must find a tangent line of 0 slope, a horizontal tangent. For non-horizontal secant lines, replace the function f by the difference of f and the secant line, that is, by

$$f(x) - \left(f(a) + (x - a)\frac{f(b) - f(a)}{b - a} \right)$$

In addition to its importance in familiar calculus, the mean value theorem is used to prove a result, Prop. 2.2.1, on the stability of fixed points.

1.4   THE FUNDAMENTAL THEOREM OF CALCULUS

The fundamental theorem of calculus (FTC) gives precise meaning to the notion that differentiation and integration are inverse operations. Suppose $F'(x) = f(x)$, then FTC has two forms

$$\int_a^b f(x)dx = F(b) - F(a) \quad \text{and} \quad \frac{d}{dx}\int_a^x f(t)dt = f(x)$$

The first form is the more commonly used. Combined with the interpretation of the definite integral as the area under the curve, the fundamental theorem of calculus states that the area under the graph of $y = f(x)$ between $x = a$ and $x = b$ is the difference $F(b) - F(a)$, where F is an antiderivative, or indefinite integral, of f. By clever geometric arguments, the ancient Greeks found the areas under a few curves. With calculus, so many areas can be found with relative ease.

1.5   DISCRETE AND CONTINUOUS MODELS

Much, but not all, of our work involves continuous models, change expressed as derivatives, dynamics as differential equations

$$\frac{dx}{dt} = f(x, t)$$

Of course, this is a fiction. The continuous distribution usually works well for quanta of electricity (electrons), but perhaps not always so well for quanta of populations (persons). Nevertheless, for large enough populations, these continuous models can give simple approaches to understanding some aspects of population dynamics.

In contrast are discrete (not discreet) models, change expressed as differences of successive generations, dynamics as iterative equations

$$x_{n+1} = f(x_n)$$

often formulated as difference equations

$$x_{n+1} - x_n = g(x_n)$$

This, too, is a simplification: it assumes non-overlapping generations, no explicit time dependence, no long-term memory effects (x_{n+1} might depend also on x_{n-1}, x_{n-2}, etc.). Nevertheless, for some systems, these models capture important aspects of the behavior.

Perhaps equally important, as pointed out so forcefully by Robert May [94], iterative equations with only slightly nonlinear functions f squash the mythology that complicated behavior must be the result of complicated models. So when we encounter complicated behavior in nature, some, perhaps much, may be explained by simple models.

May ended his paper with "an evangelical plea for the introduction of these difference equations into elementary mathematics courses, so that students' intuition may be enriched by seeing the wild things that simple nonlinear equations can do." Happily we can report that this plea, made over thirty years ago, has been answered by many, including me. Over twenty-five years ago, when I began including May's examples in freshman courses, students were amazed that a *parabola* could could produce such complicated behavior. Now the most common response is, "Oh, we saw that in high school." This is very good news, indeed.

In Chapter 2 we'll explore iterative equations through two examples, one involving population dynamics, the other cardiac dynamics. We return to discrete dynamics in Chapters 14, 16, and 19.

Chapter 2 Discrete dynamics

Most of our work will involve calculus, and although we'll learn new techniques and applications, the calculus approach to modeling already is familiar. This first chapter is different: it's about discrete models. We'll see examples of two biological processes: a model of populations with non-overlapping generations in an environment with limited resources, and some simple models of heartbeats.

We start with discrete models because they are less familiar. One of our goals is to give you some hint of the pleasure of finding connections between biology and math. If the math already is well known, the connections appear one-directional, but they aren't. Here at the start of the book, we want to establish that some math is a response to questions asked by biology. That's easier to see if the math hasn't been part of every scientist's tool box for centuries.

2.1 LOGISTIC MAP DYNAMICS

Among the simplest examples of iterative equations are models of populations with non-overlapping generations. Some insects are good examples. Suppose P_n denotes the population in generation n. The very simplest population growth model is

$$P_{n+1} = r \cdot P_n \tag{2.1}$$

where the *per capita growth rate* $r = 1 + b - d$, where b is the per capita birth rate and d the per capita death rate. The term 1 is included because if $b = d$ the population size should be constant. Sometimes this is called *Malthusian growth*, named after Thomas Malthus, who drew very gloomy conclusions from his observation that (human) populations grow geometrically ($P_{n+1} = r \cdot P_n$), while food supply grows arithmetically ($F_{n+1} = F_n + k$). Predicting the long-term behavior of populations governed by Eq. (2.1) is completely straightforward:

$$P_{n+1} \to \infty \quad \text{if } r > 1$$
$$P_{n+1} = P_n \quad \text{if } r = 1$$
$$P_{n+1} \to 0 \quad \text{if } r < 1$$

Nothing interesting here, though it is clear that the $r > 1$ result is not realistic because it does not account for population growth in an environment with (necessarily) limited resources. This can be achieved by scaling the per capita growth rate by a factor that decreases with an increase in the population size. For example

$$P_{n+1} = r \cdot (1 - P_n/K) \cdot P_n \tag{2.2}$$

where K is the maximum population supported by the environment. Instead of studying the raw population numbers P_n, we formulate the iterative equation in terms of $x_n = P_n/K$. Dividing both side of Eq. (2.2) by K, we obtain

$$x_{n+1} = r \cdot (1 - x_n) \cdot x_n \tag{2.3}$$

This function

$$L(x) = r \cdot x \cdot (1 - x)$$

is called the *logistic function*, so the iterative equation (2.3) can be written as

$$x_{n+1} = L(x_n)$$

often called the *logistic map*.

The graph of $y = L(x)$ is a parabola opening downward and passing through the points $(0,0)$ and $(1,0)$. By symmetry, the maximum of the parabola occurs at $x = 1/2$ and has value $L(1/2) = r/4$. In Exercise 2.1.6 (b), we'll see that if the logistic map is to be a plausible model of real populations, then the parameter r must lie in the range $[0,4]$.

The general problem we wish to approach is this: given an initial value x_0, can we predict anything about the sequence of iterates

$$x_1 = L(x_0), x_2 = L(x_1), x_3 = L(x_2), x_4 = L(x_3), \ldots$$

called the *orbit* of x_0, without compuing each of the x_i?

The first tool to approach this problem is a very simple and elegant way to generate the x_i geometrically. The method, called *graphical iteration* or *cobwebbing*, consists of these steps.

1. From the point $(x_0, 0)$ draw a vertical line to the graph of L, intersecting it at $(x_0, L(x_0)) = (x_0, x_1)$.
2. From the point (x_0, x_1) draw a horizontal line to the diagonal $y = x$, intersecting at (x_1, x_1).
3. From the point (x_1, x_1) draw a vertical line to the graph of L, intersecting it at $(x_1, L(x_1)) = (x_1, x_2)$.

Repeat steps 2 and 3, generating the points (x_2, x_2), (x_2, x_3), (x_3, x_3), (x_3, x_4), and so on, as long as one wishes. See Fig. 2.1.

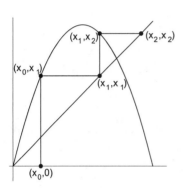

Figure 2.1. An illustration of graphical iteration.

Does this method itself reveal anything about predicting long-term behavior of the orbit? The answer depends on the value of r. In Fig. 2.2 we see the graphical iteration plots for $r = 0.9, 2.8, 3.2$ (top row, left to right), $3.5, 3.7$, and 3.83 (bottom row, left to right). How are we to understand these plots?

For $r = 0.9$ the iterates converge to the origin. The death rate exceeds the birth rate and the population marches down to extinction.

For $r = 2.8$ the iterates spiral to a constant value. The population settles down to an equilibrium, the birth rate balancing the mortality and competition rates.

For $r = 3.2$ the population cycles between a low and a high value. This is called a *2-cycle* because the population cycles between two values.

For $r = 3.5$ the population cycles between four values, a *4-cycle*.

For $r = 3.7$ we see no pattern at all, other than that the values of the iterates are constrained to lie in the range $[L^2(1/2), L(1/2)]$, where $L^2(x) = L(L(x))$. This could be a very long cycle, or it could be an instance of chaos. We'll investigate in a moment.

Just in case you think you see the pattern–as r increases the dynamics go through fixed point to 2-cycle to 4-cycle (looking more carefully still, after the 4-cycle there is an 8-cycle, then a 16-cycle, then a 32-cycle, and so on, each occupying a smaller range of r-values) to chaos–for $r = 3.83$ we find a 3-cycle. The interval of r-values that encompasses where chaos occurs contains many

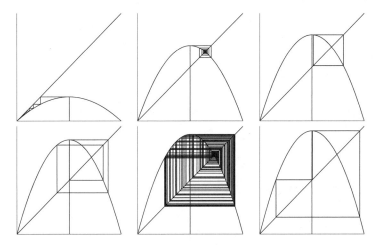

Figure 2.2. Graphical iteration for $r = 0.9, 2.8, 3.2, 3.5, 3.7$, and 3.83.

smaller intervals where periodic behavior is found. The order in which these periodic values arise is well-known. See [71], for example. The total length of the set of r-values on which chaos occurs is unknown, although M. Jakobson [68] proved this set has positive length. (Technically, positive 1-dimensional measure. This means that chaos occurs not just for a few isolated r-values, but for a substantial collection of r-values.) Beyond that, we have estimates, but few rigorous results.

How can there be questions about a parabola that no one can answer? Not just you and me, but brilliant mathematicians who have spent decades of their lives studying complex dynamics. This was the first compelling evidence [94] that we should not discount the ability of simple models to generate complicated behavior.

Despite this general lesson, some aspects of the logistic map can be understood using only tools of basic calculus, mostly just the mean value theorem and the chain rule. To illustrate this, in Sects. 2.2 and 2.3 we'll study a few properties of the logistic map.

Practice Problems

The practice problems use the *tent map*

$$T(x) = \begin{cases} r \cdot x & \text{for } x \leq 1/2 \\ r - r \cdot x & \text{for } x \geq 1/2 \end{cases} \tag{2.4}$$

In Exercise 2.1.6 (a), we'll see that to guarantee $T : [0,1] \to [0,1]$, we restrict the parameter r to the range $0 \leq r \leq 2$.

2.1.1. For the $r = 2$ tent map,

(a) compute the iterates $x_1, \ldots, x_{10}$ with $x_0 = 0.030$.

(b) Compute the iterates $y_1, \ldots, y_{10}$ with $y_0 = 0.031$.

(c) Compute $y_i - x_i$ for $i = 0, \ldots, 10$.

2.1.2. Here again we use the $r = 2$ tent map. For (a) and (b) give both graphical and algebraic solutions.

(a) Find two points that iterate to $x = 1/2$ in one step.

(b) Find four points that iterate to $x = 1/2$ in two steps.

(c) Show that 2^n points iterate to $x = 1/2$ in n steps.

Practice Problem Solutions

2.1.1. (a) The iterates are $0.030, 0.060, 0.120, 0.240, 0.480, 0.960, 0.080, 0.160, 0.320, 0.640$, and 0.720.

(b) The iterates are $0.031, 0.062, 0.124, 0.248, 0.496, 0.992, 0.016, 0.032, 0.064, 0.128$, and 0.256.

(c) The differences $y_i - x_i$ are $0.001, 0.002, 0.004, 0.008, 0.016, 0.032, -0.064, -0.128, -0.256, -0.512$, and -0.464.

2.1.2. To find points that iterate to $x = 1/2$, we'll use the left and right halves of T: $T_L(x) = 2x$ and $T_R(x) = 2 - 2x$.

(a) On the left side of Fig. 2.3 we see the graphical iteration approach to finding points that iterate to $x = 1/2$ in one step. First, drop a vertical line from $(1/2, 1)$ to $(1/2, 1/2)$. To get points that iterate to $(1/2, 1/2)$, draw horizontal lines to the left side and to the right side of the graph. Then draw vertical lines to the x-axis, finding points A and B. To find points that iterate to A and B, find the

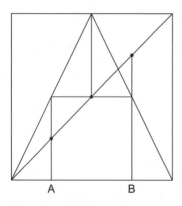
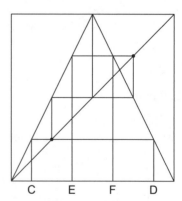

Figure 2.3. Graphical iteration plots for Practice Problem 2.1.2 (a) on the left, (b) on the right.

points where the vertical lines from A and from B intersect the diagonal line. Algebraically, $T_L(A) = 1/2$ gives $A = 1/4$ and $T_R(B) = 1/2$ gives $B = 3/4$.

(b) On the right side of Fig. 2.3 we see the graphical iteration approach to finding points that iterate to $x = A$ and to $x = B$ in one step, and consequently that iterate to $x = 1/2$ in two steps. Graphically we draw horizontal lines from (A, A) to the left and right sides of the graph of the tent map, and drop vertical lines to the x-axis, at the points C and D. Similarly, draw horizontal lines from (B, B) to the left and right sides of the graph of the tent map, and drop vertical lines to the x-axis, at the points E and F. Algebraically, $T_L(C) = 1/4$ gives $C = 1/8$, $T_R(D) = 1/4$ gives $D = 7/8$, $T_L(E) = 3/4$ gives $E = 3/8$, and $T_R(F) = 3/4$ gives $F = 5/8$.

(c) Each point a that iterates to $x = 1/2$ in $n-1$ steps gives rise to 2 points that iterate to $x = 1/2$ in n steps. Because 2 points iterate to $x = 1/2$ in 1 step, 2^n points iterate to $x = 1/2$ in n steps.

Except we must check that all of these 2^n points are distinct. Associated with each point is a sequence of n symbols, Ls and Rs, depending on whether at each point on the diagonal along the path from $1/2$ to x_i the path continues to the left or to the right in order to reach the graph of T. Call this sequence $\alpha(x_i)$. If $x_i = x_j$, then $T(x_i) = T(x_j)$, $T^2(x_i) = T^2(x_j)$, ..., $T^n(x_i) = T^n(x_j) = 1/2$, and so $\alpha(x_i) = \alpha(x_j)$. Consequently, different sequences correspond to different points and we're done.

If $r < 2$, some of these paths cannot be followed. We'll explore an example of this situation in Exercise 2.1.7.

Exercises

2.1.1. For the $r = 4$ logistic map,
(a) compute the iterates $x_1, \ldots, x_{10}$ with $x_0 = 0.300$.
(b) Compute the iterates $y_1, \ldots, y_{10}$ with $y_0 = 0.301$.
(c) Compute $y_i - x_i$ for $i = 0, \ldots, 10$.
For these calculations, carry four digits to the right of the decimal.

2.1.2. For the $r = 3.5$ logistic map,
(a) compute the iterates $x_1, \ldots, x_{10}$ with $x_0 = 0.300$.
(b) Compute the iterates $y_1, \ldots, y_{10}$ with $y_0 = 0.301$.
(c) Compute $y_i - x_i$ for $i = 0, \ldots, 10$.
(d) Compare the differences of (c) with those of Exercise 2.1.1 (c). Note any pattern you see in these differences.

2.1.3. For the $r = 4$ logistic map,
(a) compute the iterates $x_1, \ldots, x_{10}$ with $x_0 = 0.300$.

(b) Compute the iterates $y_1, \dots, y_{10}$ with $y_0 = 0.08167$.

(c) Compute $y_i - x_i$ for $i = 0, \dots, 10$.

(d) Compute $y_{i+1} - x_i$ for $i = 0, \dots, 9$.

(e) Interpret the results of (d).

2.1.4. For the $r = 2$ tent map,

(a) sketch the graphical iteration plot starting from $x_0 = 2/5$.

(b) Sketch the graphical iteration plot starting from $x_0 = 1/5$.

(c) Interpret what you see.

2.1.5. For the function

$$f(x) = \begin{cases} 9x & \text{for } 0 \le x \le 0.1 \\ 0.9 & \text{for } 0.1 \le x \le 0.6 \\ 2.25 - 2.25x & \text{for } 0.6 \le x \le 1 \end{cases}$$

(a) Plot the graph of f.

(b) Sketch the graphical iteration plot starting from $x_0 = 0.05$. Do enough steps to deduce the long-term behavior of the iterates of $x_0 = 0.05$.

(c) Sketch the graphical iteration plot starting from $x_0 = 0.9$. Do enough steps to deduce the long-term behavior of the iterates of $x_0 = 0.9$.

(d) What do you deduce? We'll come back to this function in Exercise 2.2.4.

2.1.6. (a) For all $r > 2$, show that almost all points $x_0 \in [0,1]$ iterate to $-\infty$ under the tent map (2.4). Hint: find the lengths of the intervals that iterate to $-\infty$. Show their lengths sum to 1.

(b) For $r > 4$ show that infinitely many points x_0 iterate to $-\infty$. As with the $r > 2$ tent map, almost all $x_0 \in [0,1]$ iterate to $-\infty$, but for the logistic map the proof is more subtle. We'll be content to show that infinitely many points iterate to $-\infty$.

2.1.7. Pictured on the right is a tent map with $r < 2$. In fact, for this picture $r = 1.5$.

(a) Find the coordinates of the points $A, B, \dots, G$. For instance, the coordinates of A are $(1/2, r/2)$.

(b) Find the range of r-values for which E lies to the right of G.

(c) Explain why in this range some of the paths of Practice Problem 2.1.2 cannot be constructed.

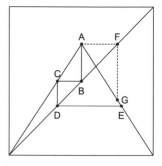

2.1.8. For a tent map with $1 < r \leq 2$, start from the top of the tent, $(1/2, r/2)$, and sketch this reverse graphical iteration path: go vertically to the diagonal, horizontally to the right branch of the tent, vertically to the diagonal, horizontally to the right branch of the tent. Continue. What do you see?

The last two problems involve using graphical iteration to plot compositions of functions. More details, and an application to finding the domain and range of a composition, can be found in [101, 102].

2.1.9. (a) Sketch the graphs of two functions, f and g, both with domain and range $[0,1]$. For any point x in $[0,1]$ on the x-axis, draw a vertical line. From the intersection of that line with the graph of g, draw a horizontal segment to the diagonal, then a vertical segment to the graph f, and finally a horizontal segment to the vertical line through x. The intersection is the point $(x, f(g(x)))$.
(b) Find an example to show that in general $f(g(x)) \neq g(f(x))$.

2.1.10. (a) Use the method of Exercise 2.1.9 to sketch the graphs of $T(T)$, which we'll denote T^2, and T^3, both for $r = 2$. Hint: for T^2, the maxima occur at those x that iterate to 1 under two applications of T, the minima occur at those x that iterate to 0 under two applications of T. Recall graphical iteration in reverse.
(b) How many maxima does the graph of T^n have?

2.2 FIXED POINTS AND THEIR STABILITY

Iterates of the logistic map are defined by $x_{n+1} = L(x_n)$. By a *fixed point* we mean a value x_n left unchanged by an application of L, so $x_{n+1} = x_n$. In the graphical iteration diagram Fig. 2.1, we can label the horizontal axis x_n and the vertical axis x_{n+1}, so the graph of the logistic function $y = L(x)$ becomes $x_{n+1} = L(x_n)$ and the diagonal line $y = x$ becomes $x_{n+1} = x_n$. Then fixed points can be identified geometrically as the intersections of the graph of the logistic function and the diagonal line. Algebraically, they are the solutions of the *fixed point equation*

$$x = L(x) \tag{2.5}$$

This gives

$$x = rx(1-x) \qquad \text{so} \qquad 0 = rx(1-x) - x = x(r-1-rx)$$

and we find two fixed points

$$x = 0 \qquad x = 1 - \frac{1}{r} = \frac{r-1}{r} \tag{2.6}$$

Note the second fixed point is in the interval $[0,1]$ only for $r \geq 1$.

The top left picture of Fig. 2.2 suggests that for *every* initial point $x_0 \in [0,1]$, graphical iteration starting from x_0 converges to 0. The middle top picture of Fig. 2.2 suggests that for many points x_0, graphical iteration converges to the non-zero fixed point. Is there a simple way to find the range of r-values for which iteration from most points converges to the one fixed point or the other?

A fixed point x_* of L is *stable* if for all x_0 close enough to x_*, $L^n(x_0) \to x_*$ as $n \to \infty$.

A fixed point x_* of L is *unstable* if every small open interval containing x_* also contains points x_0 that iterate outside of the interval. The modifier "small" is included to avoid uninteresting complications.

By an application of the mean value theorem, we can find a simple test for the stability of a fixed point.

Proposition 2.2.1. A fixed point x_* of L is stable if $|L'(x_*)| < 1$ and unstable if $|L'(x_*)| > 1$.

Proof. Suppose $|L'(x_*)| < 1$ and write $|L'(x_*)| = \delta$, so $0 \le \delta < 1$. Because L' is continuous, for x close enough to x_*, we see $|L'(x)| < 1$. For example, pick any ϵ, $\delta < \epsilon < 1$. Take $x = x_* - a$ to be the largest value of $x < x_*$ for which $|L'(x)| = \epsilon$, and take $x = x_* + b$ to be the smallest value of $x > x_*$ for which $|L'(x)| = \epsilon$. See Fig. 2.4.

Now let $c = \min\{a, b\}$ and observe that $|x - x_*| < c$ implies $|L'(x)| < \epsilon$. Suppose $|x_0 - x_*| < c$. Then

$$
\begin{aligned}
|x_1 - x_*| &= |L(x_0) - x_*| \\
&= |L(x_0) - L(x_*)| && \text{because } L(x_*) = x_* \\
&= |L'(x')||x_0 - x_*| && \text{by the mean value theorem} \\
&< \epsilon|x_0 - x_*| && \text{because } x' \text{ lies between } x_* \text{ and } x_0.
\end{aligned}
$$

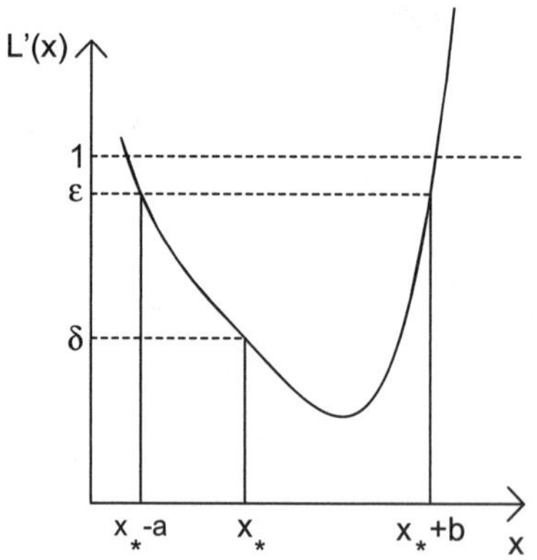

Next,

$$
\begin{aligned}
|x_2 - x_*| &= |L(x_1) - L(x_*)| \\
&= |L'(x'')||x_1 - x_*| \\
&< \epsilon|x_1 - x_*|
\end{aligned}
$$

The last inequality is because x'' lies between x_* and x_1. We see

$$|x_2 - x_*| < \epsilon|x_1 - x_*| < \epsilon^2|x_0 - x_*|$$

Continuing, we find

$$|x_n - x_*| < \epsilon^n|x_0 - x_*|$$

and so as $n \to \infty$, $x_n \to x_*$.

Figure 2.4. Bounding the derivative below 1.

The proof of the unstable part is Exercise 2.2.3. □

Note that no part of the proof of Prop. 2.2.1 depends on the specific form of $L(x)$. The argument is valid for any function $f(x)$ with $f'(x)$ continuous.

When $|f'(x_*)| = 1$, the derivative gives no information about the stability of the fixed point x_*. See Exercise 2.2.5.

Example 2.2.1. *Stability ranges for the logistic map fixed points.* For what ranges of r-values are the fixed points (2.6) of $L(x)$ stable? Note $L'(x) = r - 2rx$.

Because $L'(0) = r$, the fixed point $x = 0$ is stable for $|r| < 1$. Recalling r is non-negative, the fixed point $x = 0$ is stable for $0 \le r < 1$.

Because $L'((r-1)/r) = 2 - r$, the fixed point $x = (r-1)/r$ is stable for $|2 - r| < 1$, that is, for $1 < r < 3$. □

Fixed points also are called equilibria. Stable fixed points represent the simplest long-term behavior: the system evolves to the fixed value and stays there. In Sect. 2.3 we'll see the next-simplest behavior, cycling through a collection of values. Then in Sect. 2.4 we'll see that even the simple logistic map has behaviors far more complicated than fixed points and cycles.

Practice Problems

2.2.1. For the tent map $T(x)$,
(a) find the fixed points, and
(b) find the range of r-values for which each fixed point is stable.

2.2.2. For the function $f(x) = r \cdot x \cdot (1 - x)^2$,
(a) find the range of r-values for which $f([0,1]) \subseteq [0,1]$,
(b) find the fixed points of f, and
(c) find the range of r-values for which each of the fixed points is stable.

Practice Problem Solutions

2.2.1. (a) For all r, $x = 0$ is a fixed point of T. Note $T(1/2) = r/2$ is the highest point on the graph, and so for $r > 1$ the graph of T crosses the diagonal line at another point $x > 1/2$. That point is the solution of $r - r \cdot x = x$, that is, $x = r/(1 + r)$. Also, note that for $r = 1$, the graph of T coincides with the diagonal for $0 \le x \le 1/2$, so each of those points is a fixed point.
(b) The stability of fixed points is determined by the derivative of T

$$T'(x) = \begin{cases} r & \text{for } x < 1/2 \\ -r & \text{for } x > 1/2 \end{cases}$$

First, we see the fixed point $x = 0$ is stable for $0 \le r < 1$. On the other hand, the fixed point $x = r/(1+r)$ exists only for $r > 1$, and so is not stable for any r.

For $r = 1$ the fixed points $0 \le x \le 1/2$ are neither stable nor unstable, because near enough to any such x, points iterate neither closer to nor farther from x, but keep the same distance from x.

2.2.2. (a) The maximum value of f occurs at the x for which $f'(x) = 0$. Note $f'(x) = r \cdot (1 - 4x + 3x^2)$ and so the critical points of f are $x = 1/3$ and $x = 1$. Then $f(1/3) = 4r/27$, so the range of r-values guaranteeing $f : ([0,1]) \subseteq [0,1]$ is $0 \le r \le 27/4$.

(b) The fixed points of f are the solutions of $f(x) = x$, that is,

$$0 = f(x) - x = r \cdot x \cdot (1 - x)^2 - x = x \cdot ((r-1) - 2rx + rx^2)$$

The fixed points are $x = 0$, $x = 1 - 1/\sqrt{r}$, and $x = 1 + 1/\sqrt{r}$. The latter is not in the interval $[0,1]$, so the first two are the fixed points of f that are of interest to us.

(c) The fixed point $x = 0$ is stable for $|f'(0)| < 1$, so $|r| < 1$. The fixed point $x = 1 - 1/\sqrt{r}$ is stable for $|f'(1 - 1/\sqrt{r})| < 1$, so for $|3 - 2\sqrt{r}| < 1$. That is, $-1 < 3 - 2\sqrt{r} < 1$, so $-1 < \sqrt{r} < 2$ giving $0 \le r < 4$. Recall the fixed point $1 - 1/\sqrt{r}$ lies in $[0,1]$ only for $r \ge 1$, so this fixed point is stable for $1 \le r < 4$.

Exercises

2.2.1. For the $r = 4$ logistic map, show the non-zero fixed point $x_0 = 3/4$ is unstable. Nevertheless, show infinitely many points $x_1, x_2, x_3, \ldots$ in $[0,1]$ iterate exactly to x_0. You needn't find formulas for the x_i; an appropriate picture will suffice.

2.2.2. For the $r = 4$ logistic map, show that despite 0 being an unstable fixed point, there are points x_0, arbitrarily close to 0, for which $L^n(x_0) = 0$ for some $n > 1$. Hint: think of graphical iteration in reverse.

2.2.3. Show that if x_* is a fixed point of f with $|f'(x_*)| > 1$, then x_* is an unstable fixed point. A graphical argument suffices.

2.2.4. Recall the function of Exercise 2.1.5.
(a) Find the coordinates of the non-zero fixed point.
(b) Show this fixed point is unstable.
(c) Nevertheless, find infinitely many points that iterate to this fixed point.

2.2.5. In each of these cases, sketch a function f with fixed point x_* satisfying $|f'(x_*)| = 1$. Use graphical iteration to establish this behavior of nearby points:
(a) $f^n(x)$ goes away from x_* for $x < x_*$; $f^n(x)$ goes toward x_* for $x > x_*$,
(b) $f^n(x)$ goes toward x_* for $x < x_*$; $f^n(x)$ goes away from x_* for $x > x_*$,
(c) $f^n(x)$ goes toward x_* for $x < x_*$ and for $x > x_*$, and
(d) $f^n(x)$ goes away from x_* for $x < x_*$ and for $x > x_*$.

2.2.6. Newton's method for finding the roots of f (the x for which $f(x) = 0$) is this: start with a guess x_0 and replace it with

$$x_1 = x_0 - \frac{f(x_0)}{f'(x_0)} = N_f(x_0)$$

Then iterate

$$x_2 = N_f(x_1), x_3 = N_f(x_2), \ldots$$

The points x_0 that iterate to a root are the *basin of attraction* of that root. Use graphical iteration of N_f to find the basin of attraction for each root of $f(x) = x^2 - 1$.

2.2.7. Here we take $f(x) = r \sin(\pi x)$ with $0 \le x \le 1$ and $0 \le r \le 1$.
(a) Find the range of r-values for which the fixed point $x = 0$ is stable.
(b) Numerically approximate the range of r-values for which the non-zero fixed point is stable. Carry two digits to the right of the decimal.

The last three problems involve the iterates of the critical point(s) of the function and the stability of its fixed points. In the exercises of Sect. 2.3 we'll study the iterates of the critical points and the stability of cycles.

2.2.8. (a) For $0 < r < 1$ show the critical point $x = 1/2$ of the logistic map iterates to the stable fixed point $x = 0$.
(b) For $1 < r < 3$ show the critical point iterates to the non-zero fixed point. Appropriate drawings suffice for both (a) and (b).

2.2.9. (a) Draw a careful graph of $f(x) = x + 0.15 \sin(4\pi x)$ for $0 \le x \le 1$.
(b) Identify the fixed points and determine which are stable.
(c) Find where each of the critical points iterates.

2.2.10. For the $r = 3.4$ logistic map, show the critical point does not iterate to either fixed point. Note that neither fixed point is stable for $r = 3.4$. A sketch may guide your arguments.

2.3 CYCLES AND THEIR STABILITY

From the top right picture of Fig. 2.2 we see the logistic map has a 2-cycle, at least for some r. Can we find the 2-cycle by a geometric method similar to finding fixed points by locating the intersections of the graph and the diagonal? Can the stability of the 2-cycle be tested by conditions similar to those of Prop. 2.2.1? The answer to both is yes.

Suppose $x = a$ and $x = b$ form a 2-cycle for $L(x)$. That is,

$$L(a) = b \quad \text{and} \quad L(b) = a$$

Then both $x = a$ and $x = b$ are fixed points for $L^2 = L \circ L$:

$$L^2(a) = L(L(a)) = L(b) = a$$
$$L^2(b) = L(L(b)) = L(a) = b$$

So to find 2-cycle points, we find fixed points of L^2, that is, we locate the intersections of the graph of L^2 with the diagonal. We can sketch the graph of L^2 by the method introduced in Exercises 2.1.9 and 2.1.10. See the left side of Fig. 2.5.

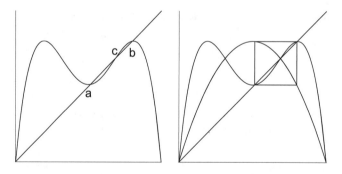

Figure 2.5. Locating a 2-cycle for the logistic map.

Here we see the graph of $L^2(x)$ and the graph of the diagonal; on the right side of Fig. 2.5 we see the graphs of both $L(x)$ and $L^2(x)$. Note the fixed points of $L(x)$, the origin and the point labeled $x = c$ on the left of the figure, also are fixed points of $L^2(x)$. This is no surprise:

$$\text{if} \quad L(x_*) = x_* \quad \text{then} \quad L^2(x_*) = L(L(x_*)) = L(x_*) = x_*$$

At least for the r-value of the logistic function plotted here, $L^2(x)$ has two additional fixed points, $x = a$ and $x = b$. These are not fixed points for $L(x)$, so they must belong to a 2-cycle for $L(x)$. This is illustrated by the graphical iteration of $L(x)$ shown on the right side of Fig. 2.5.

In general, to find points of an n-cycle for $L(x)$, locate the fixed points of $L^n(x)$ and remove from the list any that are fixed points of $L^m(x)$ for all $m < n$. We need to check only those m that divide n. Why? If a point is fixed under two applications of L, it is fixed under four, but not necessarily under three.

For example, the left side of Fig. 2.6 shows the graphs of $L(x)$ and $L^3(x)$ for $r = 3.84$. We see that the graph of $L^3(x)$ has 8 fixed points. Two of these are fixed points of $L(x)$. Because 2 does not divide 3, the 2-cycle points are not also 3-cycle points, so the remaining 6 fixed points of $L^3(x)$ must be 3-cycle points for $L(x)$. By iterating $L(x)$ we find that three of these 6 points belong to one 3-cycle for $L(x)$, the other three belong to another 3-cycle for $L(x)$. Why does graphical iteration detect one 3-cycle, but not the other? The answer lies with the stability of the cycles.

For clarity, we return to the 2-cycle. Graphical iteration shows that a 2-cycle for $L(x)$ is stable exactly when the corresponding fixed points for $L^2(x)$ are stable. Adapting Prop. 2.2.1 to $L^2(x)$, we see the 2-cycle stability is determined by $|(L^2)'(x)|$. The first question is this: for the 2-cycle points $x = a$ and $x = b$ for $L(x)$, can one be a stable fixed point for $L^2(x)$ and the other unstable? The chain rule shows us that the answer is no.

$$(L^2)'(a) = L'(L(a)) \cdot L'(a) = L'(b) \cdot L'(a) \qquad (2.7)$$
$$(L^2)'(b) = L'(L(b)) \cdot L'(b) = L'(a) \cdot L'(b)$$

So we see the fixed point $x = a$ is stable for $L^2(x)$ if and only if the fixed point $x = b$ is stable for $L^2(x)$.

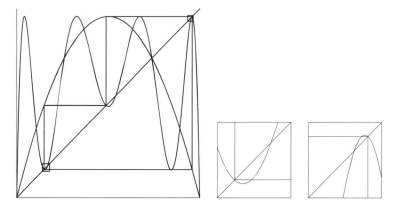

Figure 2.6. Locating a 3-cycle for the logistic map. The middle and right pictures are magnifications of the small boxes in the left picture.

To find the 2-cycle points of $L(x)$, solve

$$x = L^2(x) = L(L(x)) = L(rx(1-x)) = r(rx(1-x))(1-(rx(1-x)))$$
$$= r^2x - r^2(1+r)x^2 + 2r^3x^3 - r^3x^4$$

That is, find the roots of

$$-r^3x^4 + 2r^3x^3 - r^2(1+r)x^2 + (r^2-1)x \qquad (2.8)$$

This may look intimidating, but remember that the fixed points of $L(x)$ also are fixed points of $L^2(x)$. Also, if $x = x_*$ is a fixed point of $L^2(x)$, then $(x - x_*)$ is a factor of (2.8). The fixed points of $L(x)$ are $x = 0$ and $x = (r-1)/r$, so x and $(x - (r-1)/r)$ are factors of (2.8). Dividing these out, we find (2.8) can be factored as

$$x(x - (r-1)/r)(r^3x^2 - r^2(r+1)x + r(r+1)) \qquad (2.9)$$

The 2-cycle points for $L(x)$ are the roots of the quadratic factor of (2.9):

$$a = \frac{r+1-\sqrt{r^2-2r-3}}{2r} \quad \text{and} \quad b = \frac{r+1+\sqrt{r^2-2r-3}}{2r}$$

Applying the 2-cycle stability condition Eq. (2.7), recall $L'(x) = r - 2rx$ and we find that $L'(a)L'(b)$

$$= \left(r - \left(r+1-\sqrt{r^2-2r-3}\right)\right)\left(r + \left(r+1-\sqrt{r^2-2r-3}\right)\right)$$
$$= -r^2 + 2r + 4$$

The stability condition becomes

$$|-r^2 + 2r + 4| < 1$$

To find the range of r-values making this true, write $f(r) = -r^2 + 2r + 4$ and find all r satisfying both $f(r) \leq 1$ and $f(r) \geq -1$. In Fig. 2.7 we plot the graph

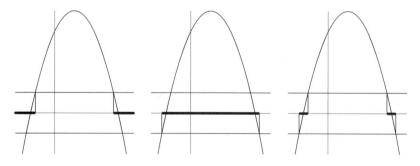

Figure 2.7. Determining the stability range of the logistic map 2-cycle.

of $y = f(r)$, together with the lines $y = 1$ and $y = -1$. On the left we show that $f(r) < 1$ for $r < -1$ and for $r > 3$. In the middle we show that $f(r) > -1$ for $1 - \sqrt{6} < r < 1 + \sqrt{6}$. On the right we see that $-1 < f(r) < 1$ for $1 - \sqrt{6} < r < -1$ and $3 < r < 1 + \sqrt{6}$. For population dynamics models we take $r \geq 0$, so the logistic map 2-cycle is stable for $3 < r < 1 + \sqrt{6}$.

Similar, though more intricate, calculations can be carried out for other cycles, including the 3-cycle pictured in Fig. 2.6.

Our final comment about cycles, continuing the idea of Exercises 2.2.8, 2.2.9, and 2.2.10, is a theorem of Fatou:

> For every stable fixed point and stable cycle of a differentiable function $f(x)$, a critical point of $f(x)$ must iterate to that point or cycle.

The logistic map $L(x)$ has only one critical point, $x = 1/2$, so for each value of r the logistic map has at most one stable fixed point or cycle. An immediate consequence is that in Fig. 2.6, only one of the two 3-cycles is stable. For some values of r, no cycle is stable. What can happen then? Is this just a mathematical curiosity, or can we observe non-repeating behavior in biological systems? We'll see.

Practice Problems

2.3.1. For the tent map T with $r = 1.75$,
(a) find the coordinates of the 2-cycle points for T.
(b) Find the coordinates of the points of the two 3-cycles for T.

2.3.2. For the function $f(x) = r\sin(\pi x)$, $0 \leq x \leq 1$, find the range of r-values for which the 2-cycle is stable. Do this numerically, correct to three digits to the right of the decimal.

Practice Problem Solutions

2.3.1. (a) From the left side of Fig. 2.8 we see that one point, a, of the 2-cycle lies in $(0, 1/2)$ and the other point, b, lies in $(1/2, 1)$. Then the point a satisfies

$$a = T(T(a)) = r - r \cdot (r \cdot a)$$

giving $a = r/(1 + r^2)$, and $b = T(a) = r \cdot a = r^2/(1 + r^2)$. (b) From the middle of Fig. 2.8 we see that two points, c and d, of the 3-cycle lie in $(0, 1/2)$ and the other point, e, lies in $(1/2, 1)$. Then the point c satisfies

$$c = T(T(T(c))) = r - r \cdot (r \cdot r \cdot c)$$

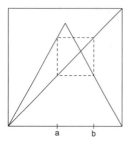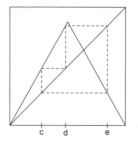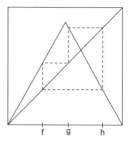

$$a \qquad b \qquad\qquad c \quad d \qquad e \qquad\qquad f \quad g \qquad h$$

Figure 2.8. The graphs for Practice Problem 2.4.2.

giving $c = r/(1+r^3)$. Then $d = T(c) = r \cdot c = r^2/(1+r^3)$ and $e = T(d) = r \cdot d = r^3/(1+r^3)$.

From the right side of Fig. 2.8 we see that one point, f, of the 3-cycle lies in $(0, 1/2)$ and the other points, g and h, lie in $(1/2, 1)$. Then the point f satisfies

$$f = T(T(T(f))) = r - r \cdot (r - r \cdot (r \cdot f))$$

giving $f = r/(1+r+r^2)$. Then $g = T(f) = r \cdot f = r^2/(1+r+r^2)$ and $h = T(g) = r - r \cdot g = (r+r^2)/(1+r+r^2)$.

2.3.2. A 2-cycle for f is two points, $x_1 \neq x_2$ with $f(x_1) = x_2$ and $f(x_2) = x_1$. We find these as fixed points of $f^2(x) = f(f(x))$. We know that fixed points of f also are solutions of $f^2(x) = x$, so we must take care to not use fixed points of f that are detected as solutions of $f^2(x) = x$. This is easy: if x is a solution of $f^2(x) = x$, then just check that $f(x) \neq x$. To test the stability of this 2-cycle, we adapt Eq. (2.7):

$$|f'(x_2) \cdot f'(x_1)| < 1$$

That is,

$$\rho(r) = |r\pi \cos(\pi x_2) r\pi \cos(\pi x_1)| < 1$$

We've given the name $\rho(r)$ to the product of the derivatives so we can refer to it without writing the whole thing again. Here's what we find, using the Mathematica code in Sect. B.1.

r	x_1	x_2	$\rho(r)$
0.7199	0.645746	0.645746	0.999476
0.7200	0.642055	0.649482	0.999349
0.8332	0.444848	0.820725	0.99878
0.8333	0.444766	0.820786	1.00062

We see that at $r = 0.7199$ the 2-cycle equation $f^2(x) = x$ has detected a fixed point, stable because

$$|f'(x_1)| = \sqrt{\rho(0.7199)} = 0.999738 < 1$$

We need the square root because

$$\rho(0.7199) = |f'(x_2)f'(x_1)| = |f'(x_1)|^2$$

At $r = 0.7200$ we have a stable 2-cycle because $x_1 \neq x_2$ and $\rho(0.7200) < 1$. At $r = 0.8332$ we have a stable 2-cycle because $\rho(0.8332) < 1$; at $r = 0.8333$ we have an unstable 2-cycle because $\rho(0.8333) > 1$. So the function f has a stable 2-cycle for $0.720 \leq r \leq 0.833$.

Exercises

2.3.1. How many 4-cycles does the $r = 2$ tent map exhibit?

2.3.2. How many 5-cycles does the $r = 2$ tent map exhibit?

2.3.3. How many 6-cycles does the $r = 2$ tent map exhibit?

2.3.4. How many 4-, 5-, and 6-cycles does the $r = 4$ logistic map exhibit?

Exercises 2.3.5, 2.3.6, and 2.3.7 involve the function $f(x) = rx(1 - x^2)$ for $0 \leq r \leq 3\sqrt{3}/2$.

2.3.5. (a) Find the coordinates of the fixed points of f.
(b) Show f has a non-zero fixed point for $r > 1$.
(c) Find the range of r-values for which this fixed point is stable.

2.3.6. Find the range of r-values for which f has a stable 2-cycle. Do this numerically, correct to four digits to the right of the decimal.

2.3.7. Find a range of r-values for which f has a stable 4-cycle. Do this numerically, correct to five digits to the right of the decimal.

Exercises 2.3.8 and 2.3.9 involve the function $f(x) = rx^2(1 - x)$ for $0 \leq r \leq 27/4$.

2.3.8. (a) Find the coordinates of the three possible fixed points of f.
(b) Show f has only one fixed point for $r < 4$, exactly two fixed points for $r = 4$,

and three fixed points for $r > 4$.

(c) Show the fixed point at $x = 0$ is stable for all r, $0 \leq r \leq 27/4$.

(d) Show the middle fixed point is unstable for all $r > 4$. Argue graphically, not algebraically.

(e) Find the range of r-values for which the right fixed point is stable. Do this numerically, correct to five digits to the right of the decimal.

(f) In the r-range of (e), the fixed point at $x = 0$ and the right fixed point both are stable. Why doesn't Fatou's theorem forbid this?

2.3.9. (a) Find the range of r-values for which f has a stable 2-cycle. Do this numerically, correct to five digits to the right of the decimal.

(b) Find the range of r-values for which f has a stable 4-cycle. Do this numerically, correct to five digits to the right of the decimal.

2.3.10. Prove that for any prime number p, p divides $2^p - 2$. Hint: think of the number of p-cycles of the $r = 2$ tent map.

2.4 CHAOS

This section is more complicated than the others of this chapter. We won't use the ideas presented here again until Chapter 13.

The basic notions of chaos have been discovered and rediscovered several times over the last hundred years or so. Late in the 19th century, Henri Poincaré [126] developed many of these ideas in his study of the stability of the solar system. Early in the 20th century, Jacques Hadamard [59] and George Birkhoff [15] found chaos in the motion of particles on negatively curved (saddle-shaped) surfaces. During World War II Lucy Cartwright and John Littlewood [24] discovered chaos in mathematical models of radar circuits. Building on the work of Barry Saltzman [138], in 1963 Edward Lorenz [87] showed long-term weather prediction is impossible by demonstrating that even simple models of atmospheric convection are chaotic. The first general appreciation of chaos came with Robert May's 1976 *Nature* paper [94] about dynamics of resource-limited populations. After that came excellent texts by Robert Devaney [31]; Heinz-Otto Peitgen, Hartmut Jürgens, and Dietmar Saupe [121]; and many others; as well as James Gleick's popular book *Chaos* [51]; then Michael Crichton's *Jurassic Park* [30]; the "Time and punishment" episode of a *Simpsons* Halloween show [151]; and on and on. This is an immense, active field with many applications. Here we sample some basics.

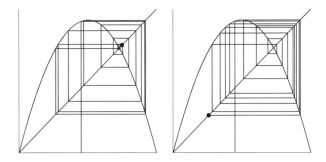

Figure 2.9. Sixteen iterates of the $r = 3.7$ logistic map, with $x_0^l = 0.450$ (left) and $x_0^r = 0.451$ (right).

Many physical and biological systems have relatively stable dynamics. Bump a grandfather clock and the motion of the pendulum stays more or less the same. Homeostasis is the mechanism by which many organisms adjust to minor perturbations in their environments. That small changes have small effects is a common observation. In contrast, chaotic systems exhibit *sensitivity to initial conditions*: under iteration of the dynamical process governing the system, very small changes in the initial conditions can produce significantly different behaviors. For example, Fig. 2.9 shows 16 iterates of the $r = 3.7$ logistic map (the model May studied in [94]), starting from $x_0^l = 0.450$ (the l superscript identifies points on the left graph) and $x_0^r = 0.451$ (r for the right graph; don't confuse this superscript label with the logistic map parameter r). The location of the 16th iterate is indicated by the dot on the diagonal line. Note how widely separated these have become after even this modest number of iterations.

In Fig. 2.10 we see another representation of this effect: for each i we plot $x_i^l - x_i^r$. Note sometimes the difference is positive, sometimes negative, sometimes large, sometimes small. These types of differences are mixed together without obvious pattern.

Figure 2.10. Differences of the $x_0^l = 0.450$ and $x_0^r = 0.451$ logistic map iterates for $r = 3.7$.

Why do we care? If no physical or biological system were chaotic, this would be a curiosity for mathematicians, but would have no impact beyond keeping some of my tribe from getting into mischief. However, for years now evidence has been building that some populations [86, 94, 95], cardiac [46, 47] and neurological [2] systems, and even some diseases [142] behave chaotically.

When in the 1980s chaos entered general scientific discourse, many people were hypnotized. Here was a path that can find simple explanations for some complicated phenomena. Who wouldn't be interested? Thousands of papers were published on chaos, spread among pretty much every field. But many problems are more complex: all systems have noise on top of whatever deterministic dynamics are present, and some systems are genuinely high-dimensional, that is, they cannot be modeled by anything as simple as a logistic map or its relatives. While the original enthusiasm about chaos has led to many interesting insights, in some cases early claims about chaos were premature. In the case of heartbeats, see [8, 49, 141, 169, 182] for a glimpse of how complicated the issues are. An appropriate balance of low-dimensional chaos and high-dimensional complexity is not always easy to find.

For now, we'll learn a bit about how chaos works, how it looks. Sensitivity to initial conditions is relevant in the world because we cannot know the initial conditions of any physical systems exactly. Often we cannot tell if, when we start measuring, the population is at 0.450 of its maximum value or at 0.451. If this is a population of agricultural pests and we are trying to schedule application of pesticides, comparing the left and right sides of Fig. 2.9 points out a problem.

On pages 68–69 of [127], Poincaré gives a clear formulation of sensitivity to initial conditions in a physical system, specifically, weather.

> We see that great disturbances are generally produced in regions where the atmosphere is in unstable equilibrium. The meteorologists see very well that the equilibrium is unstable, that a cyclone will be formed somewhere, but exactly where they are not in a position to say; a tenth of a degree more or less at any given point, and the cyclone will burst here and not there, and extend its ravages over districts it would otherwise have spared. If they had been aware of this tenth of a degree, they could have known it beforehand, but the observations were neither sufficiently comprehensive nor sufficiently precise, and that is the reason why it all seems due to the intervention of chance.

In some popular literature (and films), sensitivity to initial conditions is called the *butterfly effect*, because often the effect is misstated this way:

The flap of a butterfly wing in Mexico causes a tornado in Texas

though the locations and type of storm vary with the source. In the late 1980s, my colleague David Peak gave a clear formulation of his–and it turns out,

Poincaré's–objection. The tiny energy of the flap of a butterfly's wings cannot organize into the vast energy of a tornado. A tornado always was going to occur. The flap of the butterfly's wings introduces a small observational uncertainty which grows to swamp our ability to predict where the tornado will occur. The butterfly effect isn't about causing storms; it's about destroying our ability to predict storms.

Thus the bad news about chaos is that it limits our ability to predict. By running many simulations from different, but nearby, initial conditions, we can generate a probability distribution of outcomes. As far as our ability to predict events, in chaotic systems we see that, contrary to Einstein's dictum, god does play dice with the universe, though Einstein's comment was about quantum mechanics, not about chaotic dynamics.

But the news chaos brings is not all bad. We'll see some of the good consequences in Sect. 13.9. In the meantime, if you're impatient there's always Google. Try "controlling chaos."

For now we'll sketch two tools, Liapunov exponents and symbolic dynamics, that help us detect chaotic behavior in ways more reliable than simple visual inspection.

The Liapunov exponent quantifies the degree of chaos, expressed by sensitivity to initial conditions. In addition, the Liapunov exponent can be used to estimate the time horizon over which some prediction is possible. The idea is simple. Sensitivity to initial conditions means that nearby points have trajectories that diverge, at least for a while. In any biological system, the range of values is limited, so the differences between two trajectories cannot exceed this range; eventually trajectories must become close, only to diverge again. So we should measure the average divergence of nearby trajectories. The instantaneous divergence of infinitesimally close trajectories is measured by the magnitude of the derivative, so the *Liapunov exponent* λ is defined by

$$\lambda = \lim_{N \to \infty} \frac{1}{N} \sum_{i=1}^{N} \ln\left(\left| \frac{df}{dx} \right|_{f^i(x_0)} \right) \tag{2.10}$$

Because we expect divergence to be exponential, we average the log of the magnitude of the derivative along the trajectory. Often this calculation can be approached only by numerical approximation. We'll see some example calculations in a moment, but right away we can make two observations.

- At every point of a stable cycle, $|df/dx| < 1$, so the Liapunov exponent is negative.
- If on average nearby points diverge, then often $|df/dx| > 1$ and the Liapunov exponent is positive.

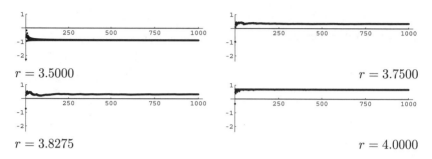

$r = 3.5000$ $r = 3.7500$

$r = 3.8275$ $r = 4.0000$

Figure 2.11. Estimating the logistic map Liapunov exponent.

So, roughly, a negative Liapunov exponent indicates a stable cycle, a positive Liapunov exponent indicates chaos. In case you wondered how we can distinguish between chaos and a stable million-cycle, which won't show a repeating pattern over a time we're likely to sample, the sign of the Liapunov exponent tells them apart.

In Fig. 2.11 we see graphs of approximations of λ, with $N = 1, \ldots, 1000$, for the logistic map with the r-value below the corresponding graph. That the graphs appear to fall along a horizontal line suggests that these numerical sequences have converged to the Liapunov exponents of these logistic maps. A bit more work is needed to establish the result conclusively, but for us, these graphs suffice. Mathematica code to approximate the Liapunov exponent by Eq. (2.10) is given in Sect. B.2.

As a check, for the $r = 3.5$ logistic map, the iterates of almost every initial point converge to a (stable) 4-cycle, so we expect that the Liapunov exponent is negative. Indeed, this is shown in the first graph of Fig. 2.11.

The magnitude of a positive Liapunov exponent indicates how rapidly iterates of nearby initial points diverge and, consequently, the time horizon of prediction of future behavior. Small positive λ signal slow growth in the divergence of nearby iterates, and consequently we can predict further into the future with confidence. Large λ mean that initially nearby iterates diverge rapidly so the reliability of a prediction based on a given set of iterates evaporates quickly.

In a few instances, we can compute the Liapunov exponent exactly. The tent map T of Eq. (2.4) is differentiable except at $x = 1/2$ and at all other points $|T'(x)| = r$. Then provided the iterates of x_0 avoid $x = 1/2$, the formula (2.10) gives

$$\lambda = \lim_{N \to \infty} \frac{1}{N} \sum_{i=1}^{N} \ln(r) = \ln(r)$$

For $r > 1$ we see $\lambda > 0$, consistent with the observation that $r > 1$ tent maps exhibit chaotic dynamics. For $r < 1$ we see $\lambda < 0$, consistent with the observation that for $r < 1$ tent maps the iterates converge to the (stable) fixed point $x = 0$.

In Sect. A.19 we show a way to adapt the Liapunov exponent calculation to trajectories of differential equations.

Another approach is called *symbolic dynamics*, which we'll think of as following the iterates when we've misplaced our glasses. An example should clarify the idea. We'll use Baker's map

$$B(x) = \begin{cases} 2x & \text{for } 0 \le x < 1/2 \\ 2x - 1 & \text{for } 1/2 \le x < 1 \end{cases}$$

The factors of 2 in the definition of B suggest that the dynamics of B might be more easily understood if we express x in its base-2 expansion. Recall that every number $x \in [0,1)$ has a base-2 expansion

$$x = \frac{x_1}{2} + \frac{x_2}{2^2} + \frac{x_3}{2^3} + \frac{x_4}{2^4} + \cdots \tag{2.11}$$

where each x_i is 0 or 1. These expansions are unique, except those that have a terminal infinite string of 1s have another expansion with a terminal infinite string of 0s. For example,

$$\frac{1}{2} = \frac{1}{2} + \frac{0}{2^2} + \frac{0}{2^3} + \frac{0}{2^4} + \cdots = \frac{0}{2} + \frac{1}{2^2} + \frac{1}{2^3} + \frac{1}{2^4} + \cdots$$

For the last equality we have used the sum of the geometric series: if $|r| < 1$, then $1 + r + r^2 + \cdots = 1/(1-r)$. This should be familiar from high school, but if you need a review, see the beginning of Chapter 10. For all x with two expansions, we'll always take the expansion with a terminal infinite string of 0s. This is why we take $[0,1)$ and not $[0,1]$ for the domain of B.

In Exercise 2.4.1 we see that

$$B(x) = \frac{x_2}{2} + \frac{x_3}{2^2} + \frac{x_4}{2^3} + \frac{x_5}{2^4} + \cdots \tag{2.12}$$

Keeping in mind the positions of all the denominators, we can use a shorthand notation: the *symbol sequence* of x is

$$x = (x_1, x_2, x_3, x_4, \ldots) \tag{2.13}$$

Then the effect of B on the symbol sequence of x is

$$B(x) = (x_2, x_3, x_4, x_5, \ldots) = \sigma(x_1, x_2, x_3, x_4, \ldots)$$

where σ, called the *shift map*, is a function on the space of all infinite base-2 sequences. This is called *sequence space*. The distance between two sequences is

$$d(x,y) = \frac{|x_1 - y_1|}{2} + \frac{|x_2 - y_2|}{2^2} + \frac{|x_3 - y_3|}{2^3} + \cdots \tag{2.14}$$

We have introduced sequence space and the shift map in order to illustrate a way to prove that a function is chaotic. First we'll give a definition of chaos, Devaney's definition 8.5 of [31]. A function $f : X \to X$ is *chaotic* if it satisfies three properties:

1. sensitivity to initial conditions,
2. density of periodic points, and
3. topological transitivity.

Now we'll give careful definitions of these three. By a *neighborhood* $N_\epsilon(x)$ of radius $\epsilon > 0$ of a point x, we mean all points y for which $d(x,y) < \epsilon$. Small ϵ means y is really near x. If the particular ϵ isn't important, we'll just say a neighborhood N of x.

1. A function f exhibits *sensitivity to initial conditions* if there is a constant $\delta > 0$ that is a lower bound on eventual divergence of iterates of some nearby points. That is, for every x, every neighborhood of x contains a point y with $d(f^n(x), f^n(y)) \geq \delta$ for some n.

   In order to prove something we must have precise definitions. This definition shows the work needed to mathematicize the simple notion "nearby points eventually iterate far away, for a while."
2. The periodic points of a function f are *dense* if every neighborhood of every point contains a periodic point.
3. A function f is *topologically transitive* if for every pair of neighborhoods M and N there is a point x in M with $f^n(x)$ in N for some n.

We already know an intuitive interpretation of sensitivity to initial conditions. The density of periodic points means that (unstable) cycles are just about everywhere, as close as you like to every point. Topological transitivity means that points in every neighborhood visit every other neighborhood; the space cannot be split into pieces that never communicate with each other.

For a concrete example, we'll show that the shift map is chaotic.

Proposition 2.4.1. The shift map exhibits sensitivity to initial conditions with $\delta = 1/2$.

Proof. Take any point $x = (x_1, x_2, x_3, \dots)$ and any neighborhood $N_\epsilon(x)$. No matter how small ϵ is, there is an n large enough that $1/2^n < \epsilon$. Then take $y = (y_1, y_2, y_3, \dots)$ with $y_1 = x_1$, $y_2 = x_2, \dots, y_n = x_n$. For $i > n$, y_i might or might not agree with x_i, but because every x_i and every y_i is 0 or 1, we see that $|x_i - y_i| \leq 1$. From the definition (2.14) of distance on sequence space, we see that

$$d(x,y) = \frac{|x_{n+1} - y_{n+1}|}{2^{n+1}} + \frac{|x_{n+2} - y_{n+2}|}{2^{n+2}} + \frac{|x_{n+3} - y_{n+3}|}{2^{n+3}} + \cdots$$

$$\leq \frac{1}{2^{n+1}} + \frac{1}{2^{n+2}} + \frac{1}{2^{n+3}} + \cdots = \frac{1}{2^{n+1}} \cdot \left(1 + \frac{1}{2} + \frac{1}{2^2} + \cdots\right)$$

$$= \frac{1}{2^{n+1}} \cdot \left(\frac{1}{1 - 1/2}\right) = \frac{1}{2^{n+1}} \cdot 2 < \epsilon$$

For the penultimate equality we have summed the geometric series. So we have shown y belongs to $N_\epsilon(x)$.

If $x \neq y$, there is some $k > n$ with $x_k \neq y_k$. Then

$$d(\sigma^k(x), \sigma^k(y)) = d((x_k, x_{k+1}, x_{k+2}, \ldots), (y_k, y_{k+1}, y_{k+2}, \ldots))$$

$$= \frac{|x_k - y_k|}{2} + \frac{|x_{k+1} - y_{k+1}|}{2^2} + \frac{|x_{k+2} - y_{k+2}|}{2^3} + \cdots$$

$$\geq \frac{|x_k - y_k|}{2} = \frac{1}{2}$$

That is, the shift map is sensitive to initial conditions with $\delta = 1/2$. $\square$

Proposition 2.4.2. The periodic points of the shift map are dense.

Proof. To show that periodic points are dense, given any point x and any $\epsilon > 0$ take n large enough that $1/2^n < \epsilon$. Now define the point y by repeating the first n terms of the symbol sequence for x

$$y = (x_1, x_2, \ldots, x_n, x_1, x_2, \ldots, x_n, x_1, x_2, \ldots, x_n, \ldots)$$

Certainly, y is periodic: $\sigma^n(y) = y$. To complete the proof, we need to show that $d(x,y) < \epsilon$.

$$d(x,y) = \frac{|x_{n+1} - y_{n+1}|}{2^{n+1}} + \frac{|x_{n+2} - y_{n+2}|}{2^{n+2}} + \frac{|x_{n+3} - y_{n+3}|}{2^{n+3}} + \cdots$$

$$\leq \frac{1}{2^{n+1}} + \frac{1}{2^{n+2}} + \frac{1}{2^{n+3}} + \cdots = \frac{1}{2^{n+1}}\left(1 + \frac{1}{2} + \frac{1}{2^2} + \cdots\right)$$

$$= \frac{1}{2^{n+1}} \cdot 2 = \frac{1}{2^n} < \epsilon$$

Here again we've summed the geometric series. $\square$

Proposition 2.4.3. The shift map is topologically transitive.

Proof. Given neighborhoods M and N take points y in M and z in N, and ϵ small enough that $d(y,x) < \epsilon$ implies x is in M. Finally, take n large enough that $1/2^n < \epsilon$. A point x that works to show topological transitivity is

$$x = (y_1, y_2, \ldots, y_n, z_1, z_2, z_3, \ldots) \tag{2.15}$$

Observe that x lies in M because $d(x, y)$

$$= \frac{|x_1 - y_1|}{2} + \cdots + \frac{|x_n - y_n|}{2^n} + \frac{|x_{n+1} - y_{n+1}|}{2^{n+1}} + \frac{|x_{n+2} - y_{n+2}|}{2^{n+2}} + \cdots$$

$$= \frac{|z_1 - y_{n+1}|}{2^{n+1}} + \frac{|z_2 - y_{n+2}|}{2^{n+2}} + \cdots \leq \frac{1}{2^{n+1}} + \frac{1}{2^{n+2}} + \cdots = \frac{1}{2^n} < \epsilon$$

where for the last equality we summed the geometric series, as in the previous two proofs.

Finally, note that $\sigma^n(x) = z$, a point in N. That is, the point x of Eq. (2.15) belongs to the neighborhood M and $\sigma^n(x)$ belongs to the neighborhood N. Because M and N can be any pair of neighborhoods, this shows that the shift map is topologically transitive. $\square$

By converting the real number x to a symbol sequence by its base-2 expansion (2.11), the Baker's map is converted into the shift map, and the chaotic dynamics of the shift map translates to chaotic dynamics of the Baker map. Some care must be exercised with "is converted into," but this is the basic idea.

Symbolic dynamics can be applied to many other functions. It's a powerful tool for keeping track of the combinatorics of how the function moves around pieces of its domain. Sections 1.6 and 1.7 of [31] are a quick introduction; the book [84] by Brian Marcus and Douglas Lind presents many more examples.

We'll visit symbolic dynamics again in Sect. 19.3.

Practice Problems

2.4.1. For the function

$$f(x) = \begin{cases} 3x & \text{for } 0 \leq x \leq 1/3 \\ (3/2)x - 1/2 & \text{for } 1/3 < x \leq 1 \end{cases} \tag{2.16}$$

(a) find upper and lower bounds for the Liapunov exponent of f.
(b) By simulating the dynamics of f, estimate the fraction of time that iterates are in the intervals $[0, 1/3]$ and $(1/3, 1]$. Use this information to estimate the Liapunov exponent of f.

2.4.2. Find a point $x \in [0, 1)$ for which the set of iterates $B^n(x)$ of the Baker's map is dense in $[0, 1)$.

Practice Problem Solutions

2.4.1. (a) For $0 \leq x \leq 1/3$, $f'(x) = 3$; for $1/3 < x \leq 1$, $f'(x) = 3/2$. If every iterate $f^i(x)$ lies in $[0, 1/3]$ (this isn't possible, but in this part of the problem we're looking just for bounds on the Liapunov exponent), then each term of Eq. (2.10) is $\ln(3)$ and so $\lambda = \ln(3)$. If every $f^i(x)$ lies in $(1/3,1]$, then every term of Eq. (2.10) is $\ln(3/2)$ and so $\lambda = \ln(3/2)$. If some iterates fall in one interval, other iterates in the other interval, then λ falls between these values. That is, $\ln(3)$ is an upper bound, and $\ln(3/2)$ is a lower bound, of the Liapunov exponent.

(b) With the code in Sect. B.2 we'll count the number I_1 of iterates in $[0,1/3]$ and the number I_2 of iterates in $(1/3,1]$. Different initial points will give different sequences.

I_1	I_2	I_1	I_2
33393	66608	33449	66552
33481	66520	33333	66668
33325	66676	33378	66623
33462	66539	33246	66755
33428	66573	33464	66537
33076	66925	33211	66790

The average value of I_1 is 33353.8; the average value of I_2 is 66647.2. We expect that the iterates land in $[0,1/3]$ about 1/3 of the time, and in $(1/3,1]$ about 2/3 of the time. Then we estimate the Liapunov exponent to be

$$\lambda = \frac{1}{3}\ln(3) + \frac{2}{3}\ln(3/2) \approx 0.6365$$

Because $\lambda > 0$ we expect the function f to support chaotic dynamics.

2.4.2. Directly from the definition of the Baker's map B this may seem daunting, so we'll translate this to a shift map problem. All we need is a sequence that contains all finite sequences, because then enough applications of the shift map will move any finite sequence to the left end of the symbol sequence. We'll find such a sequence and then argue why its iterates are dense. Here's one example, the point x with sequence

$$0,1,0,0,0,1,1,0,1,1,0,0,0,0,0,1,0,1,0,0,1,1,1,0,0,1,0,1,1,1,0,\ldots$$

To see all the finite sequences, organize the entries this way

$$(0)(1)(00)(01)(10)(11)(000)(001)(010)(011)(100)(101)(110)\ldots$$

We see both sequences of length 1, all four sequences of length 2, all eight sequences of length 3, and so on.

What number is x? We can write the base-2 expansion of Eq. (2.11)

$$x = \frac{1}{2^2} + \frac{1}{2^6} + \frac{1}{2^7} + \frac{1}{2^9} + \frac{1}{2^{10}} + \frac{1}{2^{16}} + \frac{1}{2^{18}} + \cdots$$

If we're lucky, we can find a pattern and recognize this as a geometric series or maybe a sum of geometric series. Here we don't appear to be lucky–at least, I haven't found a pattern–but we can find bounds on the decimal equivalent. The terms presented sum to about 0.27638626. The next non-zero term is $1/2^{21}$. Assuming all the remaining terms are non-zero (of course, they aren't), they would sum to the geometric series $(1/2^{21}) \cdot (1 + 1/2 + 1/2^2 + \ldots) = 1/2^{20}$ and we see $0.27638626 < x < 0.27638722$.

Back to showing that the iterates of x are dense. Because $x_1 = 0$ we see x belongs to $[0, 1/2)$. Continuing, we see $\sigma(x)$ belongs to $[1/2, 1)$, $\sigma^2(x)$ belongs to $[0, 1/4)$, $\sigma^3(x)$ belongs to $[1/4, 1/2)$, and so on. We see that the iterates visit every interval of the form $[k/2^n, (k+1)/2^n)$. Because for each n, every point of $[0, 1)$ lies in one of these intervals, by taking n large enough we can find a $\sigma^k(x)$ as close as we wish to every point of $[0, 1)$. Translating this result about σ into a property of B, we see that the iterates of x under B are dense.

The symbol sequence, an example of a symbolic dynamics argument, is very simple (though pretty clever the first time you see it), spare and elegant, devoid of the messiness you may have been expecting from the formulation of the problem.

Exercises

2.4.1. For the Baker's map, prove Eq. (2.12). Hint: consider two cases: $x < 1/2$, so $x_1 = 0$, and $x \geq 1/2$, so $x_1 = 1$.

2.4.2. For the function g given by

$$g(x) = \begin{cases} 3x & \text{for } 0 \leq x \leq 1/3 \\ (24/7)x - 8/7 & \text{for } 1/3 < x \leq 5/8 \\ (8/3)x - 5/3 & \text{for } 5/8 < x \leq 1 \end{cases}$$

(a) find upper and lower bounds for the Liapunov exponent of g.
(b) By simulating the dynamics of g, estimate the fraction of time that iterates are in the intervals $[0, 1/3]$, $(1/3, 5/8]$, and $(5/8, 1]$. Use this information to estimate the Liapunov exponent of g.

2.4.3. For the function h, given by

$$h(x) = \begin{cases} (3/2)x + 1/4 & \text{for } 0 \leq x \leq 1/2 \\ 2x - 1 & \text{for } 1/2 < x \leq 1 \end{cases}$$

(a) find upper and lower bounds for the Liapunov exponent of h.
(b) By simulating the dynamics of h, estimate the fraction of time that iterates

are in the intervals $[0,1/2]$ and $(1/2,1]$. Use this information to estimate the Liapunov exponent of h.

Exercises 2.4.4 and 2.4.5 involve the correspondence between base-2 symbol sequences and real numbers. Finding the real number represented by a symbol sequence (2.13) is just a matter of summing the series (2.11). If we recognize this as a geometric series, then the sum is straightforward. The reverse of this process, finding the base-2 expansion of a given number, is just iterated division and another search for a pattern. For example, we'll find the symbol sequence for $x = 1/3$. Because $1/3 < 1/2$, $x_1 = 0$. Because $1/3 > 1/4$, $x_2 = 1$. Then $1/3 - 1/4 = 1/12 < 1/8$, so $x_3 = 0$. Next, $1/12 > 1/16$, so $x_4 = 1$. Then $1/12 - 1/16 = 1/48 < 1/32$, so $x_5 = 0$. Because $1/48 > 1/64$, $x_6 = 1$. The pattern appears to be $x = (0,1,0,1,0,1,0,\dots)$. Summing the geometric series shows this is correct.

2.4.4. Find the rational equivalents of these symbol sequences.
(a) $(0,1,1,0,1,1,0,1,1,\dots) = (0,1,1)^\infty$
(b) $(1,1,1,0,1,1,0,1,1,0,1,1,0,1,1,\dots) = (1,1,1,(0,1,1)^\infty)$
(c) $(0,1,1,1,(0,1,1)^\infty)$
(d) $(0,1,0,0,(0,1,1)^\infty)$

2.4.5. Find the base-2 expansions of
(a) $1/5$, (b) $1/6$, and (c) $1/7$.

2.4.6. For the function

$$f(x) = \begin{cases} 3x & \text{for } 0 \le x < 1/3 \\ 3x - 1 & \text{for } 1/3 \le x < 2/3 \\ 3x - 2 & \text{for } 2/3 \le x < 1 \end{cases}$$

(a) using base-3 expansions

$$x = \frac{x_1}{3} + \frac{x_2}{3^2} + \frac{x_3}{3^3} + \cdots$$

where each x_i is equal to 0, 1, or 2, and any expansion with an infinite terminal sequence of 2s is replaced by the equivalent expansion with an infinite terminal sequence of 0s, show

$$f(x_1, x_2, x_3, \dots) = (x_2, x_3, x_4, \dots)$$

(b) Show f is chaotic.

Exercises 2.4.7–2.4.10 use the *shift-flip map* $\tilde{\sigma}$ with each $x_i = 0$ or 1

$$\tilde{\sigma}(x_1, x_2, x_3, x_4, \dots) = (1 - x_2, 1 - x_3, 1 - x_4, \dots)$$

2.4.7. Show the shift-flip map exhibits sensitivity to initial conditions with $\delta = 1/2$.

2.4.8. Show periodic points for the shift-flip map are dense. Hint: show that it suffices to consider cycles of even length.

2.4.9. Show the shift-flip map is topologically transitive.

2.4.10. (a) Deduce that the shift-flip map exhibits chaos. (Don't overthink this. The answer is immediate.)
(b) Show the function R exhibits chaos, where

$$R(x) = \begin{cases} 1 - 2x & \text{for } 0 \leq x < 1/2 \\ 2 - 2x & \text{for } 1/2 \leq x \leq 1 \end{cases}$$

Hint: use the base-2 expansion of 1 to find a base-2 expansion of $1 - (x_2/2 + x_3/2^2 + \ldots)$.

We've left out an important point: how do we know that the chaotic dynamics of $\tilde{\sigma}$ translates to chaotic dynamics of R? For that matter, how do we know that the chaotic dynamics of σ translates to chaotic dynamics of B? For this we need the notion of topological conjugacy, too far afield for us, but if you're interested, consult Sect. 1.7 of [31]. In Sect. 1.8 of that book, Devaney applies topological conjugacy to show that the chaotic dynamics of σ translates to chaotic dynamics of the $r = 2$ tent map and the $r = 4$ logistic map.

2.5 BASIC CARDIAC DYNAMICS

Cardiac dynamics is an intricate subject, dependent on a detailed understanding of networks of heart cells. Rather than focus on these complications, for now we consider very simple cartoon models, with the hope of capturing some clinically observed cardiac behaviors. (In Sect. 16.4 we'll see more biological background and a more realistic model.) If a simple cartoon can exhibit processes seen in real heartbeats, then perhaps a detailed understanding of cardiac physiology is not needed to model, or even to begin designing treatments for, *some* cardiac ailments.

Still, we must start with some information about the heart. The heart's primary pacemaker is the *sinoatrial (SA) node*, a cluster of cells in the upper part of the wall of the right atrium. The SA node is 1 to 3 cm long, half a cm wide, and 1 to 2 mm deep. At regular intervals of duration τ, the SA node sends electrical signals through the heart's electrical conduction system. These signals cause the atria to contract, pumping blood into the ventricles. The signal continues to the *atrioventricular (AV) node*, a small (1 mm by 2 mm by 3 mm) cluster of cells between the right atrium and the right ventricle. The AV node delays this signal for about 0.1 seconds, time for the atria to finish pumping blood into the ventricles. Then the AV node passes the signal on to the ventricles, causing them to contract and pump blood to the lungs and to the rest of the body. This is what we hear as a heartbeat.

The contracted state of the heart is the *systole* phase. Then the heart muscles relax, returning it to the *diastole* phase. Repeat this two and a half to three billion times and that's your life.

2.6 A SIMPLE CARDIAC MODEL

Suppose V_0 is the potential of the AV node immediately after receiving a signal from the SA node. This potential decays exponentially, so until another signal arrives from the SA node, the potential at the AV node is given by

$$V(t) = V_0 e^{-\alpha t}$$

where $\alpha > 0$ is the decay constant. The AV node initiates a heartbeat if when the next signal arrives from the SA node, the potential $V(t)$ lies below a threshold potential V_c. If a heartbeat occurs, the potential of the AV node is increased by an amount u.

Whether or not a heartbeat is initiated depends on the potential $V(t)$ when the next signal arrives from the SA node. So the dynamics of whether or not the heart beats is determined by $V(n\tau)$, denoted V_n. From the nth SA node signal to the $(n + 1)$st, the potential evolves according to

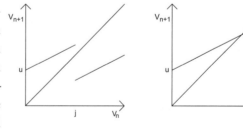

Figure 2.12. Simple cardiac dynamics models with jump at $j = V_c/e^{-\alpha\tau}$.

$$V_{n+1} = \begin{cases} e^{-\alpha\tau} V_n + u & \text{if } e^{-\alpha\tau} V_n < V_c: \text{a heartbeat occurs here} \\ e^{-\alpha\tau} V_n & \text{if } e^{-\alpha\tau} V_n \geq V_c: \text{no heartbeat occurs here} \end{cases} \qquad (2.17)$$

The graph of V_{n+1} vs V_n depends on u, $e^{-\alpha\tau}$, and V_c. Note the graph is discontinuous, with a jump at $V_n = V_c/e^{-\alpha\tau}$, and consists of two line segments, both with slope $e^{-\alpha\tau} < 1$. The left branch has V_{n+1}-intercept u, the right branch has V_{n+1}-intercept 0, though the right branch does not continue to the V_{n+1}-axis. The dynamics take two qualitatively distinct forms, depending on the location of the jump. See Fig. 2.12.

Recalling the basics of graphical iteration, we see the left image of Fig. 2.12 has no fixed point, while the right has one, stable because the slope, $e^{-\alpha\tau}$, of the tangent line (equal to the slopes of both straight line segments in the graph) satisfies $|e^{-\alpha\tau}| < 1$. Certainly, the position of the jump, the magnitude of the jump, and the slope of the line segments all influence the dynamics of this model. Are the patterns produced observed clinically?

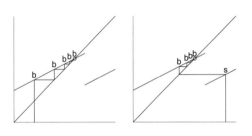

Figure 2.13. A stable heartbeat.

In Fig. 2.13 we see a model in which the left branch crosses the diagonal and so the jump occurs to the right of this crossing. We can characterize this position of the jump in terms of the model parameters. The fixed point is the intersection of

$$V_{n+1} = e^{-\alpha\tau} V_n + u \quad \text{and} \quad V_{n+1} = V_n$$

That is,

$$V_n = \frac{u}{1 - e^{-\alpha\tau}} \tag{2.18}$$

The jump occurs at $V_n = V_c/e^{-\alpha\tau}$, so the condition that the fixed point occurs, necessarily to the left of the jump, is

$$\frac{u e^{-\alpha\tau}}{1 - e^{-\alpha\tau}} < V_c \tag{2.19}$$

Note each iteration of Eq. (2.17) that occurs on the left branch corresponds to an SA node signal triggering a heartbeat. These points are labeled b. Every iteration on the right branch corresponds to the SA node signal arriving before the AV node potential has decayed adequately, so a heartbeat is skipped. These points are labeled s.

As illustrated in Fig. 2.13, parameters in this range correspond to a stable heartbeat, one beat for each pacemaker signal. Whatever the AV node potential V_0 when measurement commences, the dynamics of this model quickly leads to the stable fixed point.

What happens if condition (2.19) is violated in the model? In this case, there is no fixed point and the resulting dynamics depend delicately on the model parameters. Every iterate on the right branch corresponds to a dropped beat, a blockage that prevents the SA signal from reaching the AV node. This is called an *AV block*. These are subdivided into three classes; our examples are from the second class. On the left side of Fig. 2.14 we see a 2:1 AV block: every second signal from the AV node is dropped, resulting in a reduced heart rate. Often this ailment presents no symptoms; when symptoms occur, the most common are dizziness, light-headedness, and fainting.

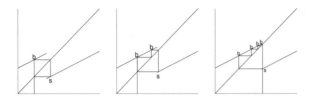

Figure 2.14. Left: a 2:1 AV block. Middle: a 3:1 AV block. Right: the Wenckebach phenomenon.

Moving the jump potential to the right (and keeping u and $e^{-\alpha\tau}$ unchanged in this figure) in the middle of Fig. 2.14 we see a similar dynamic, skipping every third beat. This is called a 3:1 AV block. On the right we see an example of the *Wenckebach phenomenon*: the interval between successive heartbeats gradually increases until a beat is dropped and the sequence repeats.

In the context of this very simple model, what can cause AV blocks? Recalling that the jump in the graph of V_{n+1} occurs at $V_c/e^{-\alpha\tau}$, we see that lowering V_c moves the jump to the left. A sufficient reduction of V_c can change the graph from the form presented in Fig. 2.13 to one of those in Fig. 2.14. Lowering V_c corresponds to increasing the refractory time between heartbeats, a result of taking calcium channel blockers, beta blockers, or cardiac glycosides. Digitalis and other drugs that increase calcium concentration decrease the refractory time.

In the Practice Problems and Exercises we'll explore parameter ranges that exhibit cycles of different lengths, and interpret these ranges. To make the formulas less cumbersome, we'll write $a = e^{-\alpha\tau}$ and we'll write Eq. (2.17) as $V_{n+1} = f(V_n)$.

Practice Problems

2.6.1. (a) Find the range of V_c/a values, bounded by functions of a and u, that give a 2-cycle for this model.

(b) Take $u = 0.5$ and plot the lower bound curve $L(a)$ and upper bound $U(a)$ for which $L(a) < V_c < U(a)$ guarantees this model exhibits a 2-cycle.
(c) Is this 2-cycle stable?

2.6.2. For $u = 0.5$ and $a = 0.5$, plot $V_{n+2} = f^2(V_n) = f(f(V_n))$
(a) for V_c below the 2-cycle range found in Practice Problem 2.6.1,
(b) for V_c in the 2-cycle range found in Practice Problem 2.6.1, and
(c) for V_c above the 2-cycle range found in Practice Problem 2.6.1.
(d) Interpret these graphs in terms of the presence or absence of a 2-cycle.

Practice Problem Solutions

2.6.1. (a) Referring to the left image of Fig. 2.14, call the potentials of the 2-cycle V_0 and V_1, with $V_0 < V_c/a < V_1$. Then

$$V_1 = f(V_0) = aV_0 + u$$
$$V_0 = f(V_1) = aV_1$$

Combining these we find

$$V_0 = a(aV_0 + u) = a^2V_0 + au$$

so

$$V_0 = \frac{au}{1 - a^2}$$

Then

$$V_1 = \frac{a^2u}{1 - a^2} + \frac{u(1 - a^2)}{1 - a^2} = \frac{u}{1 - a^2}$$

In order for this 2-cycle to exist, we must have $V_0 < V_c/a < V_1$. That is,

$$\frac{au}{1 - a^2} < V_c/a < \frac{u}{1 - a^2} \tag{2.20}$$

(b) Multiplying Eq. (2.20) by a, we obtain

$$L(a) = \frac{a^2u}{1 - a^2} \quad \text{and} \quad U(a) = \frac{au}{1 - a^2}$$

These two curves are plotted in Fig. 2.15, where both the horizontal and the vertical ranges are $[0,1]$. For V_c between these curves, the model has a 2-cycle.

(c) The 2-cycle is stable because at both 2-cycle points, the slope of the tangent line (just the slope of the straight lines that constitute the graph of f^2) has absolute value < 1.

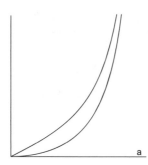

Figure 2.15. The curves $L(a)$ and $U(a)$.

2.6.2. In Practice Problem Solution 2.6.1 we see that the lower and upper bounds of the V_c/a values that admit a 2-cycle are $au/(1 - a^2) = 1/3$ and $u/(1 - a^2) = 2/3$. For (a) we'll take $V_c/a = 1/4 < 1/3$, for (b) we'll take $V_c/a = 1/2$ so $1/3 < V_c/a < 2/3$, and for (c) we'll take $V_c/a = 3/4 > 2/3$.

(a) As described in Exercises 2.1.9 and 2.1.10, we can use graphical iteration to plot the composition $f^2 = f(f)$. Here we'll use a combination of graphical iteration and algebra. First, apply the definition (2.17) to f^2,

$$f^2(V) = f(f(V)) = \begin{cases} f(V)/2 + 1/2 & \text{for } 0 \le f(V) < 1/4 \\ f(V)/2 & \text{for } 1/4 \le f(V) \le 1 \end{cases}$$

From the graph of f, we see that $0 \le f(V) < 1/4$ for $1/4 \le V < 1/2$. Then $1/4 \le f(V) \le 1$ for $0 \le V < 1/4$ and for $1/2 \le V \le 1$. Combining these, we find an expression for f^2:

$$f^2(V) = \begin{cases} f(V)/2 & \text{for } 0 \le V < 1/4 \\ f(V)/2 + 1/2 & \text{for } 1/4 \le V < 1/2 \\ f(V)/2 & \text{for } 1/2 \le V \le 1 \end{cases}$$

$$= \begin{cases} (V/2 + 1/2)/2 = V/4 + 1/4 & \text{for } 0 \le V < 1/4 \\ (V/2)/2 + 1/2 = V/4 + 1/2 & \text{for } 1/4 \le V < 1/2 \\ (V/2)/2 = V/4 & \text{for } 1/2 \le V \le 1 \end{cases}$$

This gives the left graph of the second row of Fig. 2.16.

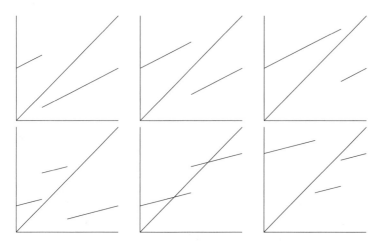

Figure 2.16. Graphs for Practice Problem 2.6.2. Top: the graphs of $f(V)$ for $V_c/a = 0.25$ (left), $V_c/a = 0.50$ (center), and $V_c/a = 0.75$ (right). Bottom: the corresponding graphs of $f^2(V)$.

(b) Why does moving V_c/a a bit to the left of $1/3$ change the number of branches of f^2 from 3 to 2? Take $V_c/a = 1/2$. Here we have

$$f^2(V) = \begin{cases} f(V)/2 + 1/2 & \text{for } 0 \le f(V) < 1/2 \\ f(V)/2 & \text{for } 1/2 \le f(V) \le 1 \end{cases}$$

From the graph of f we see that $0 \le f(V) < 1/2$ for $1/2 \le V < 1$, and $1/2 \le f(V) \le 1$ for $0 \le V \le 1/2$, so unlike in (a), neither of these regions is split into two pieces. (Okay, if we're very careful we see by graphical iteration that $f^2(1) = 1/4$. There is a second jump, but it's at the point $V = 1$, the right endpoint of the domain of f.) Then we see

$$f^2(V) = \begin{cases} (V/2 + 1/2)/2 = V/4 + 1/4 & \text{for } 0 \le V < 1/2 \\ (V/2)/2 + 1/2 = V/4 + 1/2 & \text{for } 1/2 \le V < 1 \end{cases}$$

This gives the middle graph of the second row of Fig. 2.16.

(c) The analysis is similar to that of (a). Try it and compare your answer to the right graph of the second row of Fig. 2.16 to be sure you understand the details.

(d) For $V_c/a = 1/4$ the graph of f misses the diagonal so there is no fixed point, and f^2 misses the diagonal so there is no 2-cycle. For $V_c/a = 1/2$ the graph of f misses the diagonal so there is no fixed point, but f^2 intersects the diagonal in two points so there is a 2-cycle. For $V_c/a = 3/4$ the graph of f misses the diagonal so there is no fixed point, and f^2 misses the diagonal so there is no 2-cycle.

To see how the 2-cycle appears, plot the graph of f^2 as V_c/a steps from $1/4$ to $3/4$ in small steps. Mathematica code to plot the graph of f^2 is in Sect. B.3.

Exercises

These problems use this function

$$f(V) = \begin{cases} aV + u & \text{for } V < V_c/a \\ aV & \text{for } V \ge V_c/a \end{cases}$$

where $a = e^{-\alpha\tau}$.

2.6.1. (a) Sketch the graph of f for $\alpha = \tau = 1$, $u = 0.5$, and $V_c = 0.25$.
(b) Plot f when $\alpha = 0.5$, $\tau = 1$, $u = 0.5$, and $V_c = 0.25$.
(c) Plot f when $\alpha = 1$, $\tau = 0.5$, $u = 0.5$, and $V_c = 0.25$.
(d) Explain the relation between the graphs of (a) and (b).

2.6.2. Continuing with the previous exercise
(a) Plot f when $\alpha = \tau = 1$, $u = 0.6$, and $V_c = 0.25$.
(b) Plot f when $\alpha = \tau = 1$, $u = 0.5$, and $V_c = 0.15$.

We'll label the elements of an n-cycle this way: $V_1, V_2, \ldots, V_n$ where $f(V_1) = V_2, f(V_2) = V_3, \ldots, f(V_{n-1}) = V_n$, and $f(V_n) = V_1$.

2.6.3. For the 3-cycle with $V_1 < V_2 < V_c/a < V_3$,
(a) find the $L(a)$ and $U(a)$ curves.
(b) Plot these curves for $u = 0.1, 0.5$, and 0.9.

2.6.4. For the 3-cycle with $V_1 < V_c/a < V_3 < V_2$,
(a) find the $L(a)$ and $U(a)$ curves.
(b) Plot these curves for $u = 0.1, 0.5$, and 0.9. Do some of these curves intersect at points other than the origin? Does this matter?

2.6.5. For the 4-cycle with $V_1 < V_2 < V_3 < V_c/a < V_4$,
(a) find the $L(a)$ and $U(a)$ curves.
(b) Plot these curves for $u = 0.1, 0.5$, and 0.9.

2.6.6. For the 4-cycle with $V_1 < V_c/a < V_4 < V_3 < V_2$,
(a) find the $L(a)$ and $U(a)$ curves.
(b) Plot these curves for $u = 0.1, 0.5$, and 0.9.

2.6.7. For the 5-cycle with $V_1 < V_2 < V_3 < V_4 < V_c/a < V_5$,
(a) find the $L(a)$ and $U(a)$ curves.
(b) Plot these curves for $u = 0.1, 0.5$, and 0.9.

2.6.8. For the 5-cycle with $V_1 < V_c/a < V_5 < V_4 < V_3 < V_2$,
(a) find the $L(a)$ and $U(a)$ curves.
(b) Plot these curves for $u = 0.1, 0.5$, and 0.9.

2.6.9. For the 5-cycle with $V_1 < V_4 < V_2 < V_c/a < V_5 < V_3$,
(a) find the $L(a)$ and $U(a)$ curves.
(b) Plot these curves for $u = 0.1, 0.5$, and 0.9.

2.6.10. (a) On the same graph, plot the curves $L(a)$ and $U(a)$ with $u = 0.5$ for Exercises 2.6.3–2.6.9. Shade the region between each of these pairs of curves.
(b) With these curves do you see any evidence that any parameter values support

more than one cycle? (Don't take these few examples–alone–as evidence of anything general. This is just a start. After all, the first three odd numbers > 1 support the claim that all odd numbers are prime.)

(c) To avoid AV block, what treatment strategy does this suggest?

2.7 ARNOLD'S MODEL

A related approach to studying cardiac dynamics, especially arrhythmias, was developed, but not published, by Vladimir Arnold in 1959, and developed independently in the 1970s and 80s by Leon Glass and his coworkers [29, 46, 48]. Suppose an oscillator (for example, a heart pacemaker) has period T. Relative to this oscillator, the *phase* of time t is

$$\phi(t) = t/T \text{ (mod 1)} \qquad (2.21)$$

where by t/T (mod 1) we mean the fractional part of t/T. For example,

$$7.325 \text{ (mod 1)} = 0.325$$

To understand the definition of $\phi(t)$ in Eq. (2.21), subdivide time t into intervals of length T:

$$[0, T), [T, 2T), [2T, 3T), [3T, 4T), \ldots$$

Then each time t lies in one of these intervals, and $\phi(t)$ tells us where t lies in the interval containing t. For example, suppose $T = 3$ and $t = 7.2$. Then t lies in the interval $[2T, 3T) = [6, 9)$, and in that interval, t is $1.2/3 = 0.4$ of the distance between $2T$ and $3T$. We'd say $\phi(t) = 0.4$, meaning $\phi(t)$ is 0.4 of the way between the beginnings of two successive periods of the oscillator. This is encapsulated in the definition (2.21):

$$\phi(7.2) = 7.2/3 \text{ (mod 1)} = 2.4 \text{ (mod 1)} = 0.4$$

Some cardiac arrhythmias occur when the signal from the SA node competes with another pacemaker, called an *ectopic pacemaker*, located in the ventricles. Typically, an ectopic pacemaker has a longer period; it signals more slowly than does the SA node. Heartbeats (and nerve impulses) have a refractory time: if a second signal arrives too close to the first, another heartbeat is not initiated. So if an SA signal arrives too soon after an ectopic signal, the SA signal is blocked from initiating a heartbeat. Similarly, if an ectopic signal arrives too soon after an SA signal, the ectopic signal is blocked from initiating another heartbeat.

In the model [29] of this situation, the ectopic pacemaker has period T, the SA pacemaker has period τ, and ϕ_i denotes the phase of the SA pacemaker immediately after the ith ectopic stimulus. The SA beat can reset the phase of the ectopic beat, so we are interested in how ϕ_i evolves to ϕ_{i+1}. In its most general form, Arnold's model of this evolution is given by the Arnold map

$$\phi_{i+1} = f(\phi_i) = g(\phi_i) + \tau/T \text{ (mod 1)} \tag{2.22}$$

where $g(\phi)$ is the *phase transition curve* and represents the change in the ectopic phase caused by the SA signal. In applications, often the phase transition curve is obtained from experimental data, rather than by postulating a function $g(\phi)$. Our goal is to understand the dynamics of the Arnold model, so we'll use a few simple functions $g(\phi)$.

Because $\phi = 0$ and $\phi = 1$ correspond to the same same phase of the oscillator, we think of g not as a function $g : [0,1] \to [0,1]$, but as $g : S^1 \to S^1$, where S^1 denotes a circle. The interval $[0,1]$ with its left and right endpoints identified as the same is topologically equivalent to a circle: moving left to right along the interval, when we reach the right endpoint we reappear at the left endpoint. This is the same as going once around a circle. Then $\phi(t)$ measures the fraction of the way t is around the circle. For this reason, f is called a *circle map*.

A much-studied example is

$$\phi_{i+1} = \phi_i + b\sin(2\pi\phi_i) + \tau \text{ (mod 1)} \tag{2.23}$$

In Fig. 2.17 we see three examples of Eq. (2.23), $b = -0.2, \tau = 0.4$ (left), $b = -0.2, \tau = 0.7$ (center), and $b = 0.2, \tau = 0.7$ (right).

Again we can use the ideas of graphical iteration to study the sequence of phases under iteration of Eq. (2.23). The left image of Fig. 2.18 shows 100 iterates, starting from $\phi_1 = 0.3$, for

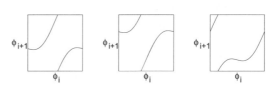

Figure 2.17. Some examples of Eq. (2.23).

$b = -0.3, \tau = 0.55$. This doesn't look like nearly 100 lines, suggesting that the ϕ_i may have converged to a cycle. The middle image of Fig. 2.18 shows iterates 100 through 200. This is an example of 2:1 *phase locking*. But nearby we can find more complicated behavior. The right side of Fig. 2.18 shows iterates 100 through 200 for $b = -0.3, \tau = 0.525$.

Regions of the $b - \tau$ plane that give rise to phase locks are called *Arnold tongues*. While the combinatorics of the relative positions of Arnold tongues is well understood, some aspects of circle maps remain unclear.

Figure 2.18. Some iterations of Eq. (2.23).

Finally, we'll consider the simpler example of the circle map

$$\phi_{i+1} = f(\phi_i) = b\phi_i + \tau \ (\text{mod } 1) \tag{2.24}$$

First we'll see how to graph $\phi_{i+1} = f(\phi_i)$ and then we'll establish some terminology for the locations of cycle points for this circle map.

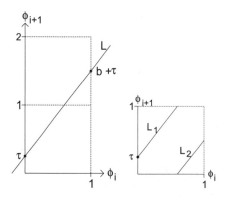

Figure 2.19. Left: the graph of $\phi_{i+1} = b\phi_i + \tau$. Right: the graph of $\phi_{i+1} = b\phi_i + \tau \ (\text{mod } 1)$.

On the left side of Fig. 2.19 we see the graph of $\phi_{i+1} = b\phi_i + \tau$, a straight line L with slope b and ϕ_{i+1}-intercept τ. Because ϕ_i lies between 0 and 1, the right-most point on the graph occurs where $\phi_i = 1$, and so $\phi_{i+1} = b + \tau$. In the case pictured, $1 < b + \tau < 2$. Because ϕ_{i+1} also must lie between 0 and 1, one more step is needed to graph $\phi_{i+1} = f(\phi_i)$. So long as $b\phi_i + \tau \leq 1$, $f(\phi_i) = b\phi_i + \tau$. The graph of this is the segment L_1 on the right of Fig. 2.19. If $1 < b\phi_i + \tau \leq 2$, $f(\phi_i) = b\phi_i + \tau - 1$, shown in the segment L_2. If $2 < b\phi_i + \tau \leq 3$ (not possible for the graph shown, but possible for some others), $f(\phi_i) = b\phi_i + \tau - 2$, and so on.

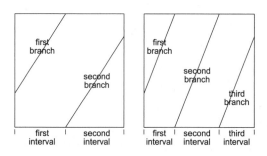

Figure 2.20. Branches and intervals of Eq. (2.24) for $1 < b\phi_i + \tau < 2$ (left graph) and $2 < b\phi_i + \tau < 3$ (right graph).

Now we establish some notation to more easily describe the locations of cycle points. On the left of Fig. 2.20 is the graph of Eq. (2.24) with $1 < b\phi_i + \tau < 2$. The graph has two branches, the first (on the left) and the second; the ϕ_i-axis is divided into two corresponding intervals, also called first and second.

On the right of Fig. 2.20 is the graph of Eq. (2.24) with $2 < b\phi_i + \tau < 3$. The graph has three branches, first, second, and third; the ϕ_i-axis is divided into three corresponding intervals, also called first, second, and third.

Phases belonging to a cycle we label ϕ_0, ϕ_1, ϕ_2, and so on, related by $f(\phi_0) = \phi_1$, $f(\phi_1) = \phi_2$, and so on. For example, a 4-cycle consists of ϕ_0, ϕ_1, ϕ_2, and ϕ_3, with

$$f(\phi_0) = \phi_1, \; f(\phi_1) = \phi_2, \; f(\phi_2) = \phi_3, \text{ and } f(\phi_3) = \phi_0$$

The examples of circle maps illustrate an important approach to using mathematics in biology. Any particular heart is immensely complicated. A detailed mathematical model that encompassed the individual nerve and muscle cells would be very difficult to build and computationally demanding, though see [54, 96] for some progress in this direction. But some aspects–maybe important aspects–of cardiac dynamics may be understood through much simpler models. Or maybe not models, which suggest some understanding of the detailed physiology, but perhaps cartoons. A simple, formal cartoon, attending to only crude physiology, yet capturing some nontrivial representative dynamics. Just a coincidence, or perhaps an example of how evolution discovers and uses laws of geometry?

Practice Problems

2.7.1. Here we use $f(\phi)$ defined by Eq. (2.24) with $1 < b + \tau < 2$.
(a) Suppose ϕ_0, ϕ_1, and ϕ_2 constitute a 3-cycle with ϕ_0 and ϕ_1 lying in the first interval and ϕ_2 in the second interval. Find the values of ϕ_0, ϕ_1, and ϕ_2 in terms of b and τ.
(b) Suppose ϕ_0, ϕ_1, ϕ_2, and ϕ_3 constitute a 4-cycle with ϕ_0, ϕ_1, and ϕ_2 lying in the first interval and ϕ_3 in the second interval. Find the values of ϕ_0, ϕ_1, ϕ_2, and ϕ_3 in terms of b and τ.
(c) Are these cycles stable?

2.7.2. Here we'll use the circle map defined by Eq. (2.24) with $1 < b + \tau < 2$. This condition is equivalent to $1 - b \leq \tau \leq 2 - b$. Together with the condition $0 \leq \tau \leq 1$, this shows the 2-cycle parameter range is contained in a parallelogram, the light gray region in the left image of Fig. 2.22.
(a) Suppose ϕ_0 and ϕ_1 form a 2-cycle with ϕ_0 in the first interval and ϕ_1 in the second interval. Find the values of ϕ_0 and ϕ_1 in terms of b and τ.
(b) In this region, $1 - b \leq \tau \leq 2 - b$ and $0 \leq \tau \leq 1$, plot the area where this circle map has a 2-cycle.

Practice Problem Solutions

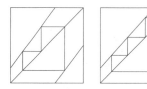

Figure 2.21. A 3- and a 4-cycle.

Here we sketch the graphical iteration paths of a 3-cycle (left) and a 4-cycle (right) for circle maps. The 3-cycle, for Practice Problem 2.7.1(a), is for the circle map with $b = 1.3$ and $\tau = 0.2$; the 4-cycle, for Practice Problem 2.7.1(b), is for the circle map with $b = 1.1$ and $\tau = 0.2$.

2.7.1. (a) As we see in the right image of Fig. 2.19, the condition $1 < b + \tau \leq 2$ means that the graph of $f(\phi)$ has two branches. From the left graph of that figure we see

$$\phi_1 = b\phi_0 + \tau \quad \phi_2 = b\phi_1 + \tau \quad \phi_0 = b\phi_2 + \tau - 1 \qquad (2.25)$$

Combining these we find ϕ_0 satisfies

$$\phi_0 = b\phi_2 + \tau - 1 = b(b\phi_1 + \tau) + \tau - 1 = b(b(b\phi_0 + \tau) + \tau) + \tau - 1$$
$$= b^3\phi_0 + b^2\tau + b\tau + \tau - 1$$

Solving for ϕ_0 gives

$$\phi_0 = \frac{1 - \tau - b\tau - b^2\tau}{b^3 - 1}$$

We can find the other 3-cycle values from Eq. (2.25):

$$\phi_1 = \frac{b - \tau - b\tau - b^2\tau}{b^3 - 1} \qquad \phi_2 = \frac{b^2 - \tau - b\tau - b^2\tau}{b^3 - 1}$$

(b) From the right graph of Fig. 2.21 we see

$$\phi_1 = b\phi_0 + \tau \quad \phi_2 = b\phi_1 + \tau \quad \phi_3 = b\phi_2 + \tau \quad \phi_0 = b\phi_3 + \tau - 1 \qquad (2.26)$$

Combining these we find ϕ_0 satisfies

$$\phi_0 = b(b(b(b\phi_0 + \tau) + \tau) + \tau) + \tau - 1$$

Solving for ϕ_0 gives

$$\phi_0 = \frac{1 - \tau - b\tau - b^2\tau - b^3\tau}{b^4 - 1}$$

We can find the other 4-cycle values from Eq. (2.26):

$$\phi_1 = \frac{b - \tau - b\tau - b^2\tau - b^3\tau}{b^4 - 1} \qquad \phi_2 = \frac{b^2 - \tau - b\tau - b^2\tau - b^3\tau}{b^4 - 1}$$
$$\phi_3 = \frac{b^3 - \tau - b\tau - b^2\tau - b^3\tau}{b^4 - 1}$$

(c) Because $0 \leq \tau \leq 1$, the condition $1 < b + \tau < 2$ implies $b > 1$. Each line segment making up the graph of f has slope b, so each line segment making up the graph of f^n has slope b^n. Consequently, all cycles are unstable.

2.7.2. (a) Because the 2-cycle has ϕ_0 in the first interval and ϕ_1 in the second interval,

$$\phi_1 = b\phi_0 + \tau \quad \text{and} \quad \phi_0 = b\phi_1 + \tau - 1$$

Solving for ϕ_0 as in the previous problem, we find

$$\phi_0 = \frac{1 - \tau - b\tau}{b^2 - 1} \quad \text{and} \quad \phi_1 = \frac{b - \tau - b\tau}{b^2 - 1}$$

(b) For $1 < b + \tau < 2$, the graph of f has a single jump at $b\phi + \tau = 1$, that is, at $\phi = (1 - \tau)/b$. Then the condition that ϕ_0 lies in the left interval and ϕ_1 lies in the right interval is

$$0 < \phi_0 = \frac{1 - \tau - b\tau}{b^2 - 1} < \frac{1 - \tau}{b} \quad \text{and} \quad \frac{1 - \tau}{b} < \phi_1 = \frac{b - \tau - b\tau}{b^2 - 1} < 1 \quad (2.27)$$

Multiplying the ϕ_0 inequalities by $b(b^2 - 1)$ gives

$$0 < b(1 - \tau - b\tau) < (b^2 - 1)(1 - \tau)$$

Simplifying and solving these inequalities for τ gives

$$\tau < \frac{1}{1 + b} = U(b) \qquad L(b) = \frac{1 + b - b^2}{1 + b} < \tau \qquad (2.28)$$

Multiplying the ϕ_1 inequalities of Eq. (2.27) by $b(b^2 - 1)$ gives

$$(b^2 - 1)(1 - \tau) < b(b - \tau - b\tau) < b(b^2 - 1)$$

Simplifying and solving these inequalities for τ gives the same conditions as those of Eq. (2.28).

On the left side of Fig. 2.22, the curve $\tau = U(b)$ is concave up, and the curve $\tau = L(b)$ is concave down. The area between these curves, and in the light gray parallelogram $1 - b \leq \tau \leq 2 - b$ and $0 \leq \tau \leq 1$, determines the parameters where this family of circle maps has a 2-cycle. Shaded in dark gray in Fig. 2.22, this region is an Arnold tongue, or more precisely, a part of an Arnold tongue.

Exercises

In these exercises use f defined by Eq. (2.24).

2.7.1. Plot the graph of f^2 for the points $(b, \tau) = (1.25, 0.42)$, $(1.5, 0.38)$, $(1.75, 0.22)$, $(1.25, 0.32)$, $(1.5, 0.13)$, and $(1.25, 0.02)$. The first three points are

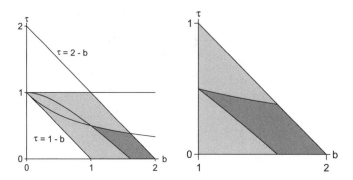

Figure 2.22. The dark region is the parameter range for $1 < b + \tau < 2$ exhibiting a 2-cycle. The right image is a magnification of part of the left.

just below the upper boundary of the shaded region of Fig. 2.22; the last three are just above the lower boundary. Do these graphs support the presence of 2-cycles for f? Why couldn't these graphs show two fixed points for f?

2.7.2. Why don't parameters in the region between the curves $\tau = L(b)$ and $\tau = U(b)$ of Fig. 2.22, with $0 \le b \le 1$, give a 2-cycle? Hint: pick a point in that region and plot f^2.

2.7.3. For $2 < b + \tau < 3$,
(a) plot the parallelogram analogous to that of Practice Problem 2.7.2 for $1 < b + \tau < 2$.
(b) Suppose this f exhibits a 2-cycle with ϕ_0 in the first interval and ϕ_1 in the second. Find the (b, τ) coordinates of these cycle points.
(c) Write the inequalities analogous to Eq. (2.27).
(d) In the parallelogram of (a), sketch the region where this circle map exhibits such a 2-cycle.

2.7.4. Repeat Exercise 2.7.3 for $3 < b + \tau < 4$.

2.7.5. Building on Exercises 2.7.3 and 2.7.4, argue that this 2-cycle Arnold tongue extends infinitely far to the right.

2.7.6. For $2 < b + \tau < 3$,
(a) find the (b, τ) coordinates for the 2-cycle with ϕ_0 in the first interval and ϕ_1 in the third interval.

(b) Find the curves bounding the region where this circle map exhibits such a 2-cycle.

(c) Sketch this region in the $2 < b + \tau < 3$, $0 < \tau < 1$ parallelogram.

2.7.7. For $1 < b + \tau < 2$,

(a) find the (b, τ) coordinates for the 3-cycle with ϕ_0 and ϕ_1 in the first interval and ϕ_2 in the second interval.

(b) Find the curves bounding the region where this circle map exhibits such a 3-cycle.

(c) Sketch this region in the $1 < b + \tau < 2$, $0 < \tau < 1$ parallelogram.

2.7.8. For $1 < b + \tau < 2$,

(a) find the (b, τ) coordinates for the 3-cycle with ϕ_0 in the first interval and ϕ_1 and ϕ_2 in the second interval.

(b) Find the curves bounding the region where this circle map exhibits such a 3-cycle.

(c) Sketch this region in the $1 < b + \tau < 2$, $0 < \tau < 1$ parallelogram.

2.7.9. Plot the graphs of Exercises 2.7.3 and 2.7.6 in the same graph. Do the regions overlap?

2.7.10. (a) Plot the graphs of Exercises 2.7.7 and 2.7.8 in the same graph. Do the regions overlap?

(b) For all $b > 1$, show every cycle is unstable.

(c) Comment on the existence of multiple cycles sharing the same parameter values. Specifically, what determines which cycle appears under iteration?

Chapter 3   Differential equations models

A man wanders around under a street lamp and looks at the ground. A cop approaches and asks the man what he's doing. Looking for my keys, replies the man. Where'd you lose them? asks the cop. Back there, says the man, pointing to a dark alley. Then why are you looking under the street lamp? asks the cop. Because the light's better here, replies the man.

The iterative equations of Ch. 2 work well in some settings, and are plausible models for systems that consist of agents which can occupy only a finite number of states: the number of open ion channels in a nerve cell at a given time and the number of Cs in the nucleotide sequence of a gene, for example. But some systems are better modeled by continuous variables: temperature, electrical current, and pH, for example. And although growing, the body of underlying theory for iterative equations still cannot address some natural questions.

In contrast to this, models with continuous variables can use the whole supply of tools from calculus and differential equations, mature subjects. To be sure, these techniques cannot solve all problems, but they can solve many, and some fairly easily. In a wide variety of settings, these solutions match observed data. In other words,

Sometimes the man is mistaken: he lost his keys under the street lamp.

For example, in Sect. 4.7 we'll see that we can solve the logistic differential equation $dx/dt = rx(1 - x)$. That is, we can find a function f with which we can write $x = f(t, x_0)$. In contrast, we cannot solve the logistic map $x_{n+1} = rx_n(1 - x_n)$, that is, we cannot find a function f that gives $x_n = f(n, x_0)$. All we can do is iterate. Right now, we have no tools comparable to calculus to study iteration. Happily, many physical and biological systems are well modeled by differential equations. Let's see some examples.

3.1 SOME TYPES OF MODELS

Instead of modeling how the value of a variable in one generation depends on its value in the previous generation by $x_{n+1} = f(x_n)$ now we turn our attention to modeling how the rate of change of a variable depends on its current value, on time, and perhaps on other factors. Because these equations specify the derivatives of the variables, they are called *differential equations*. We consider three types of differential equations:

$$dx/dt = f(t) \qquad \textit{pure-time differential equations}$$

$$dx/dt = g(x) \qquad \textit{autonomous differential equations}$$

$$dx/dt = h(x, t) \qquad \textit{non-autonomous differential equations}$$

Because the right-hand sides depend on only one variable, t for pure-time equations, x for autonomous equations, we'll call these two types *single-variable differential equations*. The contrast with non-autonomous equations should be clear: the right-hand side of a non-autonomous equation depends explicitly on both x and t.

Pure-time differential equations

Through a very clever experiment (beautifully described in [130]) by which he essentially discovered the idea of integration years before Newton was born (Galileo died in 1642, Newton was born in 1643), Galileo found that the speed with which an object falls is proportional to the time it falls. Because speed s is the rate of change of height h, this observation can be written as

$$\frac{dh}{dt} = k \cdot t$$

Galileo showed that this implies that $h = k \cdot t^2/2$. This is where he anticipated a geometric version of Newton's definite integral. Galileo could find the area under the curve only by geometry; he did not have Newton's antiderivatives.

We shall see that these general equations $dx/dt = f(t)$ can be solved by integration. But it turns out that integration is not always such an easy task. We'll study several techniques in Sects. 4.1, 4.2, 4.3, 4.4, 4.6, and 4.7. These are a tiny part of the subject of techniques of integration, but they are enough for us.

Autonomous differential equations

The *Verhulst equation* models single-species population growth in an environment with limited resources. Let $P(t)$ denote the population at time t. Verhulst postulated that the per capita growth rate R of the population is proportional to the available resources C, that is, $R = k \cdot C$. Next, suppose f units of resources are consumed in producing each new member of the population. This gives

$$\frac{dP}{dt} = R \cdot P = k \cdot C \cdot P \quad \text{and} \quad \frac{dC}{dt} = -f \cdot \frac{dP}{dt}$$

The dC/dt equation can be solved by an application of the fundamental theorem of calculus

$$\int \frac{dC}{dt} dt = \int -f \cdot \frac{dP}{dt} dt$$
$$C(t) = -f \cdot P(t) + C_0$$

Substituting this into the dP/dt equation gives

$$\frac{dP}{dt} = k \cdot (-f \cdot P + C_0) \cdot P = k \cdot C_0 \cdot \left(1 - \frac{f}{C_0} \cdot P\right) \cdot P$$

writing $r = k \cdot C_0$ and $C_0/f = K$ this is the Verhulst equation

$$\frac{dP}{dt} = rP\left(1 - \frac{P}{K}\right) \tag{3.1}$$

We interpret r as the growth rate and K the carrying capacity, the maximum population size that can be sustained long-term by the environment.

In Sects. 4.6 and 4.7 we'll learn how to solve Eq. (3.1). In this case we'll be able to find an explicit expression for $P(t)$, but often the best we'll be able to do is find a relation between t and the dependent variable, x or P or whatever we call it.

Non-autonomous differential equations

Suppose an animal population has a seasonally varying food supply, $f(t) = a + b\sin(t)$, and that the growth rate of the population P is proportional to both the population and the food supply. Then

$$\frac{dP}{dt} = r(a + b\sin(t))P$$

This equation is called non-autonomous because the right-hand side depends explicitly on P and t. We shall see that solving these equations can be easy or difficult, depending on the form of the right-hand side. For example, if $h(x,t) = h_1(x) \cdot h_2(t)$, then the method of Sect. 4.6 can be applied and with that, $dx/dt = h(x,t)$ can be solved by integration. When $h(x,t)$ does not admit a multiplicative decomposition, other methods, including the series techniques of Sect. 10.9 and the method of integrating factors of Sect. A.1 can be used.

Much of the rest of these notes will be concerned with solving these types of equations or systems of these equations, at least when finding closed-form solutions (we can write a formula for the solution) is not too much work, or in deducing general properties of solutions when finding closed-form solutions is too much work or is impossible. In Sect A.2 we'll show that solutions exist, even if we can't find them. We have little intrinsic interest in writing solutions to these equations. Rather, our real goal is to deduce some properties of the long-term behavior of the systems modeled by these equations. In addition, we are interested in the robustness of these behaviors, for as models of any physical or biological systems, our equations are not precise. Rigid behavior depending on the exact form and parameters of a model holds little hope of capturing the behavior of complex biological systems in noisy environments. Still, before getting to more realistic settings we develop some straightforward tools for solving simple differential equations. This will occupy us for a while.

3.2 EXAMPLES: DIFFUSION, GOMPERTZ, SIR

Here we build a few examples: drug concentration, diffusion across a cell membrane, the Gompertz population growth model, and the SIR (Susceptible-Infected-Recovered) model of epidemic spread. In Sects. 3.3 and 3.4 we see how Anna Laird matched the Gompertz model to observations of tumor growth, and that this was the basis for the chemotherapy scheduling regime of Larry Norton and Richard Simon.

Example 3.2.1. *Drug delivery-site concentration.* The rate at which the delivery-site concentration C of a drug declines is proportional to the concentration

$$\frac{dC}{dt} = -kC \tag{3.2}$$

where the constant k is positive, because the concentration decreases with time. A natural question is to find the *half-life*, the time $t_{1/2}$ required for the concentration to drop to $1/2$ its original value. In Exercise 3.2.1 (a) we'll verify that $C(t) = C(0)e^{-kt}$ is a solution of Eq. 3.2. Then to find the half-life we solve

$$\frac{1}{2}C(0) = C(t_{1/2}) = C(0)e^{-kt_{1/2}}$$

by canceling $C(0)$ from the left and right sides of the equation, taking the ln, and solving for $t_{1/2}$. This gives

$$t_{1/2} = \frac{\ln(1/2)}{-k} = \frac{\ln(2^{-1})}{-k} = \frac{-\ln(2)}{-k} = \frac{\ln(2)}{k}$$

and we've found the half-life in terms of the decay constant k. $\square$

Example 3.2.2. *Simple diffusion across a cell membrane.* Suppose C_i denotes the concentration of a chemical species inside a cell, and C_o the concentration outside the cell. The species exits the cell, crossing the cell membrane through a transmembrane channel, at a rate proportional to C_i, that is, at the rate αC_i where the constant of proportionality α depends on properties of the chemical species and of the cell membrane. Similarly, the species enters the cell at the rate αC_o. Combining these, with signs appropriate to whether C_i increases or decreases, we find

$$\frac{dC_i}{dt} = \alpha(C_o - C_i) \tag{3.3}$$

At least to the level of accuracy of this model, we see that if initially $C_i > C_o$, then C_i decreases monotonically until reaching an equilibrium at $C_i = C_o$. Another possibility is for C_i to undershoot C_o and then undergo a damped oscillation, approaching $C_i = C_o$. In Example 4.6.1 we'll see that damped oscillations are excluded in this simple model. $\square$

Example 3.2.3. *The Gompertz population model.* In 1825 the English mathematician Benjamin Gompertz published a populaion growth model [53] incorporating exponential growth for small populations and slower growth for larger populations. From this idea, Gompertz derived the differential equation

$$\frac{dP}{dt} = a\ln\left(\frac{K}{P(t)}\right)P(t) \tag{3.4}$$

where $a\ln(K/P(t))$ is the per capita growth rate and K is the carrying capacity, the maximum population size supported by available space and food. Here's why this is sensible. Observe that $(d/dP)a\ln(K/P) = -a/P$ so the per capita growth rate always is decreasing, as Gompertz desired. Second, if $P > K$ then $\ln(K/P)$ is negative and the population begins to decrease. We see that K is the largest population the environment can support before the population begins to decline.

In Fig. 3.1 we plot $f(x) = x\ln(1/x)$ to illustrate the general shape of the right-hand side of Eq. (3.4). We see that the increase of P eventually is overpowered by the decrease of $\ln(K/P)$.

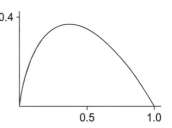

Figure 3.1. Graph of $x\ln(1/x)$.

For the moment, assuming $dP/dt \to 0$ as $t \to \infty$, Eq. (3.4) gives $P(t) \to K$ as $t \to \infty$. Fig. 3.1 is a plot of the right-hand side of Eq. (3.4) with $x = P(t)$, for $a = K = 1$. Note that rapid growth of small populations slows, eventually reversing, reflecting the negative consequences of too much growth in an environment with limited resources.

In Exercise 3.2.1 (d) we see a solution of Eq. (3.4) is a *Gompertz function*

$$P(t) = K\left(\frac{P_0}{K}\right)^{e^{-at}} \tag{3.5}$$

where $P_0 = P(0)$. A *Gompertz curve* is a plot of $(t, P(t))$.

With this solution we can find an expression for the constant K.

$$P_\infty = \lim_{t\to\infty} P(t) = \lim_{t\to\infty} K\left(\frac{P_0}{K}\right)^{e^{-at}} = K\left(\frac{P_0}{K}\right)^{e^{-\infty}} = K\left(\frac{P_0}{K}\right)^{0} = K$$

and so the Gompertz equation (3.4) can be rewritten as

$$\frac{dP}{dt} = a\ln\left(\frac{P_\infty}{P(t)}\right)P(t) \tag{3.6}$$

and Eq. (3.5) can be expressed as

$$P(t) = P_\infty\left(\frac{P_0}{P_\infty}\right)^{e^{-at}} \tag{3.7}$$

where K is replaced by P_∞. $\square$

Our final two examples are simple models of the spread of diseases.

Example 3.2.4. *The Susceptible-Infected-Recovered (SIR) model.* Divide the population into three groups: susceptible S, infected I, and recovered R. We begin with a nonlethal disease for which recovery confers immunity against future infection and infection produces permanent sterility.

The model parameters are β, δ, ϵ, and γ, where β is the per capita birth rate, δ is the per capita death rate, ϵ is the probability that a contact between a susceptible and an infected leads to infection of the susceptible per unit time, and

γ is the per capita exit rate from I to R. The reciprocal of γ has an interpretation that we shall use several times, so we state it as a proposition.

Proposition 3.2.1. The per capita exit rate from I to R, γ, is proportional to the reciprocal of the mean residence time in I.

The idea of the proof is straightforward. For example, if the mean residence time in the infected state is 7 days, then on average every day we expect 1/7 of the infecteds to leave the infected state by recovering. Note that likely this doesn't hold near the start of the infection. We must wait long enough that the average can be observed. In the case of the instance we're discussing, we'd need to wait at least 7 days, and really a bit longer, to find a mean residence time of 7 days.

The sterility assumption implies that only susceptibles give birth, and susceptibles give birth to susceptibles, so new susceptibles enter the population at the rate βS, the per capita birth rate times the size of the susceptible (reproductive) population. Similarly, at the rate γI infecteds recover; at the rates δS, δI, and δR, susceptibles, infecteds, and recovereds die.

The *principle of mass action*, that the number of encounters between two populations is proportional to the product of those populations, implies that at the rate ϵSI susceptibles become infected.

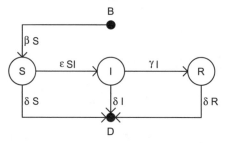

We can represent these dynamical relations by the *transition graph* between the states S, I, and R, indicated by circles in Fig. 3.2. An arrow from a circle contributes a negative term to the rate of change of the circle's variable; an arrow into a circle contributes a positive term. The states B and D, birth and death, are treated differently. Arrows from B contribute positively to the rate of

Figure 3.2. The dynamical relations between the three population groups of Ex. 3.2.4.

change of the circle to which they point, and do nothing else. Arrows into D contribute negatively to the rate of change of the circle from which they lead, and do nothing else.

From this graph we write the differential equations for this SIR model.

$$\frac{dS}{dt} = \beta S - \epsilon SI - \delta S \quad \frac{dI}{dt} = \epsilon SI - \gamma I - \delta I \quad \frac{dR}{dt} = \gamma I - \delta R \qquad (3.8)$$

Other options are investigated in Practice Problem 3.2.2, and in Exercises 3.2.4, 3.2.5, and 3.2.6. □

You might expect that these populations would settle down to equilibrium values, susceptibles, infecteds, and recovereds in balance. Sometimes this is the case, but in Sect. 13.2 we'll see that for some parameter values the populations cycle from low to high and back to low, usually a bit out of step.

Example 3.2.5. *SIR and malaria.* The SIR model was formulated in 1927 by W. O. Kermack and A. G. McKendrick [73], extending the work of Ronald Ross on the dynamics of malaria epidemics. While in India, in 1897 Ross discovered malaria parasites in the gut of mosquitoes, of a species identified as *Anopheles* by the Italian zoologist G. B. Grassi. Ross went on to uncover the complete life cycle of the parasite and to establish its role in passing malaria to birds. For this work Ross was awarded the 1902 Nobel Prize for Physiology or Medicine. The story of why Grassi did not share the prize is a chapter about greed, ego, and politics in science. Not the finest moment in our history, but one that illustrates that scientists are subject to the same foibles as are non-scientists. We're just a bit more completely distracted by our work than are most people.

Ross used his study of malaria to develop a mathematical model for the spread of the disease. Among other things, he sought to explain why the *epidemic curve*, a plot of the number of new infecteds as a function of time, was nearly symmetric about its maximum, and why epidemics terminate before all susceptibles are infected. He published three influential papers [134, 135, 136], the last two with coauthor Hilda Hudson, on the mathematics of epidemics. One of Ross's goals was to understand, from a simple model, the empirical observation (the second point mentioned above) that epidemics die out before every susceptible is infected. Common explanations offered then included that some of the susceptibles were naturally immune, or that the virulence of the infectious agent degraded with time. Ross, with some interest and talent in math, wondered if this could be a consequence of general principles. Can it be derived from a mathematical model of the epidemic?

We'll use the SIR model Eq. (3.8) for a simplification of Ross's analysis. Ross assumed that the birth rate β and the death rate δ are 0 and that the size N of the population is constant. (The "no death" condition can be relaxed if we think of the R population as "removed" rather than "recovered," where we understand that removed includes both those for whom infection has provided a perhaps short-term immunity, as well as those who have died. In neither case can members of the R population re-enter the I or the S population. Oddly, in order for this calculation to work, N must be constant and so dead members of the population still are counted as members of the population.) Simplified with these substitutions, the SIR model can be written

$$\frac{dS}{dt} = -\epsilon SI \quad \frac{dI}{dt} = \epsilon SI - \gamma I \quad \frac{dR}{dt} = \gamma I \tag{3.9}$$

As with the logistic map in Chapter 2, we are not so much interested in the number of individuals in S, I, or R, but in the fraction of the total population N in each group. So we rescale the variables $X = S/N$, $Y = I/N$, and $Z = R/N$, and obtain new, but equivalent, differential equations

$$\frac{dX}{dt} = \frac{dS}{dt}\frac{1}{N} = -N\epsilon\frac{S}{N}\frac{I}{N} = -N\epsilon XY$$

$$\frac{dY}{dt} = \frac{dI}{dt}\frac{1}{N} = N\epsilon\frac{S}{N}\frac{I}{N} - \gamma\frac{I}{N} = N\epsilon XY - \gamma Y$$

$$\frac{dZ}{dt} = \frac{dR}{dt}\frac{1}{N} = \gamma\frac{I}{N} = \gamma Y$$

Finally, the time units are arbitrary, accidents of clocks and counting. The only natural time scale in the problem is $1/\gamma$, the mean residence time in the infected state by Prop. 3.2.1. The new time parameter is

$$\tau = t/(1/\gamma) = \gamma t$$

By the chain rule,

$$\frac{dX}{d\tau} = \frac{dX}{dt}\frac{dt}{d\tau} = \frac{dX}{dt}\frac{1}{\gamma} = -\frac{N\epsilon}{\gamma}XY$$

Similarly,

$$\frac{dY}{d\tau} = \frac{dY}{dt}\frac{1}{\gamma} = \frac{N\epsilon}{\gamma}XY - Y$$

and

$$\frac{dZ}{d\tau} = \frac{dZ}{dt}\frac{1}{\gamma} = Y$$

Write $R_0 = N\epsilon/\gamma$. Then our rescaled SIR model is

$$\frac{dX}{d\tau} = -R_0 XY \quad \frac{dY}{d\tau} = R_0 XY - Y \quad \frac{dZ}{d\tau} = Y \tag{3.10}$$

At time $\tau = 0$, almost everyone in the population is susceptible (initially, no one in the population is immune), so $X(0) \approx 1$, $Y(0) \approx 0$, but $Y(0) \neq 0$ or there would be no epidemic, and $Z(0) = 0$, because you must enter the Y population before you can enter the Z population.

An epidemiological interpretation of R_0 is straightforward. At the start of the disease outbreak, few are infected so almost the entire population N is susceptible. Now ϵ is the probability per unit time that a contact between a susceptible and an infected leads to infection of the susceptible. Then $N\epsilon$ is the average number of infecteds produced by a single infected per unit time, so $R_0 = N\epsilon/\gamma$ is the average number of infecteds produced by a single infected for the duration that person is infected. We call R_0 the *basic reproductive ratio* of the disease. Then we see that if $R_0 < 1$, on average a single infected will give rise to fewer than 1 new infecteds and so the the infection will die out.

If $R_0 > 1$, the infection can spread. A natural question is how far can it spread? Will every susceptible eventually become infected, or will the epidemic subside leaving some susceptibles uninfected? We can find out.

Every infected eventually enters the recovered (or removed) population, Z. Now $Y \geq 0$, so $dZ/d\tau \geq 0$. That is, Z is a non-decreasing function of τ, and Z is bounded above by 1, so the limit $Z_\infty = \lim_{\tau \to \infty} Z(\tau)$ exists. (We'll say more about this in a moment.) We'd like to show $Z_\infty < 1$, which means that some of the population never enters the R population. Because everyone in the I population eventually enters the R population, $Z_\infty < 1$ means that some of the S population never enters the I population; that is, some susceptibles never are infected. To show this, first we'll derive a relationship between X and Z, then use $X + Y + Z = 1$ to find an expression for $dZ/d\tau$ that depends only on Z.

By the chain rule, $dX/d\tau = (dX/dZ) \cdot (dZ/d\tau)$. Solving for dX/dZ we find

$$\frac{dX}{dZ} = \frac{dX/d\tau}{dZ/d\tau} = \frac{-R_0 XY}{Y} = -R_0 X$$

where the penultimate equality follows by Eq. (3.10). In Example 4.6.4 we'll see how to solve this. For now, we'll say that the solution is

$$X(Z) = X(0)e^{-R_0 Z}$$

(It's easy to check that this function X is a solution of $dX/dZ = -R_0 X$.) Then using the third equation of Eq. (3.10), $X + Y + Z = 1$, and the expression we've just derived for $X(Z)$, we find

$$\frac{dZ}{d\tau} = Y = 1 - X - Z = 1 - X(0)e^{-R_0 Z} - Z \qquad (3.11)$$

As $\tau \to \infty$, $Z \to Z_\infty$ monotonically, that is, without wiggling through ever smaller increases and decreases. Here's why. Once you enter the R population, you can't leave, so R never decreases. And the R population is bounded above by the total population N. A function that always increases and is bounded above must approach a limit. (This seems obvious, but it's the fairly subtle result that the real numbers are *complete*, roughly that there are no gaps in the real numbers.) Then as $\tau \to \infty$ we also have $dZ/d\tau \to 0$. Because $X(0) \approx 1$, Eq. (3.11) shows

$$0 = 1 - X_0 e^{-R_0 Z} - Z \approx 1 - e^{-R_0 Z} - Z$$

That is, Z_∞ is (approximately) a solution of

$$Z_\infty = 1 - e^{-R_0 Z_\infty} \tag{3.12}$$

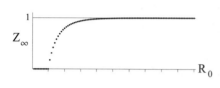

Figure 3.3. Z_∞ as a function of R_0.

In Fig. 3.3 we plot approximate solutions of Eq. (3.12) for R_0 ranging from 0 to 10. Note the graph shows that the infection dies out quickly if $R_0 < 1$, which we've seen before. And the graph shows that for all R_0, the epidemic disappears before all the susceptibles are infected. Not surprisingly, the larger the value of R_0, the larger is the fraction of infecteds, as measured by Z_∞.

We'll make two more observations from this formulation of Ross's analysis. The first is that while we cannot analytically solve Eq. (3.12) for Z_∞, we can solve it for R_0,

$$R_0 = -\frac{\ln(1 - Z_\infty)}{Z_\infty} \tag{3.13}$$

Although measuring $1/\gamma$, the mean time an infected patient remains infected, can be straightforward, some complications can occur with trying to measure ϵ, the probability that a susceptible-infected encounter will produce a new infected. But public health records can be used to approximate Z_∞, and we can use Eq. (3.13) to estimate R_0. Recall that $R_0 = N\epsilon/\gamma$, so if we can measure R_0, N, and $1/\gamma$, we can estimate ϵ.

The second observation is that we can use (numerical) solutions of Eq. (3.10) to plot the epidemic curves, the number of new infecteds, measured by $-dX/d\tau$, as a function of the rescaled time τ. But shouldn't the number of new infecteds be measured by $dY/d\tau$? No: $dY/d\tau$ measures the number of new infecteds minus the number of new recovereds. Because every new infected is a susceptible who just has become infected, $-dX/d\tau$ is the measure of new infecteds. By the first equation of Eq. (3.10), we'll plot $R_0 X(\tau) Y(\tau)$. Three examples are shown in Fig. 3.4 for $R_0 = 2, 4$, and 6, for $0 \leq \tau \leq 10$.

Each curve is approximately symmetric about its maximum value, a feature observed in epidemic curves constructed from public health records. One of Ross's goals was to derive this symmetry as a consequence of the general principles of the model and not of the particulars of the disease. In addition, we see that larger R_0 produce infections that peak at a higher rate and die out sooner. This reinforces our intuition. □

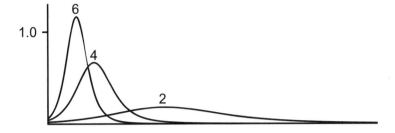

Figure 3.4. Epidemic curves for $R_0 = 2, 4$, and 6.

In Sect. A.3 we use the SIR model to derive the epidemic threshold theorem of Kermack and McKendrick.

Practice Problems

3.2.1. In the Verhulst equation (Eq. (3.1)) take $r = 1$ and $K = 1$. Plot $y = f(P) = rP(1 - P/K)$ as a function of P, for $0 \leq P \leq 1$. Deduce the shape of the $P(t)$ curve, the *solution curve* for all $t \geq 0$.

3.2.2. Modify the SIR model of Ex. 3.2.4, dropping the sterility assumption. Write the differential equations of the model and explain the role of each term on the right-hand side of the equations.
(a) Assume susceptibles, infecteds, and recovereds all give birth to susceptibles.
(b) Assume susceptibles and recovereds give birth to susceptibles, but infecteds give birth to infecteds.

3.2.3. For the rescaled SIR model (Eq. (3.10)), show that if the limiting fraction of recovereds, Z_∞, for an epidemic in population A is larger than Z_∞ for an epidemic in population B, then the basic reproductive ratio R_0 for A is larger than R_0 for B.

Practice Problem Solutions

3.2.1. First we see that the graph of $f(P)$ is a parabola, passing through $(0,0)$ and $(1,0)$. See the left side of Fig. 3.5. From this we see that, for example, if $P(t)$ is small and positive, $P(t)$ grows slowly. Then $P(t)$ grows most quickly for $P(t)$ near $P(t) = 1/2$, and ever more slowly as $P(t) \to 1$. This we sketch on the right side of Fig. 3.5, the graph of Eq. (3.5) with $P(0) = 1000$.

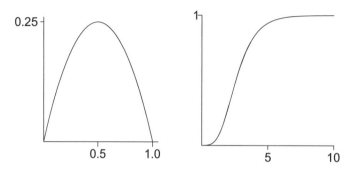

Figure 3.5. A sketch of $f(P)$ (left) and $P(t)$ (right) for Practice Problem 3.2.1. Note the scales on the axes.

3.2.2. (a) The graph is shown on the left side of Fig. 3.6. From this we see the differential equations are

$$\frac{dS}{dt} = \beta(S+I+R) - \delta S - \epsilon SI \quad \frac{dI}{dt} = \epsilon SI - \gamma I - \delta I \quad \frac{dR}{dt} = \gamma I - \delta R$$

Because S, I, and R give birth to S, dS/dt contains the positive term $\beta(S+I+R)$. The negative term δS represents the decrease in S from death; the negative term ϵSI is the decrease in S when susceptibles become infected. Here we have applied the principle of mass action.

Susceptibles who become infected leave the susceptible population and enter the infected population, hence the positive term ϵSI in dI/dt. The negative term δI represents the decrease in I from death; the negative term γI the decrease in I by recovery of infecteds.

Infecteds who recover enter the the recovered population, hence the positive term γI in dR/dt. The negative term δR represents the decrease in R from death. (b) The graph is shown on the right side of Fig. 3.6. From this we see the differential equations are

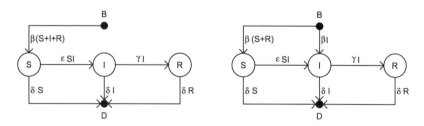

Figure 3.6. The transition graphs for Practice Problem 3.2.2 (a) (left) and (b) (right).

$$\frac{dS}{dt} = \beta(S+R) - \delta S - \epsilon SI \qquad \frac{dI}{dt} = \epsilon SI + \beta I - \gamma I - \delta I \qquad \frac{dR}{dt} = \gamma I - \delta R$$

The changes from part (a) are these. The positive term in dS/dt is $\beta(S+R)$ because only susceptibles and recovereds give birth to susceptibles. The dI/dt has another positive term, βI, because now infecteds give birth to infecteds.

3.2.3. The simplest approach to this problem is to plot R_0 as a function of Z_∞ using Eq. (3.13), which we have done in Fig. 3.7. Because the graph is increasing, a higher Z_∞ value means a higher R_0 value.

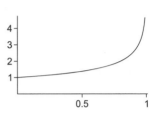

Figure 3.7. A plot of R_0 as a function of Z_∞.

We can try to do this without graphing, by computing the derivative of $f(x) = -\ln(1-x)/x$ and showing it is positive, so the graph is increasing.

$$f'(x) = \frac{\dfrac{x}{1-x} + \ln(1-x)}{x^2} = \frac{x + (1-x)\ln(1-x)}{x^2(1-x)}$$

In the range $0 < x < 1$, the denominator of f' is positive and x is positive, but $(1-x)\ln(1-x)$ is negative. In order to show $f' > 0$, we must show that $x > -(1-x)\ln(1-x)$. This may seem challenging, but there's a trick that helps. Write $g(x) = -(1-x)\ln(1-x)$ and calculate

$$g'(x) = \ln(1-x) + 1 \qquad g''(x) = \frac{1}{x-1}$$

Certainly, for $0 < x < 1$, $g''(x) < 0$, so the graph of g is concave down and so lies below all its tangents. Next, $g(0) = 0$ and $g'(0) = 1$, so $y = x$ is the tangent line to $y = g(x)$ at $x = 0$. Consequently, the graph $y = g(x)$ lies below the graph $y = x$. In other words, $x > g(x) = -(1-x)\ln(1-x)$ and so $f' > 0$. This is what we needed to show.

Exercises

3.2.1. (a) Verify that $C(t) = C(0)e^{-kt}$ is a solution of Eq. (3.2).
(b) In Eq. (3.2) with $k = 10$, for what times $t_{1/4}$ and $t_{1/8}$ are $C(t_{1/4}) = (1/4)C(0)$ and $C(t_{1/8}) = (1/8)C(0)$?
(c) Find the pattern for the times required for these concentration drops

$$C(0) \to \frac{1}{2}C(0) \to \frac{1}{4}C(0) \to \frac{1}{8}C(0)$$

(d) Verify that Eq. (3.5) is a solution of Eq. (3.4).

3.2.2. For the system

$$\frac{dx}{dt} = -2x + y^3 \qquad \frac{dy}{dt} = -0.0001y$$

(a) for small values of y, show $y(t)$ changes very slowly.
(b) Show that over short time scales and for small x, the growth rate of x is approximately constant.

3.2.3. Suppose dx/dt varies directly with x and inversely with y, with a proportionality constant $a > 0$. Suppose dy/dt varies directly with y^2, with a proportionality constant $b = -c$ with $c > 0$. Assume $x(t) > 0$ and $y(t) > 0$ for all t.
(a) Write the equations for this system.
(b) As t increases, show that $x(t)$ increases more and more rapidly.

3.2.4. Write the equations for a simplified SIR model (Example 3.2.4) with constant population size $N = S + I + R$. Let μ denote the per capita birth rate, equal to the per capita death rate, ϵ the fraction of S and I encounters that result in a susceptible becoming infected, and γ the per capita recovery rate of infecteds. Assume susceptibles, infecteds, and recovereds all give birth to susceptibles. Explain the terms of your equations.

3.2.5. Suppose $N = S + I + R$ is constant and infection gives no immunity, that is, infecteds who recover re-enter the susceptible population. Define μ, ϵ, and γ as in Exercise 3.2.4. This is called the SIS model.
(a) Assuming birth rate = death rate = 0, write the differential equations for this model. Explain the terms of your equations.
(b) Assuming a constant population $N = S + I$, and that both infecteds and susceptibles give birth to susceptibles, write the differential equations for this model. Explain the terms of your equations.

3.2.6. In addition to S, I, and R, add a state E, the latent state, infecteds who cannot yet infect others. Suppose ϵ denotes the fraction of SI encounters that give rise to a susceptible entering the latent state, η denotes the per capita rate at which latents enter state I, and μ and γ are as in Exercise 3.2.4. This is called the SEIR model.

(a) Assuming birth rate $=$ death rate $= 0$, write the differential equations for this model. Explain the terms of your equations.

(b) Assuming a constant population $N = S + E + I + R$, and that all four states give birth to susceptibles, write the differential equations for this model. Explain the terms of your equations.

3.2.7. Suppose the basic reproductive ratio of a disease is $R_0 = 2$. Estimate the fraction of the susceptible population uninfected when the disease dies out. Express your answer by carrying two digits to the right of the decimal.

3.2.8. Suppose the half-life of ^{18}F, the radioisotope in fludeoxyglucose used in PET scans, is 110 minutes. How long must the patient wait for 95% of the ^{18}F to decay?

3.2.9. For the rescaled SIR model Eq. (3.10),

(a) If the epidemic terminates at $Z_\infty = 0.7$, find the basic reproductive ratio R_0 of the disease.

(b) If the expected time in the infected state is 2 days, estimate the average number of new cases per unit time, per infected individual.

3.2.10. For the SIS model of Exercise 3.2.5 with birth rate $=$ death rate $= 0$,

$$S' = -\epsilon SI + \gamma I \qquad I' = \epsilon SI - \gamma I$$

(a) Rescale this model using $X = S/N$, $Y = I/N$, and replacing d/dt with $d/d\tau$ where $t = \tau/\gamma$. Recall we write $R_0 = \epsilon N/\gamma$ and that $X + Y = 1$.

(b) Find the value of Y for which Y is increasing most rapidly as a function of τ.

3.3 LAIRD'S GROWTH LOGIC

The Gompertz equation (3.4) exhibits growth rates that increase with size for small populations and decrease with size for large populations. While Gompertz used this equation for his actuarial studies of population growth, the characteristic sigmoid curve, similar to the right of Fig. 3.5, has appeared in a variety of fields, including the growth of normal embryos, animal populations, visceral organs, and neural nets, and the dynamics of enzymes with cooperative binding. In 1964, Anna Laird used the Gompertz curve to fit tumor growth data.

In this and the next section we will explore the logic of Laird's mathematical argument that some tumors follow a Gompertz curve and see how her work influenced Larry Norton and Richard Simon in the 1970s in their design of cancer treatment schedules.

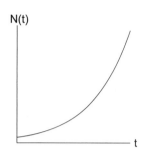

Figure 3.8. Exponential growth: $N(t) = be^{at}$.

When Laird looked into the dynamics of tumor growth [78, 79, 80], the conventional model was a simple exponential process that would terminate when the tumor exhausted the host's nutritional resources or available space. These truncated exponential models match the growth of bacteria colonies and fit the observations of certain types of cancer, including L1210 leukemia in mice [153]. Laird pointed out that this behavior is rare and observed for only a brief duration. Observation of tumor growth over a long period of time (spanning an increase in tumor size by a factor of 100 to 1000) led Laird to conclude that most tumors grow more slowly as they get larger. For (untruncated) exponential growth, illustrated in Fig. 3.8, the growth rate (the slope of the tangent line) increases with time.

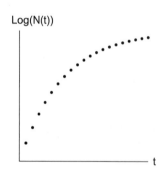

Figure 3.9. Cartoon of Laird's plot of $\log(N(t))$ vs t for the Ehrlich ascites tumor.

From data for an Ehrlich ascites tumor (an undifferentiated tumor common in laboratory studies), Laird plotted $\log(N(t))$ vs t, where $N(t)$ denotes the number of tumor cells, a measure of the tumor size, at time t. In Fig. 3.9 we show a cartoon of Laird's data.

If the tumor were growing exponentially, this plot would have been a straight line. Just take the log of both sides of $N(t) = be^{at}$ to get $\log(N(t)) = at + \log(b)$. Then the points $(t_i, \log(N(t_i)))$ would lie on a straight line of slope a. Instead, the curve suggested to Laird that the growth rate of the ascites tumor decreased with time. The equation that fits this tumor growth data has this form

$$N(t) = N_0 e^{(c/a)(1-e^{-at})} \tag{3.14}$$

This is another Gompertz function, with $N_0 = N(0)$ the initial tumor size, and a and c are constants that determine the growth behavior. Below we'll see that $a\ln(N_\infty/N)$ is the per capita growth rate and $c = a\ln(N_\infty/N_0)$.

To justify calling $N(t)$ defined by Eq. (3.14) a Gompertz function, we'll compare it with $P(t)$ defined by Eq. (3.5).

$$N(t) = N_0 e^{(c/a)(1-e^{-at})} = N_0 e^{c/a} e^{-(c/a)e^{-at}} = N_0 e^{c/a} (e^{-c/a})^{e^{-at}}$$

$$= N_0 \frac{N_\infty}{N_0} \left(\frac{N_0}{N_\infty}\right)^{e^{-at}} = K \left(\frac{N_0}{N_\infty}\right)^{e^{-at}}$$

where the penultimate equality comes from writing $e^{-c/a} = N_0/N_\infty$. This is just Eq. (3.7), with P replaced by N. Compare this with Example 3.2.3, to see the interpretations of a and c given above.

When t is small, the growth function is approximately exponential. To see this, in Sect. 10.8 we'll show

$$e^x = 1 + x + \frac{x^2}{2!} + \frac{x^3}{3!} + \frac{x^4}{4!} + \cdots$$

If x is small, x^2 is small^2, that is, really small; x^3 is small^3, really really small; and so on. Consequently, for small x, we can use the approximation $e^x \approx 1 + x$. Then for small t, $e^{-at} \approx 1 - at$ and so

$$N(t) = N_0 e^{(c/a)(1-e^{-at})} \approx N_0 e^{(c/a)(1-(1-at))} = N_0 e^{(c/a)(at)} = N_0 e^{ct}$$

On the other hand, as $t \to \infty$, $1 - e^{-at} \to 1$ and the Gompertz function approaches the horizontal asymptote

$$N_\infty = \lim_{t \to \infty} N(t) = N_0 e^{c/a} \tag{3.15}$$

From this we see that

$$e^{c/a} = \frac{N_\infty}{N_0} \tag{3.16}$$

If you're worried that this is inconsistent with the calculation $P_\infty = K$ from Example 3.2.3, compare Eqs. (3.7)

$$P(t) = P_\infty \left(\frac{P_0}{P_\infty}\right)^{e^{-at}}$$

and (3.14)

$$N(t) = N_0 e^{(c/a)(1-e^{-at})} = N_0 e^{c/a} e^{-(c/a)e^{-at}} = N_\infty e^{-(c/a)e^{-at}}$$

$$= N_\infty \left(\frac{1}{e^{c/a}}\right)^{e^{-at}} = N_\infty \left(\frac{N_0}{N_\infty}\right)^{e^{-at}} \tag{3.17}$$

where the third equality comes by Eq. (3.15), and the last by Eq. (3.16). So we see that the calculation for N_∞ is consistent with that of P_∞.

Thus, c and a determine the asymptotic behavior of particular tumors. Laird used the *method of least squares* with rat, mouse, and rabbit data to determine the values for c, a, and N_0, starting with arbitrarily small values for all three and then computing the three parameters simultaneously with successive approximations. The approach Laird used to determine these values required a computer, but the method of least squares itself isn't difficult. Given a set of n experimental data points

$$\{(x_1, y_1), (x_2, y_2), \ldots, (x_n, y_n)\},$$

the slope m is

$$\frac{n(x_1 y_1 + x_2 y_2 + \cdots + x_n y_n) - (x_1 + x_2 + \cdots + x_n)(y_1 + y_2 + \cdots + y_n)}{n(x_1^2 + x_2^2 + \cdots + x_n^2) - (x_1 + x_2 + \cdots + x_n)^2}$$

Given m, the y-intercept b is

$$b = \frac{(y_1 + y_2 + \cdots + y_n) - m(x_1 + x_2 + \cdots + x_n)}{n}$$

Although these equations may look complicated, the derivation is rather straightforward, which you will see in Sect. 9.1.

While the biological reason for this type of Gompertzian growth isn't established, Laird proposed two possible explanations:

1. there is a rapid increase in the mean duration times of successive generations, and
2. there is a rapid loss in the number of viable tumor cells.

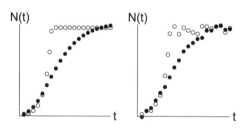

Figure 3.10. Truncated exponential vs Gompertz; right, with added noise.

She also argued that a simple passive failure of the host to provide nutritional support was not an adequate explanation because if that were the case, the graphs of tumor growth would look more like bacterial growth graphs up to a point, followed by an abrupt transition to a flat, horizontal plot. In tumor size curves the convergence to the horizontal asymptote is much more gradual. In the left image of Fig. 3.10 we see Gompertzian growth (filled circles) and truncated exponential growth (open circles). No one could fail to distinguish these graphs, but of course they are too clean. The right image of Fig. 3.10 replots the points of the left graph, with some noise added to the vertical coordinate of each point. Measurements in the physical world include some noise-with the obvious exception: counting rather than measuring; there may be 3 or 4 raccoons in my back yard, but not 3.5-so the right image, or something more complex, is closer to what we'll see.

In the next section we'll see how Laird's application of the Gompertz model to tumor growth was instrumental in designing a more effective chemotherapy schedule.

Practice Problems

In problems 3.3.1 and 3.3.2 for the given lists of points
(a) plot the points.
(b) Find the best-fitting line.
(c) Comment on whether the line is a reasonable fit for the data. One way to measure goodness of fit is the average difference between the actual y-values y_i and the predicted values $mx_i + b$. But the average of the differences isn't such a good measure, because positive and negative differences can subtract out. We could take the average of the absolute values of the differences, but more commonly used is the square root of the average of the square of the differences.

3.3.1. (1, 2.1), (2, 2.77), (3, 3.44), (4, 4.11), (5, 4.78), (6, 5.45), (7, 6.12), (8, 6.79), (9, 7.46), (10, 8.13)

3.3.2. (1, 0), (2, 3), (3, 2), (4, 5), (5, 4), (6, 7), (7, 6), (8, 9), (9, 8), (10, 11)

3.3.3. One of these lists of points lies on the graph of $y = 1 + ax^2$, the other on the graph of $y = 1 + bx^3$, for some values of a and b. Say which list goes with which graph. Give a reason to support your choice. Your justification need not include a plot of the points or an estimation of the parameters a and b.
(a) (0, 1.00), (0.2, 1.04), (0.4, 1.17), (0.6, 1.40), (0.8, 1.70), (1.0, 2.10), (1.2, 2.58), (1.4, 3.16), (1.6, 3.82), (1.8, 4.56), (2.0, 5.40)
(b) (0, 1.00), (0.2, 1.01), (0.4, 1.05), (0.6, 1.17), (0.8, 1.41), (1.0, 1.81), (1.2, 2.38), (1.4, 3.20), (1.6, 4.28), (1.8, 5.57), (2.0, 7.40)

Practice Problem Solutions

3.3.1. (a) The points are plotted on the left of Fig. 3.11.
(b) We compute

$$x_1 + x_2 + \cdots + x_{10} = 55$$
$$y_1 + y_2 + \cdots + y_{10} = 51.15$$
$$x_1 y_1 + x_2 y_2 + \cdots x_{10} y_{10} = 336.6$$
$$x_1^2 + x_2^2 + \cdots x_{10}^2 = 385$$

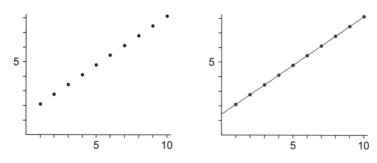

Figure 3.11. Points (left), points and best-fitting line (right) for Practice Problem 3.3.1.

From this we find the slope is $m = 0.67$ and the y-intercept is $b = 1.43$. That is, the equation of the best-fitting line is

$$y = 0.67x + 1.43$$

(c) On the right of Fig. 3.11 the points are superimposed on the line. To test the goodness of fit, we'll compute

$$\sqrt{\frac{(y_1 - (mx_1 + b))^2 + \cdots + (y_{10} - (mx_{10} + b))^2}{10}} \approx 1.4 \times 10^{-16}$$

So indeed this appears to be a perfect fit.

3.3.2. (a) The points are plotted on the left of Fig. 3.12.
(b) We compute

$$x_1 + x_2 + \cdots + x_{10} = 55$$
$$y_1 + y_2 + \cdots + y_{10} = 55$$
$$x_1y_1 + x_2y_2 + \cdots x_{10}y_{10} = 390$$
$$x_1^2 + x_2^2 + \cdots x_{10}^2 = 385$$

From this we find the slope is $m = 1.06$ and the y-intercept is $b = -0.33$. That is, the equation of the best-fitting line is

$$y = 1.06x - 0.33$$

(c) On the right of Fig. 3.12 the points are superimposed on the line. To test the goodness of fit, we'll compute

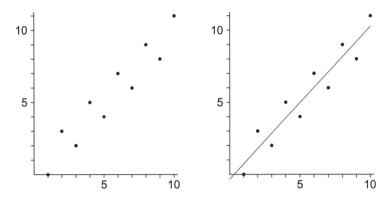

Figure 3.12. Points (left), points and best-fitting line (right) for Practice Problem 3.3.2.

$$\sqrt{\frac{(y_1 - (mx_1 + b))^2 + \cdots + (y_{10} - (mx_{10} + b))^2}{10}} \approx 1.54$$

The points appear to lie along two parallel lines, with the best-fitting line passing through the middle of the points. This line does not appear to be a good fit to the data. A first impression might be the line is a good model, with the data points scattered above and below the line, perhaps due to some environmental perturbations. But such perturbations are unlikely to be so symmetrical. A better model would be a *stair-step*, connecting the dots in the order they occur along the x-axis. This might be explained by alternating rapid increases, followed by slow recoveries. Learn to read the story the data are telling. Symmetry usually has a source that is not random.

3.3.3. Remember the graphs of $y = x^2$ and $y = x^3$. For $0 \le x < 1$ we have $x^3 < x^2$, while $x^3 > x^2$ for $x > 1$. The y-values of list (b) are lower than those of list (a) for $x \le 1.2$ and greater than those of list (a) for $x \ge 1.4$. Does it matter that the points of list (b) jump above those of list (a) around $x = 1.2$, while the graph of $y = x^3$ jumps above the graph of $y = x^2$ at $x = 1$? No, varying the coefficients a and b can move the crossover point left or right, but regardless of the (positive) coefficients, for small enough x we see $1 + bx^3 < 1 + ax^2$ and for large enough x $1 + bx^3 > 1 + ax^2$. So we expect (a) is the quadratic data and (b) is the cubic data.

Exercises

In Probs. 3.3.1–3.3.4, the data sets correspond to the y-values for eleven values of x from 0 to 10 in steps of 1.

(a) Sketch the graph.

(b) Find the best-fitting line.

(c) Comment on whether or not the line is a reasonable fit for the data. Use the goodness of fit data in the practice problems.

3.3.1. 0.20, 0.96, 2.57, 4.11, 6.51, 7.53, 8.49, 10.40, 12.49, 13.87, 14.65

3.3.2. -0.27, -1.87, -3.82, -5.51, -7.64, -9.28, -10.93, -12.14, -14.93, -15.78, -18.03

3.3.3. -0.10, 0.06, 0.13, 0.16, 0.11, 0.34, 1.24, 1.46, 1.78, 2.03, 2.32

3.3.4. -0.30, 1.16, 4.19, 9.23, 16.14, 25.11, 36.27, 48.78, 63.84, 81.48, 99.61

In Probs. 3.3.5–3.3.8, the data sets correspond to the y-values for eleven values of x from 0 to 2 in steps of 0.2. Are these better fit by $y = a + x^2$ or by $y = ae^x$? Estimate the parameter a. Explain your choice.

3.3.5. 1.20, 1.47, 1.79, 2.19, 2.67, 3.26, 3.98, 4.87, 5.94, 7.26, 8.87

3.3.6. 1.20, 1.24, 1.36, 1.56, 1.84, 2.2, 2.64, 3.16, 3.76, 4.44, 5.20

3.3.7. 0.80, 0.84, 0.96, 1.16, 1.44, 1.80, 2.24, 2.76, 3.36, 4.04, 4.08

3.3.8. 0.90, 1.10, 1.34, 1.64, 2.00, 2.45, 2.99, 3.65, 4.46, 5.44, 6.65

3.3.9. For Gompertzian growth of Eq. (3.14) with $t \geq 0$,

(a) show the tangent of $N(t)$ is not horizontal for any t,

(b) find the value of t for which $N'(t)$ takes its maximum value,

(c) find this maximum value of $N'(t)$, and

(d) show this maximum value of $N'(t)$ goes to ∞ as $c \to \infty$, but for any finite value of c, the maximum of $N'(t)$ is finite.

3.3.10. For the function $f(t) = a(1 - \cos(t))$, $0 \leq t \leq \pi$,

(a) sketch the graph of $f(t)$ and verify that it is a sigmoid, as is the Gompertz function $N(t)$.

(b) Could $f(t)$ be another model for Gompertzian growth? Specifically, does $f(t)$ satisfy property (a) of Exercise 3.3.9?

3.4 THE NORTON-SIMON MODEL

Laird's work that showed some tumor growth is fit better by a Gompertz curve rather than by a truncated exponential did not have much influence until the 1970s, when oncologist Larry Norton and bioinformatician Richard Simon applied Laird's findings to the design of chemotherapy treatment schedules [104, 105, 106, 107, 108, 109, 144]. The mid-1970s marked a puzzling time for oncologists and an unsettling time for cancer patients. By the end of a chemotherapy regimen, tumors would appear to be eradicated completely, only to return later in full force. Strangely, the regrown tumors often even had the same drug sensitivity as the originals. Doctors, including Norton and Simon, who began to investigate the problem put forth two possible explanations: either tumors were developing drug-resistant cells (a scary thought! truly drug-resistant tumors would bring a whole host of problems), or there was an inadequate killing of the drug-sensitive cells, or both.

At the time, chemotherapy regimens frequently employed intense doses initially, with a tapering off as the tumor decreased in size. This pattern had many logical reasons: to curb the depletion of bone marrow reserves late in treatment (a side effect of chemotherapy), to reduce toxicity, and to acknowledge the possible long-term adverse consequences of chemotherapy. In addition, the widely held exponential growth model implies that as a tumor grows, its growth rate is slowed by resource limitation. Consequently, the smaller the tumor, the more rapidly it grows. Chemotherapy agents are poisons; a chemotherapy schedule is a plan to deliver the poison at a level sufficient to kill the cancer, but not to kill, or to damage badly, the patient. The rate at which a cancer cell absorbs nutrients and chemotherapy agents is proportional to the cell growth rate, so fast-growing cancer cells absorb proportionally more poison. A lower dose, less damaging to the patient, still could be effective against the cancer. These reasons standardized chemotherapy regimens with tapered dosing schedules. Some cancers like colon carcinoma and malignant melanomas did appear, after surgical resection, to retain cells that were drug-resistant enough to escape death from low doses of chemotherapy. Norton and Simon also observed that most clinical evidence suggested some tumors were less sensitive to therapy when they were very small or very large than when they were of intermediate size. This is a characteristic of Gompertzian growth; under exponential growth the sensitivity decreases with increasing tumor size. Consequently, Norton and Simon deduced that inadequate killing is the culprit of many observed recurrences of cancers including Hodgkin's lymphoma and Burkitt's lymphoma. These observations pointed to the inadequacy of the *small is sensitive* model. We'll explore how

Norton and Simon's incorporation of Laird's use of the Gompertz model for tumor growth translated to more effective designs of chemotherapy schedules.

First, we introduce some foundation for their model. Start with the differential equation that corresponds to exponential growth

$$\frac{dN(t)}{dt} = a \cdot N(t) \tag{3.18}$$

Eq. (3.18) has a growth rate that increases in direct proportion to tumor size, N. The larger the tumor gets, the faster it grows, which makes sense only if the number of cells undergoing division is the sole factor in tumor growth. Here the growth factor a is positive because cell reproductive rate exceeds cell death rate. Skipper's L1210 leukemia in mice had a growth curve approximated by Eq. (3.18), to which we return later.

Norton and Simon modified Eq. (3.18) by introducing the instantaneous growth factor $GF(N)$, which is the fraction of the tumor that doubles in the time dt. Because most anticancer agents attack dividing cells, this fraction will be important in the tumor's responsiveness to therapy. For the exponential model, the GF is constant. In the Gompertz model (3.6), we have

$$GF(N) = \ln(N_\infty/N(t)) \tag{3.19}$$

where $N_\infty = \lim_{t\to\infty} N(t)$, called the *plateau tumor size*. The Gompertz version of Eq. (3.18) is

$$\frac{dN(t)}{dt} = a \cdot \ln\left(\frac{N_\infty}{N(t)}\right) \cdot N(t) \tag{3.20}$$

The solution of Eq. (3.20) is a Gompertz curve. In Exercise 3.2.1 (d) we saw that a Gompertz curve is described by Eq. (3.17), restated here.

$$N(t) = N_\infty \left(\frac{N_0}{N_\infty}\right)^{e^{-at}} \tag{3.21}$$

A Gompertz curve has these three properties, which we'll use in this section.

Proposition 3.4.1. The Gompertz curve $N(t)$ of Eq. (3.21)
(i) has a horizontal asymptote at $N = N_\infty$,
(ii) has an inflection point at $N = N_\infty/e$, and
(iii) has the maximum of dN/dt occuring at this inflection point.

Proof. (i) Because a is positive, as $t \to \infty$ we see $e^{-at} \to 0$ and so

$$\lim_{t\to\infty} N(t) = \lim_{t\to\infty} N_\infty \left(\frac{N_0}{N_\infty}\right)^{e^{-at}} = N_\infty \left(\frac{N_0}{N_\infty}\right)^0 = N_\infty$$

That is, $N = N_\infty$ is a horizontal asymptote.

(ii) Differentiating the expression for $N(t)$ twice we find

$$\frac{d^2 N(t)}{dt^2} = a^2 N_\infty \left(\frac{N_0}{N_\infty}\right)^{e^{-at}} \ln\left(\frac{N_0}{N_\infty}\right) \left(e^{at} + \ln\left(\frac{N_0}{N_\infty}\right)\right) e^{-2at}$$

Then $d^2 N/dt^2 = 0$ at t that satisfies $e^{at} + \ln(N_0/N_\infty) = 0$. Recalling that $N_\infty = N_0 e^{c/a}$ (Eq. (3.15)), we see that this t is

$$t = t_{\text{infl}} = \frac{1}{a} \ln\left(\frac{c}{a}\right) \tag{3.22}$$

Moreover, because $\ln(N_0/N_\infty) < 0$, the second derivative has the opposite sign of the factor

$$S(t) = e^{at} + \ln\left(\frac{N_0}{N_\infty}\right)$$

For $0 \le t < \ln(c/a)/a$, with a bit of algebra we see that $S(t) < 0$ and so $d^2 N/dt^2 > 0$ and the Gompertz curve is concave up. For $t > \ln(c/a)/a$, the Gompertz curve is concave down. (In Exercise 3.4.3 you'll work through the steps.) Consequently, at $t = t_{\text{infl}}$ the Gompertz curve has its inflection point. The N value for this t is

$$N(t_{\text{infl}}) = N_\infty \left(\frac{N_0}{N_\infty}\right)^{e^{-a(1/a)\ln(c/a)}} = N_\infty \left(\frac{N_0}{N_\infty}\right)^{e^{-\ln(c/a)}}$$

$$= N_\infty \left(\frac{N_0}{N_\infty}\right)^{a/c} = N_\infty \left(\frac{1}{e^{c/a}}\right)^{a/c} = \frac{N_\infty}{e}$$

where the penultimate equality comes from applying Eq. (3.15) again. That is, $N = N_\infty/e$ is the inflection point.

(iii) To see that the maximum of dN/dt occurs at the inflection point, note that the inflection point is a critical point of dN/dt because at the inflection point $0 = d^2 N/dt^2 = d(dN/dt)/dt$. To show this is a maximum, we'll use the second-derivative test.

$$\frac{d^2}{dt^2} \frac{dN}{dt} = \frac{d^3 N}{dt^3} = -a^3 e^{-3at} N_\infty \left(\frac{N_0}{N_\infty}\right)^{e^{-at}} \ln\left(\frac{N_0}{N_\infty}\right) \cdot$$

$$\left(e^{2at} + 3e^{at} \ln\left(\frac{N_0}{N_\infty}\right) + \left(\ln\left(\frac{N_0}{N_\infty}\right)\right)^2\right) \tag{3.23}$$

The sign of $d^3 N/dt^3$ evaluated at $t = t_{\text{infl}}$ is the same as the sign of the bracketed expression (3.23) evaluated at $t = t_{\text{infl}}$. Simple substitution and yet another application of Eq. (3.15) gives

$$e^{at} = \frac{c}{a}, \ e^{2at} = \left(\frac{c}{a}\right)^2, \text{ and } \ln\left(\frac{N_0}{N_\infty}\right) = -\frac{c}{a}$$

Combining these, we find that (3.23) equals $-(c/a)^2$. Consequently, the second derivative of dN/dt is negative at the inflection point of N and so at this point dN/dt has its maximum. $\square$

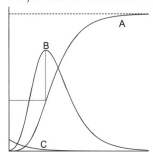

Figure 3.13. Tumor size (A), growth rate (B), and GF (C).

An example of a Gompertz curve is curve A of Fig. 3.13. The curve B is the growth rate, the right-hand side of Eq. (3.20). As we showed in Prop. 3.4.1 (iii), the maximum of this growth curve B occurs at the inflection point of the Gompertz curve A. As indicated by the gray line, the maximum growth rate occurs when the tumor has grown to about 37% of its maximum value, the plateau tumor size N_∞, indicated by the horizontal dashed line. This is a graphical expression of Prop. 3.4.1 (ii), that the maximum growth rate occurs at N_∞/e. The curve C is the instantaneous growth factor $GF(N)$. As illustrated by this figure, Norton and Simon pointed out that even when the instantaneous growth factor is large, the growth rate need not be large.

With tapered dosing, typically when the tumor has shrunk to about 37% of its maximum size, the dosage is at a lower level, not so effective when a (Gompertz) tumor is at its maximum growth rate. (Recall that some cancers are well modeled by truncated exponential growth. This argument doesn't apply to those.) In order to find a better chemotherapy schedule, we need to model how therapy inhibits tumor growth. That was Norton and Simon's next step.

For treated tumors, Norton and Simon introduced a function $D(N,t)$, which represents the *growth-inhibitory influence* of the anticancer agents. This is a function of two variables, N and t, since it depends on the tumor size N and explicitly on t through $L(t)$, the level of therapy at time t. Then the growth equation becomes

$$\frac{dN(t)}{dt} = a \cdot GF(N) \cdot N(t) - D(N,t)$$

Norton and Simon next investigated alternative expressions for $D(N,t)$ to see which is the most consistent with clinical observations. They explored three alternatives:

$$D(N,t) = \begin{cases} D_1(N,t) = K \cdot L(t) \\ D_2(N,t) = K \cdot L(t) \cdot N(t) \\ D_3(N,t) = K \cdot L(t) \cdot N(t) \cdot GF(N) \end{cases} \qquad (3.24)$$

Now we'll look at each alternative in some detail, examining the biological implications of each model.

The function $D_1(N,t)$ represents a treatment for which the growth-inhibitory effect is independent of N, the size of the tumor. That is, a fixed number of tumor cells is killed by a given dose of therapy. With this *fixed-number kill* hypothesis, the value $D_1(N,t)$ is propor-

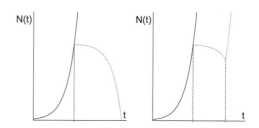

Figure 3.14. Solutions of Eq. (3.25) with exponential $GF(N)$.

tional only to $L(t)$, the level of therapy. Then the growth equation becomes

$$\frac{dN(t)}{dt} = a \cdot GF(N) \cdot N(t) - K \cdot L(t) \qquad (3.25)$$

The exponential case is illustrated in Fig. 3.14. The dark curve represents unchecked exponential growth. The inhibitory factor $L(t)$ we'll take to be 0 before the onset of the treatment, 1 during the treatment, and 0 after the treatment. In the left graph of the figure, the dashed vertical line marks the beginning of the treatment. The gray curve represents tumor growth after the start of therapy. Note that treatment for an adequate duration kills the tumor. The right graph shows that if treatment stops (second dashed line) before the tumor is eradicated, then exponential tumor growth resumes. Recall that for exponential growth, the right-hand side of Eq. (3.25) is

$$a \cdot N - K \cdot L(t) = a \cdot \left(N - \frac{K}{a} \cdot L(t) \right) \qquad (3.26)$$

We see that the relative influence of $L(t)$ is greater for smaller tumor size N. This implies that for any level of $L(t)$ large enough to initiate tumor regression–that is, if the bracketed factor on the right side of Eq. (3.26) is negative–the tumor size $N(t)$ would decrease quickly to 0. Norton and Simon noted that this is widely regarded as unrealistic, since it suggests that most therapy that causes tumor regression would automatically result in cure if the treatment is applied for sufficient time.

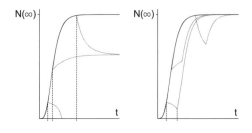

Figure 3.15. Solutions of Eq. (3.25) with Gompertz $GF(N)$.

This model is also inadequate for Gompertzian growth. In Fig. 3.15 the dark curve is the solution of Eq. (3.25) with $GF(N) = \ln(N_\infty/N(t))$. We note the plateau at $N = N_\infty$. The dashed vertical lines in the left graph indicate the onset of treatment, that is, the transition from $L(t) = 0$ to $L(t) = 1$. With the parameters of our simulation, if chemotherapy begins below the inflection point of the Gompertz curve (a noticeable bit below, according to our simulations (see Exercise 3.4.4)), $N(t) \to 0$, while if therapy is turned on when the tumor size is a bit higher, $N(t)$ approaches a plateau strictly below N_∞. In the right graph we see that regardless of whether the tumor size was headed for 0 or for this lower plateau, when chemotherapy is stopped the tumor resumes its climb to N_∞. To avoid crowding the diagram, we have shown the therapy-on and therapy-off dashed vertical lines for only one curve.

Since the maximum of dN/dt occurs at the inflection point of the graph of $N(t)$ (Prop. 3.4.1 (iii)), the relative influence of $L(t)$ would be least at this value, and tumors of intermediate size would be least sensitive. For tumors a bit below the inflection point, therapy leads to eradication, and for larger tumors, regression. Once chemotherapy begins, regression increases with the difference of tumor size from the inflection point size. (See Exercise 3.4.5.) In other words, small and large tumors would have maximum sensitivity, contradicting observation. Clearly $D_1(N,t)$ is not a realistic growth inhibition function, but it was an instructive wrong turn.

So, what now? If $D(N,t)$ dependent on only $L(t)$ is insufficient, why not introduce dependence also on $N(t)$? This is precisely what Norton and Simon did. The second alternative $D_2(N,t)$ of Eq. (3.24) gives the growth equation

$$\frac{dN(t)}{dt} = a \cdot GF(N) \cdot N(t) - K \cdot L(t) \cdot N(t) \qquad (3.27)$$

While no one subscribed to the fixed-number kill hypothesis, introducing this small change happened to correspond with the popular and accepted *percentage-kill*, or *log-kill*, hypothesis. It had been observed that for exponentially growing tumor populations, a fixed percentage of the tumor mass is killed for a given $L(t)$. The left graph of Fig. 3.16 illustrates this effect, precisely what

Skipper [153] observed with murine leukemia L1210. For exponential growth, Eq. (3.27) can be rewritten as

$$\frac{dN(t)}{dt} = (a - K \cdot L(t)) \cdot N(t)$$
(3.28)

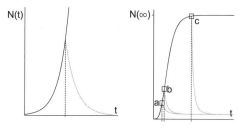

In Eq. (3.28), complete eradication is assured under treatment with

$$K \cdot L(t) > a$$

Figure 3.16. Solutions of Eq. (3.27) with exponential (left) and Gompertz (right) $GF(N)$.

for a sufficient period of time. But note: in the left graph of Fig. 3.14 we see that the decline to tumor eradication occurs along a concave down curve, leading to eradication at a finite (and short, in that graph) time. With the percentage-kill growth inhibition shown in the left graph of Fig. 3.16, the decline to tumor eradication will occur along a concave up curve, giving a slower approach to eradication, which is more consistent with observation.

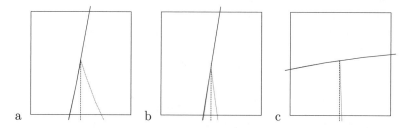

Figure 3.17. Magnifications of regions a, b, and c of the right graph of Fig. 3.16.

For Gompertzian growth kinetics, the percentage-kill function $D_2(N,t)$, illustrated in the right graph of Fig. 3.16, is not what we want. One problem is that chemotherapy appears to reduce the tumor to an equilibrium size, smaller than N_∞ to be sure, but still not eradicated. Another problem with this model appears when we recall Laird's observation that tumors grow most rapidly when they are of moderate size, and consequently, chemotherapy should cause the fastest regression when applied to tumors of moderate size. Graphically, regression speed is indicated by the slope of the growth curve after chemotherapy is turned on. In Fig. 3.17 we find magnifications of the three boxes of the right graph of Fig. 3.16 and see that the regression speed increases with the tumor size, behavior inconsistent with Laird's observation.

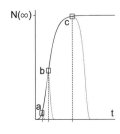

N(∞)

c

b

a

t

Figure 3.18.
Solutions of
Eq. (3.29) with
Gompertz *GF*.

The second alternative, $D_2(N,t)$, fits observations for exponentially growing tumors, but not Gompertzian ones. Can we do better? Norton and Simon hypothesized that the regression rate of a Gompertzian tumor is proportional to the growth rate of the same tumor if no therapy had been introduced. This must mean $D(N,t)$ includes the instantaneous growth factor. So we introduce the *GF* into $D_3(N,t)$, and we have the growth equation

$$\frac{dN(t)}{dt} = a \cdot GF(N) \cdot N(t) - K \cdot L(t) \cdot N(t) \cdot GF(N)$$

(3.29)

illustrated in Fig. 3.18.

In Fig. 3.19 we see the regression rate, revealed by the slope when chemotherapy is turned on, is fastest for moderate-size tumors (b).

Analytically we can see this by observing that when the (constant-level) chemotherapy is turned on, the inhibitory factor is $f(N) = N \ln(N_\infty/N)$. We'll show that this inhibitory factor takes its largest value at the inflection point of the tumor growth curve $N(t)$. We compute $f'(N) = -1 + \ln(N_\infty/N)$ and see that the critical point, the solution of $f'(N) = 0$, is $N = N_\infty/e$, the inflection point of the tumor growth curve by Prop. 3.4.1 (ii). The critical point is a maximum by the second-derivative test, because $f''(N) = -1/N < 0$. This shows that for Gompertzian tumors, the inhibitory factor $GF(N) \cdot N(t)$ takes its maximum at the inflection point of the growth curve, so $D_3(N,t)$ also has the largest effect around the inflection point when tumors are of intermediate size, and the inhibitory effect decreases with increasing size and with decreasing size. Thus, it fits clinical data for the previously unaccounted-for Gompertzian tumors.

But does this model now neglect the exponential cases? Did we just trade one incomplete model for another? No: for the exponential case we have $GF(N) = 1$

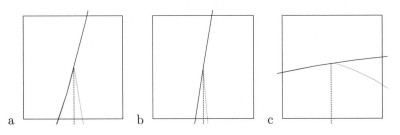

a   b   c

Figure 3.19. Magnifications of regions a, b, and c of the graph of Fig. 3.18.

and so Eq. (3.29) reduces to Eq. (3.28), which we already saw fit with clinical data for exponential tumors. Skipper was not wrong, but simply too narrow, or rather his results were applied too broadly.

In the redesign of chemotherapy schedules, Norton and Simon point out that Eq. (3.29) suggests that higher doses and prolonged therapy have a better chance of tumor eradication. They advocated for doses *as large as possible for as long as possible*. Since the rate of depletion decreases as the tumors get smaller, lower doses of chemotherapy may have been simply lengthening the disease-free interval prior to recurrence, but not affecting the eventual rate of recurrence at all.

Fig. 3.20 simulates a Gompertzian tumor under constant chemotherapy doses ($L(t)$ described by light graph). Even in the absence of biochemical resistance to therapy, some cancers may be difficult to cure solely on kinetic grounds. Norton and Simon point out that maximizing $L(t)$ for each chemotherapy agent is key, with the optimal schedule for chemotherapy having very high doses early on during induction. While toxicity prevents oncologists from using the optimal schedule, Norton and Simon suggest a method called intensification in which the chemotherapy dose levels increase after their initial induction.

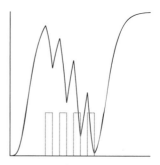

Figure 3.20. Gompertz tumor size under pulsed chemotherapy (light curve).

To this end, Norton and Simon predicted that decreasing the interval between doses to every two weeks (*high-density dosing*) would hold tumor regrowth in check. During clinical trials, this schedule decreased the odds of disease recurrence by 26% and increased survival by 31%, a significant
advancement.

According to Vincent DeVita (page 1492 of [144]) at the Yale School of Medicine, "What Norton did was take information about biological growth and integrate it into treatment scheduling. [...] We give chemotherapy on days one and eight and we give radiotherapy five days out of seven. Why? Because that's the schedule that conforms to the five day work weeks. But, frankly, tumors are smarter than that. We need to think about what's driving the growth of the tumor. And if that means giving the treatment at 2 a.m., that's what we have to do."

In other words, the lessons from the success of the Norton-Simon hypothesis include the need for physicians who design clinical trials to attend more to mathematical models. But this is an uphill battle. The first publications of the Norton-Simon hypothesis in the 1970s received such a hostile response that Norton considered leaving oncology. But he stuck with his belief that mathematical models can give useful insights into designing cancer treatments. His persistence paid off. The main reward is the improved lives of so many patients. That Norton received the 2004 David A. Karnovsky award, the highest honor given by the American Society of Clinical Oncology, was recognition that the role of mathematics in cancer treatment has gained some acceptance. But this journey is long and still we have far to go.

Practice Problems

In problems that ask for plots of differential equation solution curves, you can use the Mathematica code in Sect. B.4. This is just one approach. If you have a favorite piece of software to plot solutions of differential equations, by all means use that.

3.4.1. For the Gompertz fixed-number kill model of Eq. (3.25),
(a) before chemotherapy is turned on (so $L(t) = 0$), show that $N = N_\infty$ is a fixed point, and
(b) once chemotherapy is turned on (so $L(t) = 1$), show that $N = N_\infty$ is not a fixed point.

3.4.2. Plot the solution of Eq. (3.29) with $a = 1$, $K = 2$, $N_\infty = 100$, and $N_0 = 0.01$ for $0 \leq t \leq 15$, when chemotherapy is turned on at $t = 6$. Comment on the symmetry of the curve. How would the shape of the curve change if instead of $K = 2$ we had $K = 1$ or $K = 3/2$?

3.4.3. For the Gompertz percentage-kill model of Eq. (3.27) with $a = 1$, $K = 2$, $N_\infty = 100$, and $N_0 = 0.01$, when chemotherapy starts at $t = 5$, estimate the time when $N(t)$ drops below $N = 20$.

Practice Problem Solutions

3.4.1. (a) In Sect. 4.5 we'll show that the fixed points of a differential equation $x' = f(x)$ are the solutions of $f(x) = 0$. This is easy to see: if $f(x) = 0$, then $x' = 0$ and so x doesn't change. It is a fixed point. Before chemotherapy is turned on, Eq. (3.25) becomes

$$dN/dt = a\ln(N_\infty/N) \cdot N$$

Then to show $N = N_\infty$ is a fixed point, we must show that the right-hand side of this equation then is 0. This is easy: $a\ln(N_\infty/N_\infty) \cdot N_\infty = 0$ because $\ln(1) = 0$.

(b) When chemotherapy is turned on, Eq. (3.25) becomes

$$dN/dt = a\ln(N_\infty/N) \cdot N - k$$

If $N = N_\infty$, then $dN/dt = -k$. Consequently, $N = N_\infty$ is not a fixed point. In fact, once chemotherapy is turned on, at $N = N_\infty$ we have shown that dN/dt is negative. Turning on chemotherapy forces N to drop below N_∞.

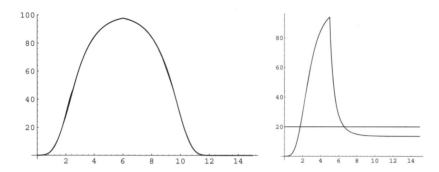

Figure 3.21. Left: the solution curve for Practice Problem 3.4.2. Right: the graph of N for Practice Problem 3.4.3.

3.4.2. With these parameters, the chemotherapy function is

$$L(t) = \begin{cases} 0 & \text{if } t < 6 \\ 1 & \text{if } t \geq 6 \end{cases} \tag{3.30}$$

A Mathematica definition of this function is presented in Sect. B.4.

The left graph of Fig. 3.21 is a plot of the solution curve, which appears to be symmetric about a vertical line through its maximum point. To see why this curve is symmetric, rewrite Eq. (3.29), grouping the common factors of the two terms on the right side of the equation.

$$\frac{dN(t)}{dt} = GF(N) \cdot N(t) \cdot (a - k \cdot L(t)) = GF(N) \cdot N(t) \cdot \begin{cases} 1 & \text{if } t < 6 \\ -1 & \text{if } t \geq 6 \end{cases}$$

That is, once $t \geq 6$, N decreases in just the same way it increased for $t < 6$. We can express this mathematically as

$$\left.\frac{dN}{dt}\right|_{t=s} = -\left.\frac{dN}{dt}\right|_{t=12-s} \quad \text{for } 6 \leq s \leq 12$$

If we change k from 2 to 1, then for $t \geq 6$ we have $a - k \cdot L(t) = 0$. This means that the curve will grow just as in the graph above up to $t = 6$ and then remain constant, because $dN/dt = 0$ for $t \geq 6$.

If we change k to $3/2$, then after $t = 6$, $a - k \cdot L(t) = -1/2$. Thus dN/dt is negative, but N decreases more slowly than it increased for corresponding times $t < 6$. The graph isn't symmetrical: the part to the right of $t = 6$ is a stretched copy of the reflection of the part to the left of $t = 6$.

3.4.3. The Matematica code to plot the graph of $N(t)$ and $N = 20$ is in Sect. B.4. (In Example 4.6.2 we'll see how to solve this differential equation.) From the right graph of Fig. 3.21 we see that $N(t)$ drops below $N = 20$ at about $t = 6.5$. This level of chemotherapy drops the tumor size very quickly, ...if this model were accurate.

Exercises

3.4.1. If we rewrite Eq. (3.20) by using the division property of logarithms we have

$$\frac{dN(t)}{dt} = a \cdot \ln(N_\infty) \cdot N(t) - a \cdot N(t) \cdot \ln(N(t))$$

Plot $f(x) = a \cdot \ln(N_\infty) \cdot x - a \cdot x \cdot \ln(x)$ for $1 \le x \le N_\infty$, for
(a) $a = 0.40, N_\infty = 100$,   (b) $a = 0.25, N_\infty = 100$,
(c) $a = 0.40, N_\infty = 75$,   (d) $a = 0.25, N_\infty = 75$.
What parts of the graph do a and N_∞ control?

3.4.2. In Eq. (3.28), how long is a sufficient period of time for the tumor to be eradicated? To answer this question, we have to define what we mean by eradication. Since we aren't given units for N, we could choose units of volume, mass, or number of cells. Let's choose the number of cells. In this case, it seems appropriate to define eradication as the point at which the number of cells is less than 1. Let's take $N(t) = 0.9$ to indicate eradication of the tumor. Using this convention, find the time it would take for the tumor to be eradicated, when
(a) $K \cdot L(t) = a + 0.1$ (b) $K \cdot L(t) = a + 0.01$ (c) $K \cdot L(t) = a$

3.4.3. For $t > \ln(c/a)/a$ show the Gompertz curve given by Eq. (3.21) is concave down.
(a) Show $t > \ln(c/a)/a$ implies $e^{at} > c/a$.
(b) Show $N_\infty/N_0 = e^{c/a}$ (Eq. (3.15)) implies $\ln(N_0/N_\infty) = -c/a$.
(c) Deduce that $d^2N/dt^2 < 0$.

3.4.4. For the Gompertzian growth of Eqs. (3.19) and (3.25), find a relation between K and a guaranteeing that $dN/dt < 0$ at the inflection point given by Eq. (3.22). Assume chemotherapy has just been turned on, so $L(t) = 1$.

3.4.5. For the Gompertz fixed-number kill model of Eqs. (3.19) and (3.25),
(a) during the administration of chemotherapy (so we may take $L(t) = 1$), there is one value of N for which $dN/dt = 0$. Find the equation that this value $N = N_*$ satisfies.
(b) For $N_\infty = 10$, $a = 1$, and $K = 2$, estimate the value of N_*.

3.4.6. For the Gompertz percentage-kill model of Eqs. (3.19) and (3.27), assume chemotherapy has just begun, so $L = 1$.
(a) Show $N = 0$ is a solution of $dN/dt = 0$. (Hint: you'll need to compute $\lim_{N \to 0} \ln(N_\infty/N) \cdot N$.)
(b) Find the non-zero solution $N = N_*$ of $dN/dt = 0$.
(c) Determine the signs of dN_*/dK and of dN_*/da.

3.4.7. For the Gompertz percentage-kill model of Eqs. (3.19) and (3.27) with $a = 1$, $K = 2$, $N_\infty = 100$, and $N_0 = 0.01$, when chemotherapy starts at $t = 5$, in Practice Problem Solution 3.4.3 we saw that $N(t)$ drops below $N = 20$ at about $t = 6.5$.
(a) When does $N(t)$ drop below $N = 20$ if we change K to $K = 3$?
(b) When does $N(t)$ drop below $N = 20$ if we return K to $K = 2$ and change a to $a = 0.5$?
(c) Are the signs of these changes from the $t = 6.5$ value of Practice Problem Solution 3.4.3 consistent with the movement of the fixed point N_* found in Exercise 3.4.6 (b)?

3.4.8. For the model of Eqs. (3.19) and (3.29) with $a = 1$, $K = 2$, $N_\infty = 100$, and $N_0 = 0.01$, when chemotherapy starts at $t = 5$,
(a) estimate the time when $N(t)$ drops below $N = 5$.
(b) Estimate the time when $N(t)$ drops below $N = 1$.

So far, we have considered models with chemotherapy taking one of two levels, 0 and a positive constant. In the next two problems we'll consider a model with variable drug levels.

3.4.9. Use the Gompertz model of Eqs. (3.19) and (3.29) with $a = 1$, $K = 2$, $N_\infty = 100$, and $N(0) = 0.01$. Plot the curve $N(t)$, $0 \le t \le 20$, for

(a) constant-dose therapy

$$L(t) = \begin{cases} 0 & \text{for } 0 \le t < 5 \\ 1 & \text{for } 5 \le t < 10 \\ 0 & \text{for } 10 \le t \le 20 \end{cases} \tag{3.31}$$

Mathematica code for $L(t)$ is in Sect. B.4.

(b) variable-dose therapy

$$L(t) = \begin{cases} 0 & \text{for } 0 \le t < 5 \\ (15 - t)/10 & \text{for } 5 \le t < 15 \\ 0 & \text{for } 15 \le t \le 20 \end{cases}$$

(c) Verify that the total dose of schedule (b) equals the total dose of schedule (a). That is, compute the area under both $L(t)$ curves.

(d) Tapering down the dose presents the patient with fewer challenges, but is it as effective? What does this example tell you? The rise of $N(t)$ at $t = 10$ in (a) is clear because that is the time the therapy ends, but can you explain the rise of $N(t)$ at $t = 10$ in (b)?

(e) In (a) turn the therapy on at $t = 5$ and off at $t = 12$. Does this eliminate the tumor? In (b) replace the middle part of the definition of L with $L(t) = (19 - t)/14$ for $5 \le t < 19$. (And consequently the last part of the definition is $L(t) = 0$ for $19 \le t$.) Does this eliminate the tumor?

3.4.10. So maybe a linear decline of drug dosage reduces the therapy level too quickly. Again, we'll use the Gompertz model of Eqs. (3.19) and (3.29) with $a = 1$, $K = 2$, $N_\infty = 100$, and $N(0) = 0.01$.

(a) Plot the curve $N(t)$, $0 \le t \le 20$, for

$$L(t) = \begin{cases} 0 & \text{for } 0 \le t < 5 \\ 1 - (2(t - 5)/15)^2 & \text{for } 5 \le t < 25/2 \\ 0 & \text{for } 25/2 \le t \le 20 \end{cases}$$

(b) Verify that the total dose of this schedule equals the total dose of schedules (a) and (b) of Exercise 3.4.9. Don't be intimidated by the integral. Expand the square and see it's just a polynomial.

(c) Compare the graph of this exercise with the graphs of Exercise 3.4.9 (a) and (b). What conclusions can you draw?

Chapter 4   Single-variable differential equations

In this chapter we'll learn ways to solve some pure-time differential equations $(dx/dt = f(t))$ in Sects. 4.1–4.4, and some autonomous differential equations $(dx/dt = g(x))$ in Sects. 4.5–4.7. Pure-time differential equations first.

In principle, solving equations of the form

$$\frac{dx}{dt} = f(t) \tag{4.1}$$

is easy. Integrate both sides

$$\int \frac{dx}{dt} dt = \int f(t) dt = F(t) + C$$

obtaining

$$x(t) = F(t) + C$$

where F is an antiderivative of f, that is, $F'(t) = f(t)$.

The differential equation (4.1) does not specify the solution completely, because of the constant of integration C. In general, we need another piece of information, the value of $x(t)$ for some time t, to solve the equation completely.

4.1 SIMPLE INTEGRATION

First, though, we need some rules for finding antiderivatives. While relatively few rules suffice for finding derivatives, for antiderivatives the situation is much murkier, largely because there is no antiderivative version of the product rule. Sects. 4.2, 4.3, 4.4, and 4.7 are devoted to strategies for finding antiderivatives. In this section we present a few simple rules, and investigate the effect of the constant of integration. Suppose a is any constant.

$$\int a \cdot f(t)\, dt = a \cdot \int f(t)\, dt \qquad\qquad \text{constant multiplier rule}$$

$$\int (f(t) \pm g(t))\, dt = \int f(t)\, dt \pm \int g(t)\, dt \qquad \text{sum and difference rule}$$

$$\int t^n\, dt = \begin{cases} t^{n+1}/(n+1) + C & \text{for } n \neq -1 \\ \ln(|t|) + C & \text{for } n = -1 \end{cases} \qquad \text{constant exponent rule}$$

A few others can be obtained from familiar differentiation formulas

$$\int e^t\, dt = e^t + C \qquad\qquad \text{because } (e^t)' = e^t$$

$$\int \cos(t)\, dt = \sin(t) + C \qquad \text{because } (\sin(t))' = \cos(t)$$

$$\int \sin(t)\, dt = -\cos(t) + C \qquad \text{because } (\cos(t))' = -\sin(t)$$

$$\int \sec^2(t)\, dt = \tan(t) + C \qquad \text{because } (\tan(t))' = \left(\frac{\sin(t)}{\cos(t)}\right)' = \sec^2(t)$$

where we used the quotient rule for the last equality.

With just these few rules, we can solve some problems.

Example 4.1.1. *Position, speed, and acceleration.* Suppose $y(t)$ denotes the height (position) of an object at time t, $v(t) = y'(t)$ its speed, and $a(t) = v'(t)$ its acceleration. Knowing $a(t) = -10$, find expressions for $v(t)$ and $y(t)$. First note that

$$v(t) = \int a(t)\, dt = \int -10\, dt = -10t + C_1$$

To evaluate C_1, substituting in any value for t will do, but then we must ask, "For which t do we know the speed?" Assuming the object is launched (vertically) at $t = 0$ with speed $v(0)$, we obtain

$$v(t) = -10t + v(0)$$

Next,

$$y(t) = \int (-10t + v(0))\, dt = -10 \int t\, dt + \int v(0)\, dt = -5t^2 + v(0)t + C_2$$

To evaluate C_2, again any value of t will work, so long as we know the position at that time t. Under the launch at $t = 0$ assumption, $y(0)$ is the initial position. That is,

$$y(t) = -5t^2 + v(0)t + y(0)$$

For instance, suppose $y(0) = 0$ and $v(0) = 100$. (Because acceleration due to gravity is directed downward and we took $a(t)$ to be negative, positive speeds mean up, negative speeds mean down.) When will the object hit the ground? Assume the object is thrown from the ground.   From the information about $y(0)$ and $v(0)$, we know

$$y(t) = -5t^2 + 100t$$

The object hits the ground when $y(t) = 0$. That is,

$$0 = y(t) = -5t^2 + 100t = t(-5t + 100)$$

The solutions are $t = 0$ (this is the launch from the ground) and $t = 20$.   How fast is the object traveling when it hits the ground?

$$v(20) = -10 \cdot 20 + 100 = -100$$

Note the impact speed equals the launch speed. If we change the model to include air resistance, decelerating proportional to v^2, this no longer is the case: the impact speed will be lower than the launch speed. On the other hand, the resulting differential equation is harder to solve, but its solution has the realistic characteristic that falling objects do not accelerate without bound, but rather reach a terminal velocity, determined by the object's weight and shape, and the viscosity of the medium through which it falls. For a person falling spread out as a skydiver falls, terminal velocity is about 54 m/sec. □

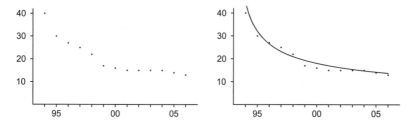

Figure 4.1. New disease cases per 100,000.

Example 4.1.2. *The rate of new disease cases.* On the left side of Fig. 4.1 we see a graph of data on new cases of a disease per 100,000 population. On the right we have fit the curve

$$y(t) = 2 + \frac{42.43}{\sqrt{t}}$$

where we measure time so the year 1994 corresponds to $t = 1$. Later we discuss how to match curves with data points. Many texts just present an equation and ask you to make deductions, but all such equations come from data in one way or another. To remind you that this subject is grounded in experimental data, we will continue to illustrate the source of equations from time to time.

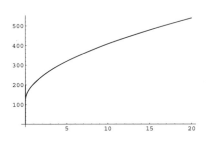

Figure 4.2. Cases per 100,000.

Thinking of the annual rate of new cases as the rate of change of $x(t)$ = number of cases per 100,000 at time t, we have this model for the spread of the disease

$$\frac{dx}{dt} = 2 + \frac{42.43}{\sqrt{t}}$$

We begin by integrating both sides with respect to t,

$$x(t) = \int (2 + 42.43 t^{-1/2})\, dt = 2t + 84.86 t^{1/2} + C$$

where C represents the number of cases per 100,000 in 1993 ($t = 0$), estimated at 120. □

Practice Problems

4.1.1. Suppose $x'(t) = t^2 + 2$ for all t.
(a) If $x(0) = 1$, find $x(t)$ for all t.
(b) If $x(1) = 1$, find $x(t)$ for all t.

4.1.2. (a) For $x'(t) = \cos(t) + \sin(t)$ and $x(0) = 2$, find $x(t)$ for all t.
(b) For $x'(t) = t^2 + t + 2 - t^{-1} - t^{-2}$ and $x(1) = 1$, find $x(t)$ for all t.
(c) For $x'(t) = at^2 + t + 2$ and $x(0) = 1$, find a so that $x(1) = 2$.

4.1.3. An object is thrown vertically from a height of $y(0) = 0$ with speed $v(0) = 200$. At what time does the object reach its maximum height?

Practice Problem Solutions

4.1.1. (a) First, $x(t) = \int x'(t)\, dt = \int (t^2+2)\, dt = t^3/3 + 2t + C$. To find the
value of C, observe that $1 = x(0) = C$, so $x(t) = t^3/3 + 2t + 1$.
(b) Again, $x(t) = t^3/3 + 2t + C$. Then $1 = x(1) = 1/3 + 2 + C$, giving $C = -4/3$.

4.1.2. (a) First, $x(t) = \int x'(t)\, dt = \int (\cos(t) + \sin(t))\, dt = \sin(t) - \cos(t) + C$.
To evaluate C, use $2 = x(0) = \sin(0) - \cos(0) + C = 0 + 1 + C$, so $C = 1$.
(b) First, $x(t) = \int x'(t)\, dt = \int (t^2 + t + 2 - 1/t - 1/t^2)\, dt = t^3/3 + t^2/2 + 2t -$
$\ln(t) + 1/t + C$. To evaluate C, use $1 = x(1) = 1/3 + 1/2 + 2 - \ln(1) + 1 + C$,
so $C = -17/6$.
(c) First, $x(t) = \int x'(t)\, dt = \int (at^2 + t + 2)\, dt = at^3/3 + t^2/2 + 2t + C$. To
evaluate C, use $1 = x(0) = C$, so $x(t) = at^3/3 + t^2/2 + 2t + 1$. Then $2 = x(1) =$
$a/3 + 1/2 + 2 + 1$, giving $a = -9/2$.

4.1.3. Recall acceleration is $a = -10$. Then $v(t) = \int -10\, dt = -10t + v(0) =$
$-10t + 200$. Next, $y(t) = \int v(t)\, dt = -10t^2/2 + 200t + y(0) = -5t^2 + 200t$.
The maximum height occurs when the vertical speed is 0. That is, $0 = -10t +$
200, so $t = 20$.

Exercises

4.1.1. At time $t = 0$ an object is thrown vertically from a height $y(0) = 5$ with
speed $v(0) = 100$. When will the object hit the ground ($y(t) = 0$)? How fast is
the object traveling when it hits the ground?

4.1.2. Suppose $x'(t) = 2t + \sqrt{t}$ and $x(1) = 2$. Find $x(t)$ for all t.

4.1.3. Suppose $x'(t) = t^3 - t + 1$ and $x(0) = 1$. Find $x(t)$ for all t.

4.1.4. Suppose $x''(t) = t^3 - t + 1$, $x'(0) = 1$, and $x(0) = 1$. Find $x(t)$ for all t.

4.1.5. Suppose $x'(t) = at^2 + 2t + 1$ for some constant a. Find the value of a so
that $x(0) = 1$ and $x(1) = 0$.

4.1.6. Find all the functions $x(t)$ with $x''(t) = 2t^2 - t + 2$ and $x(1) = 1$.

4.1.7. Find all the functions $x(t)$ with $x''(t) = 2t^2 - t + 2$ and $x'(0) = 1$.

4.1.8. Find all the functions $x(t)$ with $x'''(t) = 2t^2 - t + 2$ and $x(0) = 1$.

4.1.9. Find all the functions $x(t)$ with $x'''(t) = 2t^2 - t + 2$, $x(0) = 1$, and $x'(0) = 1$.

4.1.10. Find all the functions $x(t)$ with $x'''(t) = 2t^2 - t + 2$, $x(0) = 1$, and $x(1) = 1$.

4.2 INTEGRATION BY SUBSTITUTION

Now we begin our study of more complicated techniques of integration. First, recall the chain rule for derivatives

$$(f(g(t)))' = f'(g(t)) \cdot g'(t)$$

Integrating, we obtain

$$\int f'(g(t)) \cdot g'(t)\, dt = \int (f(g(t)))'\, dt = f(g(t)) + C$$

When facing an integral of a complicated function, we try to parse it as a composition of functions times the derivative of the inner function of the composition. Keeping track of the details is made more straightforward by the substitution of a new variable, hence this technique's name, "integration by substitution." This appears to be most easily grasped through examples. We give two simple examples, then go through a list of more complicated practice problems.

Example 4.2.1. *A simple example of integration by substitution.* Evaluate $\int e^{2t}\, dt$. We know $\int e^t\, dt = e^t + C$, so substitute

$$w = 2t \quad \text{then} \quad dw = 2\, dt \quad \text{and so} \quad \frac{1}{2}\, dw = dt$$

The last step is included because we must match scrupulously each factor that involves t with a factor that involves w. This gives

$$\int e^{2t}\, dt = \int e^w \frac{1}{2}\, dw = \frac{1}{2} \int e^w\, dw = \frac{1}{2} e^w + C = \frac{1}{2} e^{2t} + C$$

The last step expresses the solution in terms of the original variable. $\square$

Example 4.2.2. *A slightly less simple integration by substitution.* Evaluate $\int t^2 \cos(t^3)\, dt$. Substitute

$$w = t^3 \quad \text{then} \quad dw = 3t^2\, dt \quad \text{and so} \quad \frac{1}{3}\, dw = t^2\, dt$$

The substitution gives

$$\int t^2 \cos(t^3)\, dt = \int \cos(w) \frac{1}{3}\, dw = \frac{1}{3}\sin(w) + C = \frac{1}{3}\sin(t^3) + C$$

Differentiate this solution to understand how integration by substitution is the inverse operation of the chain rule of differentiation. □

Because computer algebra systems can evaluate all these integrals, there must be some sort of decision tree or heuristic to figure out what to substitute, or even if substitution alone suffices or if another technique is required. Rather than try to formalize this, we'll rely on intuition and practice. The more you practice, the better your intuition becomes. If you want more sample problems, Google can provide tons.

Practice Problems

4.2.1. Evaluate these integrals.

(a) $\displaystyle\int \cos(t)(1+\sin(t))^2\, dt$ (b) $\displaystyle\int \sec^2(t)\tan^2(t)\, dt$

(c) $\displaystyle\int \frac{t^2}{(1+t^3)^4}\, dt$ (d) $\displaystyle\int \frac{t^2}{1+t^3}\, dt$

4.2.2. Evaluate these integrals.

(a) $\displaystyle\int \tan(t)\, dt$ (b) $\displaystyle\int t\sqrt{1+t^2}\, dt$

(c) $\displaystyle\int t^3\sqrt{1+t^2}\, dt$ (d) $\displaystyle\int e^{4t}\sqrt{1+e^{2t}}\, dt$

Practice Problem Solutions

4.2.1. (a) The most complicated factor in the integrand is $(1+\sin(t))^2$. If we substitute $w = 1+\sin(t)$, then this factor becomes w^2 and $dw = \cos(t)\, dt$. Now we can evaluate the integral.

$$\int \cos(t)(1+\sin(t))^2\, dt = \int w^2\, dw = \frac{w^3}{3} + C = \frac{(1+\sin(t))^3}{3} + C$$

(b) Recalling that $(\tan(t))' = \sec^2(t)$, substitute $w = \tan(t)$ so $dw = \sec^2(t)\, dt$ and

$$\int \sec^2(t)\tan^2(t)\, dt = \int w^2\, dw = \frac{w^3}{3} + C = \frac{\tan^3(t)}{3} + C$$

(c) The denominator $(1+t^3)^4$ is the trickiest part of the integral so we'll substitute $w=1+t^3$. Then $dw=3t^2\,dt$ and $t^2\,dt=\frac{1}{3}\,dw$. We obtain

$$\int \frac{t^2}{(1+t^3)^4}\,dt = \int \frac{1}{3}w^{-4}\,dw = \frac{1}{3}\frac{w^{-3}}{-3}+C = \frac{1}{9}\frac{-1}{w^3}+C = \frac{-1}{9(1+t^3)^3}+C$$

(d) Use the substitution of (c), obtaining

$$\int \frac{t^2}{1+t^3}\,dt = \int \frac{1}{3}w^{-1}\,dw = \frac{1}{3}\ln|w|+C = \frac{1}{3}\ln|1+t^3|+C$$
$$= \ln(|1+t^3|^{1/3})+C$$

4.2.2. (a) What to substitute isn't so obvious here, until we recall that $\tan(t) = \sin(t)/\cos(t)$. Substitute $w=\cos(t)$ so $dw=-\sin(t)\,dt$ and $\sin(t)\,dt=-dw$. Then

$$\int \frac{\sin(t)}{\cos(t)}\,dt = \int -\frac{dw}{w} = -\ln|w|+C = -\ln|\cos(t)|+C$$
$$= \ln|\sec(t)|+C$$

The substitution $w=\sin(x)$ works, too, but the calculation is quite a bit more involved.

(b) The factor $\sqrt{1+t^2}$ suggests that we substitute $w=1+t^2$. Then $dw=2t\,dt$ and $\frac{1}{2}\,dw=t\,dt$, obtaining

$$\int t\sqrt{1+t^2}\,dt = \int \frac{1}{2}w^{1/2}\,dw = \frac{1}{2}\frac{w^{3/2}}{3/2}+C = \frac{1}{3}(1+t^2)^{3/2}+C$$

(c) This looks like (b), but substituting $w=1+t^2$ we obtain $dw=\frac{1}{3}t\,dt$. This leaves a factor of t^2 still to be treated. From the original substitution, note that $t^2=w-1$, so this substitution gives

$$\int t^3\sqrt{1+t^2}\,dt = \int \frac{1}{2}(w-1)\sqrt{w}\,dw = \frac{1}{2}\int (w-w^{3/2})\,dw$$
$$= \frac{1}{2}\left(\frac{w^{5/2}}{5/2}-\frac{w^{3/2}}{3/2}\right)+C$$
$$= \frac{1}{5}(1+t^2)^{5/2}-\frac{1}{3}(1+t^2)^{3/2}+C$$

(d) Recall that $e^{4t}=(e^t)^4$ and $e^{2t}=(e^t)^2$. This suggests the substitution $w=e^t$, so $dw=e^t\,dt$ and we have

$$\int e^{4t}\sqrt{1+e^{2t}}\,dt = \int (e^t)^3\sqrt{1+(e^t)^2}e^t\,dt = \int w^3\sqrt{1+w^2}\,dw$$

This is just the integral of (c), so using that result we have

$$\int e^{4t}\sqrt{1+e^{2t}}\,dt = \frac{1}{5}(1+e^{2t})^{5/2}-\frac{1}{3}(1+e^{2t})^{3/2}+C$$

Exercises

4.2.1. Evaluate these integrals.

(a) $\displaystyle\int \frac{\cos(t)}{\sin^3(t)}\, dt$ (b) $\displaystyle\int \csc^2(t)\cot(t)\, dt$ (c) $\displaystyle\int e^t\sqrt{1+e^t}\, dt$

(d) $\displaystyle\int e^{4t}\sqrt{1+e^{2t}}\, dt$ (e) $\displaystyle\int \frac{\cos(t)}{1+\sin(t)}\, dt$ (f) $\displaystyle\int t^3(t^2+3)^{3/2}\, dt$

4.2.2. Evaluate these integrals.

(a) $\displaystyle\int (t+1)(t^2+2t)^{3/2}\, dt$ (b) $\displaystyle\int \cos(t)\sqrt{1+\sin(t)}\, dt$

(c) $\displaystyle\int \cos(t)\sin(t)\sqrt{1+\sin^2(t)}\, dt$ (d) $\displaystyle\int \frac{\sqrt{1+\sqrt{t}}}{\sqrt{t}}\, dt$

4.2.3. Evaluate these integrals.

(a) $\displaystyle\int -\cos(t)\sin(\sin(t))\, dt$ (b) $\displaystyle\int -\cos(t)\cos(\sin(t))\, dt$

(c) $\displaystyle\int (t+1/2)\cos(t^2+t)\, dt$ (d) $\displaystyle\int_0^{\pi/2} \frac{\cos(t)\sin(t)}{1+\cos^2(t)}\, dt$

4.2.4. Evaluate these integrals.

(a) $\displaystyle\int_0^1 t^3\sqrt{t^4+1}\, dt$ (b) $\displaystyle\int (t^7-t^3)\sqrt{t^4+1}\, dt$

(c) $\displaystyle\int \sin(t)\cos(t)\sqrt{\cos^2(t)+1}\, dt$ (d) $\displaystyle\int \sin^3(t)\cos(t)\sqrt{\sin^2(t)+1}\, dt$

4.2.5. Evaluate these integrals.

(a) $\displaystyle\int \frac{1}{\frac{1}{t}+1}\frac{1}{t^2}\, dt$ (b) $\displaystyle\int_1^2 \frac{1}{t+t^2}\, dt$

(c) $\displaystyle\int e^{\tan(t)}\sec^2(t)\, dt$ (d) $\displaystyle\int \cos(t)\cos(\sin(t))\cos(\sin(\sin(t)))\, dt$

4.3 INTEGRATION BY PARTS

Integration by parts is based on the product rule for derivatives, and is as close as we can come to a product rule for integrals, but it's not all that close. For differentiable functions $u(t)$ and $v(t)$, recall

$$(uv)' = u'v + uv'$$

Rearrange to

$$uv' = (uv)' - u'v$$

and integrate both sides

$$\int uv'\, dt = \int (uv)'\, dt - \int vu'\, dt = uv - \int vu'\, dt$$

or in its more familiar form,

$$\int u\, dv = uv - \int v\, du \qquad (4.2)$$

In order to apply integration by parts, the integrand must be split into two factors, u and dv. Generally, for dv first try the most complicated factor that we can integrate: to fill in the uv part of Eq. (4.2), we need v and so dv must be a function that we can integrate. The hope with the "most complicated" guide is that this choice will make $\int v\, du$ simple enough to evaluate. We'll see this isn't always true, but then a few tricks may help. We begin with two examples, then some practice problems.

Example 4.3.1. *A simple integration by parts example.* Evaluate $\int te^t\, dt$. The most complicated factor we can integrate is $dv = e^t\, dt$, leaving $u = t$. Usually we write this in a table, with u and dv in the first row, du and v in the second.

$$u = t \qquad dv = e^t\, dt$$
$$du = dt \qquad v = e^t$$

Then Eq. (4.2) gives

$$\int te^t\, dt = te^t - \int e^t\, dt = te^t - e^t + C$$

Check by differentiating, of course. □

Example 4.3.2. *A less simple integration by parts example.* Evaluate $\int \ln(t)\, dt$. There appears to be only one factor, $\ln(t)$, and integrating that is the whole problem. So how can integration by parts help? Looking more closely, we see there are two factors, but one, dt usually is unnoticed, being part of every integral. Nevertheless, let's apply integration by parts with $dv = dt$ and see what happens

$$u = \ln(t) \qquad dv = dt$$
$$du = (1/t)\, dt \qquad v = t$$

Then

$$\int \ln(t)\, dt = t\ln(t) - \int t(1/t)\, dt = t\ln(t) - \int dt = t\ln(t) - t + C$$

Although it may appear quite specialized, this trick is useful in other situations. Keep your eyes open. □

Practice Problems

4.3.1. Evaluate these integrals.

(a) $\int t\sin(t)\,dt$ (b) $\int \dfrac{\ln(t)}{t^2}\,dt$ (c) $\int t^2 e^t\,dt$ (d) $\int \tan^{-1}(t)\,dt$

4.3.2. Evaluate these integrals.

(a) $\int (\ln(t))^2\,dt$ (b) $\int \cos(\ln(t))\,dt$ (c) $\int e^t\cos(t)\,dt$ (d) $\int t\sec^2(t)\,dt$

Practice Problem Solutions

4.3.1. (a) Taking $dv = \sin(t)\,dt$ seems to be the natural choice. Then

$$u = t \qquad\qquad dv = \sin(t)\,dt$$
$$du = dt \qquad\qquad v = -\cos(t)$$

and

$$\int t\sin(t)\,dt = -t\cos(t) - \int -\cos(t)\,dt = -t\cos(t) + \sin(t) + C$$

(b) If we remember how to integrate $\ln(t)$, we'd take $dv = \ln(t)dt$. If we don't remember that integral, then we take $dv = t^{-2}\,dt$. This gives

$$u = \ln(t) \qquad\qquad dv = t^{-2}\,dt$$
$$du = t^{-1}\,dt \qquad\qquad v = -t^{-1}$$

and

$$\int \frac{\ln(t)}{t^2}\,dt = -\frac{1}{t}\ln(t) - \int \frac{1}{t}\cdot\frac{-1}{t}\,dt = -\frac{\ln(t)}{t} - \frac{1}{t} + C$$

Taking $dv = \ln(t)\,dt$ works, but the calculation is more complicated.

(c) The most complicated factor we can integrate is e^t, so

$$u = t^2 \qquad\qquad dv = e^t\,dt$$
$$du = 2t\,dt \qquad\qquad v = e^t$$

We obtain

$$\int t^2 e^t\,dt = t^2 e^t - \int 2t e^t\,dt$$

Now either apply integration by parts again, or recall Example 4.3.1. In either case,

$$\int t^2 e^t\,dt = t^2 e^t - (te^t - e^t) + C$$

Note that if we had made the other choice, $dv = t^2\, dt$, we would obtain

$$\int t^2 e^t\, dt = \frac{t^3}{3} e^t - \int \frac{t^3}{3} e^t\, dt$$

We have made the integral more complicated, going from a quadratic times an exponential to a cubic times an exponential. This kind of behavior usually signals that we should make a different choice for dv.

(d) This is similar to the Example 4.3.2. If we could take $dv = \tan^{-1}(t)\, dt$, we wouldn't need integration by parts because we could just do the integral. So we are left with $dv = dt$ and

$$u = \tan^{-1}(t) \qquad dv = dt$$

$$du = \frac{dt}{1+t^2} \qquad v = t$$

This gives

$$\int \tan^{-1}(t)\, dt = t\tan^{-1}(t) - \int \frac{t}{1+t^2}\, dt = t\tan^{-1}(t) - \frac{1}{2}\ln(1+t^2) + C$$

where the last equality is obtained by substituting $w = 1 + t^2$, so $dw = 2t\, dt$, in $\int t/(1+t^2)\, dt$.

4.3.2. (a) As in the solution of Practice Problem 4.3.1 (d), we take $dv = dt$ and so

$$u = (\ln(t))^2 \qquad dv = dt$$

$$du = 2\ln(t)\frac{1}{t}\, dt \qquad v = t$$

This gives

$$\int (\ln(t))^2\, dt = t(\ln(t))^2 - \int 2\ln(t)\frac{1}{t} t\, dt = t(\ln(t))^2 - 2\int \ln(t)\, dt$$

Applying integration by parts again, or recalling Ex. 4.3.2, we find

$$\int (\ln(t))^2\, dt = t(\ln(t))^2 - 2(t\ln(t) - t) + C$$

(b) If we are to use integration by parts, here again we have only one choice: $dv = dt$.

$$u = \cos(\ln(t)) \qquad dv = dt$$

$$du = -\sin(\ln(t))\frac{1}{t}\, dt \qquad v = t$$

Then

$$\int \cos(\ln(t))\, dt = t\cos(\ln(t)) - \int -\sin(\ln(t))\frac{1}{t} t\, dt$$

$$= t\cos(\ln(t)) + \int \sin(\ln(t))\, dt \qquad (4.3)$$

This doesn't appear to have helped at all. We've gone from integrating $\cos(\ln(t))$ to integrating $\sin(\ln(t))$. This certainly isn't any simpler, but on the other hand, it isn't any more complicated, either. Let's try to integrate $\sin(\ln(t))$ by parts and see what happens.

$$u = \sin(\ln(t)) \qquad\qquad dv = dt$$

$$du = \cos(\ln(t))\frac{1}{t}\, dt \qquad v = t$$

so

$$\int \sin(\ln(t))\, dt = t\sin(\ln(t)) - \int \cos(\ln(t))\frac{1}{t} t\, dt$$

$$= t\sin(\ln(t)) - \int \cos(\ln(t))\, dt \qquad (4.4)$$

Combining Eqs. (4.3) and (4.4), we find

$$\int \cos(\ln(t))\, dt = t\cos(\ln(t)) + t\sin(\ln(t)) - \int \cos(\ln(t))dt$$

an equation we can solve for $\int \cos(\ln(t))\, dt$:

$$\int \cos(\ln(t))\, dt = \frac{1}{2}(t\cos(\ln(t)) + t\sin(\ln(t))) + C$$

(c) This is not so straightforward for a reason different from that of the previous problem. Namely, which of e^t and $\cos(t)$ is the more complicated factor? Having no clear guidance, pick one, say $dv = \cos(t)\, dt$. Then

$$u = e^t \qquad\qquad dv = \cos(t)\, dt$$

$$du = e^t\, dt \qquad v = \sin(t)$$

giving

$$\int e^t \cos(t)\, dt = e^t \sin(t) - \int e^t \sin(t)\, dt \qquad (4.5)$$

We've replaced one integral by another of about equal complexity, so recalling (b) we'll integrate by parts again. We must be careful to make similar choices for u and dv

$$u = e^t \qquad\qquad dv = \sin(t)\, dt$$

$$du = e^t\, dt \qquad v = -\cos(t)$$

and so

$$\int e^t \sin(t)\, dt = -e^t \cos(t) - \int -e^t \cos(t)\, dt \qquad (4.6)$$

Combining Eqs. (4.5) and (4.6), we find an equation

$$\int e^t \cos(t)\, dt = e^t \sin(t) - \left(-e^t \cos(t) - \int -e^t \cos(t)\, dt \right)$$

which we can solve for $\int e^t \cos(t)\, dt$:

$$\int e^t \cos(t)\, dt = \frac{1}{2}(e^t \sin(t) + e^t \cos(t)) + C$$

Try taking $du = e^t dt$ in the second integration by parts. You will obtain a well-known, and not at all useful, equation.

On the other hand, taking $dv = e^t\, dt$ in both integrations by parts works just fine.

(d) We do know how to integrate $\sec^2(t)$, so that's the obvious choice for dv

$$u = t \qquad\qquad dv = \sec^2(t)\, dt$$
$$du = dt \qquad\qquad v = \tan(t)$$

Then integration by parts gives

$$\int t\sec^2(t)\, dt = t\tan(t) - \int \tan(t)\, dt$$

Do we know how to integrate $\tan(t)$? Yes, if we remember Practice Problem 4.2.2 (a). Then

$$\int t\sec^2(t)\, dt = t\tan(t) - \ln|\sec(t)| + C$$

If we don't remember the integral of $\tan(t)$, probably we could derive it quickly, if we saw the right substitution again. Keeping a list of simple integrals we know is a good idea. More on this in Sect. 4.4.

Exercises

4.3.1. Evaluate these integrals.

(a) $\int t\cos(t)\, dt$ (b) $\int t^3 e^t\, dt$ (c) $\int e^t \sin(t)\, dt$ (d) $\int t^5 \ln(t)\, dt$

4.3.2. Evaluate these integrals. Do not multiply out the integrand of (a).

(a) $\int t(t-2)^{99}\, dt$ (b) $\int \sin^{-1}(t)\, dt$ (c) $\int t^2 \sin(t)\, dt$ (d) $\int (t\ln(t))^2\, dt$

4.3.3. Evaluate these integrals. Do not multiply out the integrand of (a).

(a) $\int (t-1)^2(t-2)^{10}\, dt$ (b) $\int \sin(\ln(t))\, dt$

(c) $\int t\ln(t^4+2t^2+1)\, dt$ (d) $\int e^t\ln(e^{2t}+4e^t+4)\, dt$

4.3.4. Evaluate these integrals.

(a) $\int \ln(t^3+3t^2+3t+1)\, dt$ (b) $\int \dfrac{\ln(\ln(t)^2+2\ln(t)+1)}{t}\, dt$

(c) $\int -\dfrac{\ln\left(\dfrac{1}{t^2}+\dfrac{2}{t}+1\right)}{t^2}\, dt$ (d) $\int \ln(t^2+1)\, dt$

4.3.5. Where's the mistake with this argument? Taking $u=1/t$ and $dv=dt$, we have $du=(-1/t^2)dt$ and $v=t$, so integration by parts gives

$$\int \frac{1}{t}dt = \frac{1}{t}t - \int t\frac{-1}{t^2}dt = 1+\int \frac{1}{t}dt$$

Subtracting $\int (1/t)dt$ from both sides, we see $0=1$.

4.4 INTEGRAL TABLES AND COMPUTER INTEGRATION

So far the principal techniques of integration that we've introduced are substitution (Sect. 4.2) and integration by parts (Sect. 4.3). In Sect. 4.7 we introduce one more, partial fractions expansion. These are the only techniques we study in any detail, but some additional integrals arise in interesting situations. Common approaches involve using a table of integrals or a computer algebra system, Mathematica or Maple, for example.

A table of integrals is given in Appendix C. In the practice problems we illustrate its use with some examples, and then show how Mathematica handles these examples. One challenge of computer integration is translating the result into more familiar form. We illustrate this as well. In Mathematica, the indefinite and definite integrals

$$\int f(t)\, dt \quad \text{and} \quad \int_a^b f(t)\, dt$$

are evaluated by

Integrate[f[t],t]   and   Integrate[f[t],{t,a,b}]

Note also that function names including Sin and Exp and Log (the natural logarithm, ln) are capitalized, multiplication is denoted by * and exponentiation by ^, and functional arguments are enclosed in square brackets, [ ], not round brackets ( ). For example, in Mathematica $\sin(t^2 + \sqrt{t})$ is written this way

$$\text{Sin}[t^2 + \text{Sqrt}[t]]$$

The syntax of Mathematica is more intricate, but this is enough to start.

Practice Problems

4.4.1. Use the table of integrals in Appendix C to evaluate these integrals.

(a) $\displaystyle\int \sin^2(e^t)e^t\, dt$

(b) $\displaystyle\int \frac{1}{t\sqrt{1-\ln^2(t)}}\, dt$

(c) $\displaystyle\int \frac{\cos(t)}{\sqrt{1-\sin^2(t)}}\, dt$

(d) $\displaystyle\int \frac{1}{\sqrt{9+4t^2}}\, dt$

4.4.2. Evaluate the integrals of Practice Problem 4.4.1 (a)–(d) using Mathematica. Compare these answers with those obtained using the table of integrals.

Practice Problem Solutions

4.4.1. (a) This looks similar to rule 1 in the table of integrals in Appendix C, except for the e^t. The substitution $w = e^t$ gives $dw = e^t\, dt$ and so

$$\int \sin^2(e^t)e^t\, dt = \int \sin^2(w)\, dw = \frac{w}{2} - \frac{\sin(2w)}{4} + C = \frac{e^t}{2} - \frac{\sin(2e^t)}{4} + C$$

where the second equality follows by 1 of the integral table. Nothing new here, except we need to be familiar with the integrals in the table.

(b) Rule 30 in the table might look like the best starting point, because of the t multiplying the $\sqrt{1-\ln^2(t)}$. However, the $\ln^2(t)$ shows this cannot work: the t in rule 30 cannot be both t and $\ln(t)$. A better match is rule 28, substituting $w = \ln(t)$ so $dw = (1/t)\, dt$. This gives

$$\int \frac{1}{t\sqrt{1-\ln^2(t)}}\, dt = \int \frac{1}{\sqrt{1-w^2}}\, dw = \sin^{-1}(w) + C = \sin^{-1}(\ln(t)) + C$$

where the second equality follows by 28 of the integral table.

(c) This looks similar to Practice Problem 4.4.1 (b), so try rule 28 again, along with the substitution $w = \sin(t)$ so $dw = \cos(t)\, dt$.

$$\int \frac{\cos(t)}{\sqrt{1-\sin^2(t)}}\, dt = \int \frac{1}{\sqrt{1-w^2}}\, dw = \sin^{-1}(w) + C$$

$$= \sin^{-1}(\sin(t)) + C = \pm t + n\pi + C$$

where the second equality follows by rule 28 of the integral table. What happened here? Look again at the integrand:

$$\sqrt{1-\sin^2(t)} = \sqrt{\cos^2(t)} = \pm\cos(t)$$

We see the integral really is $\int \pm 1 dx$. What about the $n\pi$? That can be absorbed into the constant C. Sometimes a bit of algebraic simplification before integrating can save work.

(d) This looks like rule 20 from the table, except for the 4 multiplying t^2. Substituting $w = 2t$ so $dw = 2\, dt$ and $dt = (1/2)\, dw$, we see

$$\int \frac{1}{\sqrt{9+4t^2}}\, dt = \int \frac{1}{2}\frac{1}{\sqrt{9+w^2}}\, dw = \frac{1}{2}\ln\left|w + \sqrt{9+w^2}\right| + C$$

$$= \frac{1}{2}\ln\left|2t + \sqrt{9+4t^2}\right|$$

where the second equality follows by 20 of the integral table.

4.4.2. Now let's integrate these using Mathematica.

(a) Integrate[(Sin[Exp[t]]^2)*Exp[t],t] returns (2*E^t - Sin[2*E^t])/4.

(b) Integrate[1/(t*Sqrt[1 - (Log[t])^2]),t] returns ArcSin[Log[t]]. Recall that Mathematica uses Log to denote the natural logarithm So far, so good.

(c) Integrate[Cos[t]/(Sqrt[1 - (Sin[t])^2]),t] returns ArcSin[Sin[t]]. Again, this agrees with our use of substitution and the integral table. So what's the point of warning about abstract integration and machine integration giving different answers?

(d) The Mathematica command Integrate[1/(Sqrt[9 + 4*(t^2)]),t] returns ArcSinh[2*t/3]/2. The hyperbolic sine, sinh(t), is defined by

$$\sinh(t) = \frac{e^t - e^{-t}}{2}$$

and ArcSinh[t] is the inverse of sinh(t). For $t = 0$,

$$\text{arcsinh}(2t/3) = 0$$

and

$$(1/2)\ln\left|2t + \sqrt{9+4t^2}\right| = \ln(3)/2 \approx 0.549306$$

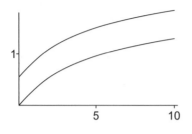

so these functions are not identical. In Fig. 4.3 we plot these functions, and see that they appear to differ by a constant, in this case, $\ln(3)/2$. How can we see that this is true? Check that both answers have derivatives equal to $1/\sqrt{9+4t^2}$.

Figure 4.3. Plots of two solutions of Practice Problem 4.4.1 (d).

$$\left(\frac{1}{2}\ln|2t+\sqrt{9+4t^2}|\right)' = \frac{1}{2}\frac{2+(1/2)(9+4t^2)^{-1/2}(8t)}{2x+\sqrt{9+4t^2}}$$

$$= \frac{1+2t(9+4t^2)^{-1/2}}{2t+\sqrt{9+4t^2}} \cdot \frac{2t-\sqrt{9+4t^2}}{2t-\sqrt{9+4t^2}} = \frac{1}{\sqrt{9+4t^2}}$$

Recalling (or looking it up if you don't recall) $(\operatorname{arcsinh}(t))' = 1/\sqrt{1+t^2}$, we see

$$\left(\operatorname{arcsinh}\left(\frac{2t}{3}\right)\right)' = \frac{1}{2}\cdot\frac{1}{\sqrt{1+(2t/3)^2}}\cdot\frac{2}{3} = \frac{1}{3\sqrt{1+(4t^2/9)}}$$

and the derivatives are identical. Computer algebra systems may express integrals in unfamiliar forms. We view this as an opportunity to learn new functional relations.

Exercises

4.4.1. Evaluate these integrals using the table in Appendix C. If you have a computer algebra system, check the integrals with that system and compare that answer with the one you found with the table.

(a) $\displaystyle\int \frac{e^t}{\sqrt{4-e^{2t}}}\, dt$

(b) $\displaystyle\int \frac{e^t}{\sqrt{4+e^{2t}}}\, dt$

(c) $\displaystyle\int \frac{\sec(1/t)}{t^2}\, dt$

(d) $\displaystyle\int \frac{t-1}{t^2+3t+2}\, dt$

4.4.2. Same instructions as the preceding problem.

(a) $\displaystyle\int \frac{2t-1}{t^2+5t+4}\, dx$

(b) $\displaystyle\int t\sin^{-1}(t)\, dt$

(c) $\displaystyle\int \frac{\cos(t)}{\sqrt{5+\sin^2(t)}}\, dt$

(d) $\displaystyle\int \frac{\sin(t)\cos(t)}{\sqrt{5+\sin^4(t)}}\, dt$

4.4.3. Same instructions as the preceding problem. For one of these problems, you might want to review completing the square.

(a) $\displaystyle\int \frac{1}{\sqrt{t}\sqrt{9+t}}\,dt$

(b) $\displaystyle\int \frac{e^t}{\sqrt{2e^t + e^{2t}}}\,dt$

(c) $\displaystyle\int \frac{1/t^3}{1/t^4 + 1}\,dt$

(d) $\displaystyle\int \frac{\cos(t)}{\sqrt{4 - \sin^2(t)}}\,dt$

4.5 FIXED POINT ANALYSIS

Now we move on to autonomous differential equations. Recall these have the form

$$\frac{dx}{dt} = g(x) \tag{4.7}$$

That is, the rate of change of x depends on x, but not explicitly on t. In this section we'll learn how to find the fixed points of (4.7) and determine their stability. Next, in Sect. 4.6 we introduce separation of variables, a simple method for solving some autonomous equations. Finally, in Sect. 4.7 we add partial fractions expansions to our list of techniques of integration. This method is useful if the function $g(x)$ on the right-hand side of Eq. (4.7) is a ratio of polynomials.

Even when we cannot find an explicit solution of Eq. (4.7), some general properties of the long-term behavior of the solution can be obtained by modifying the fixed point analysis of Sect. 2.2. In this section we sketch several examples of Eq. (4.7) for biological systems, and develop a graphical method for finding fixed points and determining their stability.

Example 4.5.1. *Diffusion of chemicals across a cell membrane.* Denote by I the concentration of a chemical inside a cell, and by O the concentration of this chemical in the medium surrounding the cell, that is, outside the cell. Assuming the volume of the medium surrounding the cell is much larger than the volume of the cell, we can take O to be approximately constant. The cell is separated from the medium by the cell membrane. Denote by p the membrane permeability, taken to be constant for the moment, of this chemical. Then

$$\frac{dI}{dt} = \text{the rate the chemical enters the cell} - \text{the rate the chemical leaves}$$

$$= pO - pI = p(O - I)$$

where we have assumed that the rate at which the chemical enters the cell is proportional to the concentration outside the cell, and in fact is equal to pO, and that the rate the chemical leaves the cell is pI. $\square$

Example 4.5.2. *A continuous model for population growth with limited resources.* Denote by $P(t)$ the population at time t, by r the growth rate of the population assuming unlimited resources, and by K the carrying capacity of the environment. Recall that the Verhulst equation (3.1) for population growth is

$$\frac{dP}{dt} = r\left(1 - \frac{P}{K}\right)P \tag{4.8}$$

Rescale the population as a fraction of the carrying capacity, $x(t) = P(t)/K$. Then Eq. (4.8) becomes the *logistic differential equation*

$$\frac{dx}{dt} = r(1-x)x \tag{4.9}$$

A medical variation on Verhulst's equation is to let $P(t)$ denote the volume of a tumor at time t. The volume obviously has an upper bound, hence the growth rate r is limited by a carrying capacity, most simply modeled through multiplication of r by $(1 - P(t)/K)$. Destroying tumor cells by a chemotherapy agent with per capita death rate c, the percentage-kill model described in Sect. 3.4, Eq. (4.8) becomes

$$\frac{dP}{dt} = r\left(1 - \frac{P}{K}\right)P - cP \tag{4.10}$$

For therapy agents administered in a time-varying fashion, $c = c(t)$ and Eq. (4.10) becomes a non-autonomous differential equation. We study equations of this type in Sects. 10.9 and A.1. $\square$

What can we say about Eq. (4.7) without solving it? The fixed points, or equilibria, of a system described by Eq. (4.7) are those points x_* for which $g(x_*) = 0$, because at these points, $dx/dt = 0$ and so the value of x does not change.

For the diffusion model of Ex. 4.5.1, the fixed point I_* is the solution of $0 = p(O - I_*)$, that is, $I_* = O$. This is no surprise: the equilibrium occurs when both sides of the membranes have the same concentration.

For the Verhulst model of Ex. 4.5.2, the fixed points are the solutions, P_*, of

$$0 = r\left(1 - \frac{P_*}{K}\right)P_*$$

That is, there are two fixed points: $P_* = 0$ and $P_* = K$. If P_* is exactly 0 or K, then $P(t)$ will stay at that value. What if $P(t_0)$ is near 0 or is near K? To answer this question, we graph the right-hand side of Eq. (4.8), or more generally, the right-hand side of Eq. (4.7), as illustrated in Fig. 4.4. The fixed points are A and B.

Let's consider A first. For $x < A$, $g(x) > 0$ and so $dx/dt > 0$. That is, $x(t)$ increases toward A. For $x > A$ and near A, $g(x) < 0$ and so $dx/dt < 0$. That is, $x(t)$ decreases toward A. Combining these observations, we see that for $x(t_0)$ near A, $x(t)$ moves toward A as t increases, so A is a stable fixed point.

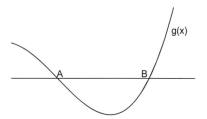

Figure 4.4. Determining the stability of the fixed points A and B.

Now consider B. For $x < B$ and x near B, $g(x) < 0$ and so $dx/dt < 0$. That is, x decreases away from B. For $x > B$, $g(x) > 0$ and so $dx/dt > 0$. That is, $x(t)$ increases away from B. Combining these observations, we see that for $x(t_0)$ near B, $x(t)$ moves away from B, so B is an unstable fixed point.

We have assumed nothing beyond the differentiability of g, so the arguments of the preceding two paragraphs prove

Theorem 4.5.1. *For the system (4.7) with $g(x)$ differentiable, a fixed point x_* is stable if $dg/dx|_{x_*} < 0$ and is unstable if $dg/dx|_{x_*} > 0$.*

For the diffusion model, $g(I) = p(O - I)$. Then $dg/dI = -p$ and the fixed point $I = O$ is stable.

For the Verhulst model, $g(P) = r(1 - P/K)P$. Then $dg/dP = r - 2rP/K$. The fixed point $P = 0$ has $dg/dP|_0 = r > 0$, so is unstable. The fixed point $P = K$ has $dg/dP|_K = -r < 0$, so is stable. In fact, $g(P) > 0$ for all P, $0 < P < K$, so every $P > 0$ grows monotonically until it reaches $P = K$. This is an example of a differential equation for which we can determine the qualitative long-term behavior without solving the differential equation.

Theorem 4.5.1 does not address one case: $dg/dx|_{x_*} = 0$. Here the derivative gives no information about the dynamics near the fixed point. Rather, we must investigate the graph of $g(x)$ near the fixed point and adapt the reasoning of the theorem to these graphs. Four possibilities arise. Fig. 4.5 illustrates these cases.

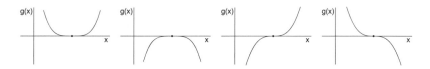

Figure 4.5. Examples with $dg/dx|_{x_*} = 0$, with x_* marked by the dot.

In the first image of Fig. 4.5, $dx/dt > 0$ for $x < x_*$ and for $x > x_*$, so for $x < x_*$, x increases toward x_*; for $x > x_*$, x increases away from x_*. Thus the fixed point x_* is neither stable nor unstable. We might say x_* is stable on the left and unstable on the right, but this sensible language is uncommon.

Arguing similarly, going from left to right we see that the fixed point of the second graph is neither stable nor unstable (we could say it is unstable on the left and stable on the right), the fixed point of the third graph is unstable, and the fixed point of the fourth (right) graph is stable.

Finding fixed points and determining their stability may not seem like much, but for single-variable differential equation (4.7) the only options for a trajectory are to converge to a fixed point, run away from a fixed point, or diverge to infinity. Without solving the differential equation, we can't tell how fast these things happen, but we can tell which happen.

Practice Problems

4.5.1. For the differential equation

$$\frac{dx}{dt} = (1-x)\sin(x) = g(x)$$

find the fixed points and determine their stability.

4.5.2. For constant k, the differential equation

$$\frac{dx}{dt} = x^3 - x + k = g(x)$$

has one, two, or three fixed points. Find the range of k-values for which the differential equation has exactly one fixed point, exactly two fixed points, and exactly three fixed points. In each case, determine the stability of every fixed point.

Practice Problem Solutions

4.5.1. The fixed points are the solutions of $(x-1)\sin(x) = 0$, that is, $x = 1$ and $x = n\pi$ for every integer n. Next,

$$\frac{dg}{dx} = \cos(x) - \sin(x) - x\cos(x)$$

Evaluating dg/dx at the fixed points, we find

- $x = 1$ is stable,
- $x = 2n\pi$ are stable for $n > 0$,
- $x = (2n+1)\pi$ are unstable for $n > 0$,

- $x = 2n\pi$ are unstable for $n \leq 0$, and
- $x = (2n+1)\pi$ are stable for $n < 0$.

Remembering the $x = 1$ fixed point, note that the stability of the fixed points alternates.

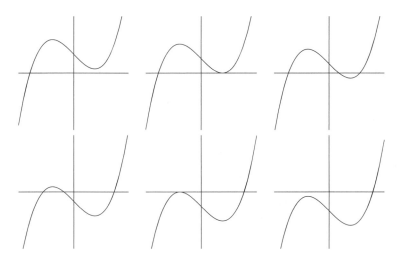

Figure 4.6. Graphs of $g(x) = x^3 - x + k$ for $k = 1/2, 2\sqrt{3}/9, 1/4, -1/4, -2\sqrt{3}/9$, and $-1/2$. The plot range is $-1.5 \leq x \leq 1.5$ and $-1.5 \leq y \leq 1.5$.

4.5.2. By inspecting the graphs for sample values of k, we see that the number of fixed points changes at those k for which the local max or the local min of $g(x)$ occurs on the x-axis. The critical points of $g(x)$ are the solutions of $0 = g'(x) = 3x^2 - 1$, that is, $x_\pm = \pm 1/\sqrt{3}$; $g(x_\pm) = \mp 2\sqrt{3}/9 + k$ and the number of critical points changes at $k = \pm 2\sqrt{3}/9$. This gives five ranges of k-values. Inspecting the graph of $g(x)$ in each, here we list the range of k-values, the number of fixed points, and the stability of those fixed points, ordered from left to right.

k	number	stability
$k > 2\sqrt{3}/9$	one	unstable
$k = 2\sqrt{3}/9$	two	unstable, neither
$-2\sqrt{3}/9 < k < 2\sqrt{3}/9$	three	unstable, stable, unstable
$k = -2\sqrt{3}/9$	two	neither, unstable
$k < -2\sqrt{3}/9$	one	unstable

Exercises

4.5.1. Find the fixed points and determine their stability for $dx/dt = g(x)$ with $g(x) = x^3 + 3x^2 + 2x$.

4.5.2. Find the fixed points and determine their stability for $dx/dt = g(x)$ with
(a) $g(x) = x^3 - 3x^2 + 3x - 1$
(b) $g(x) = x^4 + 4x^3 + 6x^2 + 4x + 1$
(c) $g(x) = x^5 + 5x^4 + 10x^3 + 10x^2 + 5x + 1$

4.5.3. Find the fixed points and determine their stability for $dx/dt = g(x)$ with
(a) $g(x) = x^3 + 6x^2 + 11x + 6$ (Hint: factor the polynomial. Remember division of polynomials.)
(b) $g(x) = x^3 + 2x^2 - x - 2$
(c) $g(x) = x^4 - 5x^2 + 4$
(d) $g(x) = x^4 + 4x^3 + 3x^2$

4.5.4. Suppose $dx/dt = g(x)$ for some polynomial $g(x)$, and $x_1 < x_2 < \cdots < x_n$ are the fixed points of the differential equation. Can both x_i and x_{i+1} be stable? Can both x_i and x_{i+1} be unstable? Explain.

4.5.5. Suppose $dI/dt = pO - pI^2$. Assume O and p are positive constants. Find the fixed points and test their stability.

4.5.6. Suppose $g(x)$ is a polynomial of degree n (that is, the highest power of x is n). Consider the differential equation $dx/dt = g(x)$.
(a) If n is even, what is the maximum number of distinct fixed points this equation can have? What is the minimum number? Explain.
(b) If n is odd, what is the maximum number of distinct fixed points this equation can have? What is the minimum number? Explain.

4.5.7. Suppose $g(x)$ is a polynomial with zeros $x_1 < x_2 < \cdots < x_m$. Describe the dynamics of the solutions of dx/dt at every point except the fixed points. We already know what happens at those points.

4.5.8. For the differential equation $dx/dt = \sin(x) + k$, $0 \le x \le 2\pi$, as k ranges from -1.1 to 1.1, count the number of fixed points as a function of k. Determine the stability of each of these fixed points as a function of k itself. Values of k where the number of fixed points changes are called *bifurcation points*.

4.5.9. For the differential equation $dx/dt = \sin(x) + ax + b$, $0 \leq x \leq 2\pi$,
(a) show that for $a = 1/2$ the differential equation exhibits bifurcations as b increases from -1.4 to 1.4.
(b) Show that for $a = 3/2$ the differential equation does not exhibit bifurcations as b increases from -1.4 to 1.4.
(c) Find the value of a for which the differential equation stops exhibiting bifurcations as b increases from -1.4 to 1.4.

4.5.10. For the differential equation $dx/dt = x^3 + ax + b$, $-2 \leq x \leq 2$,
(a) show that for $a = -1/2$ the differential equation exhibits bifurcations as b increases from -1 to 1.
(b) Show that for $a = 1/2$ the differential equation does not exhibit bifurcations as b increases from -1 to 1.
(c) Find the value of a for which the differential equation stops exhibiting bifurcations as b increases from -1 to 1.

4.6 SEPARATION OF VARIABLES

In principle, solving an autonomous differential equation is simple, but there are two sometimes significant, occasionally insurmountable, bumps along that road. Because we can always check a putative solution of a differential equation by differentiating, any protocol, no matter how mathematically nonsensical, is worth a try. We'll use one called *separation of variables*. (This name also refers to an approach for solving partial differential equations, a subject we do not investigate here.) Separation of variables is based on the absurd predicate that the derivative dx/dt can be treated as a quotient. So from

$$\frac{dx}{dt} = g(x)$$

we pretend that

$$\frac{1}{g(x)} \, dx = dt$$

Integrate both sides (integrating the left side is the first potential bump)

$$G(x) = \int \frac{1}{g(x)} \, dx = \int dt = t + C \tag{4.11}$$

and solve

$$G(x) = t + C \tag{4.12}$$

for x. This last step is the second, and possibly much larger, bump.

To validate this approach, suppose we have found $G(x)$ by Eq. (4.11). Now take d/dt of both sides of Eq. (4.12). With the chain rule we obtain

$$\frac{dG}{dx}\frac{dx}{dt} = 1 \tag{4.13}$$

Applying the fundamental theorem of calculus to Eq. (4.11) we rewrite Eq. (4.13) as

$$\frac{1}{g(x)}\frac{dx}{dt} = 1$$

and we recover the original autonomous differential equation.

Example 4.6.1. *Simple diffusion across a cell membrane, revisited.* Apply separation of variables to Eq. (3.3) of Example 3.2.2 to obtain

$$\frac{dC_i}{C_o - C_i} = \alpha \, dt$$

Integrate both sides and take C_o to be a constant.

$$-\ln|C_o - C_i| = \alpha t + b$$

Then exponentiate both sides and solve for C_i,

$$C_i = C_o - e^{-\alpha t} e^{-b} = C_o - e^{-\alpha t}(C_o(0) - C_i(0))$$

As t increases, C_i moves monotonically toward C_o, increasing if $C_o(0) > C_i(0)$, decreasing if $C_o(0) < C_i(0)$. So we see that as claimed in Example 3.2.2, there are no oscillations. □

Example 4.6.2. *The Gompertz percentage-kill model.* Now we'll solve the differential equations of Practice Problem 3.4.3. There are two:

$$\frac{dN}{dt} = aN \ln\left(\frac{N_\infty}{N}\right) \qquad\qquad \text{for } t < 5$$

$$\frac{dN}{dt} = aN \ln\left(\frac{N_\infty}{N}\right) - KN \qquad\qquad \text{for } t \geq 5$$

Noting that $\ln(N_\infty/N) = -\ln(N/N_\infty)$, dividing both sides of both equations by N_∞, and substituting $z = N/N_\infty$, these equations become

$$\frac{dz}{dt} = -az \ln(z) \text{ for } t < 5 \quad \text{and} \quad \frac{dz}{dt} = -az \ln(z) + Kz \text{ for } t \geq 5$$

Apply separation of variables to the first equation and integrate

$$\int \frac{dz}{z \ln(z)} = \int -a \, dt$$

$$\int \frac{dw}{w} = \int -a \, dt \qquad\qquad\qquad \text{substituting } w = \ln(z)$$

$$\ln(\ln(z)) = -at + b$$
$$\ln(z) = e^{-at+b} = e^{-at}e^{b} = e^{-at}B$$
$$z = e^{e^{-at}B} = e^{Be^{-at}} = \left(e^{B}\right)^{e^{-at}}$$

and so we see that

$$N = N_{\infty}\left(e^{B}\right)^{e^{-at}}$$

Substituting $t = 0$ we find $N(0) = N_{\infty}\left(e^{B}\right)^{e^{0}}$ and so $e^{B} = N(0)/N_{\infty}$. That is, the solution of the first equation is

$$N(t) = N_{\infty}\left(\frac{N(0)}{N_{\infty}}\right)^{e^{-at}}$$

Apply separation of variables to the second equation and integrate

$$\int \frac{dz}{z(\ln(z) + K/a)} = \int -a dt$$

with the substitution $w = \ln(z) + K/a$, then simplify. Again using $B = e^{b}$, this gives

$$N(t) = N_{\infty}\left(e^{B}\right)^{e^{-at}}e^{-K/a}$$

Take $t = 0$ and solve for e^{B} to obtain

$$N(t) = N_{\infty}e^{-K/a}\left(\frac{N(0)e^{K/a}}{N_{\infty}}\right)^{e^{-at}}$$

This is a bit messy, but not bad if we are careful with the algebra. $\square$

We'll end with a simple example in two forms, solving a simple growth equation and solving a differential equation of Example 3.2.5.

Example 4.6.3. *The simple growth equation.* Solve

$$\frac{dP}{dt} = \lambda P \qquad (4.14)$$

By separation of variables we obtain

$$\frac{dP}{P} = \lambda dt$$

Integrate and then exponentiate both sides to obtain

$$\ln|P| = \lambda t + c \qquad \text{and then} \qquad P(t) = e^{\lambda t}e^{c}$$

Note that

$$P(0) = e^{0}e^{c} = e^{c}$$

so we can rewrite the solution of Eq. (4.14) as

$$P(t) = P(0)e^{\lambda t} \tag{4.15}$$

We'll use this often in later sections. □

Example 4.6.4. *Solve the equation $dX/dZ = -R_0 X$. This is an easy variation of Example 4.6.3:*

$$\int \frac{dX}{X} = \int -R_0 \, dZ$$

$$\ln(X) = -R_0 Z + C$$

$$X(Z) = e^{-R_0 Z} e^C$$

$$X(Z) = X(0)e^{-R_0 Z} \qquad \text{Take } Z = 0 \text{ in the previous equation.}$$

This is as we claimed in Example 3.2.5. □

Practice Problems

4.6.1. (a) Suppose $x(0) = 10$. Solve $dx/dt = \cos^2(x)$.
(b) Solve $dx/dt = x/(x^2 + 1)$ by finding an explicit expression for x as a function of t, or if that is impossible, find a relation between x and t.

4.6.2. Separation of variables also can solve some non-autonomous equations. For example, solve $dx/dt = t/(tx + x)$.

Practice Problem Solutions

4.6.1. (a) By separation of variables we obtain

$$\frac{dx}{\cos^2(x)} = dt \qquad \text{That is,} \qquad \sec^2(x)dx = dt$$

Integrate both sides.

$$\tan(x) = t + c \qquad \text{and so} \qquad x(t) = \arctan(t + c)$$

Note that

$$x(0) = \arctan(c), \qquad \text{so} \qquad c = \tan(x(0))$$

and we see

$$x(t) = \arctan(t + \tan(x(0)))$$

Because $x(0) = 10$, the solution is $x(t) = \arctan(t + \tan(10))$.
(b) By separation of variables and some algebra we obtain

$$\left(x + \frac{1}{x}\right)dx = dt$$

Integrate both sides.

$$\frac{x^2}{2} + \ln(x) = t + C$$

Because x occurs both inside the ln and in a term outside the ln, we cannot solve this relation for x.

4.6.2. By separation of variables we find

$$2x\,dx = \frac{t}{t+1}\,dt = \left(1 - \frac{1}{t+1}\right)dt$$

Division of polynomials gives the last equality. Integrate both sides.

$$x^2 = t - \ln|t+1| + C$$

Take the square root of both sides to solve for x.

Sometimes we can solve complicated relations for x, sometimes not. Whether or not we can depends on practice, patience, and luck. For example, can we solve $e^{2x} + 2e^x + 1 = t + C$ for x? Once we recall $e^{2x} = (e^x)^2$, then we see that the left-hand side of the equation is a quadratic in e^x and factors as $(e^x + 1)^2$. A bit of algebra gives $x = \ln(\sqrt{t+C} - 1)$. If you don't see an obvious way to solve the relation for x, just experiment a bit.

Much of our work will focus on what we can find without solving differential equations, but every once in a while we encounter equations that can be solved. When this happens, we should be able to solve them.

Exercises

4.6.1. For the functions $g(x)$ in (a)–(i), solve the equation $dx/dt = g(x)$ explicitly for $x(t)$. In (e), take $x(0) = 3$.

(a) $\dfrac{1}{1+x}$ (b) $\dfrac{1}{4x + ((x^3)/3)}$ (c) $\dfrac{1 + e^{2x}}{e^x}$

(d) $e^{-x} + e^x$ (e) $\sqrt{9 - x^2}$ (f) $x\sqrt{1 - x^2}$

(g) $x\sqrt{x^2 - 1}$ (h) $x^2\sqrt{x^2 - 9}$ (i) $\sqrt{x^2 - 9}$

4.6.2. For the functions $g(x)$ in (a)–(f), find a relation between x and t determined by the equation $dx/dt = g(x)$. You need not solve this relation explicitly for x.

(a) $\dfrac{1}{\sqrt{x^2 - 4}}$ (b) $\dfrac{x}{\sqrt{x^2 - 1}}$ (c) $\dfrac{1}{\sqrt{1 - (1/x^2)}}$

(d) $\dfrac{\sqrt{1 - (1/x^2)}}{x}$ (e) $\dfrac{x^2}{\sqrt{x^2 - 1}}$ (f) $\dfrac{\sqrt{x^2 + 1}}{x + 1}$

4.6.3. This is a step in an analytic approach to Exercise 3.4.8. For positive constants a and K, solve $z' = -az \ln(z) + Kz \ln(z)$.

4.6.4. For the functions $h(x, t)$ in (a)–(d), solve the equation $dx/dt = h(x, t)$ explicitly for $x(t)$.

(a) $\dfrac{t^3 - 1}{x^2}$ (b) $\dfrac{\cos(t)}{\sin(x)}$ (c) $x(t^4 + t^2)$ (d) $\dfrac{\sqrt{t+1}}{x\sqrt{x^2 + 1}}$

4.6.5. For the functions $h(x, t)$ in (a)–(d), solve the equation $dx/dt = h(x, t)$ explicitly for $x(t)$, or find a relation between x and t if you cannot solve for x.

(a) $\dfrac{\sin(x)}{\cos(t)}$ (b) $\dfrac{\cos(t)}{\ln(x)}$ (c) $\dfrac{te^t}{1 + \ln(x)}$ (d) $\dfrac{\sin(t)}{\sin^{-1}(x)}$

4.6.6. Find an expression for $p(t)$, the size of a population with a seasonally varying food supply governed by $dp/dt = \alpha(1 + \sin(t))p$ for some positive constant α. Express $p(t)$ in terms of $p(0)$.

4.7 INTEGRATION BY PARTIAL FRACTIONS

Recall the rescaled version of Verhulst's equation, Eq. (4.9), the logistic differential equation

$$\frac{dx}{dt} = rx(1 - x)$$

Separation of variables gives

$$\frac{dx}{x(1 - x)} = r\,dt$$

We know $\int 1/x\,dx = \ln|x| + C$ and $\int 1/(1 - x)\,dx = -\ln|1 - x| + C$. Can we combine these in some fashion to evaluate $\int 1/x(1 - x)\,dx$? There is no product rule for integrals, but there is a sum rule. Can we write $1/x(1 - x)$ as a sum of $1/x$ and $1/(1 - x)$? That is, can we find constants a and b for which

$$\frac{1}{x(1 - x)} = \frac{a}{x} + \frac{b}{1 - x}$$

is true for all x in the domains of these functions? Finding a common denominator for the terms of the right,

$$\frac{1}{x(1 - x)} = \frac{a(1 - x)}{x(1 - x)} + \frac{bx}{x(1 - x)}$$

Equating the numerators we find

$$1 = a(1 - x) + bx = (b - a)x + a \tag{4.16}$$

In order for this to be true for all x, the coefficients of like powers of x must be equal.

coefficients of x	$0 = b - a$
coefficients of the constant	$1 = a$

That is, $a = b = 1$ and we have derived the algebraic equality

$$\frac{1}{x(1-x)} = \frac{1}{x} + \frac{1}{1-x}$$

Then

$$\int \frac{dx}{x(1-x)} = \int \frac{dx}{x} + \int \frac{dx}{1-x} = \ln|x| - \ln|1-x| + C = \ln\left|\frac{x}{1-x}\right| + C$$

So from the separation of variables formalism we deduce

$$\ln\left|\frac{x}{1-x}\right| = rt + C$$

Exponentiating both sides gives

$$\left|\frac{x}{1-x}\right| = e^{rt+C} = Ke^{rt}$$

Assuming $0 < x < 1$ so $|x/(1-x)| = x/(1-x)$, and solving for x, we obtain

$$x(t) = \frac{Ke^{rt}}{1 + Ke^{rt}}$$

Finally, substituting $t = 0$ and solving for K we find $K = x(0)/(1-x(0))$.

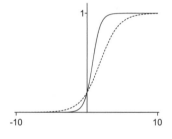

Figure 4.7. Two solution curves of Eq. (4.9).

In Fig. 4.7 we plot two solutions, with different r-values and both with $x(0) = 0.2$. Note the curves cross the vertical axis (the $t = 0$ line) at $x(0) = 0.2$. For the dashed curve, $r = 0.75$; for the solid curve, $r = 1.75$. Also, these curves agree with the stability analysis of Sect. 4.5, because all solutions converge to the stable fixed point $x = 1$.

This method of converting a ratio of polynomials into a sum of polynomials with simple denominators is called finding a *partial fractions expansion*. Four more points need to be covered.

1. There is a computationally simpler way to find the coefficients.
2. What happens if the highest power of the numerator is greater than or equal to the highest power of the denominator?
3. What if the denominator contains repeated factors?

4. What if the denominator contains a quadratic factor that does not have (real) linear factors? We shall not present the most general result, but give examples sufficient to solve all problems we'll encounter.

Example 4.7.1. *A simpler approach to finding the coefficients.* Return to the left equation in Eq. (4.16), $1 = a(1-x) + bx$. Even though this comes from an equation of functions whose domains exclude $x = 0$ and $x = 1$, such exclusions are not implied by this equation. What can we find from these substitutions?

Substituting $x = 0$ gives $1 = a$.   Substituting $x = 1$ gives $1 = b$.

This is only a bit simpler than the original calculation. For more complicated polynomials, this method can be substantially quicker. $\square$

Example 4.7.2. *Too high power of x in the numerator.* For any polynomial $p(x) = a_n x^n + a_{n-1} x^{n-1} + \cdots + a_1 x + a_0$, with $a_n \neq 0$, we say n is the degree $\deg(p(x))$ of the polynomial. In order to apply partial fractions to the ratio $p(x)/q(x)$ of polynomials with $\deg(p(x)) \geq \deg(q(x))$, first we must divide (by long division of polynomials or synthetic division), obtaining

$$\frac{p(x)}{q(x)} = r(x) + \frac{s(x)}{q(x)}$$

with $\deg(s(x)) < \deg(q(x))$. Integrating $r(x)$ is straightforward, then apply partial fractions to integrate $s(x)/q(x)$. An example calculation is given in Practice Problem 4.7.1 (b). $\square$

Example 4.7.3. *A power of a linear factor in the denominator.* If the denominator contains $(x-1)^2$, then the partial fractions expansion will contain the terms (in general, we do need both)

$$\frac{a}{x-1} \quad \text{and} \quad \frac{b}{(x-1)^2}$$

An example calculation is given in Practice Problem 4.7.2 (a). $\square$

Example 4.7.4. *An irreducible quadratic factor in the denominator.* For example, if the denominator contains the factor $(x^2 + 2x + 2)$ (to see this has no real factors, apply the quadratic formula to obtain complex roots), then the partial fractions expansion will contain a term

$$\frac{ax + b}{x^2 + 2x + 2}$$

An example calculation is given in Practice Problem 4.7.2 (b). $\square$

Although the name may sound a bit imposing, a partial fractions expansion is nothing more than simple algebra. To be sure, some calculations can be tedious, but the method is straightforward after some practice.

Practice Problems

Solve explicitly for $x(t)$, if you can. If not, you always can find a relation between x and t.

4.7.1. (a) $\dfrac{dx}{dt} = \dfrac{x^2 - 7x + 12}{2x + 1}$ (b) $\dfrac{dx}{dt} = \dfrac{x^2 + 3x + 2}{x^3 + 1}$

4.7.2. (a) $\dfrac{dx}{dt} = \dfrac{(x-1)(x+2)^2}{3x - 9}$ (b) $\dfrac{dx}{dt} = \dfrac{x^3 + 4x}{2x^2 - x + 4}$

Practice Problem Solutions

4.7.1. (a) We must integrate $(2x+1)/(x^2 - 7x + 12)$. The denominator factors as $x^2 - 7x + 12 = (x-3)(x-4)$. Then we have

$$\frac{2x+1}{(x-3)(x-4)} = \frac{a}{x-3} + \frac{b}{x-4}$$

so

$$2x + 1 = a(x-4) + b(x-3)$$

Taking $x = 4$ gives $b = 9$; taking $x = 3$ gives $a = -7$. Then

$$\int \frac{2x+1}{x^2 - 7x + 12}\, dx = \int \frac{-7}{x-3}\, dx + \int \frac{9}{x-4}\, dx$$
$$= -7\ln|x-3| + 9\ln|x-4|$$

so we conclude

$$\ln\left(\frac{|x-4|^9}{|x-3|^7}\right) = t + c \quad \text{That is,} \quad \frac{|x-4|^9}{|x-3|^7} = e^{t+c}$$

We cannot solve this explicitly for $x(t)$.

(b) Here we must integrate $(x^3 + 1)/(x^2 + 3x + 2)$. The degree of the numerator exceeds the degree of the denominator, so divide the polynomials. Here's a reminder, using long division of polynomials. If you learned synthetic division, that works just fine, too.

$$\begin{array}{r} x \quad -3 \\ \hline x^2 +3x +2\,\big|\, x^3 \qquad\qquad\qquad +1 \\ x^3 +3x^2 +2x \\ \hline -3x^2 -2x +1 \\ -3x^2 -9x -6 \\ \hline 7x +7 \end{array}$$

That is,

$$\frac{x^3 + 1}{x^2 + 3x + 2} = x - 3 + \frac{7x + 7}{x^2 + 3x + 2}$$

Next, use a partial fractions expansion for $(7x+7)/(x^2+3x+2)$:

$$\frac{7x+7}{x^2+3x+2} = \frac{a}{x+2} + \frac{b}{x+1} \quad \text{so} \quad 7x+7 = a(x+1) + b(x+2)$$

Take $x = -2$ to find $a = 7$; take $x = -1$ for $b = 0$. Really, $b = 0$?. Sure:

$$\frac{7x+7}{(x+2)(x+1)} = \frac{7(x+1)}{(x+2)(x+1)} = \frac{7}{x+2}$$

We could have gotten this by canceling a factor of $x+1$, once we view the numerator as $7x+7 = 7(x+1)$. Integrating, we conclude

$$\frac{x^2}{2} - 3x + 7\ln|x+3| = t + C$$

Again, we cannot solve this for $x(t)$.

4.7.2. (a) We must integrate $(3x-9)/((x-1)(x+2)^2)$. Use the partial fractions expansion

$$\frac{3x-9}{(x-1)(x+2)^2} = \frac{a}{x-1} + \frac{b}{x+2} + \frac{c}{(x+2)^2}$$

Putting all the terms over a common denominator and equating the numerators, we obtain

$$3x - 9 = a(x+2)^2 + b(x-1)(x+2) + c(x-1)$$

Taking $x = 1$ gives $a = -2/3$. Taking $x = -2$ gives $c = 5$. Substituting in these values of a and c and taking any other value of x gives $b = 2/3$. Then

$$\int \frac{3x-9}{(x-1)(x+2)^2} dx = -\frac{2}{3} \int \frac{dx}{x-1} + \frac{2}{3} \int \frac{dx}{x+2} + 5 \int \frac{dx}{(x+2)^2}$$

and so

$$-\frac{2}{3}\ln|x-1| + \frac{2}{3}\ln|x+2| - 5\frac{1}{x+2} = t + C$$

Yet another we cannot solve explicitly for x.

(b) We must integrate $(2x^2 - x + 4)/(x(x^2 + 4))$. Use the partial fractions expansion

$$\frac{2x^2 - x + 4}{x(x^2 + 4)} = \frac{a}{x} + \frac{bx + c}{x^2 + 4}$$

Putting all the terms over a common denominator and equating the numerators, we obtain

$$2x^2 - x + 4 = a(x^2 + 4) + (bx + c)x \qquad (4.17)$$

In this case, substituting in values of x doesn't provide any simplification, because the coefficient of a has no real roots. So instead we find a, b, and c by equating coefficients of like powers of x. Regrouping the right side of Eq. (4.17),

$$2x^2 - x + 4 = (a + b)x^2 + cx + 4a$$

we find

$$2 = a + b, \quad -1 = c, \quad \text{and} \quad 4 = 4a$$

and so

$$a = 1, \quad b = 1, \quad \text{and} \quad c = -1$$

Then

$$\int \frac{2x^2 - x + 4}{x(x^2 + 4)} dx = \int \frac{1}{x} dx + \int \frac{x - 1}{x^2 + 4} dx$$

The last integral is more easily done if it is split into two terms,

$$\int \frac{x - 1}{x^2 + 4} dx = \int \frac{x}{x^2 + 4} dx - \int \frac{1}{x^2 + 4} dx$$

The first is a simple substitution, the second can be gotten from rule 41 of the integral table, using $x^2 + 4 = 4((x/2)^2 + 1)$. Putting all this together gives

$$\int \frac{2x^2 - x + 4}{x(x^2 + 4)} dx = \ln|x| + \frac{1}{2} \ln(x^2 + 4) - \frac{1}{2} \arctan\left(\frac{x}{2}\right)$$

As usual, the right-hand side equals $t + C$. Also as usual, we cannot solve this explicitly for $x(t)$.

By this point you may be thinking that partial fractions expansions are not often useful, at least as far as finding explicit solutions, x as a function of t, for many autonomous differential equations. However, we shall see that the solutions we obtain do provide useful information. Patience is the key to paradise.

Exercises

Solve these differential equations, explicitly for $x(t)$ if possible. If not, find a relation between x and t.

4.7.1.

(a) $\dfrac{dx}{dt} = \dfrac{x^2 + 5x + 6}{2x + 5}$

(b) $\dfrac{dx}{dt} = \dfrac{x^2 + 3x + 2}{x^2 + 5x + 5}$

(c) $\dfrac{dx}{dt} = \dfrac{x^3 + 6x^2 + 11x + 6}{x^2 + 2x + 2}$

(d) $\dfrac{dx}{dt} = \dfrac{x^3 + 4x^2 + 5x + 2}{2x^2 + 6x + 5}$

4.7.2.

(a) $\dfrac{dx}{dt} = \dfrac{x^2 - 1}{x^3 + x}$

(b) $\dfrac{dx}{dt} = \dfrac{x^2 + 1}{x^3}$

(c) $\dfrac{dx}{dt} = \dfrac{x^2 + 3x + 2}{2x + 3}$

(d) $\dfrac{dx}{dt} = \dfrac{x^3 - 4x}{x^2 - 3x + 2}$

4.7.3.

(a) $\dfrac{dx}{dt} = \dfrac{x^3 + 2x^2 + x}{x^2 - 3x + 2}$

(b) $\dfrac{dx}{dt} = \dfrac{x^2 + 2x - 3}{x + 1}$

(c) $\dfrac{dx}{dt} = \dfrac{x^2 + 2x + 3}{x + 1}$

(d) $\dfrac{dx}{dt} = \dfrac{x^3 + 6x^2 + 11x + 6}{x + 1}$

4.7.4.

(a) $\dfrac{dx}{dt} = \dfrac{e^{2x} + 3e^x + 2}{e^{2x} + 3e^x}$

(b) $\dfrac{dx}{dt} = \dfrac{e^{2x} + 5e^x + 6}{e^{2x} + e^x}$

(c) $\dfrac{dx}{dt} = \dfrac{\sin^2(x) + 5\sin(x) + 6}{\sin(x)\cos(x) + \cos(x)}$

(d) $\dfrac{dx}{dt} = \dfrac{2\tan^2(x) - 3\tan(x) - 2}{\tan(x)\sec^2(x) + \sec^2(x)}$

4.7.5.

(a) $\dfrac{dx}{dt} = \dfrac{x^4 + 5x^2 + 4}{2x^3 + x^2 + 8x + 1}$

(b) $\dfrac{dx}{dt} = \dfrac{x^4 - 1}{2x^3 + x^2 - 2x + 1}$

(c) $\dfrac{dx}{dt} = \dfrac{x^3 - x^2 + x - 1}{x^2 + x}$

(d) $\dfrac{dx}{dt} = \dfrac{x^3 - x^2 + 2x - 2}{x^2 - 2x - 2}$

Chapter 5 Definite integrals and improper integrals

Many problems in biology and medicine involve diffusion across membranes and throughout volumes. In this chapter we develop some simple tools to calculate areas and volumes by definite integrals. In addition, we'll learn how to handle infinite limits of integration, and find an example of what infinite limits can tell us about the biological world.

In the introduction to integrals in calculus 1, you saw that the value of the definite integral $\int_a^b f(x)dx$ is the area under the curve $y = f(x)$ and above the x-axis, between $x = a$ and $x = b$. By the fundamental theorem of calculus, this is evaluated by

$$\int_a^b f(x)dx = F(x)\Big|_a^b = F(b) - F(a)$$

where $F(x)$ is any antiderivative of $f(x)$, that is, $F'(x) = f(x)$.

The only point requiring some care is that where the graph of $f(x)$ lies below the x-axis, the area is negative. For example,

$$\int_0^{2\pi} \sin(x)dx = -\cos(x)\Big|_0^{2\pi} = -\cos(2\pi) - (-\cos(0)) = -1 + 1 = 0$$

This is expected, because between $x = 0$ and $x = 2\pi$, half the graph of $\sin(x)$ lies above the x-axis, and half lies below.

This area computation has an obvious extension: to find the area of the region bounded between two curves, always subtract the curve below from the curve above. An important step in this is locating all the points of intersection of the curves. For example, find the area of the region bounded by the curves $f(x) = x$ and $g(x) = x^3 - x$. To find the intersections, solve

$$x = x^3 - x \quad \text{that is,} \quad 0 = x^2 - 2x = x(x^2 - 2)$$

The points of intersection are $(-\sqrt{2}, -\sqrt{2})$, $(0,0)$, and $(\sqrt{2}, \sqrt{2})$. Because the functions are continuous, whichever lies above the other at one point of $(-\sqrt{2}, 0)$ lies above throughout that interval. Because $f(-1) = -1 < 0 = g(-1)$, on the interval $(-\sqrt{2}, 0)$ the curve $g(x)$ lies above $f(x)$. Similarly for $(0, \sqrt{2})$. Because $f(1) = 1 > 0 = g(1)$, on the interval $(0, \sqrt{2})$ the curve $f(x)$ lies above $g(x)$. See Fig. 5.1.

The area of the region bounded by these curves is

$$\int_{-\sqrt{2}}^{0} ((x^3 - x) - x) \, dx + \int_{0}^{\sqrt{2}} (x - (x^3 - x)) \, dx$$

$$= \left(\frac{x^4}{4} - x^2 \right) \Big|_{-\sqrt{2}}^{0} + \left(-\frac{x^4}{4} + x^2 \right) \Big|_{0}^{\sqrt{2}} = 1 + 1$$

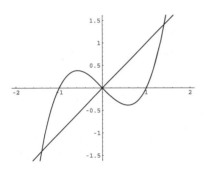

Note if we simply integrate $f(x) - g(x)$ or $g(x) - f(x)$ between $x = -\sqrt{2}$ and $x = \sqrt{2}$, we obtain 0. This occurs for the same reason that $\int_{0}^{2\pi} \sin(x) dx = 0$: for half of the interval, $f(x)$ lies above $g(x)$, for the other half, $g(x)$ lies above $f(x)$, and these two pieces have the same areas.

Figure 5.1. The graphs of $y = x$ and $y = x^3 - x$.

If we modify the original problem of this chapter by asking for the area of the region bounded by $y = \sin(x)$ and $y = 0$, between $x = 0$ and $x = 2\pi$, the appropriate integrals are

$$\int_{0}^{\pi} (\sin(x) - 0) \, dx + \int_{\pi}^{2\pi} (0 - \sin(x)) \, dx = 2 + 2 = 4$$

5.1 DEFINITE INTEGRALS BY SUBSTITUTION

When we use substitution, definite integrals can be evaluated in two ways. Both give the same result, of course.

Example 5.1.1. *Original variable limits.* Evaluate the integral by substitution, express the result in terms of the original variable, and apply the fundamental theorem. To evaluate

$$\int_0^1 \frac{2x}{(1+x^2)^3}\, dx$$

substitute $w = 1 + x^2$ so $dw = 2x\, dx$ and

$$\int \frac{2x}{(1+x^2)^3}\, dx = \int \frac{1}{w^3}\, dw = -\frac{1}{2w^2} = -\frac{1}{2(1+x^2)^2}$$

Then

$$\int_0^1 \frac{2x}{(1+x^2)^3}\, dx = -\frac{1}{2(1+x^2)^2}\Big|_0^1 = \frac{3}{8}$$

That is, substitute, integrate, express the integral in the original variable, and use the given limits of integration. $\square$

Example 5.1.2. *New variable limits.* Evaluate the integral of Example 5.1.1 by substitution, and apply the fundamental theorem with limits in terms of the new variable. With the substitution $w = 1 + x^2$, observe that $x = 0$ gives $w = 1$ and $x = 1$ gives $w = 2$. Then

$$\int_0^1 \frac{2x}{(1+x^2)^3}\, dx = \int_1^2 \frac{1}{w^3}\, dw = -\frac{1}{2w^2}\Big|_1^2 = -\frac{1}{8} - \left(-\frac{1}{2}\right) = \frac{3}{8}$$

This approach usually is shorter, but we must remember to use the values of the new variable when evaluating the integral. $\square$

Practice Problems

Use the method of Ex. 5.1.2 to solve these problems. For each, find the area under the curve $y = f(x)$ between the given bounds on x.

5.1.1. $f(x) = \cos(x)/(2 + \sin(x))^2$ between $x = 0$ and $x = \pi/2$.

5.1.2. $f(x) = (2 - e^{2x})e^x$ between $x = 0$ and $x = 1$.

5.1.3. $f(x) = \sin(\sin(x))\cos(x)$ between $x = 0$ and $x = \pi/2$.

Practice Problem Solutions

5.1.1. The substitution $w = \sin(x)$ gives $dw = \cos(x)\, dx$ and so

$$\int_0^{\pi/2} \frac{\cos(x)}{(2+\sin(x))^2}\, dx = \int_0^1 \frac{dw}{(2+w)^2} = -\frac{1}{2+w}\Big|_0^1 = \frac{1}{6}$$

5.1.2. The substitution $w = e^x$ gives $dw = e^x\, dx$ and so

$$\int_0^1 (2 - e^{2x})e^x\, dx = \int_0^e (2 - w^2)\, dw = \left(2w - \frac{w^3}{3}\right)\Big|_1^e = 2e - \frac{e^3}{3} - \frac{5}{3}$$

5.1.3. The substitution $w = \sin(x)$ gives $dw = \cos(x)\, dx$ and so

$$\int_0^{\pi/2} \sin(\sin(x)) \cos(x)\, dx = \int_0^1 \sin(w)\, dw = -\cos(w)\Big|_0^1$$

$$= 1 - \cos(1) \approx 0.459698$$

Exercises

In Exercises 5.1.1, 5.1.2, and 5.1.3 use the method of Example 5.1.2 to evaluate the integrals. If you use the table of integrals in Appendix C, give the number of the integral you use.

5.1.1.

(a) $\displaystyle\int_1^2 x^2\sqrt{x^3 + 1}\, dx$

(b) $\displaystyle\int_1^2 \frac{x}{x^2 + 1}\, dx$

(c) $\displaystyle\int_0^1 xe^{-x^2}\, dx$

(d) $\displaystyle\int_0^{\pi/2} \sin^5(x)\cos(x)\, dx$

5.1.2.

(a) $\displaystyle\int_0^1 \frac{x+1}{x^2 + 2x + 3}\, dx$

(b) $\displaystyle\int_0^{\pi^{1/3}} x^2 \sin(x^3)\, dx$

(c) $\displaystyle\int_2^3 \frac{\ln(x)}{x}\, dx$

(d) $\displaystyle\int_5^6 \frac{1}{x\ln(x)}\, dx$

5.1.3.

(a) $\displaystyle\int_0^{\pi/2} \cos(x)\sin(x)e^{\sin(x)}\, dx$

(b) $\displaystyle\int_{\ln(1/2)}^{\ln(\sqrt{2}/2)} \frac{e^x}{\sqrt{1 - e^{2x}}}\, dx$

(c) $\displaystyle\int_2^3 \frac{\ln(x)}{x}\, dx$

(d) $\displaystyle\int_{\pi/6}^{\pi/3} \cos(x)\ln(\sin(x))\, dx$

5.1.4. Find the area between the curves $y = f(x)$ and $y = g(x)$, between the given x-values.
(a) $f(x) = \tan^2(x)\sec^2(x)$ and $g(x) = 0$, for $0 \le x \le \pi/4$.
(b) $f(x) = (1 + \ln(x)^2)/x$ and $g(x) = 0$, for $1 \le x \le 2$.

5.1.5. Find the area of the region bounded by $f(x)$ and $g(x)$.
(a) $f(x) = 4\ln(x)(1 - \ln(x))/x$ and $g(x) = \ln(x)/x$.
(b) $f(x) = (1 - \ln(x)^2)/x$ and $g(x) = \ln(x)^2/x$.

5.2 VOLUME BY INTEGRATION

When finding the area between $y = f(x)$ and $y = g(x)$, $f(x) \geq g(x)$, $a \leq x \leq b$, with the integral $\int_a^b (f(x) - g(x))\, dx$, we integrate the length of the segment from $g(x)$ to $f(x)$ for $a \leq x \leq b$. The integral of a length can give an area. Similarly, the integral of an area can give a volume. Let's see an example.

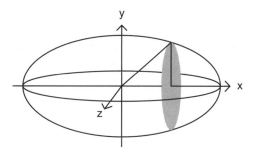

First we find the volume of an ellipsoid that intersects the xy-plane in the ellipse $x^2/4 + y^2 = 1$. The range of x-values is $-2 \leq x \leq 2$. Cross-sections perpendicular to the x-axis are circles of radius $y = \sqrt{1 - x^2/4}$. See Fig. 5.2. Then the volume of the ellipsoid is the integral of the area of these circles:

Figure 5.2. How to find the volume of an ellipsoid.

$$\text{volume} = \int_{-2}^{2} \pi y^2 \, dx = \pi \int_{-2}^{2} \pi \left(1 - \frac{x^2}{4}\right) dx = \frac{8\pi}{3}$$

This seems straightforward. Just to be sure, we test this approach with a case for which we already know the answer. For a sphere of radius r, cross-sections perpendicular to the x-axis are circles of radius $y = \sqrt{r^2 - x^2}$. Then the volume is

$$\text{volume} = \int_{-r}^{r} \pi y^2 \, dx = \pi \int_{-r}^{r} \pi \left(r^2 - x^2\right) dx = \frac{4\pi r^3}{3}$$

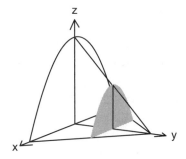

This method to find the volume by integration of the area of cross-sections can be applied to any cross-section, not just circles. For example, suppose the cross-section in the plane $y = y_0$ is the portion of the parabola $z = (1 - y_0) - x^2/(1 - y_0)$ with $z \geq 0$. See Fig. 5.3. For each of these parabolas, the range of x-values is $y_0 - 1 \leq x \leq 1 - y_0$. The area of the $y = y_0$ cross-section is

Figure 5.3. How to find the volume of a solid with parabolic cross-sections.

$$\text{area}(y_0) = \int_{y_0-1}^{1-y_0} ((1-y_0) - x^2) \, dx = \frac{4}{3}(y_0 - 1)^2$$

and the volume is

$$\text{volume} = \int_0^1 \text{area}(y) \, dy = \int_0^1 \frac{4}{3}(y-1)^2 \, dy = \frac{4}{9}$$

Practice Problems

5.2.1. Find the volume of the solid with square cross-section perpendicular to the x-axis and with side length x^2 for $0 \le x \le 1$.

5.2.2. Find the volume of the solid with annular cross-section perpendicular to the x-axis with outside radius x^2 and inside radius x, for $1 \le x \le 2$.

Practice Problem Solutions

5.2.1. The cross-section at x has area $(x^2)^2 = x^4$, so the volume is

$$\int_0^1 x^4 \, dx = \frac{x^5}{5}\Big|_0^1 = \frac{1}{5}$$

5.2.2. The cross-section at x has area $\pi(x^2)^2 - \pi x^2$, so the volume is

$$\int_1^2 (\pi x^4 - \pi x^2) \, dx = \pi \left(\frac{x^5}{5} - \frac{x^3}{3}\right)\Big|_1^2 = \pi \frac{58}{15}$$

Exercises

5.2.1. Find the volume of the solid with right triangular cross-section perpendicular to the x-axis, with base x and altitude x^2, for $0 \le x \le 1$.

5.2.2. Find the volume of the solid with elliptical cross-section perpendicular to the x-axis, with semi-major axis x and semi-minor axis x^2, for $0 \le x \le 1$. (The area of an ellipse of semi-axes a and b is πab.)

5.2.3. Find the volume of the solid with circular cross-section of radius $\sin(x)$, $0 \le x \le \pi$. (Hint: remember $\sin^2(x) = (1 - \cos(2x))/2$.)

5.2.4. Find the volume of the solid with circular cross-section of radius $\sin^{3/2}(x)$, $0 \le x \le \pi$. (Hint: remember $\sin^3(x) = \sin(x)(1 - \cos^2(x))$.)

5.2.5. Find the volume of the solid with cross-section an annulus (region between concentric circles) of outer radius x and inner radius x^2, $0 \le x \le 1$.

5.2.6. Find the volume of the solid with cross-section a rectangle of base x and height $1 - x^2$, $0 \le x \le 1$.

5.2.7. Find the volume of the solid obtained when the curve $y = x^2$, $-1 \le x \le 1$, is rotated about the x-axis.

5.2.8. Find the volume of the solid obtained when the curve $y = x^2$, $-1 \le x \le 1$, is rotated about the line $y = 1$.

5.2.9. Find the volume of the cone obtained when $z = x$, $0 \le z \le 1$, is rotated about the z-axis.

5.2.10. (a) Find the volume of the solid obtained when the curve $y = x - x^2$, $0 \le x \le 1$, is rotated about the x-axis.
(b) Find the volume of the solid formed when the curve of (a) is rotated about the y-axis.

5.3 LENGTHS OF CURVES

The rates of many physiological processes depend on the length of a capillary, an axon, a muscle cell. In this section we introduce a simple method for computing the length of a curve, given a functional representation of that curve. Apply the usual warning: the large part of the real work of any problem is finding a mathematical model. For now, we assume the model already has been found.

What do we know about finding lengths? For straight lines the answer comes from Pythagoras: the segment between (x_1, y_1) and (x_2, y_2) is the hypotenuse of the right triangle with base length $|x_1 - x_2|$ and altitude length $|y_1 - y_2|$, so the line segment length is

$$\sqrt{(x_1 - x_2)^2 + (y_1 - y_2)^2}$$

Now suppose we want to find the length of the graph of a function $y = f(x)$ between $x = a$ and $x = b$. We can pick sample points on the graph, connect successive points with straight line segments, and add the lengths of the segments. As we take more sample points and as these points become closer together, does the sum of the length approach a limit? If it does, then to call this limit the length of the graph is sensible. Can we find a simple expression for this limit?

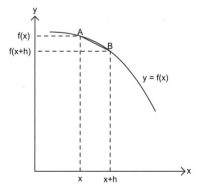

Figure 5.4. Approximating a curve by line segments.

Fig. 5.4 is our guide in this calculation. The straight line segment joining points A and B has length

$$\sqrt{((x+h)-x)^2 + (f(x+h)-f(x))^2} = \sqrt{1 + \left(\frac{f(x+h)-f(x)}{h}\right)^2}\, h \quad (5.1)$$

In the limit as $h \to 0$, the sum of the terms of Eq. (5.1) becomes an integral, and, if $f(x)$ is differentiable, the integrand Eq. (5.1) becomes

$$\sqrt{1 + (f'(x))^2}\, dx$$

Then the length of the curve is

$$L = \int_a^b \sqrt{1 + (f'(x))^2}\, dx \quad (5.2)$$

To illustrate this method, we find the length of the quarter of the circle $x^2 + y^2 = r^2$ in the first quadrant. This is the graph of $y = f(x) = \sqrt{r^2 - x^2}$ for $0 \le x \le r$. Applying Eq. (5.1) we find the length is

$$\int_0^r \sqrt{1 + \left(\frac{x}{\sqrt{r^2-x^2}}\right)^2}\, dx = \int_0^r \frac{r}{\sqrt{r^2-x^2}}\, dx = r\sin^{-1}\left(\frac{x}{r}\right)\bigg|_0^r = \frac{r\pi}{2}$$

where the penultimate equality follows from rule 28 of the integral table in Appendix C. Note this is one-quarter of the circumference of the circle of radius r.

Practice Problems

5.3.1. Find the arclength of $f(x) = x^2/2$ from $x = 0$ to $x = 1$.

5.3.2. Find the arclength of $f(x) = \ln(x)$ from $x = 1$ to $x = 2$.

Practice Problem Solutions

5.3.1. Because $f'(x) = x$, the arclength is $\displaystyle\int_0^1 \sqrt{1 + x^2}\, dx$

$$= \left(\frac{x}{2}\sqrt{1+x^2} + \frac{1}{2}\ln|x + \sqrt{1+x^2}|\right)\bigg|_0^1 = \frac{\sqrt{2}}{2} + \frac{1}{2}\ln(1+\sqrt{2})$$

where the penultimate equality follows from rule 16 of the integral table.

5.3.2. Because $f'(x) = 1/x$, the arclength is $\displaystyle\int_1^2 \sqrt{1 + \frac{1}{x^2}}\, dx$

$$= \int_1^2 \frac{\sqrt{x^2+1}}{x}\, dx = \left(\sqrt{1+x^2} - \ln\left(\frac{1+\sqrt{1+x^2}}{x}\right)\right)\Bigg|_1^2$$

$$= \left(\sqrt{5} - \ln\left(\frac{1+\sqrt{5}}{2}\right)\right) - \left(\sqrt{2} - \ln(1+\sqrt{2})\right)$$

where the penultimate equality follows from rule 18 of the integral table.

Exercises

5.3.1. Find the arclength of $f(x) = mx$ from $x = 0$ to $x = 1$. Does this agree with basic geometry?

5.3.2. Find the arclength of $f(x) = x^3/6 + 1/(2x)$ from $x = 1$ to $x = 2$.

5.3.3. Find the arclength of $f(x) = x^{3/2}$ from $x = 0$ to $x = 1$.

5.3.4. Find the arclength of $f(x) = x^2/2 - \ln(x)/4$ from $x = 2$ to $x = 3$.

5.3.5. Find the arclength of the parabola $f(x) = (1/2)x^2$ from $x = 0$ to $x = 1$. This should be simple, right?

5.3.6. Find the arclength of the parabola $f(x) = x^2$ from $x = 0$ to $x = 1$. Is this twice the length you found in Exercise 5.3.5?

5.3.7. Find the arclength of $f(x) = x + x^2/2$ from $x = 0$ to $x = 1$.

5.3.8. Find the arclength of $f(x) = \ln(1+x)$ from $x = 0$ to $x = 1$.

5.3.9. Find the arclength of $f(x) = x^4/8 + 1/(4x^2)$ from $x = 1$ to $x = 2$.

5.3.10. Find the arclength of $f(x) = (e^x + e^{-x})/2$ from $x = 0$ to $x = 1$.

5.4 SURFACE AREA BY INTEGRATION

Many body components exhibit an approximate axial symmetry of varying radius. Diffusion into or out of such a component occurs across the surface, so knowing the (approximate) area of these surfaces is important in calculating the total diffusion. In this section we learn to compute the areas of surfaces of revolution, which are idealized models of these membranes. Again, remember that hard work goes into finding an approximate representation of biological components. For now, we assume that work has been done by others.

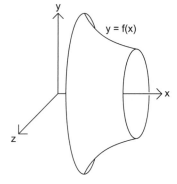

Figure 5.5. The surface obtained by revolving the graph $y = f(x)$ about the x-axis

Fig. 5.5 illustrates the construction of a surface of revolution: revolve the graph of $y = f(x)$ about the x-axis.

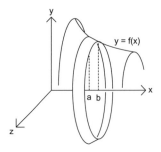

A surface of revolution is made up of a family of circles, one of radius $f(x)$ for each x. Guided by our calculation of the length of a curve, we approximate portions of the curve by straight line segments. As illustrated in Fig. 5.6, revolving each of these segments about the x-axis gives a portion of a cone, called a frustrum. So we need to find the area of these frustra, add them up, and take the limit as width of the frustra goes to 0.

Figure 5.6. Calculating surface area with frustra.

From geometry recall that the surface area of the cone with base radius r and height h is

$$\pi r \sqrt{r^2 + h^2} = \pi rL$$

where L is called the slant height of the cone. See the left side of Fig. 5.7. Then the area of the frustrum on the right side of Fig. 5.7 is the difference of the areas of the large and small cones:

$$\text{frustrum area} = \pi r'(L + L') - \pi rL$$
$$= \pi(r' - r)L + \pi r'L'$$
$$= \pi rL' + \pi r'L' = \pi(r + r')L'$$

The last line follows from the similarity of the triangles $\triangle ABC$ and $\triangle ADE$: corresponding sides have the same ratios, that is,

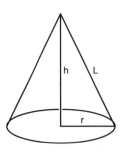
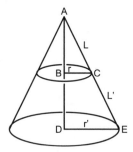

Figure 5.7. The area of a cone (left); the area of a frustrum (right).

$$\frac{r'}{L+L'} = \frac{r}{L}$$
$$r'L = r(L+L')$$
$$(r'-r)L = rL'$$

Now rewrite the frustrum area as

$$2\pi \frac{r+r'}{2} L'$$

In the limit as $r \to r'$, the average of r and r' approaches $r' = f(x)$, the length L' approaches the arclength $\sqrt{1+(f'(x))^2}\,dx$, and the sum of the areas of the frustra approaches the integral

$$\text{area} = \int_a^b 2\pi f(x)\sqrt{1+(f'(x))^2}\,dx \tag{5.3}$$

Example 5.4.1. *Area of a surface of revolution.* Compute the area of the surface obtained by revolving the curve $y = x^3$, $0 \le x \le 1$, about the x-axis.
   With Eq. (5.3) we find

$$\text{area} = \int_0^1 2\pi x^3 \sqrt{1+(3x^2)^2}\,dx = \frac{\pi}{18}\int_1^{10}\sqrt{w}\,dw = \frac{\pi}{27}(10^{3/2}-1)$$

where we have used the substitution $w = 1+9x^4$. □

Practice Problems

5.4.1. Find the area of the sphere of radius r.

5.4.2. Find the area of the surface obtained by revolving $y = \sin(x)$, $0 \le x \le \pi$, about the x-axis.

Practice Problem Solutions

5.4.1. The sphere of radius r, centered at the origin, is obtained by revolving $y = f(x) = \sqrt{r^2-x^2}$, $-r \le x \le r$, about the x-axis. Then $f'(x) = x/\sqrt{r^2-x^2}$ and so the area is

$$\text{area} = \int_{-r}^r 2\pi\sqrt{r^2-x^2}\sqrt{1+\frac{x^2}{r^2-x^2}}\,dx$$
$$= \int_{-r}^r 2\pi\sqrt{r^2-x^2}\,\frac{r}{\sqrt{r^2-x^2}}\,dx = 2\pi\int_{-r}^r r\,dx = 4\pi r^2$$

5.4.2. Because $f(x) = \sin(x)$ and $f'(x) = \cos(x)$, the area is given by

$$\text{area} = \int_0^\pi 2\pi \sin(x)\sqrt{1+\cos^2(x)}\ dx = -2\pi \int_1^{-1} \sqrt{1+u^2}\ du$$

$$= -2\pi \left(\frac{u}{2}\sqrt{1+u^2} + \frac{1}{2}\ln\left(u+\sqrt{1+u^2}\right)\right)\Big|_1^{-1}$$

$$= 2\pi \left(\sqrt{2} + \frac{1}{2}\ln(1+\sqrt{2}) - \frac{1}{2}\ln(-1+\sqrt{2})\right)$$

where the second equality uses the substitution $u = \cos(x)$, and the third uses rule 16 of the integral table.

Exercises

5.4.1. Find the area of the surface obtained when $y = \sqrt{x}$, $0 \le x \le 1$, is revolved about the x-axis.

5.4.2. Find the area of the surface obtained when $y = e^x$, $0 \le x \le 1$, is revolved about the x-axis.

5.4.3. Find the area of the surface obtained when $y = e^{-x}$, $0 \le x \le 1$, is revolved about the x-axis.

5.4.4. Find the area of the surface obtained when $y = \sqrt{x+1}$, $0 \le x \le 1$, is revolved about the x-axis.

5.4.5. Find the area of the surface obtained when $y = \sqrt{2x - x^2}$, $0 \le x \le 2$, is revolved about the x-axis.

5.4.6. Show that the surface obtained when $x = g(y)$, $c \le y \le d$, is revolved about the y-axis has area

$$\int_c^d 2\pi g(y)\sqrt{1+(g'(y))^2}\ dy$$

5.4.7. Find the area of the surface obtained when $x = y^3$, $0 \le y \le 1$, is revolved about the y-axis.

5.4.8. For the surface obtained when $x = y^4$, $0 \le y \le 2$, is revolved about the y-axis, show the area is $\ge 64/5$.

5.4.9. For the surface obtained when $x = 1/y$, $1 \leq y \leq 2$, is revolved about the y-axis, show the area is $\geq 2\pi \ln(2)$.

5.4.10. Find the area of the surface obtained when $x = 1/y$, $1 \leq y \leq 2$, is revolved about the y-axis. This is tricky. I used integration by parts, a substitution, and the integral table.

5.5 IMPROPER INTEGRALS

A definite integral $\int_a^b f(x)\, dx$ has limits that are real numbers and an implicit assumption that the domain of f includes the interval $[a, b]$. An integral is called *improper* if $a = -\infty$, $b = \infty$, or if for some $c \in [a, b]$, $\lim_{x \to c^+} f(x) = \pm\infty$ or $\lim_{x \to c^-} f(x) = \pm\infty$. In this section we'll give examples of improper integrals and determine their convergence or divergence (these terms mean just what you think they mean) by direct calculation. A way to determine convergence even if we can't find an antiderivative of the function is presented in Sect. 5.6. So far as we know, no physical object is infinite in size or in duration, so why should we care about improper integrals, beyond their mathematical interest? Some settings to apply improper integrals are shown in Sect. 5.7.

Examples illustrate the types of behaviors of improper integrals.

Example 5.5.1. *A convergent improper integral.*

$$\int_1^\infty \frac{1}{x^2}\, dx = \lim_{T \to \infty} \int_1^T \frac{1}{x^2}\, dx = \lim_{T \to \infty} -\frac{1}{x}\Big|_1^T = \lim_{T \to \infty} \left(-\frac{1}{T} + 1\right) = 1$$

Because the limit is a number, we say this improper integral *converges*. □

What, besides a number, could the limit be? How about ∞?

Example 5.5.2. *A divergent improper integral.*

$$\int_1^\infty \frac{1}{x}\, dx = \lim_{T \to \infty} \int_1^T \frac{1}{x}\, dx = \lim_{T \to \infty} \ln(x)\Big|_1^T = \lim_{T \to \infty} \left(\ln(T) - \ln(1)\right) = \infty$$

Because the limit is ∞, we say this improper integral *diverges*. □

What about this general case?

$$\int_1^\infty \frac{1}{x^n}\, dx$$

We have already seen that this integral converges for $n = 2$ and diverges for $n = 1$. For all $n \neq 1$,

$$\int_1^\infty \frac{1}{x^n} dx = \lim_{T \to \infty} \frac{x^{-n+1}}{-n+1}\bigg|_1^T = \lim_{T \to \infty} \left(\frac{T^{-n+1}}{-n+1} - \frac{1}{-n+1} \right)$$

This integral converges if the exponent of T is negative, i.e., if $n > 1$. The integral diverges if the exponent of T is positive, i.e., if $n < 1$. Putting this all together with Ex. 5.5.2, we see

$$\int_1^\infty \frac{1}{x^n} dx \begin{cases} \text{converges} & \text{if } n > 1 \\ \text{divergres} & \text{if } n \leq 1 \end{cases} \tag{5.4}$$

Even when both integration limits are finite, some integrals are improper.

Example 5.5.3. *An improper integral with finite limits of integration.*

$$\int_0^1 \frac{1}{x} dx$$

Here the problem occurs at the lower limit of integration: the integrand is undefined at $x = 0$. Guided by our treatment of infinite limits of integration, we replace the $x = 0$ limit with $x = T$ and take the limit as $T \to 0$.

$$\int_0^1 \frac{1}{x} dx = \lim_{T \to 0} \int_T^1 \frac{1}{x} dx = \lim_{T \to 0} \ln(x)\bigg|_T^1 = \lim_{T \to 0} (\ln(1) - \ln(T))$$

Because $\lim_{T \to 0} \ln(T) = -\infty$, we see this improper integral diverges. $\square$

What about other powers of x? For $n \neq 1$,

$$\int_0^1 \frac{1}{x^n} dx = \lim_{T \to 0} \frac{x^{-n+1}}{-n+1}\bigg|_T^1 = \lim_{T \to 0} \left(\frac{1}{-n+1} - \frac{T^{-n+1}}{-n+1} \right)$$

This limit exists if $-n+1 > 0$. Combined with Ex. 5.5.3, we see

$$\int_0^1 \frac{1}{x^n} dx \begin{cases} \text{converges} & \text{if } n < 1 \\ \text{diverges} & \text{if } n \geq 1 \end{cases} \tag{5.5}$$

Now we'll combine the convergence conditions (5.4) and (5.5), with substitution and perhaps other techniques to evaluate the integral.

Practice Problems

Determine if the improper integral converges or diverges.

5.5.1.

(a) $\displaystyle\int_1^\infty \frac{t^2}{(1+t^3)^{3/2}}\,dt$ (b) $\displaystyle\int_1^\infty \frac{\cos(x)\sin(x)}{(1+\cos^2(x))^2}\,dx$ (c) $\displaystyle\int_1^2 \frac{1}{x-1}\,dx$

5.5.2.

(a) $\displaystyle\int_{\pi/2}^\pi \frac{\sin(x)}{\cos^2(x)}\,dx$ (b) $\displaystyle\int_{\pi/2}^\pi \tan(x)\sec(x)\,dx$ (c) $\displaystyle\int_0^2 \frac{1}{(t-1)^3}\,dt$

Practice Problem Solutions

5.5.1. (a) Substitute $w = 1+t^3$ so $dw = 3t^2\,dt$. The limits $t=1$ and $t=\infty$ become $w=2$ and $w=\infty$, so the integral is

$$\int_1^\infty \frac{t^2}{(1+t^3)^{3/2}}\,dt = \int_2^\infty \frac{1}{3}\frac{1}{w^{3/2}}\,dw$$

Because $3/2 > 1$, we see this integral converges by (5.4).

(b) Substitute $w = 1+\cos^2(x)$ so $dw = -2\cos(x)\sin(x)dx$. The lower limit becomes $1+\cos^2(1)$, the upper limit becomes ...what? Let's try a different approach. Evaluate this as an indefinite integral, substitute back for x, and then take the $x \to \infty$ limit.

$$\int \frac{\sin(x)\cos(x)}{1+\cos^2(x)}\,dx = \int -\frac{1}{2}\frac{1}{w^2}\,dw = \frac{1}{2w} = \frac{1}{2(1+\cos^2(x))}$$

As $x \to \infty$, $1/(2(1+\cos^2(x)))$ oscillates between $1/4$ and $1/2$, so this improper integral does not converge, illustrating another way in which an improper integral can diverge.

(c) The integrand is undefined at the $x=1$ limit, so this is an improper integral. The substitution $w = x-1$ converts this integral to $\int_0^1 dw/w$, which diverges by (5.5).

5.5.2. (a) The integrand is undefined at the $x=\pi/2$ limit, so this is an improper integral. With the substitution $w = \cos(x)$, we find $dw = -\sin(x)dx$, $x = \pi/2$ becomes $w=0$, and $x=\pi$ becomes $w=-1$. This translates the integral to

$$\int_{\pi/2}^\pi -\frac{\sin(x)}{\cos^2(x)}\,dx = \int_0^{-1} -\frac{dw}{w^2}$$

which diverges by (5.5).

(b) Note that $\tan(x)\sec(x) = \sin(x)/\cos^2(x)$, so this problem is equivalent to Practice Problem 5.5.2 (a).

(c) The integrand is defined at both limits of integration, but *not* throughout the interval of integration: $x = 1$ is the problem. So we'll split this into two integrals, both improper.

$$\int_0^2 \frac{1}{(t-1)^3}\, dt = \lim_{T\to 1^-} \int_0^T \frac{1}{(t-1)^3}\, dt + \lim_{S\to 1^+} \int_S^2 \frac{1}{(t-1)^3}\, dt$$

We use a separate variable for each limit to emphasize that both must be evaluated independently, that if either diverges then the original integral diverges. A substitution of $w = t - 1$ turns both these into integrals that diverge by (5.5).

Exercises

In Probs. 5.5.1–5.5.5, do these improper integrals converge or diverge?

5.5.1.

$$\text{(a)}\ \int_1^\infty \frac{1}{\sqrt{t}(1+\sqrt{t})^{3/2}}\, dt \qquad \text{(b)}\ \int_1^\infty \frac{1}{\sqrt{t}(1+\sqrt{t})^{1/2}}\, dt$$

5.5.2.

$$\text{(a)}\ \int_0^1 \cot^2(t)\sec^2(t)\, dt \qquad \text{(b)}\ \int_0^1 \frac{1}{\sin^2(t)}\, dt$$

5.5.3.

$$\text{(a)}\ \int_0^2 \frac{t}{t^2-1}\, dt \qquad \text{(b)}\ \int_0^1 \cot(t)\, dt$$

5.5.4.

$$\text{(a)}\ \int_0^\pi \sec(t)\, dt \qquad \text{(b)}\ \int_0^\infty \sin(t)\, dt$$

5.5.5.

$$\text{(a)}\ \int_0^\infty \frac{\cos(t)}{1+\sin^2(t)}\, dt \qquad \text{(b)}\ \int_0^\infty \frac{\cos(t)}{e^{\sin(t)}}\, dt$$

5.6 IMPROPER INTEGRAL COMPARISON TESTS

Does $\int_1^\infty \frac{1}{x+x^{3/2}}\, dx$ converge or diverge? While this integral can be evaluated, finding an antiderivative is tricky. When we are interested only in convergence, must we evaluate this integral, or can we determine convergence by evaluating a simpler integral?

Recall that an integral of a non-negative function represents the area of the region bounded by the graph of the function and the x-axis. The comparison test for improper integrals is the interpretation of convergence or divergence when the area under one curve is contained in the area under another.

Theorem 5.6.1. *Comparison Test.* Suppose $0 \le f(x) \le g(x)$ for all $x \ge a$. Then
(a) $\int_a^\infty f(x)dx$ converges if $\int_a^\infty g(x)dx$ converges, and
(b) $\int_a^\infty g(x)dx$ diverges if $\int_a^\infty f(x)dx$ diverges.

The work, of course, is finding an effective function for comparison. For example, consider $1/(x+x^{3/2})$. First choices for comparison are $1/x$ and $1/x^{3/2}$. By (5.4), $\int_1^\infty 1/x\, dx$ diverges and $\int_1^\infty 1/x^{3/2}\, dx$ converges. Because $1/(x+x^{3/2})$ is less than both $1/x$ and $1/x^{3/2}$, comparison to $1/x^{3/2}$ is useful, showing that $\int_1^\infty 1/(x+x^{3/2})\, dx$ converges.

The point of this example is to illustrate that two of the four possible comparisons ((a) and (b) of the theorem) are useful, and these two are useless:

(c) being less than a function with diverging integral tells us nothing, and
(d) being greater than a function with converging integral tells us nothing.

Sometimes trickier comparisons are necessary. For example, consider

$$\int_1^\infty \frac{1}{(x+\sqrt{x})}dx \qquad (5.6)$$

Does this improper integral converge or diverge? Obvious comparisons are to $1/x$ and $1/\sqrt{x}$. By (5.4) both improper integrals $\int_1^\infty 1/x\, dx$ and $\int_1^\infty 1/\sqrt{x}\, dx$ diverge, but these comparisons don't help: $1/(x+\sqrt{x})$ is less than both $1/x$ and $1/\sqrt{x}$, and we've just mentioned that being smaller than a divergent integral tells us nothing. We need to find another comparison. Note that for $x \ge 1$, we have $x \ge \sqrt{x}$. Then

$$\sqrt{x}+x \le x+x \qquad \text{and so} \qquad \frac{1}{2x} \le \frac{1}{x+\sqrt{x}}$$

Again by (5.4), $\int_1^\infty 1/2x\, dx$ diverges, so by the Comparison test, the improper integral (5.6) diverges.

The Comparison Test can be adapted to integrals that are improper because the integrand diverges to infinity at a point of the interval of integration. For example, $1/\sin(x)$ diverges at $x=0$. Does the integral

$$\int_0^1 \frac{1}{\sin(x)}dx \qquad (5.7)$$

converge or diverge? Recall that for $x \ge 0$,

$$\sin(x) \le x \quad \text{and so} \quad \frac{1}{\sin(x)} \ge \frac{1}{x}$$

Then the divergence of $\int_0^1 1/x\,dx$ by (5.5) implies the divergence of the improper integral (5.7).

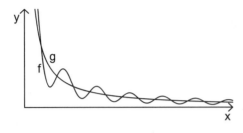

Figure 5.8. The integrals of f and g should both converge or both diverge.

The reason that the Comparison Test works is evident, but with some more effort, the hypotheses for the comparison can be relaxed considerably. Fig 5.8 illustrates a possibilty. Neither $f(x)$ nor $g(x)$ ever stays above the other, no matter how large the initial value of the interval over which the improper integral is evaluated. Yet it appears that about as much of the graph of f lies above the graph of g as lies below it. Very roughly, then, the improper integrals $\int_1^\infty f(x)\,dx$ and $\int_1^\infty g(x)\,dx$ should both converge or both diverge. The precise condition is given in the Limit Comparison Test, Thm. 5.6.2.

Theorem 5.6.2. *Limit Comparison Test.* Suppose $f(x)$ and $g(x)$ are positive functions for $a \leq x < \infty$, and

$$\lim_{x\to\infty} \frac{f(x)}{g(x)} = L \quad \text{with} \quad 0 < L < \infty.$$

Then $\int_a^\infty f(x)\,dx$ and $\int_a^\infty g(x)\,dx$ both converge or both diverge.

For example, does the improper integral

$$\int_1^\infty \frac{1}{x^2 - \sin^2(x)}\,dx \tag{5.8}$$

converge or diverge? Comparison with $\int_1^\infty 1/x^2\,dx$ doesn't work, because the integrand of (5.8) is greater than $1/x^2$. But we can apply the Limit Comparison Test:

$$\lim_{x\to\infty} \frac{1/x^2}{1/(-\sin^2(x) + x^2)} = \lim_{x\to\infty}\left(-\left(\frac{\sin(x)}{x}\right)^2 + 1\right) \tag{5.9}$$

Now a bit of care is needed. We are accustomed to thinking of the limit of $\sin(x)/x$ as the $x \to 0$ limit, which is 1. This would give a value of 0 for the limit (5.9), and so the Limit Comparison Test cannot be applied. However, in (5.9) we are taking the $x \to \infty$ limit, not the $x \to 0$ limit. Certainly, $\lim_{x\to\infty}\sin(x)/x = 0$. The limit of (5.9) is 1 and so the improper integral (5.8) converges by the Limit Comparison Test.

We conclude this section with a puzzling geometric example. Suppose S is the surface formed when $y = 1/x$, $1 \le x < \infty$, is revolved about the x-axis, and E is the solid enclosed by capping S with the unit disc D in the plane $x = 1$. When we extend the method of Sect. 5.2 to improper integrals, we compute the volume of E:

$$\text{volume} = \int_1^\infty \pi \left(\frac{1}{x}\right)^2 dx = \lim_{T\to\infty} -\pi \frac{1}{x}\Big|_1^T = \lim_{T\to\infty}\left(-\frac{\pi}{T}+\frac{\pi}{1}\right) = \pi$$

When we extend the method of Sect. 5.4 to improper integrals, we compute the surface area of S:

$$\text{area} = \int_1^\infty 2\pi \frac{1}{x}\sqrt{1+\left(\frac{1}{x^2}\right)^2}\,dx$$

Now the antiderivative of this function is the goal of Exercise 5.4.10, and I don't want to deprive you of the pleasure of solving that problem if you haven't already. Instead, we'll ask a question more modest than to find the area. Let's just check if the surface area integral converges. Why should we check this? Surely if the volume is finite, then the area must also be finite. Let's see.

$$\frac{1}{x}\sqrt{1+\frac{1}{x^4}} > \frac{1}{x}$$

We know $\int_1^\infty 1/x\,dx$ diverges, so by Theorem 5.6.1 the surface area integral diverges. That is, the surface S has infinite area, yet it (along with the disc D) encloses a finite volume. This is pretty strange.

Practice Problems

Determine whether these improper integrals converge or diverge.

5.6.1.

(a) $\displaystyle\int_1^\infty \frac{e^{-x}}{1+x^2}\,dx$ (b) $\displaystyle\int_1^\infty \frac{1}{\ln(x)}\,dx$

5.6.2.

(a) $\displaystyle\int_2^\infty \frac{1}{x^3-1}\,dx$ (b) $\displaystyle\int_1^\infty \frac{\cos(1/x)}{x}\,dx$

Practice Problem Solutions

5.6.1. (a) For $x \ge 1$, $e^{-x} < 1$ and $1+x^2 > x^2$, so

$$\frac{e^{-x}}{1+x^2} < \frac{1}{1+x^2} < \frac{1}{x^2}$$

Because $\int_1^\infty 1/x^2\,dx$ converges, the integral of this problem converges. Alternatively, $\int 1/(1+x^2)\,dx$ can be evaluated using rule 41 from the table of integrals.
(b) Note $\ln(x) < x$, so $1/\ln(x) > 1/x$, so the integral of this problem diverges by the Comparison test.

5.6.2. (a) The first comparison to come to mind is $1/x^3$. By (5.4), $\int_2^\infty 1/x^3\,dx$ converges. Unhappily, $1/(x^3 - 1) > 1/x^3$, so the Comparison Test cannot be applied. Instead, we apply the Limit Comparison Test:

$$\lim_{x\to\infty} \frac{1/x^3}{1/(x^3-1)} = \lim_{x\to\infty} \frac{x^3-1}{x^3} = \lim_{x\to\infty} \left(1 - \frac{1}{x^3}\right) = 1$$

Because the limit exists and is a positive number, the improper integrals $\int_2^\infty 1/x^3\,dx$ and $\int_2^\infty 1/(x^3 - 1)\,dx$ have the same convergence behavior, and by (5.4), we see $\int_2^\infty 1/x^3\,dx$ converges.
(b) Use the Limit Comparison Test with $1/x$

$$\lim_{x\to\infty} \frac{\cos(1/x)/x}{1/x} = \lim_{x\to\infty} \cos(1/x) = 1$$

So by the Limit Comparison test, $\int_1^\infty \cos(1/x)/x\,dx$ diverges because $\int_1^\infty 1/x\,dx$ diverges by (5.4).

Exercises

Determine whether these improper integrals converge or diverge.

5.6.1.

$$\text{(a)} \int_1^\infty \frac{1}{e^x + 1}\,dx \quad \text{(b)} \int_0^1 \frac{x^4}{\sqrt{1+x^2}}\,dx$$

5.6.2.

$$\text{(a)} \int_1^\infty \frac{1+\cos(x)}{1+x^2}\,dx \quad \text{(b)} \int_1^\infty \frac{\cos^2(x)}{1+x^3}\,dx$$

5.6.3.

$$\text{(a)} \int_2^\infty \frac{\sqrt{x}}{x^2 - \sqrt{x}}\,dx \quad \text{(b)} \int_2^\infty \frac{\sqrt{x}}{x - \sqrt{x}}\,dx$$

5.6.4.

$$\text{(a)} \int_1^\infty \frac{1+\cos(x)}{x^3 - x}\,dx \quad \text{(b)} \int_1^\infty \frac{\sin(1/x)}{x}\,dx$$

5.6.5.

$$\text{(a) } \int_0^3 \frac{x}{\sqrt{9-x^2}}\, dx \quad \text{(b) } \int_0^\infty \frac{1}{\sqrt{x+x^4}}\, dx$$

5.7 STRESS TESTING GROWTH MODELS

Improper integrals may appear to be useless for any problem in biology, or, for
that matter, in physics, because so far as we know all biological and physical
systems are finite in extent, in duration, and in levels of substructure. A
consequence of this finiteness is that every mathematical model necessarily is
based on a limited amount of data leading us to ask how well does the model
predict behavior outside the range of data from which it was built? An example
may clarify this point.

Suppose measurements suggest that members of an immortal (really–see the
end of this section) species of jellyfish grow with a rate

$$\frac{dR}{dt} = \frac{1}{(1+t)^{3/2}}$$

where $R(t)$ denotes the radius of a jellyfish of age t, so

$$R(T) = \int_0^T \frac{1}{(1+t)^{3/2}}\, dt$$

A crude test of the plausibility of this model is to check its predictions in the
$T \to \infty$ limit. We do not observe really, really big jellyfish, so $R(T)$ should not
increase without bound.

$$R(T) = \int_0^T \frac{1}{(1+t)^{3/2}}\, dt = -2\frac{1}{(1+t)^{1/2}}\Big|_0^T = 2 - \frac{2}{(1+T)^{1/2}}$$

As $T \to \infty$, $R(T)$ approaches a limiting value of 2, so at least in this regard, the
growth model is plausible.

If the measurements suggested

$$\frac{dR}{dt} = \frac{1}{(1+t)^{1/2}}$$

does this model pass the same test? Integrating we obtain

$$R(T) = \int_0^T \frac{1}{(1+t)^{1/2}}\, dt = 2(1+t)^{1/2}\Big|_0^T = 2(1+T)^{1/2} - 2$$

As $T \to \infty$, evaluation of the improper integral shows this model predicts
$R(T) \to \infty$. This suggests either that the model is completely wrong, or perhaps
that the model is valid over only a limited range of time, or maybe a jellyfish the

size of Michigan lurks in the depths of the ocean. Probably not. With a bell diameter of 10 m and tentacle length of 30 m, the *Cyanea capillata* (Lion's mane) is the largest known species.

Yes, some jellyfish do appear to be immortal. See pg. 45 of [75] and pgs. 72–75 of [45]. When stressed to the point of death, the tiny (about 1 cm) Mediterranean species *Turritopsis dohrnii* reverts to its immature polyp stage by a process called transdifferentiation and repeats its life cycle. So far, observation suggests that this process can continue forever. But this particular jellyfish has not been seen to grow arbitrarily large. This immortal species of jellyfish does not fit into our example.

To be sure, this application of improper integrals is a very limited test, but it can save us from making some mistakes.

Practice Problems

5.7.1. Pictured on the right is a plot of growth rate data for a long-lived species; the coordinates of the points are in the table below. Plot the growth curves $g_1(t) = 1/(1 + t)$ and $g_2(t) = 1/(1+t)^{3/2}$. If the data points

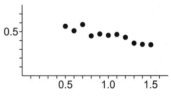

do not give a clear match with one of these curves, use the long t size projections to select a model.

(0.5, 0.561)	(0.6, 0.509)	(0.7, 0.580)	(0.8, 0.452)
(0.9, 0.474)	(1.0, 0.458)	(1.1, 0.468)	(1.2, 0.435)
(1.3, 0.369)	(1.4, 0.356)	(1.5, 0.352)	

5.7.2. For a disease without recovery in a constant population of size $10,000$, suppose the infected population grows at the rate of $dI/dt = 300/(1+t)^{1.1}$, and the initial infected population is 1000.
(a) Will the infected population ever exceed 20% of the total population?
(b) If so, estimate the time until 20% of the total population is infected.

Practice Problem Solutions

5.7.1. On the right we plot the curves $g_1(t)$ and $g_2(t)$ and the data points. The data do not appear to give a clear choice of one growth rate curve, so we compute the sizes attained in long-time.

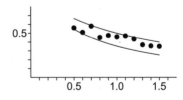

$$R_1(T) = \int_0^T g_1(t)dt = \ln(1+T)$$

$$R_2(T) = \int_0^T g_2(t)dt = 2 - \frac{2}{\sqrt{1+T}}$$

As $T \to \infty$, the growth rate $g_1(t)$ predicts unbounded length, while the growth rate $g_2(t)$ predicts length bounded above by 2.

5.7.2. To find the limiting value of the population, evaluate the integral

$$I(t) = \int \frac{300}{(1+t)^{1.1}} \, dt = \frac{-3000}{(1+t)^{0.1}} + c$$

With $I(0) = 1000$ we have $c = 4000$.
(a) Then the $t \to \infty$ limit of $I(t)$ is 4000, which is twice 20% of the population.
(b) To estimate the time when $I(t)$ first exceeds 2000, solve

$$4000 - \frac{3000}{(1+t)^{0.1}} = 2000 \quad \text{and we find} \quad \left(\frac{3000}{2000}\right)^{10} = 1+t$$

Then $t = (3/2)^{10} - 1 \approx 57$.

Exercises

5.7.1. The table shows growth rate data of a long-lived species for $2.0 \le t \le 2.95$. Plot these points and the growth curves $g_1(t) = 1/t$, $g_2(t) = 1/t^2$, and $g_3(t) = 1/t^3$. Predict the long-term behavior based on the best-matching curve.

(2.00, 0.388)	(2.05, 0.401)	(2.10, 0.327)	(2.15, 0.384)
(2.20, 0.372)	(2.25, 0.303)	(2.30, 0.304)	(2.35, 0.227)
(2.40, 0.208)	(2.45, 0.243)	(2.50, 0.280)	(2.55, 0.218)
(2.60, 0.239)	(2.65, 0.278)	(2.70, 0.278)	(2.75, 0.226)
(2.80, 0.245)	(2.85, 0.224)	(2.90, 0.222)	(2.95, 0.170)

5.7.2. The table shows growth rate data of a long-lived species for $2.0 \le t \le 2.95$. Plot these points and the growth curves $g_1(t) = 1/t$, $g_2(t) = 1/t^2$, and $g_3(t) = 1/t^3$. Predict the long-term behavior based on the best-matching curve.

(2.00, 0.242)	(2.05, 0.378)	(2.10, 0.334)	(2.15, 0.304)
(2.20, 0.375)	(2.25, 0.281)	(2.30, 0.296)	(2.35, 0.118)
(2.40, 0.283)	(2.45, 0.115)	(2.50, 0.153)	(2.55, 0.082)
(2.60, 0.182)	(2.65, 0.156)	(2.70, 0.089)	(2.75, 0.288)
(2.80, 0.119)	(2.85, 0.119)	(2.90, 0.084)	(2.95, 0.235)

5.7.3. Can a bacteria growth rate of $dP/dt = 100/(1+2t)^{1.2}$, with t measured in minutes, be maintained indefinitely in a body with a volume of 95 liters? Suppose the volume of an individual bacterium is 7×10^{-13} milliliters and the average lifetime of the host is about 40,000,000 minutes.

5.7.4. Suppose the bacteria growth rate is $dP/dt = 100/(1 + 2t)^{0.9}$. Can this growth rate be maintained indefinitely? Use the volume parameters of Exercise 5.7.3.

5.7.5. Suppose the population of infecteds grows at a rate of $dI/dt = 10e^{-t/10}$ starting from an initial incursion of 100 infecteds. Will the population of infecteds ever exceed 150? If so, when?

Chapter 6   Power laws

Functional dependence in nature takes many forms. The entropy of an ideal gas is proportional to the log of the number of corresponding microstates (Boltzmann's law); the seasonal variation of daylight time is approximately sinusoidal; gravitational attraction between two objects varies with the reciprocal of the square of the distance between their centers. This last is an example of a power law, in which one variable is proportional to a power of another variable. In this case that power is -2.

Specifically, we say f exhibits a *power law* dependence on x if $f(x) = k \cdot x^d$ for some constants d and k. Because they appear in many biological settings, in this chapter we learn how to recognize power laws and also learn some of the implications of power law scalings.

6.1   THE CIRCUMFERENCE OF A CIRCLE

How did the ancient Greeks calculate the circumference of a circle? Put another way, because π is defined as the ratio of circumference to diameter of a circle (why this ratio is the same for all circles is a cute problem to ponder for a bit), one way to calculate the value of π is to calculate the circumference of the unit circle.

In Sect. 5.3 we estimated the length of a curve by finding the length of polygonal approximations. Successive approximations used ever shorter segments, and if the curve is well-behaved (specifically, if it is differentiable everywhere, or almost everywhere), then with decreasing segment size, the sum of the segment lengths converges to a limit, which we take to be the length of the curve.

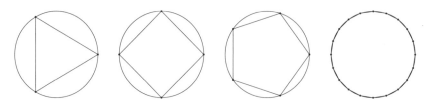

Figure 6.1. Approximating the unit circle with polygons.

We avoided calculating the limit by expressing it as an integral and evaluating the integral. Without this option, some messy calculations are unavoidable. Of course, many people know the answer to a few digits, and Kate Bush [21] sang a lot of digits in her song "π." In Fig. 6.1 we see the unit circle approximated by polygons with 3, 4, 5, and 20 sides. As the number of sides grows, the length of the sides decreases and the polygons appear to better approximate the circle.

Do the lengths of these polygons approach a limit? For each polygon, the sides have the same length, so the polygon perimeter satisfies

$$\text{perimeter} = \text{number of sides} \cdot \text{length of a side}$$

We've oriented the polygons so for each the point $(1,0)$ is a vertex. Above the x-axis, the polygon side with one vertex $(1,0)$ has its other vertex at $(\cos(2\pi/n),\sin(2\pi/n))$, where n is the number of sides. Then the perimeter of the polygon with n sides is

$$n\sqrt{(\cos(2\pi/n)-1)^2+\sin^2(2\pi/n)} \qquad (6.1)$$

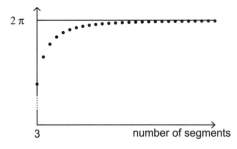

For each n this gives a lower bound for 2π. In Fig. 6.2 we plot the perimeters of the polygons approximating the unit circle, for $n = 3, \ldots, 30$. Note how closely the perimeters approach 2π.

The Greeks were a bit more careful than this: they also circum-scribed polygons around the circle,

Figure 6.2. Perimeters of some polygons approximating the circle.

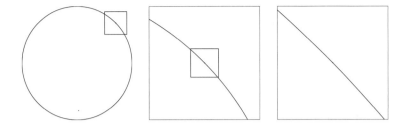

Figure 6.3. The image to the left of each is a magnification of the box to the right.

so for each number n of polygon sides, they had both upper and lower bounds on the circumference of the circle.

This approach works because as we zoom in on an ever-smaller portion of the circle, that portion more closely resembles a straight line segment. See Fig. 6.3. This is just a visualization of the fact that with increasing magnification portions of circles are better approximated by their tangent lines. That is, the curve is differentiable, or smooth.

Practice Problem

6.1.1. By dividing the domain $[0, 1]$ into segments of length $1/n$, find the straight line approximation of the length of the portion of the curve $y = x^3$, between $x = 0$ and $x = 1$. Express your answer as a sum; you need not simplify the result. Numerically estimate this sum for $n = 10$, 50, and 100.

Practice Problem Solution

6.1.1. The endpoints of the sides of the approximation are 0, $(1/n)^3$, $(2/n)^3$, ..., $(n/n)^3$. Then the length of the approximation is

$$\sqrt{\left(0 - \frac{1}{n}\right)^2 + \left(0 - \left(\frac{1}{n}\right)^3\right)^2} + \sqrt{\left(\frac{1}{n} - \frac{2}{n}\right)^2 + \left(\left(\frac{1}{n}\right)^3 - \left(\frac{2}{n}\right)^3\right)^2}$$

$$+ \cdots + \sqrt{\left(\frac{n-1}{n} - \frac{n}{n}\right)^2 + \left(\left(\frac{n-1}{n}\right)^3 - \left(\frac{n}{n}\right)^3\right)^2}$$

Numerically we have $L \approx 1.54682$ for $n = 10$, $L \approx 1.54782$ for $n = 50$, and $L \approx 1.54786$ for $n = 100$.

Exercises

6.1.1. Using Eq. (6.1), how many sides are needed to show $\pi > 3.14$? How many are needed to show $\pi > 3.141$?

In Exercises 6.1.2–6.1.4 approximate the length of the given curve by dividing the domain into segments of length $1/n$. Numerically estimate this sum for $n = 10$, 50, and 100.

6.1.2. Approximate the length of $y = x^2$ between $x = 0$ and $x = 1$.

6.1.3. Approximate the length of $y = x^4$ between $x = 0$ and $x = 1$.

6.1.4. Approximate the length of $y = x^{10}$ between $x = 0$ and $x = 1$.

6.1.5. By plotting the graphs of the curves in Exercises 6.1.2–6.1.4, explain why the lengths approach 2.

6.2 SCALING OF COASTLINES, LOG-LOG PLOTS

In contrast to familiar smooth mathematical models, much of nature is rough: coastlines, profiles of mountain ranges, tree bark, a cat's tongue (if you have little hair on top of your head and have been groomed there by a cat, you'll know that cat's tongue is better described as a field of little cacti than as a bit of sandpaper), many body membranes, fitness landscapes in abstract evolutionary space, and on and on. Nature–both animate and inanimate–exhibit an array of rough surfaces.

Benoit Mandelbrot's insight was to go beyond agreeing that bark is rough, and to ask how is it rough. Can we measure how some surfaces are rougher than others? Do rough surfaces share any trait? The answer is that, in contrast to smooth curves and surfaces, zooming in on rough objects reveals more or less the same structures. Rough curves are not well approximated by their tangent lines, or may have tangent lines nowhere or on only a sparse set of points, no matter how closely we look. In the same manner, rough surfaces are not well approximated by their tangent planes.

Mandelbrot began his presentation of ways to characterize and quantify roughness with a study of coastlines [91], so we'll start there, too. Do a simple experiment with Google Maps: zoom in on a portion of a coastline. It is no surprise that as we look ever more closely, we see ever more detail. Bays and promontories are decorated with ever smaller bays and promontories, down to individual rocks and grains of sand, themselves far from smooth. Surely this signals something interesting.

In Fig. 6.4 we see a NASA photograph of Britain, overlaid with
two polygonal approximations of
the coastline. Clearly, smaller sides
of the polygon will pick up more
features of the coastline and so
the polygon length increases as
the side length decreases. What is
interesting–and we shall see this is
very interesting, indeed–is the *way*
in which the polygon length increases.

Figure 6.4. Approximating the coastline of
Britain by line segments. Shorter segments
better approximate the coastline.

So far as we know, for coastlines this was observed first by Lewis Fry
Richardson [133]. Using a very accurate atlas, Richardson measured the lengths
of polygonal approximations of several coastlines, including the west coast of
Britain. His plot of polygon length vs scale (polygon side length) reveals an
approximately L-shaped curve, not useful in any obvious way.

In Fig. 6.5 Richard-
son plotted log(length)
vs log(scale). Except for
the point farthest to the
right, the points lie close
to a straight line, con-
sistent with a power law
relation between length
and scale. Here's why. If

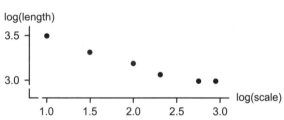

Figure 6.5. Richardson's power law scaling.

$L(\epsilon)$ denotes the length measured at scale ϵ, then a power law relation means
that there are constants a and b for which

$$L(\epsilon) = a\epsilon^b \tag{6.2}$$

Taking the log of both sides of Eq. (6.2) gives

$$\log(L(\epsilon)) = \log(a) + b\log(\epsilon) \tag{6.3}$$

That is, power laws can be recognized by a linear trend in log-log plots.

Another common way to present coastline data is found by noting $L(\epsilon)$, the
coastline length measured at the resolution ϵ, is given by

$$L(\epsilon) = N(\epsilon) \cdot \epsilon \tag{6.4}$$

where $N(\epsilon)$ is the number of length ϵ segments used in the polygonal
approximation of the coastline. Substitute Eq. (6.4) into Eq. (6.3) and simplify
to obtain

$$\log(N(\epsilon)) = \log(a) + (b-1)\log(\epsilon)$$

Eqs. (6.2) and (6.4) imply a power law scaling for $N(\epsilon)$, namely $N(\epsilon) = c\epsilon^d$. Mandelbrot's insight was that the exponent d is a dimension, and so coastline length measurements diverge as ϵ decreases because coastlines have a dimension greater than 1. We'll explore this broadening of our familiar notion of dimension in Sect. 6.4.

Practice Problems

For these problems, plot the points $(\log_{10}(x), \log_{10}(y))$. From visual inspection of these log-log plots, determine if the data are generated by a power law $y = a \cdot x^b$. For those that are, estimate a and b. In all these problems, $x = 1, 2, \ldots, 20$. The values given are $y(1), \ldots, y(20)$.

6.2.1. (a) 1.1, 1.61, 2.01, 2.36, 2.67, 2.95, 3.21, 3.45, 3.68, 3.90, 4.11, 4.31, 4.51, 4.70, 4.88, 5.05, 5.23, 5.39, 5.56, 5.71
(b) 6, 4.10, 3.28, 2.80, 2.48, 2.24, 2.06, 1.91, 1.79, 1.69, 1.60, 1.53, 1.46, 1.41, 1.35, 1.31, 1.26, 1.22, 1.19, 1.16

6.2.2. (a) 1.1, 1.21, 1.33, 1.46, 1.61, 1.77, 1.95, 2.14, 2.36, 2.59, 2.85, 3.14, 3.45, 3.80, 4.18, 4.59, 5.05, 5.56, 6.12, 6.73
(b) 1.2, 2.24, 3.23, 4.18, 5.11, 6.02, 6.91, 7.80, 9.00, 9.49, 9.95, 10.39, 10.82, 11.23, 11.62, 12.00, 12.37, 12.73, 13.08, 13.42

Practice Problem Solutions

6.2.1. (a) The left graph of Fig. 6.6 shows the log-log plot of the data of Practice Problem 6.2.1 (a). The points appear to fall along a straight line, indicating a power law dependence. The Mathematica code to find the best-fitting line is presented in Sect. B.5. The line is $y = 0.550x + 0.041$, so $b = 0.550$ and $a = 10^{0.041} = 1.099$.
(b) The right graph of Fig. 6.6 shows the log-log plot of the data of Practice Problem 6.2.1 (b). The points appear to fall along a straight line, indicating a power law dependence. The best-fitting line is $y = -0.550x + 0.778$, so $b = -0.550$ and $a = 10^{0.778} = 5.998$.

6.2.2. (a) The left graph of Fig. 6.7 shows the log-log plot of the data of Practice Problem 6.2.2 (a). The points do not fall along a straight line, so a power law relation is not supported.
(b) The right graph of Fig. 6.7 shows the log-log plot of the data of Practice Problem 6.2.2 (b). Here the situation is more complicated. The first 8 points appear to lie on one straight line, the last 12 on another. One interpretation of

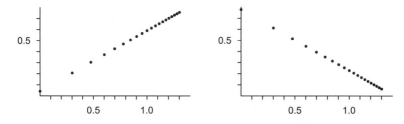

Figure 6.6. Log-log plots for Practice Problem 6.2.1 (a) on the left, and (b) on the right.

this situation is the presence of two power laws, one for smaller x, the other for larger. The best-fitting line through the first 8 points is $y = 0.900x + 0.079$, so $b = 0.900$ and $a = 10^{.079} = 1.200$. The best-fitting line through the last 12 points is $y = 0.167x + 0.930$, so $b = 0.167$ and $a = 10^{0.930} = 8.511$.

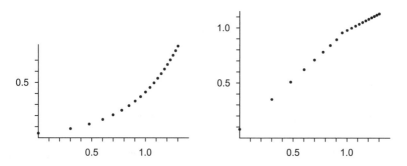

Figure 6.7. Log-log plots for Practice Problem 6.2.2 (a) on the left, and (b) on the right.

Exercises

For the data given in Probs. 6.2.1–6.2.9, plot $(\log_{10}(x), \log_{10}(y))$. From visual inspection of the graph, determine if the data are generated by a power law $y = a \cdot x^b$. For those that are, estimate a and b. In all these problems, $x = 1, 2, \ldots, 10$. The values given are $y(1), \ldots, y(10)$.

6.2.1. $0.94, 1.61, 2.36, 3.05, 3.40, 3.72, 4.12, 5.00, 5.11, 5.42$

6.2.2. $1.50, 2.12, 2.60, 3.00, 3.35, 3.67, 3.97, 4.24, 4.50, 4.74$

6.2.3. $2.00, 4.29, 6.70, 9.19, 11.75, 14.35, 17.01, 19.70, 22.42, 25.18$

6.2.4. $3.12, 5.82, 8.39, 10.86, 13.28, 15.65, 17.98, 20.27, 22.54, 24.78$

6.2.5. $3.39, 2.95, 2.72, 2.57, 2.46, 2.37, 2.30, 2.24, 2.18, 2.14$

6.2.6. $10.21, 7.74, 6.58, 5.86, 5.36, 4.99, 4.69, 4.44, 4.24, 4.06$

6.2.7. $1.068, 1.204, 1.366, 1.510, 1.585, 1.758, 1.984, 2.158, 2.335, 2.611$

6.2.8. $1.268, 1.684, 2.232, 2.902, 3.687, 4.814, 6.310, 8.171, 10.582, 13.803$

6.2.9. $2.470, 3.994, 5.535, 7.046, 8.474, 9.187, 9.435, 9.614, 9.778, 10.017$

6.2.10. In [157] Donald Turcotte presents data suggesting that in southern California the annual number N of earthquakes of magnitude $\geq m$ scales as $\log(N) = -0.89m + a$, for some constant a. The magnitude scale already is logarithmic, explaining the presence of m instead of $\log(m)$. If magnitude ≥ 6 earthquakes occur about once per year, about how many years pass between magnitude 8 earthquakes in southern California?

6.3 ALLOMETRIC SCALING

Very roughly, the surface area, A, of an animal scales as its linear size, L, squared: $A \propto L^2$, by which we mean there is a constant k_1 with $A = k_1 L^2$, at least approximately. Similarly, the mass, M, of an animal scales as $M = k_2 L^3$. Heat dissipation occurs across the surface, so the total metabolic rate of an animal is proportional to L^2, hence to $M^{2/3}$, and the metabolic rate per unit mass then is proportional to $M^{-1/3}$, or so argued Max Rubner [137] in 1883. Pulse rate is related to metabolic rate per unit mass, so smaller animals should have faster pulse rates and larger animals slower. Indeed, this is observed, familiar even. A mouse's heart beats very rapidly, a whale's heart very slowly. Add in the observation that most mammal hearts beat 1 to 2 billion times during the animal's life and we understand that in the absence of external perturbation (early death due to predation or disease, among others), a mouse has a shorter life than a person, who in turn has a shorter life than at least some whales, for example, the bowhead. See [89].

This makes perfect sense, but careful measurements by Max Kleiber [76] (yes, another Max–what are the odds?) in 1932 revealed something different: for most animals, over a range of sizes spanning 21 orders of magnitude, the metabolic rate per unit mass varies as $M^{-1/4}$ rather than as $M^{-1/3}$. Why is this true?

Adding to the interest of this question, plants exhibit this $M^{-1/4}$ metabolic rate per unit mass scaling. Power laws have been observed in not only the vascular structures of animals and plants, but also the internal structures of cells and the abstract networks of resource distribution in forests and other ecosystems. Also, more biological variables exhibit power law scalings with mass: life-span scales as $M^{1/4}$, age of first reproduction as $M^{3/4}$, the time of embryonic development as $M^{-1/4}$, and the diameters of tree trunks and of aortas as $M^{3/8}$, for example.

Maybe first we should answer the question "Why do we care?" For one thing, the amazing range of validity for this power law suggests some sort of underlying universal mechanism. Suggests only; does not prove. Two recent, and different, explanations are given by Geoffrey West, James Brown, and Brian Enquist [171, 172], and by Jayanth Banavar, Amos Maritan, and Andrea Rinaldo [9]. West et al. argue that metabolisms

Figure 6.8. A cast of a dog's lungs.

use hierarchical resource distribution networks, all of which terminate in a smallest scale, capillaries or the terminal level of lung airways, for example. See Fig. 6.8, a photograph of a cast of a dog's lung.

For complicated biological systems, area and volume should not refer to Euclidean area and volume, but to *biological area* and *biological volume*. These are the total effective surface area across which nutrients and energy are exchanged between inside and outside, and the volume of biologically active material. The biological volume often is less than the Euclidean volume enclosed by the material boundary because this boundary encloses empty regions, usually exhibiting a range of sizes. Examples of biological area include the total area of the lungs and the gastrointestinal tract, the total area of a leaf, and the area of mitochondrial membranes.

West's group suggests that the biological area A depends in a complicated way on several, perhaps many, length scales $A = A(l_0, l_1, \ldots, l_N)$, where l_0 is the smallest length scale. The central point is this: that the smallest scale is approximately constant for many, many species. Whale capillaries are not vastly magnified versions of mouse capillaries, any more than whale blood cells are vastly magnified versions of mouse blood cells. Now compare the areas for two species, one a linear factor λ times the size of the other. If *all* the lengths scaled by λ, then the larger area A' and the smaller A are related by

$$A' = A(\lambda l_0, \lambda l_1, \ldots, \lambda l_N) = \lambda^2 A(l_0, l_1, \ldots, l_N) = \lambda^2 A$$

However, being constant across species, the smallest length l_0 does not scale, so

$$\begin{aligned} A' &= A(l_0, \lambda l_1, \ldots, \lambda l_N) \\ &= \lambda^{2+\epsilon_A(l_0, l_1, \ldots, l_N)} A(l_0, l_1, \ldots, l_N) = \lambda^{2+\epsilon_A(l_0, l_1, \ldots, l_N)} A \end{aligned} \tag{6.5}$$

where the exponent $\epsilon_A(l_0, l_1, \ldots, l_N)$ accounts for the fact that l_0 is not scaled, while the remaining l_i are.

For example, take $N = 1$ and suppose $A(l_0, l_1) = l_0 l_1 + l_1^2$. If only λ_1 is scaled, we see

$$A(l_0, \lambda l_1) = \lambda l_0 l_1 + \lambda^2 l_1 = \lambda^{2+\epsilon_A(l_0, l_1)} A(l_0, l_1)$$

where $\epsilon_A(l_0, l_1) = \log_\lambda((l_0 + \lambda l_1)/(l_0 + l_1)) - 1$.

Similarly, the volume influenced by this network depends on the length factors $l_0, l_1, \ldots, l_N$. Again, scaling by λ leaves l_0 unchanged, so

$$\begin{aligned} V' &= V(l_0, \lambda l_1, \ldots, \lambda l_N) \\ &= \lambda^{3+\epsilon_V(l_0, \lambda l_1, \ldots, \lambda l_N)} V(l_0, l_1, \ldots, l_N) = \lambda^{3+\epsilon_V(l_0, \lambda l_1, \ldots, \lambda l_N)} V \end{aligned} \tag{6.6}$$

Combining Eqs. (6.5) and (6.6), and omitting the arguments of ϵ_A and ϵ_V, we find

$$A = kV^{(2+\epsilon_A)/(3+\epsilon_V)}$$

Finally, $V = A \cdot L$, where $L = L(l_0, l_1, \ldots, l_N)$ is a length factor depending on all the individual l_i. As with A and V, L exhibits scaling behavior, this time of the form

$$\begin{aligned} L' &= L(l_0, \lambda l_1, \ldots, \lambda l_N) \\ &= \lambda^{1+\epsilon_L} L(l_0, l_1, \ldots, l_N) = \lambda^{1+\epsilon_L} L \end{aligned} \tag{6.7}$$

Combining Eqs. (6.5) and (6.7) we find

$$V' = A'L' = \lambda^{2+\epsilon_A} A \lambda^{1+\epsilon_L} L = \lambda^{3+\epsilon_A+\epsilon_L} V$$

From this and Eq. (6.6), we see $\epsilon_V = \epsilon_A + \epsilon_L$, and so the area-volume relation is

$$A = kV^{(2+\epsilon_A)/(3+\epsilon_A+\epsilon_L)} = \hat{k} M^{(2+\epsilon_A)/(3+\epsilon_A+\epsilon_L)}$$

West, Brown, and Enquist conjecture that natural selection has maximized fitness by maximizing metabolic capacity, the rate at which energy and matter are taken from the environment and distributed within the organism. They argue maximizing metabolic capacity is equivalent to maximizing the exponent b in the scaling relation $A \propto M^b$. That is, we maximize

$$b(\epsilon_A, \epsilon_L) = \frac{2 + \epsilon_A}{3 + \epsilon_A + \epsilon_L}$$

subject to the constraints $0 \leq \epsilon_A \leq 1$ and $0 \leq \epsilon_L \leq 1$, resulting from the observation that the exponent of λ in Eq. (6.5) must lie between 2 and 3, and the exponent of λ in Eq. (6.7) must lie between 1 and 2. We could plot $b(\epsilon_A, \epsilon_L)$ in this region to find its maximum, but consider this reasoning. When ϵ_L is held constant, b is an increasing function of ϵ_A; when ϵ_A is held constant, b is a decreasing function of ϵ_L. Then the scaling exponent is maximized with $\epsilon_A = 1$ and $\epsilon_L = 0$, giving $b = 3/4$ and so the metabolic rate per unit mass scales as $M^{-1/4}$.

Banavar et al. [9] take a different approach. They assume that microscopic nutrient transfer is independent of organism size and occurs at L^D sites, where L is expressed as a multiple of the average distance between transfer sites and D is the dimension of the organism. They argue that natural selection has minimized C, the total volume of blood; they give the name *efficient* to those organisms that minimize C, and in a moment we'll see how they show that for efficient organisms,

$$C \propto L^{D+1} \qquad (6.8)$$

From this it follows that the metabolic rate B scales as

$$B \propto M^{D/(D+1)}$$

For $D = 3$ we recover the familiar $3/4$ power law. How do Banavar, Maritan, and Rinaldo obtain Eq. (6.8)? They show that for any spanning network (that is, a network reaching all the nutrient transfer sites),

$$L^{D+1} \leq C \leq L^{2D}$$

for large L. The efficiency assumption of minimum C gives Eq. (6.8). They have made no assumption of any hierarchical structure of the distribution network, though they note that such structures would shorten the network length, thus increasing its viability and efficiency.

Both approaches give the same allometric scaling law, but they use different methods and make different assumptions. They do show that it is possible to find a universal explanation for this metabolic scaling law, though of course neither explanation may be correct, and there may be no single explanation at all. There may be some very deep science here, waiting to be discovered, or maybe not. In a 1983 *NOVA* episode, reported in [38], Richard Feynman remarked

People say to me, "Are you looking for the ultimate laws of physics?" No, I'm not. I'm just looking to find out more about the world. And if it turns out there is a simple ultimate answer which explains everything, so be it. That would be very nice to discover.

If it turns out it's like an onion with millions of layers and we're just sick and tired of looking at the layers, then that's the way it is.

For those of you more interested in practical medicine than in underlying science, here's another reason to understand allometric scaling. We need to get the metabolic power law right in order to properly scale the doses of experimental drugs from lab rats to human test subjects.

But we can't let this be our last word on the subject. In their study of a Costa Rican rain forest, West, Enquist, and Brown [35, 173] found two power law relations, one between the diameters of the branches of a tree and the number of branches having that diameter, the other between the trunk diameter of a tree and the number of trees having that trunk diameter. Astonishingly, both power laws have the same exponent. In this specific sense, branch is to tree as tree is to forest. The resource distribution pathways of a tree, discovered and polished by evolution, match those of the forest. This could be a coincidence, but also it could point out an interesting connection.

6.4 POWER LAWS AND DIMENSIONS

Some power laws have an interesting geometric interpretation. For example, suppose we take an arc of a circle, or any other smooth curve, and cover it with boxes of side length r. Let $N(r)$ denote the minimum number of boxes of side length r needed to cover the curve. A power law relation between $N(r)$ and r would take the form

$$N(r) = k\left(\frac{1}{r}\right)^d \tag{6.9}$$

where the r dependence is $(1/r)^d$ instead of r^d because as the box side length decreases, the number of boxes increases. If this relation holds, then

$$\log(N(r)) = d\log(1/r) + \log(k) \tag{6.10}$$

So finding $N(r_i)$ for a collection of values $r_1, r_2, \ldots, r_N$, the points

$$(\log(1/r_i), \log(N(r_i)))$$

should lie along a straight line with slope d. What is d?

In Fig. 6.9 we see one-eighth of the unit circle covered with boxes of side length $r = 0.05, 0.02, 0.008$, and 0.004. The number of boxes needed is

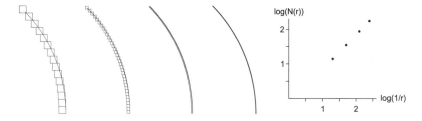

Figure 6.9. Boxes covering a circle arc, and a log-log plot.

14, 35, 88, and 176. The right side of the figure shows the log-log plot. The points do indeed appear to fall on a straight line; the slope of the line is 1, the dimension of the circle.

A similar calculation for a filled-in circle or square or triangle, etc., gives points on a line of slope 2. This motivates an interpretation of the exponent. If a shape is covered with $N(r)$ squares of side length r, and if the points $(\log(1/r_i), \log(N(r_i)))$ lie on (or pretty close to) a straight line, then the slope of the line is the *box-counting dimension* of the shape.

Must the box-counting dimension always be an integer? After all, interpreted as the number of independent directions in which we can move, the familiar notion of dimension must take on integer values. For Euclidean shapes, the box-counting dimension agrees with the familiar notion of dimension. The more closely we look at a smooth curve, no matter how wiggly, the more it looks like its tangent line. But some–many, maybe most–natural shapes are more complex. When magnified they do not appear simpler, but rather small pieces look like scaled copies of the whole. This observation has been familiar to artists and to children for centuries, but it was Benoit Mandelbrot's genius to develop a mathematics to analyze these shapes. This is fractal geometry. Mandelbrot's book [90] is fundamental. Abandoning modestly, [41] is a more accessible source. A shape is a *fractal* if it is made up of pieces, each of which resembles the whole in some fashion. Fractals arise so often in nature because the same growth process operates over many size scales. While mathematical fractals can exhibit similar structures on arbitrarily small scales, natural fractals exhibit this self-similarity on only a limited range of sizes, called the *scaling range*. Whether or not a fractal description is useful depends on the extent of the scaling range.

We'll give a simple illustration, not pushing the scaling range to its limits. In Fig. 6.10 we see a portion of a deer skull suture, with the coverings by the largest and the smallest of the five box sizes, 1, 1/2, 1/4, 1/8, and 1/16. This range is not at all sufficient to take seriously the dimension calculation, but this does

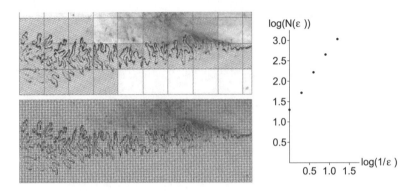

Figure 6.10. Approximating skull suture box-counting dimension.

illustrate that smaller boxes pick up finer details. For this example, the numbers of boxes are 20, 52, 165, 448, and 1080. In the log-log plot, the best-fitting line has slope about 1.462, and indeed this suture does look quite wiggly. The higher the dimension, the closer the curve comes to filling the plane.

Often linear regression is used to find the slope of the best-fitting line, although this method must not be used blindly. For example, applying linear regression to the data points of the right graph of Fig. 6.7 would not give the dimension of anything related to the data.

Some care must be exercised here. As pointed out on page 298 of [34], small errors in values of x or y near 0 can produce large changes in $\log(x)$ or $\log(y)$. If some values of either x or y are near 0, fitting Eq. (6.9) directly may be preferable to fitting Eq. (6.10).

Another issue is that an accurate estimate of the box-counting dimension may require using very, very small boxes. Many published papers report box-counting dimension calculations based on too small a range of boxes. The examples presented here, including the skull suture picture, are to illustrate the method, not to be treated as templates of serious calculations. Any estimate of dimension based on a small collection of boxes must be regarded as suspect.

For mathematical examples, sometimes we know a formula for $N(r_i)$. In this case, the box-counting dimension can be computed by a limit

$$d = \lim_{r_i \to 0} \frac{\log(N(r_i))}{\log(1/r_i)} \tag{6.11}$$

Examples of applications of dimension and scaling are given in Sect. 6.5.

Practice Problems

6.4.1. Compute the box-counting dimension for this data, if the data support such a computation.

r_i	0.5	0.25	0.125	0.0625	0.03125
$N(r_i)$	3	9	27	81	243

6.4.2. Suppose two objects, A and B, are covered with boxes of side lengths $r_1 > r_2 > \cdots$, requiring $N_A(r_i)$ boxes to cover A and $N_B(r_i) = 2N_A(r_i)$ boxes to cover B. Suppose d_A is the box-counting dimension of A, computed by Eq. (6.11). How is the box-counting dimension of B related to that of A?

Practice Problem Solutions

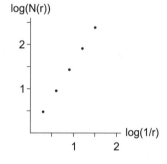

6.4.1. In Fig. 6.11 we plot the points $(\log(1/r_i), \log(N(r_i)))$. These appear to fall along a straight line, supporting the computation of the box-counting dimension. Applying the method sketched in Sect. B.5, the best-fitting line through these data points is

$$\log(N(r_i)) = 1.585 \log(1/r_i) + 0$$

Figure 6.11. The log-log plot of Practice Problem 6.4.1.

The box-counting dimension is the slope of this line, so $d \approx 1.585$. Because all the points do appear to lie on a line, we can estimate the slope with any pair of points. For example, the last two data points give the same value for the dimension

$$d \approx \frac{\log(243) - \log(81)}{\log(1/.03125) - \log(1/.0625)} = 1.585.$$

6.4.2. By Eq. (6.11) the dimension d_A is

$$d_A = \lim_{r_i \to 0} \frac{\log(N_A(r_i))}{\log(1/r_i)} \quad \text{and} \quad d_B = \lim_{r_i \to 0} \frac{\log(N_B(r_i))}{\log(1/r_i)}$$

Then

$$d_B = \lim_{r_i \to 0} \frac{\log(2N_A(r_i))}{\log(1/r_i)} = \lim_{r_i \to 0} \frac{\log(2)}{\log(1/r_i)} + \lim_{r_i \to 0} \frac{\log(N_A(r_i))}{\log(1/r_i)} = d_A$$

In terms of the log-log plot, the points $(\log(r_i), \log(N_B(r_i)))$ are the points $(\log(r_i), \log(N_A(r_i)))$ each shifted up by $\log(2)$, so lying on a line parallel to the line through $(\log(r_i), \log(N_A(r_i)))$.

Exercises

In Exercises 6.4.1, 6.4.2, and 6.4.3 estimate the box-counting dimension if the log-log plot of the data points appear to lie on a line.

6.4.1.

r_i	0.5	0.25	0.125	0.0625	.003125
$N(r_i)$	6	18	54	162	486

6.4.2.

r_i	0.5	0.25	0.125	0.0625	0.03125
$N(r_i)$	4	16	64	256	1024

6.4.3.

r_i	0.5	0.25	0.125	0.0625	0.03125
$N(r_i)$	2	8	32	128	512

6.4.4. Suppose two objects, A and B, are covered with $N_A(r_i)$ and $N_B(r_i)$ boxes of side lengths $r_1 > r_2 > \cdots > r_m$, and that $N_B(r_i) = N_A(r_i)^2$. Suppose the log-log plot for A consists of points on a straight line of slope d_A. How is the box-counting dimension of B related to that of A?

Figure 6.12. Steps in constructing the Cantor middle-thirds set of Exercise 6.4.5.

6.4.5. The *Cantor middle-thirds set* C is constructed iteratively. Begin with $C_0 = [0,1]$. Then $C_1 = [0,1/3] \cup [2/3,1]$ is formed by removing $(1/3,2/3)$, the middle third of C_0. Next, $C_2 = [0,1/9] \cup [2/9,1/3] \cup [2/3,7/9] \cup [8/9,1]$ is formed by removing the middle third of each interval of C_1, and so on. See Fig. 6.12. The Cantor middle-thirds set is $C = \cap_{n=0}^{\infty} C_n$. Using boxes of side length $r_n = 1/3^n$, for $n = 1, 2, 3, \ldots$, find a formula for $N(r_n)$ and compute the box-counting dimension of the Cantor middle-thirds set.

G_0 G_1 G_2 G_4

Figure 6.13. Steps in constructing the Sierpinski gasket of Exercise 6.4.6.

6.4.6. The *Sierpinski gasket* G is constructed iteratively. Begin with G_0, the filled-in triangle with vertices $(0,0)$, $(1,0)$, and $(0,1)$. Then G_1 is formed from G_0 by removing the interior of the triangle with vertices $(1/2,0)$, $(1/2,1/2)$, and $(0,1/2)$, the *middle triangle* of G_0. Note G_1 consists of three filled-in triangles. Next, G_2 is formed by removing the middle triangles of each of the three filled-in triangles of G_1, and so on. See Fig. 6.13. The Sierpinski gasket is $G = \bigcap_{n=0}^{\infty} G_n$. Using boxes (squares) of side length $1/2^n$, for $n = 1,2,3,\ldots$, find a formula for $N(r_n)$ and compute the box-counting dimension of the Sierpinski gasket.

6.4.7. (a) Suppose A is the product of two Cantor middle-thirds sets. That is,

$$A = C \times C = \{(x,y) : x \in C \text{ and } y \in C\}$$

See the left side of Fig. 6.14. Using boxes of side length $r_n = 1/3^n$, for $n = 1,2,3,\ldots$, find a formula for $N(r_n)$ and compute $d(A)$, the box-counting dimension of A.

(b) Suppose B is the product of a Cantor middle-thirds set and the interval $[0,1]$. That is,

$$B = C \times [0,1] = \{(x,y) : x \in C \text{ and } y \in [0,1]\}$$

See the right side of Fig. 6.14. Using boxes of side length $r_n = 1/3^n$, for $n = 1,2,3,\ldots$, find a formula for $N(r_n)$ and compute $d(B)$, the box-counting dimension of B.

(c) From the results of (a) and (b), speculate on the relation between $d(X \times Y)$, $d(X)$, and $d(Y)$.

6.4.8. (a) Suppose A is the union of a Cantor middle-thirds set (viewed as a subset of $[0,1]$) and the interval $[1,2]$. See the left side of Fig. 6.15. Using boxes of side length $r_n = 1/3^n$, for $n = 1,2,3,\ldots$, find a formula for $N(r_n)$ and compute $d(A)$, the box-counting dimension of A. Hint: in general $\log(a+b) \neq \log(a) + \log(b)$, but note $\log(a+b) = \log(a(1+b/a)) = \log(a) + \log(1+b/a)$.

(b) Suppose B is the union of a Sierpinski gasket (with vertices $(0,0)$, $(1,0)$, and

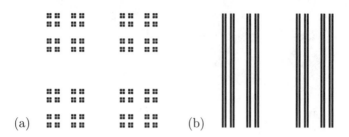

Figure 6.14. Product constructions for Exercise 6.4.7.

$(0,1))$ and the interval $\{(x,0) : 1 \le x \le 2\}$. See the right side of Fig. 6.15. Using boxes of side length $r_n = 1/2^n$, for $n = 1, 2, 3, \ldots$, find a formula for $N(r_n)$ and compute $d(B)$, the box-counting dimension of B.

(c) From the results of (a) and (b), speculate on the relation between $d(X \cup Y)$, $d(X)$, and $d(Y)$.

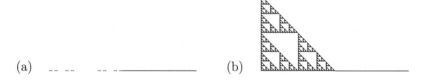

Figure 6.15. Union constructions for Exercise 6.4.9.

6.4.9. Compute the box-counting dimensions of the shapes (a) and (b) in Fig. 6.16. Do these agree with your answer to Exercise 6.4.8 (c)?

(a) The union of a Cantor middle-thirds set and a line segment different from that of the left side of Fig. 6.15.

(b) The union of a Sierpinski gasket and a filled-in square.

Figure 6.16. (a) and (b), constructions for Exercise 6.4.9; (c) and (d), constructions for Exercise 6.4.10.

6.4.10. Compute the box-counting dimensions of the shapes (c) and (d) in Fig. 6.16.

(a) A union of two Sierpinski gaskets.

(b) A union of two different Sierpinski gaskets.

6.5 SOME BIOLOGICAL EXAMPLES

In biology power laws arise in the surface area of mitochondrial membranes, in the branching structures of the pulmonary, circulatory, and nervous systems, in ion channel kinetics, in fossil data, in the distribution of epidemics, and on and on and on. We shall sketch one of these examples in a bit of detail and mention a few others.

Our lungs exchange oxygen from the air for carbon dioxide from the blood. During heavy exercise, this must be done quickly, in under 1 second. To achieve this, the lungs must have a large surface area contained in a small volume. Typical human lungs have an area of about 130 m^2, three-quarters that of a tennis court, and are contained in a volume of 5 or 6 L. The physiological problem is not just to fold the area of the lungs into their volume, but to do this in a way that makes the lung membrane efficient for exchanging gases.

Measuring the lung volume is straightforward; but measuring the surface area? Slice open and spread out all the branches and air sacks? Not a chance. The solution is indirect and beautiful. We'll get to it in a moment, after we've understood a bit more about the lung structure. Gas exchange occurs in what looks like a foam, consisting of close to half a billion alveoli, little sacks each about 1/4 mm in diameter. What architecture can bring air to such a crowd of alveoli so quickly and evenly?

Starting with the trachea, the airway splits into two branches, one of which we see on the left side of Fig. 6.17. Each of these branches splits into two, then again and again and again, between 18 and 30 times, with an average of 23 splittings. The first 15 branchings constitute the *conducting airways*. In this part of the lungs, at each branching the diameters d_1 and d_2 of the two daughter branches are related to d_0, the diameter of the mother branch, by

$$d_1 = d_2 = 2^{-1/3} \cdot d_0 \approx 0.79 \cdot d_0$$

This relation can be understood by the work of W. R. Hess [62] and C. D. Murray [98], who showed that the energy dissipation of air flow at branchings is minimized if $d_0^3 = d_1^3 + d_2^3$, so the geometry of the conducting airways has evolved to maximize the ease of air flow. Evolution is very good at discovering subtle geometry.

 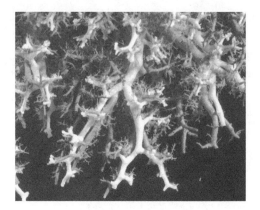

Figure 6.17. Left: a cast of a lung, together with the arterial and venous circulatory system. Right: a magnification of a lung.

Generation 15 is called the *transitional bronchiole*; all successive branching generations are decorated with alveoli. This is the *air exchange zone*, where the diameters of successive generations change very little, slowing the air flow to expedite diffusion of oxygen and carbon dioxide across the alveolar membrane. After about three generations, all branches are completely covered with alveoli. All the branches, and the approximately 10,000 alveoli that cover them, from a single transitional bronchiole, is a *pulmonary acinus*, a unit of the gas exchanger of the lungs.

On the right side of Fig. 6.17 we see the airway branching is accompanied by the branching of pulmonary arteries and veins. After following the bronchial branching all along the conducting airways, the arteries branch more frequently, on average 28 times, putting the number of capillaries at about 30 times the number of terminal airway branches. The capillaries fill the spaces between alveoli, two sides of each capillary contacting alveoli. The lungs have evolved three interlocking trees–pulmonary, arterial, and venous–getting air and blood to all parts of the volume of the lungs in such a way that every part of the gas exchange membrane lies within about 20 cm of the trachea. Air need not move very far to get into and out of the lungs.

Another design feature is of interest. The alveoli are spread along the last 8 generations of branching, each of which is surrounded by its own network of arteries. To reach the alveoli of the last branching generation, which constitute about half the total gas exchange membrane area, air must pass through all the earlier generations of alveoli. Ewald Weibel describes it this way: the air exchange units are perfused in parallel and ventilated in series. Oxygen is absorbed all along the acinus, causing the oxygen concentration to drop as we look ever more deeply into the lungs. If the oxygen concentration drops too much, blood would not be oxygenated at the terminal alveoli. This would initiate the formation of venous shunts, screening the terminal alveoli. Because these account for about half the lung area, such screening would be a physiological disaster. To avoid screening, a ratio of physical factors must be close to a ratio of morphological factors:

$$\frac{O_2 \text{ diffusion}}{O_2 \text{ permeability}} \approx \frac{\text{gas exchange area}}{\text{acinus size}} \tag{6.12}$$

For mammal lungs these ratios are about equal. Screening is avoided and the lungs function efficiently.

Emphysema causes the acinus to enlarge, and also reduces the gas exchange surface area, invalidating (6.12). The screening from the resulting venous shunts explain why the effects of emphysema are more severe than predicted from the loss of surface area alone.

Careful measurements by B. Mauroy, M. Filoche, E. Weibel, and B. Sapoval [93] show that the mother-daughter diameter ratio is closer to 0.85 than to the optimum value 0.79, calculated by Hess and Murray. How could evolution get this factor so wrong? Maybe purposefully: this ratio may be a safety factor, a fail-safe against airway constriction due to asthma. The cost, the reduction of lung volume available for air exchange, is small, because the volume of the airways themselves increases as the square of their diameter (assuming the lengths are unchanged), while by Poiseuille's law the air resistance varies as the fourth power of the diameter. So even the departures from optimality serve a purpose. Evolution does discover nuance.

Now we address how the lung surface area is measured. The method was inspired by the *Buffon needle problem*. In 1777 the French mathematician Georges-Louis Leclerc, Comte de Buffon, posed, and answered, this question to the French Academy of Sciences: what is the probability that a needle of length ℓ, when dropped on a floor of parallel boards separated by a distance d, will cross one of the joints between the boards? Buffon answered his own question: the probability is $(2/\pi) \cdot (\ell/d)$. To find the surface area of the lungs, Weibel [114, 166, 167, 168] applied Buffon's approach in a 3-dimensional setting. Thin cross-sections are sliced from the lungs, producing curves that Weibel overlaid with a grid of test lines of length L. This process is illustrated in Fig. 6.18. Weibel showed that if I lines cross the curves, the surface area A per unit volume V satisfies

$$\frac{A}{V} = 2 \cdot \frac{I}{L}$$

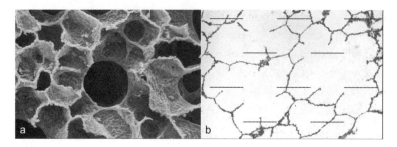

Figure 6.18. Measuring the alveolar area.

In 1963, Weibel used this approach, with images from a light microscope, to measure the alveolar surface area, obtaining a value of 80 m². In the 1970s, repeating the measurements with an electron microscope, Weibel obtained the value of 130 m² for the area. Weibel attributed this to the finer resolution of the electron microscope, and of course this is correct. After hearing Mandelbrot lecture about Richardson's work on measuring the length of the coastline of Britain, he performed a more detailed calculation, quantifying the fractality of the lungs by finding the dimension is about 2.2.

The problem of ventillating and perfusing a large area in a small volume, and with no part of the gas exchange surface more than 20 cm from the air source, was solved elegantly by evolution: three interlocking fractal trees. Weibel speculates that the genetic instructions to build the lung likely are similar to simple geometric rules to build fractals. Fractals may be more than a convenient language for us to discover patterns; they may be the way our DNA builds at least parts of our bodies.

The circulatory system exhibits a similar branching structure, though often the largest arteries conform to the shape of the organs involved. With this scaling structure, the circulatory system, which occupies only about 3% of the body's volume, is able to reach every cell. By far the simplest genetic programming to achieve this structure is to encode the branching and shrinking instructions, rather than the specific design of each artery, vein, and capillary.

From his work on the ability of fractal drums to damp vibrations, Bernard Sapoval [140] and coworkers deduced another advantage of the fractal character of the circulatory system: "the fractal structure of the human circulatory system damps out the hammer blows that our heart generates. The heart is a very violent pump, and if there were any resonance in blood circulation, you would die."

Marc-Olivier Coppens [28] has adapted these ideas to design low-turbulence chemical mixers modeled on the branching architecture of the circulatory system. This is part of his Nature-Inspired Chemical Engineering (NICE, that's the name, really) program.

Power laws appear in many other biological settings. We'll mention two that involve DNA, one about the sequences, one about folding. Sequence correlations were studied by Richard Voss [160, 161] and others. (We'll say more about this in Sect. A.22.2.) For a DNA sequence $z_1 z_2 \cdots z_N$, where each z_i is A, C, G, or T, the *length-w correlation* is $C(w) = \langle z_i z_{i+w} \rangle$. Here the product $z_i z_{i+w}$ is 1 if z_i and z_{i+w} are the same nucleotide, and 0 otherwise. The pointy brackets indicate the average. For example,

$$C(1) = \frac{1}{N}(z_1 z_2 + z_2 z_3 + \cdots z_{N-1} z_N)$$

If we think of w as a wavelength, then long w corresponds to low frequency f. For any signal, a plot of the amplitude^2 of the signal as a function of f is called the *power spectrum*, $P(f)$. Early in his career, Voss [162] studied power spectra in acoustic signals. Later he found a relation between power spectra of the form $P(f) = 1/f$ and fractal structures. More recently, Voss found that while individual DNA sequences have fairly noisy power spectra, averages over GenBank classifications exhibit relatively smooth long-range power spectra $P(f) = 1/f^\beta$ with these values of β. These long-range correlations must encode

bacteria	1.16	organelle	0.71	virus	0.82
plant	0.86	invertebrate	1.00	vertebrate	0.87
mammal	0.84	rodent	0.81	primate	0.77

something, but what? As gene therapy becomes more common, we need to unpack all the ways information is stored in the sequence.

But the sequence is not all that's important. The DNA of a single human cell would unwind into a linear strand nearly two meters long; in the cell nucleus the DNA is wound into a ball about five millionths of a meter of diameter. One folded configuration, called an *equilibrium globule*, is based on statistical mechanics arguments, but the folding process introduces knots and tangles in the sequence, and these would cause significant problems when unfolding the globule to make accessible parts of the sequence during gene activation, gene inactivation, and cell replication. A hierarchical folding mechanism, introduced by Alexander Grosberg and coworkers [56, 57], avoids the knotting problems so any portion of the globule can be unfolded as needed. Strong experimental support for the fractal globule hypothesis is presented in [83]. We'll say more about this in Sect. A.22.3.

This is nowhere near the end. Physiological scaling and fractals can be found in the linings of the intestines [52], the internal structure of the kidneys [148] and liver [36], cross-section of bones [175, 22], dynamics of some EEG [1, 7, 165] and ECG records [100], and the ways familiar memories are stored [43, 152]. Many other examples are discussed in [10, 52, 88, 103, 170], for instance. On May 15, 2018, a Google search for fractals in biology returned about 1,370,000 matches. I wonder how many you would find when you read this.

Chapter 7 Differential equations in the plane

So far we have studied differential equations with one independent variable, single-species populations as a function of time, for example. However, often we find situations in which several species interact. Representing each species as an independent variable, we find systems of this form

$$\frac{dx}{dt} = f(x,y) \qquad \frac{dy}{dt} = g(x,y) \qquad\qquad (7.1)$$

for two species, with obvious generalizations for more species. Still, we are studying autonomous differential equations; $f(x,y)$ and $g(x,y)$ are not explicit functions of t. In this setting, the xy-plane is called the *phase plane*.

If the functions f and g of Eq. (7.1) are linear, we can solve this system. The long-term behavior of solutions of linear systems falls into one of three cases: converges toward the origin, diverges to infinity, orbits in a simple closed path about the origin. We verify that these are the only possibilities by finding the solutions of all linear systems. If the world were linear, it would be easy to understand, but so much less interesting.

Many aspects of the world are nonlinear. For example, predator and prey (including T cells and viruses) interact according to the principle of mass action, that the likelihood of interaction is proportional to the products of the densities of the species, certainly a nonlinear function. To prepare an approach to non-linear differential equations, we develop some geometric tools (first, eigenvalue analysis in Sect. 8.5) that characterize the long-term behavior of linear system solutions. Appropriately modified, these tools work in many, but not all, non-linear differential equations (Sect. 9.2). We spend some time to find ways to deal with these other systems (Sect. 9.4), and to understand limit cycles (Sects. 9.6 and 9.7), periodic solutions that nearby solutions converge to or diverge from.

First, we must understand some basic geometry of differential equations in the plane.

7.1 VECTOR FIELDS AND TRAJECTORIES

Systems with n independent variables are called n-dimensional. For some of the biological processes we'll study, 2-dimensional systems will be sufficient, and they're the simplest higher-dimensional systems, so this is where we'll start. How different can 2-dimensional systems be from the 1-dimensional systems

$$\frac{dx}{dt} = h(x) \tag{7.2}$$

we have studied so far?

The *asymptotic* behavior of a system is what it does as $t \to \infty$ or as $t \to -\infty$. For 1-dimensional systems, we have seen two types of asymptotic behavior, $x(t) \to \pm\infty$ and $x(t) \to x_*$, exemplified by $dx/dt = x$ and $dx/dt = x(1-x)$, respectively. For the latter, the fixed points are $x_* = 0$ and $x_* = 1$, the solutions of $0 = h(x) = x(1-x)$. The fixed point $x_* = 0$ is unstable because $h'(0) = 1 > 0$; the fixed point $x_* = 1$ is stable because $h'(1) = -1 < 0$. How much of this remains true in 2-dimensional systems? Are any different behaviors possible?

We shall see that finding the fixed points of a 2-dimensional system is the obvious generalization of the 1-dimensional case. A point (x_*, y_*) is a fixed point of the system (7.1) if

$$f(x_*, y_*) = 0 \quad \text{and} \quad g(x_*, y_*) = 0$$

Stability of the fixed points is a little trickier. First we generalize the derivative of $h(x)$ to $f(x, y)$ and $g(x, y)$, then extend the conditions $|h'(x)| < 1$ and $|h'(x)| > 1$ to 2 dimensions. How to handle the extension to 2 dimensions of some of the conditions $|h'(x)| = 1$ is trickier. Taken together, these will occupy Ch. 9.

However, converging to fixed points or diverging to ∞ do not exhaust the possibilities. Some 2-dimensional systems exhibit periodic behavior. For example, the predator-prey populations we'll describe in Sect. 7.3 oscillate in time. In Sect. A.4 we'll give a geometric construction, developed by the mathematician Vito Volterra, showing this really is periodic behavior and not a very slow spiral converging to a fixed point.

Can anything else happen? For 2-dimensional systems, the answer is no. By a beautiful theorem of Poincaré and Bendixson (Sect. 9.7), the bounded behavior (neither variable runs away to ∞) consists of fixed points, solutions that connect fixed points, and periodic dynamics (closed curves and limit cycles). For 3-dimensional systems, the possibilities are infinitely richer. Solutions can stay bounded but never repeat, or converge to a periodic solution. Solutions starting from two points arbitrarily close together can stay nearby for some time, but eventually diverge into wildly different behavior. The chaos we saw in Sect. 2.4 for discrete models, impossible in 2-dimensional autonomous differential equations, can and often does occur for differential equations of 3 or more dimensions. Also, we can find chaotic behavior for non-autonomous 2-dimensional systems, by converting these into 3-dimensional autonomous systems taking t as the third variable. But these are complications we'll study later. For now, let's understand how to visualize solutions of the system (7.1).

First, associated with the system (7.1) we have a *vector field*. To each point (x_0, y_0) assign the vector $\langle f(x_0, y_0), g(x_0, y_0) \rangle$. We can visualize this as the vector from the point (x_0, y_0) to the point $(x_0, y_0) + (f(x_0, y_0), g(x_0, y_0))$. Drawing these vectors for every point in the phase plane would fill up the plane completely leaving nothing to be seen, a picture of a panther with eyes closed in a forest on a moonless night. Rather, we select some sample points, usually on a grid, and draw the vectors at the sample points.

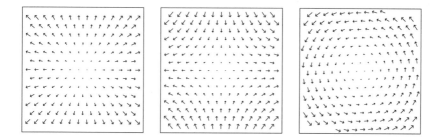

Figure 7.1. Examples of vector fields.

Consider the three vector fields

$$\langle f_1, g_1 \rangle = \langle x, y \rangle \quad \langle f_2, g_2 \rangle = \langle x, -y \rangle \quad \langle f_3, g_3 \rangle = \langle -y, x \rangle \qquad (7.3)$$

For example, at the point $(1,1)$ for $\langle f_1, g_1 \rangle$ we draw the vector from $(1,1)$ to $(1,1) + (1,1) = (2,2)$, for $\langle f_2, g_2 \rangle$ we draw the vector from $(1,1)$ to $(2,0)$, and

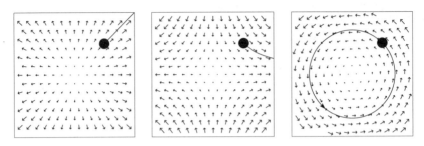

Figure 7.2. Some trajectories of vector fields. The starting point, (1,1), is shown as a dot.

for $\langle f_3, g_3 \rangle$ we draw the vector from (1,1) to (0,2). Vectors at other points are drawn similarly. In Fig. 7.1 we have scaled the lengths of the vectors so they do not overlap.

Now to be precise, a *solution curve* or *trajectory* of the system (7.1) is a function $\vec{r}(t) = \langle x(t), y(t) \rangle$ satisfying the system (7.1), so $x'(t) = f(x, y)$ and $y'(t) = g(x, y)$. The tangent vector to the curve $\vec{r}(t)$ is $\vec{r}'(t) = \langle x'(t), y'(t) \rangle$, so we see the tangent vector to each point on a trajectory is the vector field vector at that point. In Fig. 7.2 we see solution curves to the systems of Fig. 7.1, each starting at the point (1,1).

For example, consider the third system $x' = -y$, $y' = x$. In Ch. 8 we solve this system as part of a more general approach, but for now we use an old and often effective method: guessing. We're looking for two functions, the first is the derivative of the second, the second is the negative of the derivative of the first. The functions $\cos(t)$ and $\sin(t)$ are familiar examples. In order to be able to fill in the whole phase plane, take

$$\vec{r}(t) = \langle r_0 \cos(t), r_0 \sin(t) \rangle$$

where r_0 is the radius of the circle. Then

$$\vec{r}'(t) = \langle -r_0 \sin(t), r_0 \cos(t) \rangle$$

so indeed $x' = -y$ and $y' = x$. The solution curves are circles, as seen in the third graph of Fig 7.2.

Not so fast. The third image of Fig. 7.2 shows a solution curve satisfying $\vec{r}(0) = (1,1)$, indicated by the dot on the circle. Remember, by the chain rule adding a constant to the variable t doesn't alter the trajectory. It changes only where on the trajectory we are at any given time. So the solution we want is

$$\vec{r}(t) = \langle \sqrt{2}\cos(t + \pi/4), \sqrt{2}\sin(t + \pi/4) \rangle$$

If you wish, check that $\vec{r}(0) = \langle 1, 1 \rangle$.

In the next two sections we develop two examples from population dynamics, a predator-prey system and two species that compete for the same resources. Biology contains many, many other examples, some of which we shall see. For others, remember that Google is our friend.

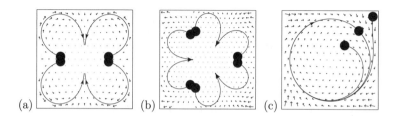

(a)   (b)   (c)

Figure 7.3. More vector fields, with some solution curves.

To illustrate that complicated behavior can arise even in 2 dimensions, in Fig. 7.3 we see three more vector fields, (a) $\langle x^3 - 3xy^2, 3x^2y - y^3 \rangle$, (b) $\langle x^4 - 6x^2y^2 + y^4, 4x^3y - 4xy^3 \rangle$, and (c) $\langle x + y - x(x^2 + y^2), -x + y - y(x^2 + y^2) \rangle$. Some solution curves are indicated; initial points appear as dots. We'll explain some of the behavior of these differential equations in Exercises 7.1.8, 7.1.9, and 7.1.10. For a hint, look at the exponents, the coefficients, and the signs in (a) and (b). Do you see a pattern?

In Sect. B.6 we'll see how to plot vector fields using Mathematica. This is useful for building intuition about how the terms of a vector field are represented in the patterns of the vectors.

Practice Problems

7.1.1. Match the vector field plots with the corresponding trajectory plots in Fig. 7.4.

7.1.2. Match the vector field plots of Fig. 7.5 with the vector fields. Explain your choices.

(a) $\langle y^2, x \rangle$ (b) $\langle x, y^2 \rangle$ (c) $\langle y, x \rangle$

Practice Problem Solutions

7.1.1. Tangents to the trajectories coincide with vectors of the vector field, so the vector field (a) has trajectories (e), the vector field (b) has trajectories (d), and the vector field (c) has trajectories (f).

7.1.2. Vector field (a) corresponds to plot (f) because $f(x,y) = y^2$ means all the vectors point to the right, and only plot (f) has this property.

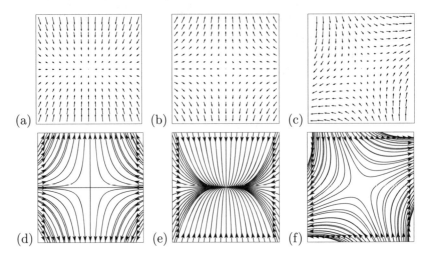

Figure 7.4. Vector fields (top) and trajectories (bottom) for Practice Problem 7.1.1.

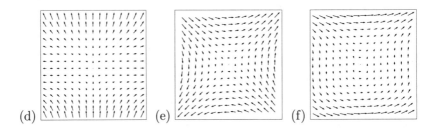

Figure 7.5. Vector fields for Practice Problem 7.1.2.

Vector field (b) corresponds to plot (d) because $g(x,y) = y^2$ means all the vectors point up, and only plot (d) has this property.

Vector field (c) corresponds to plot (e) because $f(x,y) = y$ means vectors in the upper half of the square point to the right, while vectors in the lower half of the square point to the left. Only plot (e) has this property.

Exercises

In Exercises 7.1.1–7.1.4 match the vector fields with the vector field plots of Figs. 7.6–7.9. Explain your choices.

7.1.1.   (a) $\langle x^2, y^2 \rangle$ (b) $\langle x^2, y^3 \rangle$ (c) $\langle x^2, -y^2 \rangle$

7.1.2.   (a) $\langle y^2, x^2 \rangle$ (b) $\langle -y^2, x \rangle$ (c) $\langle -y^2, x^2 \rangle$

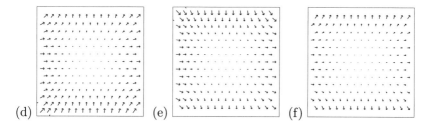

Figure 7.6. Vector fields for Exercise 7.1.1.

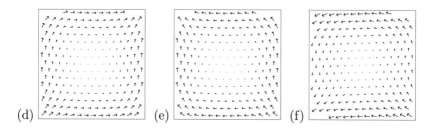

Figure 7.7. Vector fields for Exercise 7.1.2.

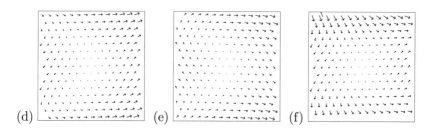

Figure 7.8. Vector fields for Exercise 7.1.3.

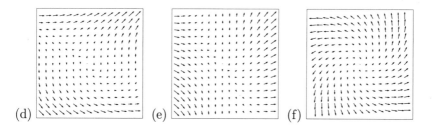

Figure 7.9. Vector fields for Exercise 7.1.4.

7.1.3. (a) $\langle x + y^2, x \rangle$ (b) $\langle x + y^2, -x \rangle$ (c) $\langle x + y^2, x - y^2 \rangle$

7.1.4. (a) $\langle y^2, x + y \rangle$ (b) $\langle x - y, x + y \rangle$ (c) $\langle x^2, x + y \rangle$

7.1.5. Match the vector fields to the trajectories in Fig. 7.10. Explain your choices.

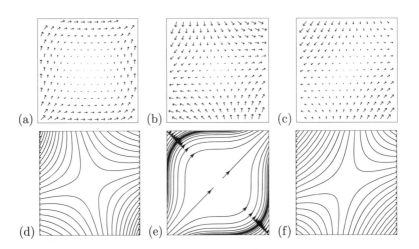

Figure 7.10. Vector fields and trajectories for Exercise 7.1.5.

7.1.6. Match the vector fields to the trajectories in Fig. 7.11. Explain your choices.

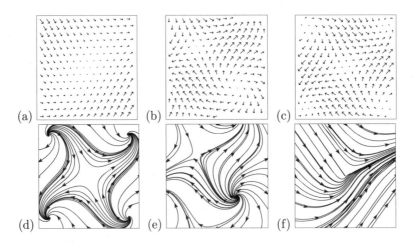

Figure 7.11. Vector fields and trajectories for Exercise 7.1.6.

7.1.7. Does either of these vector fields have trajectories that are unbounded, that is, $|x(t)| \to \infty$ or $|y(t)| \to \infty$ as $t \to \infty$? Explain your answer.

$$\text{(a)} \left\langle \frac{y}{\sqrt{x^2 + y^2}}, \frac{x}{\sqrt{x^2 + y^2}} \right\rangle \quad \text{(b)} \left\langle \frac{-y}{\sqrt{x^2 + y^2}}, \frac{x}{\sqrt{x^2 + y^2}} \right\rangle$$

In the next three problems we'll unpack some of the properties of vector fields (a) and (b) of Fig. 7.3.

7.1.8. We can represent a complex number $z = x + iy$ by the point (x, y) in the plane. The distance between the origin and the point (x, y) is $\|z\| = \sqrt{x^2 + y^2}$, called the *modulus* of z. The angle θ between the x-axis and the line from $(0, 0)$ to (x, y) is called the *argument* of z, $\arg(z)$, and satisfies $\tan(\theta) = y/x$. Suppose $z = a + ib$ and $w = c + id$.
(a) Show that $\|z \cdot w\| = \|z\| \cdot \|w\|$.
(b) Show that $\arg(z \cdot w) = \arg(z) + \arg(w)$.

7.1.9. For a complex number $z = x + iy$, the number x is called the *real part* of z, $\mathrm{Re}(z)$, and the number y is called the *imaginary part* of z, $\mathrm{Im}(z)$. (Note that the imaginary part is *not* iy.)
(a) Show the real part of z^3 is $x^3 - 3xy^2$ and the imaginary part of z^3 is $3x^2y - y^3$.
(b) Show the real part of z^4 is $x^4 - 6x^2y^2 + y^4$ and the imaginary part of z^4 is $4x^3y - 4xy^3$.
(c) From part (b) of Exercise 7.1.8, we see that $\arg(z^n) = n \arg(z)$. Using the code in Sect. B.7, verify that the trajectories of Fig. 7.3 (a) are solutions of $x' = \mathrm{Re}(z^3)$, $y' = \mathrm{Im}(z^3)$, and the trajectories of Fig. 7.3 (b) are solutions of $x' = \mathrm{Re}(z^4)$, $y' = \mathrm{Im}(z^4)$. If you're puzzled that $n = 3$ gives a four-lobed pattern and $n = 4$ gives a six-lobed pattern, we'll return to this topic in Sect. 17.7.

7.1.10. As $n \to \infty$, show that at some points on trajectories $\vec{r}(t)$ of the vector field determined by z^n, the magnitude of $\vec{r}'(t)$ is arbitrarily large.

7.2 DIFFERENTIAL EQUATIONS SOFTWARE

Chapters 8 and 9 deal with solutions of differential equations in the plane. In Chapter 8 we find these solutions; in Chapter 9 and later chapters we find solutions only occasionally, but we can understand some geometric and qualitative properties of solutions. Experimentation is an effective avenue for building intuition about how terms of a differential equation alter the vector field and trajectories. In Sect. B.7 we describe how to generate these plots using Mathematica, which is reasonably widely available in colleges and universities. Special-purpose codes are presented when we need them.

All the vector field plots and solution curves presented in this text were produced by Mathematica. Rather than discuss generalities here, we'll consider specific examples in the following sections and chapters. But we'll describe the general code in the appendix entry, B.7, for this section.

7.3 PREDATOR-PREY EQUATIONS

Our first example involves an interaction of two species, x and y. The species x is the predator and consumes y; y is the prey and has (for the moment) an unlimited food supply. (Why use x and y instead on variables that suggest the populations they represent? Because both predator and prey begin with "P.") How do x and y change with time? In the absence of prey, the predators will die at a rate proportional to their population; in the absence of predators, the prey will grow at a rate proportional to their population. If a is the predator per capita death rate and c is the prey per capita birth rate, these populations grow according to

$$x' = -ax \quad \text{and} \quad y' = cy$$

How do the predators and prey interact? Recall the principle of mass action: the probability of an encounter between a predator and a prey is proportional to xy. These encounters are good for the predators and bad for the prey. Predators eat prey, obtaining energy for, among other things, reproducing more predators. Say the effect on the predator population growth rate is $+bxy$. Prey are consumed, reducing their population. This is seen as a negative term, $-dxy$, in the prey growth rate. Then the predator-prey equations are

$$x' = -ax + bxy \tag{7.4}$$

and

$$y' = cy - dxy \tag{7.5}$$

Eqs. (7.4) and (7.5) are called the *Lotka-Volterra predator-prey equations*. As mentioned above, in Sect. A.4 we present Volterra's elegant approach to show that the trajectories are periodic.

Eq. (7.5) requires that the population of prey is small compared to the maximum supported by its food supply, assumed constant on the time scales studied. If the population of prey can affect its food supply, the cy term in Eq. (7.5) must be replaced by something like the Verhulst equation, Eq. (3.1), giving

$$y' = cy\left(1 - \frac{y}{K}\right) - dxy = cy - ey^2 - dxy \qquad (7.6)$$

Just how do the dynamics of the system Eqs. (7.4) and (7.5) differ from those of the system Eqs. (7.4) and (7.6)? Fig. 7.12 shows examples. On the left are the vector field and some periodic trajectories of the system (7.4) and (7.5) with $a = b = c = d = 1$. The middle and right images show the vector field and some trajectories of the system (7.4) and (7.6) with $a = b = c = d = 1$ and $K = 5$ (middle), $K = 1$ (right). Here we appear to see the trajectories converge to a fixed point, so the addition of the logistic term, at least for these parameter values, has a substantial effect on the predator-prey dynamics. Can you explain why for $K = 5$ the populations spiral in to a fixed point with both predator (x) and prey (y) populations non-zero, while for $K = 1$ the predator population drops to 0 and the prey population converges to its Verhulst limit?

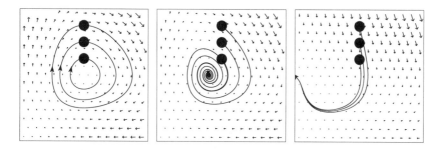

Figure 7.12. Trajectories of predator-prey systems.

An alternate approach is to plot x and y along the vertical axis and time t along the horizontal axis. We call these *population curves*. The left graph of Fig. 7.13 shows predator x and prey y curves as a function of time for the system of Eqs. (7.4) and (7.5) with $a = b = c = d = 1$. The right graph shows the $x-$ and $y-$curves for Eqs. (7.6) and (7.4) with $a = b = c = d = 1$ and $K = 5$. Fig. 7.13 shows the same information as the first two graphs of Fig. 7.12, except that if we

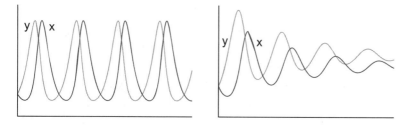

Figure 7.13. Plots of population curves.

put a scale on the time axis of Fig. 7.13, we could tell how much time the predator and the prey populations spend at each level. Both trajectories and population curves are useful.

Practice Problems

7.3.1. In Fig. 7.14, which Lotka-Volterra trajectory (plot of the curve $(x(t), y(t))$), (a) or (b), corresponds to which population curves (plot of the curves $x(t)$ and $y(t)$ vs t), (c) or (d))? Explain your answer.

7.3.2. Suppose a Lotka-Volterra system of Eqs. (7.4) and (7.5) has closed trajectories. How does increasing a alter these trajectories? Explain your reasoning. You can use the Mathematica code in Sects. B.7 and B.8 to run test cases, if you wish.

Practice Problem Solutions

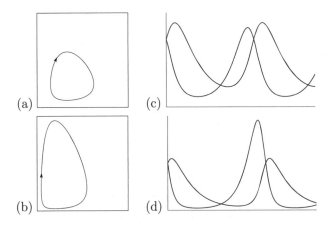

Figure 7.14. Trajectories and population curves for Practice Problem 7.3.1. The vertical scales of (a) and (b) are different from those of (c) and (d).

7.3.1. In (a) the x- and y- values have similar ranges. This can be seen by noting that the curve projects to about the same intervals on the x- and y-axes. In (b) y has a considerably larger range than x. In (c) the x- and y-values have similar ranges. In (d) the y-values have a larger range than the x-values. Consequently, plot (a) corresponds to graph (c), and plot (b) to graph (d).

Alternately, consider the intersections of the trajectories in (a) and (b) with the diagonal lines. At these intersections, $x = y$; in (c) and (d) these correspond to intersections of the $x(t)$ and $y(t)$ curves. In (b), one intersection with the diagonal occurs for small x and y, and for (d) the $x(t)$ and $y(t)$ curves cross for small values of $x(t)$ and $y(t)$, again suggesting that (b) corresponds to (d).

7.3.2. In Eq. (7.4), increasing a increases the per capita rate at which the predators x die. This lowers the x-min of the curve. A deeper dip in the predator population reduces the magnitude of the negative term in Eq. (7.5), causing a higher y-max in the curve.

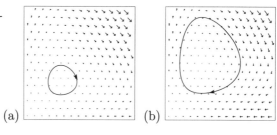

(a)   (b)

Figure 7.15. Trajectories for Practice problem 7.3.2.

A higher maximum in the prey population increases the positive term of Eq. (7.4), causing a higher x-max in the curve. Because we have made no direct change to the prey equation, any change in the y-min of the curve must result from increased predation. Because the maximum prey population is higher, we see increased predation of a higher prey population, so we expect only minor changes in y-min. Fig. 7.15 illustrates this effect. On the left we see a closed trajectory for $a = b = c = d = 1$, on the right for $a = 1.5$, $b = c = d = 1$. The x-axes and the y-axes of both graphs are plotted on the same scales.

Exercises

7.3.1. In Fig. 7.16, which trajectory, (a) or (b), corresponds to which population curves, (c) or (d)? Explain your answer.

In Exercises 7.3.2 through 7.3.6, suppose the Lotka-Volterra system of Eqs. (7.4) and (7.5) has closed trajectories. Explain your reasoning. If you do a simulation, start with $a = b = c = d = 1$ and in the first three problems, change the appropriate parameter to 2.

7.3.2. How does increasing c alter these trajectories?

7.3.3. How does increasing d alter these trajectories?

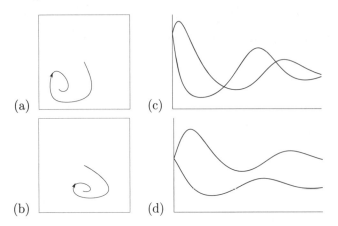

(a)

(c)

(b)

(d)

Figure 7.16. Trajectories and population curves for Exercise 7.3.1.

7.3.4. How does increasing b alter these trajectories?

7.3.5. (a) Show that the highest point (maximum y) and lowest point (minimum y) of a closed trajectory occur at the same x-value.
(b) Find the x-value of these maximum and minimum y-values.
(c) Show that the right-most point (maximum x) and left-most point (minimum x) of a closed trajectory occur at the same y-value.
(d) Find the y-value of these maximum and minimum x-values.

7.3.6. (a) At the points where a trajectory intersects the diagonal $y = x$, find dy/dx, the slope of the tangent to the trajectory. Hint: remember the chain rule.
(b) Find the points (x, y) where both $x' = 1$ and $y' = 1$. Express the values of x and y in terms of the system parameters a, b, c, and d.

7.3.7. What is the difference between the middle and right graphs of Fig. 7.12 in terms of carrying capacity and population size?

7.3.8. Show that Eq. (7.6) reduces to Eq. (7.5) when $K \to \infty$. Assume x and y remain bounded. Interpret this in terms of unlimited resources.

7.3.9. We have argued that if the prey population can approach the carrying capacity of its environment, then we must add a logistic term to limit the prey growth. Why don't we need to add a logistic term to the predator population?

7.3.10. For the predator-prey equations with logistic limits on the prey growth, Eqs. (7.4) and (7.6), take $a = b = c = d = 1$.

(a) Find the coordinates of the fixed points, the points (x, y) at which $x' = 0$ and $y' = 0$. We found those points for the Lotka-Volterra system in Exercise 7.3.5. We'll study fixed points extensively, starting in Sect. 7.5.

(b) Explore how the fixed points move and how the trajectories evolve as K goes from 5 to 1. That is, explore the transition between the middle and right figures of Fig. 7.12.

7.4 COMPETING POPULATIONS

Now suppose we have two species, again with populations denoted by x and y. Here neither preys upon the other; both compete for the same limited food supply. As small populations these grow approximately as

$$x' = ax \quad \text{and} \quad y' = dy$$

where a and b are positive constants, the per capita growth rates. As the populations increase, competition for food slows growth rates. We'll use Eq. (3.1), the Verhulst formulation of the effect of limited resources, as a guide for our model of populations that compete for limited resources

$$x' = a\left(1 - \frac{x+y}{K_x}\right)x \quad \text{and} \quad y' = d\left(1 - \frac{x+y}{K_y}\right)y \qquad (7.7)$$

where K_x and K_y are the carrying capacities of x and y. Because both x and y compete for the same resources, if $x + y > K_x$, then $x' < 0$, and if $x + y > K_y$, then $y' < 0$.

The model (7.7) assumes x and y consume resources at the same rate. If this is not the case, we can replace $(x + y)/K_x$ by $(Bx + Cy)/K_x = bx + cy$. Then the more general equations for two competing populations are

$$x' = a(1 - bx - cy)x \quad \text{and} \quad y' = d(1 - ex - fy)y \qquad (7.8)$$

In Fig 7.17 we see three examples of the system (7.7) with

left	$a = 1$, $d = 1$, $K_x = 2$, $K_y = 1$
center	$a = 10$, $d = 1$, $K_x = 2$, $K_y = 1$
right	$a = 1$, $d = 1$, $K_x = 1$, $K_y = 2$

The arrows on the solution curves have been removed to make clearer the behavior of these curves along the x and y axes.

What do these graphs tell us? Left and center seem to show that, at least for the initial populations indicated, y becomes extinct and x approaches a fixed value,

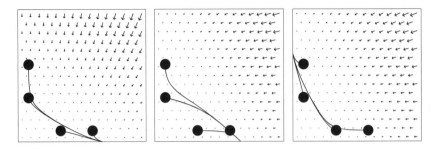

Figure 7.17. Examples of competing species systems. Dots show starting points of the trajectories.

independently of the initial population. In both of these, $K_x > K_y$. In the right graph, x appears to become extinct and y approaches a fixed value. Here $K_x < K_y$. As a first guess, it appears that the population with the larger carrying capacity forces the other to extinction and then settles down to a fixed value. Can this be true? How can we tell? As with all phenomena explored by experiments, a single set of model parameters and initial values can invalidate this hypothesis, but millions of examples supporting it provide no proof, though they can increase our confidence in the hypothesis. Still, we need to find a general approach. Be prepared to wait: the full result will take some time to develop. We'll get there in Sect. 9.2. Remember that patience is the key to paradise.

A final comment, for now, on this topic. Ways to combine this section's competing species model with the predator-prey model of Sect. 7.3 should be clear. If we have, say, one predator and two competing prey species, writing the equations is a simple extension of what we've done in this section and in Sect. 7.3, but will the dynamics still be limited to fixed points and cycles? In Sect. 13.6 we'll see that the behavior of such three-species systems can be substantially more complex; in some sense, infinitely more complex because three-species models can exhibit chaos. Whenever we think we understand everything nature can do, nature reminds us that we should be more humble.

Practice Problems

7.4.1. Suppose Eq. (7.8) is modified to take this form

$$x' = f(x,y) = x(2 - x - y) \qquad y' = g(x,y) = y(1 - (x/3) - y) \qquad (7.9)$$

(a) Show the vectors are vertical along the line $y = 2 - x$. Show the vectors are horizontal along the line $y = 1 - x/3$.
(b) Show that at all points (x,y) above the line $y = 2 - x$ the vector field points to the left, and to the right at all points below $y = 2 - x$.

(c) Can this system have a periodic trajectory that lies entirely above, or entirely below, the line $y = 2 - x$?

7.4.2. Continuing with the system (7.9),
(a) show that at all points (x,y) above the line $y = 1 - x/3$ the vectors point down, and at all points below the line, the vectors point up.
(b) Find the fixed points of this system.

Practice Problem Solutions

7.4.1. (a) Along the line $y = 2 - x$ we see $2 - x - y = 0$ and so $x' = f(x,y) = 0$. That is, along this line the vectors of the vector field have 0 horizontal component, hence must be vertical. Similarly, along the line $y = 1 - x/3$, $y' = g(x,y) = 0$ and so the vectors must be horizontal.
(b) Recall that $x \geq 0$, so the sign of $x(2 - x - y)$ is determined by the sign of $2 - x - y$. For points above the line, $x' < 0$ and so the vectors point to the left. For points below the line, $x' > 0$ and so the vectors point to the right.
(c) A periodic trajectory must have tangent vectors that point NE, other vectors that point NW, and SW, and SE. Because all the vectors above the line $y = 2 - x$ point to the left (W), no periodic trajectory can lie entirely above that line. Similarly, because all vectors below the line $y = 2 - x$ point to the right (E), no periodic trajectory can lie entirely below the line.

7.4.2. (a) Because $y \geq 0$, the sign of y' is the sign of $1 - x/3 - y$. If $y > 1 - x/3$ we see $0 > 1 - x/3 - y$ and so the vectors point down. Similarly, if $y < 1 - x/3$ we see $0 < 1 - x/3 - y$ and so the vectors point up.
(b) First, observe $x' = 0$ at $x = 0$ and along the line $y = 2 - x$. Next, $y' = 0$ at $y = 0$ and along the line $y = 1 - x/3$. The fixed points of the system occur where both $x' = 0$ and $y' = 0$. (This observation is explored in Sect. 7.5.) That is, the fixed points are given by four combinations of these conditions:

$\quad$ $x = 0$ and $y = 0$, so $(x,y) = (0,0)$,
$\quad$ $x = 0$ and $y = 1 - x/3$, so $(x,y) = (0,1)$,
$\quad$ $y = 2 - x$ and $y = 0$, so $(x,y) = (2,0)$, and
$\quad$ $y = 2 - x$ and $y = 1 - x/3$, so $(x,y) = (3/2, 1/2)$.

Exercises

In Exercises 7.4.1, 7.4.2, and 7.4.3,
(a) find the lines on which the vector field vectors are horizontal,
(b) find the lines on which the vector field vectors are vertical, and

(c) find the fixed points.

(d) The lines found in (a) and (b) cut the first quadrant into several regions. In each of these regions, determine in which direction, NE, SE, SW, or NW, the vector field vectors point.

(e) Do the considerations of (d) preclude the presence of periodic trajectories for the vector field? Explain your reasoning.

7.4.1. $x' = 2x(1 - (x+y)/4)$, $y' = 2y(1 - (x+y)/3)$

7.4.2. $x' = 2x(1 - (x+y)/6)$, $y' = 2y(1 - (x+y)/3)$

7.4.3. $x' = 2x(1 - (x+y)/4)$, $y' = 2y(1 - (x+y)/5)$

7.4.4. From the results of Exercises 7.4.1, 7.4.2, and 7.4.3, comment on the effects on the fixed points of increasing each carrying capacity K_x and K_y separately.

7.4.5. For the differential equations of Exercises 7.4.1, 7.4.2, and 7.4.3, plot some trajectories. For each differential equation, take one starting point in each region determined by $x' = 0$ and $y' = 0$.

In Exercises 7.4.6, 7.4.7, and 7.4.8, use Eq. (7.7) with $K_x = K_y = k$.

7.4.6. Can any trajectory cross the line $y = k - x$? Explain your answer.

7.4.7. For (x,y) above the line $y = k - x$, show the vector field always points down and to the left. For (x,y) below the line $y = k - x$, show the vector field always points up and to the right.

7.4.8. Using the results of Exercises 7.4.6 and 7.4.7, show that Eq. (7.7) with $K_x = K_y$ can have no periodic trajectories.

In Exercises 7.4.9 and 7.4.10, suppose both species are predators, and that the prey supply is unlimited, at least over the time scale of our analysis. Plot population curves for $(x(0),y(0)) = (0.8,0.7)$, $(0.8,0.1)$, $(0.2,0.1)$, and $(0.11,0.1)$. Interpret these plots.

7.4.9. Suppose the x predators must hunt in pairs, so the terms bx and ex of Eq. (7.8) become bx^2 and ex^2.

7.4.10. Now suppose both predators x and y must hunt in pairs, so the terms bx, ex, cy, and fy of Eq. (7.8) become bx^2, ex^2, cy^2, and fy^2.

7.5   NULLCLINE ANALYSIS

Recall the general 2-dimensional system (7.1)

$$x' = f(x,y) \qquad y' = g(x,y)$$

In Sect. 7.1 we saw that the fixed points (x_*, y_*) are the solutions of

$$f(x_*, y_*) = 0 \quad \text{and} \quad g(x_*, y_*) = 0 \qquad (7.10)$$

Sometimes, as in the examples of Sects. 7.3 and 7.4, we can solve $f(x,y) = 0$ and $g(x,y) = 0$ and obtain equations for lines or curves.

For the predator-prey model of Sect. 7.3, Eq. (7.4) gives

$$0 = -ax + bxy = x(-a + by)$$

That is

$$x = 0 \quad \text{or} \quad y = a/b \qquad (7.11)$$

These constitute the *x-nullcline* of the predator-prey model, because all along these lines, $x' = 0$, that is, the vector field vectors are vertical. These are the solid gray lines in Fig. 7.18. Similarly, the *y*-nullcline (the dashed gray lines in Fig. 7.18) for Eq. (7.5) is defined by

$$y = 0 \quad \text{or} \quad x = c/d \qquad (7.12)$$

The first observation is to be careful when finding the fixed points. From Eq. (7.10) we see the fixed points are the points at which the nullclines intersect. For this predator-prey system, the nullclines are the four lines given by Eqs. (7.11) and (7.12). Too hastily we might conclude this system has four fixed points, $(0,0)$, $(c/d, a/b)$, $(0, a/b)$, and $(c/d, 0)$. The first two are fixed points, the third is the intersection of the two lines constituting the *x*-nullcline, and the fourth is the intersection of the two lines constituting the *y*-nullcline. Consequently, $(0, a/b)$ and $(c/d, 0)$ are not fixed points. In Fig. 7.18 circles indicate the fixed points.

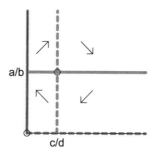

Figure 7.18. Nullclines in a predator-prey system with $a = b = c = 1$, $d = 2$.

Second, the x- and y-nullclines are all the points at which the vector field vectors are either vertical or horizontal. When the nullclines are removed, the phase plane (the first quadrant in the case of populations) is divided into regions on which the vector field points NE (Northeast), SE, SW, or NW. So long as the functions f and g are continuous, testing a single point in each region determines if the vector field is NE, SE, SW, or NW throughout that region. Sometimes this is easy enough to determine directly from the functions $f(x,y)$ and $g(x,y)$. For example, $y < a/b$ means $-a + by < 0$ and so $x' < 0$. That is, below the line $y = a/b$ the vector field points W, above this line the vector field points E. Similarly, the vector field points N to the left of $x = a/b$ and S to the right of $x = a/b$. Combining these, we obtain the direction vectors in Fig. 7.18. From these, we see the trajectories go around the fixed point $(c/d, a/b)$, but we cannot tell if the trajectories are periodic or spiral in to the fixed point. These two possibilities have substantially different biological interpretations. We'll have to do a bit of work to see which is the case. Yet again, we mention that in Sect. A.4 we'll see an elegant geometrical way to show that the trajectories are periodic.

Now we analyze the nullclines for the two competing species model of Sect. 7.3. In Eq. (7.7), we see the x-nullcline is the pair of lines

$$x = 0 \quad \text{and} \quad y = -x + K_x$$

and the y-nullcline is the pair of lines

$$y = 0 \quad \text{and} \quad y = -x + K_y$$

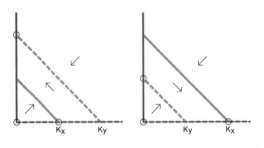

Figure 7.19. Nullclines in two competing species systems.

In Fig. 7.19 we see two examples. The fixed points are $(0,0)$, $(K_x, 0)$, and $(0, K_y)$. The nullclines divide the first quadrant into three regions: upper, middle, and lower. As with the predator-prey equations, we use the signs of x' and y' in each region to determine the vector field direction in that region. For both graphs of Fig. 7.19, the vector field points SW in the upper region and NE in the lower region. In the left graph of Fig. 7.19, with $K_y > K_x$, the vector field points NW in the middle region, so it appears that all off-axis trajectories converge to the fixed point $(0, K_y)$. In the right graph of Fig. 7.19, with $K_y < K_x$, the vector field points SE in the middle region, so it appears that all off-axis trajectories converge to the fixed point $(K_x, 0)$.

When we combine these observations, we see that whichever species has the larger carrying capacity appears to force the other to extinction, and then settles down to a population of constant size. Though plausible, this conclusion is not an obvious outcome for the system, but is a simple consequence of analysis of the nullclines.

Practice Problems

7.5.1. Sketch the nullclines and locate the fixed points of the system

$$x' = x - y^2 \qquad y' = y - x^3$$

7.5.2. Sketch the nullclines and locate the fixed points of the predator-prey equations with logistic prey limitations, given by (7.4) and (7.6). Do two examples, one with $K > a/b$ and one with $K < a/b$.

7.5.3. Recall the two competing species model (7.7).
(a) Explain why the growth factors a and d do not alter the positions of the fixed points.
(b) Suppose $K_x = K_y$. Locate the fixed points of the system.

Practice Problem Solutions

7.5.1. The x-nullcline is the curve $x = y^2$ (the solid gray curve in Fig. 7.20). The y-nullcline is the curve $y = x^3$ (the dashed gray curve in Fig. 7.20). The fixed points are the intersections of the nullclines, so

$$x = y^2 = (x^3)^2 = x^6 \quad \text{so} \quad 0 = x^6 - x = x(x^5 - 1)$$

This gives $x = 0$ and $x = 1$. (The equation $x^5 = 1$ has five solutions: $x = 1$ and two pairs of complex conjugate solutions. We are interested in only the real solution.) The fixed points are $(0, 0)$ and $(1, 1)$. See Fig. 7.20.

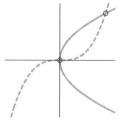

Figure 7.20.
Nullclines and fixed points for Practice Problem 7.5.1.

7.5.2. Given by Eq. (7.4), the x-nullcline is the pair of lines $x = 0$ and $y = a/b$, the solid gray lines in Fig. 7.21. Given by Eq. (7.6), the y-nullcline is the pair of dashed gray lines $y = 0$ and $x = (c/d)(1 - y/K)$.

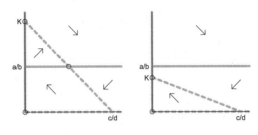

Figure 7.21. Nullclines in predator-prey systems with limited prey resources, Practice Problem 7.5.2.

On the left side of Fig. 7.21 we see the nullclines for the parameter values $a = d = b = 1$, $c = K = 2$. Note this system has three fixed points: $(0,0)$, $(0,K)$, and $(c(bK - a)/(dbK), a/b)$. The nullclines divide the phase plane into four regions. In the upper right region, the right-hand side of (7.4) is positive and the right-hand side of (7.6) is negative, so the vector field points SE in that region. Similarly, the vector field points SW in the lower right region, NW in the lower left region, and NE in the upper left region. This is consistent with either clockwise periodic trajectories around the off-axis fixed point, or trajectories that spiral clockwise into that fixed point.

On the right side of Fig. 7.21 we see the nullclines for the parameter values $a = d = b = 1$, $c = 2$, $K = 0.75$. This system has only two fixed points, $(0,0)$ and $(0,K)$. The nullclines divide the phase plane into three regions: upper, middle, and lower. The vector field points SE in the upper region, SW in the middle region, and NW in the lower region. This suggests that all off-axis trajectories converge to the fixed point $(0,K)$. So it seems that adding a resource limitation to the prey forces the predators into extinction and the prey population settles down to a constant value.

7.5.3. (a) The fixed points occur at the intersections of the nullclines, and the growth factors a and d do not occur in the nullcline equations.
(b) If $K_x = K_y$, then the line $y = K_x - x$ is both an x-nullcline and a y-nullcline. Every point on this line is a fixed point.

Exercises

For these systems, sketch the nullclines, locate the fixed points on the nullcline graph, and find numerical values of the fixed points. For Exercises 7.5.4, 7.5.5, 7.5.7, 7.5.8, 7.5.9, and 7.5.10, in each region determined by the nullclines, determine whether the vector field vectors point NE, SE, SW, or NW.

7.5.1. $x' = y - yx^2$ $y' = x - xy^2$

7.5.2. $x' = y^2 - y + yx^2$ $y' = x^2 - x + xy^2$

7.5.3. $x' = 1 - x^2 - y^2 \qquad y' = x^2 - y^2$

7.5.4. $x' = (x-1)^2 + y^2 - 1 \qquad y' = x^2 + (y-1)^2 - 1$

7.5.5. $x' = y - x^2 + 1 \qquad y' = y + x^2 - 1$

7.5.6. $x' = x^2 + y^2 - 1 \qquad y' = y + x^2 - 1$

7.5.7. $x' = x + y^2 - 1 \qquad y' = x - y^2 + 1$

7.5.8. $x' = y - \sin(x) \qquad y' = y - x^2$

7.5.9. $x' = y + x^3 \qquad y' = y - x$

7.5.10. $x' = x^2 + y^2 - 1 \qquad y' = x^2 + y^2 - 4$

7.6 THE FITZHUGH-NAGUMO EQUATIONS

The Hodgkin-Huxley equations are the first comprehensive mathematical model for nerve impulse transmission. This is a 4-dimensional system; we'll discuss it in Sect. 13.6. In 1960 Richard Fitzhugh [39] analyzed some properties of solutions of the Hodgkin-Huxley equations by reducing it to 2 dimensions. This was a sensible approach, because two of the four variables on the Hodgkin-Huxley equations change much more slowly than the other two. So he held the slow variables constant at their equilibrium values and study the two fast variables. While this did reproduce some of the behaviors of the full Hodgkin-Huxley equations, it missed some important features, trains of impulses, for example.

We should menton that all of Fitzhugh's simulations were performed on electronic analog computers. In the late 1950s and early 1960s, digital computers were huge, slow, and difficult to use. If you're unfamiliar with analog computers, Google can give you an illuminating hour or so.

In 1961 Fitzhugh [40] developed another 2-dimensional model that captured more of the Hodgkin-Huxley dynamics. In 1962 J. Nagumo, S. Arimoto, and S. Yoshizawa [99] designed an electronic circuit that exhibits the same dynamics as the model now named for both Fitzhugh and Nagumo.

Here's how Fitzhugh derived his model. Rather than simplify the Hodgkin-Huxley equations, he thought of them as one member of a family of nonlinear differential equations that exhibit certain types of complex dynamics, then tried to find a 2-dimensional system with similar dynamics. He began with the van der Pol oscillator

$$x'' + c(x^2 - 1)x' + x = 0 \tag{7.13}$$

proposed and studied in the 1920s by the physicist Balthasar van der Pol. This equation is known to exhibit some of the dynamics seen in the Hodgkin-Huxley equations, so Fitzhugh took this for a starting point. Then Fitzhugh converted the van der Pol equation into a Lienard system (Sect. 9.8) and obtained the equivalent system

$$x' = c(y + x - x^3/3) \qquad \text{and} \qquad y' = -x/c \tag{7.14}$$

We'll derive this in Sect. 9.8. The final modifications are these

$$x' = c(y + x - x^3/3 + z) \qquad \text{and} \qquad y' = -(x - a + by)/c \tag{7.15}$$

Here z is the imposed current. This model is in a *current clamp* situation, that is, the imposed current z can be set however we wish–a step function or a rectangular pulse, for example–and the variations in x and y won't influence z.

For $z = 0$ the nullclines are

$$x\text{-nullcline: } y = -x + x^3/3 \qquad y\text{-nullcline: } y = -x/b + a/b \tag{7.16}$$

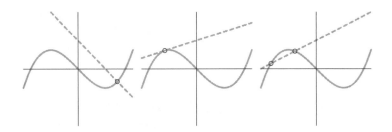

Figure 7.22. Nullclines for the Fitzhugh-Nagumo equations (7.15).

In Fig. 7.22 we see plots of Eqs. 7.16, the x- and y-nullclines of Eq. (7.15) for $a = b = 1$ (left), $a = b = -3.221954$ (middle), and $a = b = -2$ (right). In all these we've taken $a = b$ so the y-nullcline intersects the y-axis at $y = 1$. The system with nullclines in the left image has one fixed point; the middle has two. Hold on, you may say, I see only one fixed point. But for large enough x, the cubic curve grows faster than any (non-vertical) line and so the nullclines intersect at a point to the right of the image. In the right image, the nullclines intersect in three points. As far as the number of fixed points is concerned, these are all the possibilities.

In Figs. 7.23–7.26 we plot the population curves $x(t)$ (solid) and $y(t)$ (dashed) on the left and a trajectory in the xy plane on the right.

Figure 7.23. Population curves and a trajectory for Eq. (7.15) for $a = 0.8, b = 0.35$, and $c = 1$. The x population curve is solid; the y curve is dashed.

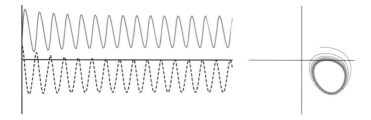

Figure 7.24. Population curves and a trajectory for $a = 0.8, b = 0.15$, and $c = 1$.

For the parameters $a = 0.8, b = 0.35$, and $c = 1$ of Fig. 7.23 the x- and y-nullclines intersect in a single point, so the Fitzhugh-Nagumo equations have a single fixed point. The population curves show oscillatory convergence to constant values; the phase plane shows a trajectory that spirals down to a fixed point.

For the parameters $a = 0.8, b = 0.15$, and $c = 1$ of Fig. 7.24 the x- and y-nullclines intersect in a single point, so again the Fitzhugh-Nagumo equations have a single fixed point. Unlike in Fig. 7.23, here the fixed point appears to be unstable and instead the population curves are roughly sine curves, though out of phase: the local maxima of the y-curve are a bit to the right of the local minima of the x-curve. The curves have similar shapes, and the trajectory is roughly a circle.

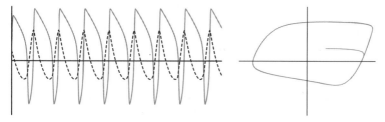

Figure 7.25. Population curves and a trajectory for $a = 0.8, b = 0.15$, and $c = 3$.

For the parameters $a = 0.8, b = 0.15$, and $c = 3$ of Fig. 7.25 the x- and y-nullclines intersect in a single point, so again the Fitzhugh-Nagumo equations have a single fixed point. Like in Fig. 7.24, the fixed point appears to be unstable and the population curves are periodic, though not symmetrical. Both the x and y population curves can be described as trains of impulses. The trajectory still is periodic, but is very far from a circle. Also note that the convergence to this periodic trajectory appears to be much more rapid than that in Fig. 7.24.

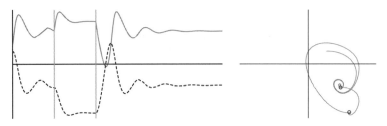

Figure 7.26. Population curves and a trajectory for $a = 0.8, b = 0.35$, and $c = 1$, with a pulse of $z = 1$ between $t = 10$ and $t = 20$, these times signaled by vertical gray lines.

In Fig. 7.26 we've returned the parameters to those of Fig. 7.23 where we saw convergence to a fixed point. In this case, at $t = 10$ we change z from 0 to 1, and at $t = 20$ we change z back to 0. In the population curve graphs, these times are marked by vertical gray lines. In the phase plane plot, the higher small circle shows that by $t = 10$ the trajectory is near the fixed point. When $z = 1$, the x-nullcline is $y = -x + x^3/3 - 1$ and so the fixed point shifts down by 1 and to the right by 0.15. Here's why: the y-nullcline has slope $-1/b$, so $-b\Delta y = \Delta x$. The lower small circle, marking $t = 20$, is near this shifted fixed point. After z returns to 0, the trajectory returns to the unshifted fixed point.

In Sect. 9.2 we'll determine the stability of these fixed points. The observed periodic trajectory we'll treat in Chapter 9. For now, though, we are content to observe that Fitzhugh achieved his goal: a 2-dimensional system that exhibits much of the rich dynamics of a neuron.

Practice Problems

In these practice problems we'll study a variation of the Fitzhugh-Nagumo equations

$$x' = -x^3 + (a+1)x^2 - ax - y + z \qquad \text{and} \qquad y' = (x - by)/c \qquad (7.17)$$

Some signs and coefficients differ from those of Eq. (7.15). Think of this as just another member of the family of differential equations that Fitzhugh envisioned.

For $z = 0$ the nullclines are

$$x\text{-nullcline: } y = -x^3 + (a+1)x^2 - ax \qquad y\text{-nullcline: } y = x/b \qquad (7.18)$$

Then the fixed points are the solutions of

$$x/b = -x^3 + (a+1)x^2 - ax$$

That is,

$$0 = -x^3 + (a+1)x^2 - (a+1/b)x = -x(x^2 - (a+1)x + (a+1/b))$$

So the solutions are

$$x_0 = 0, \quad x_\pm = \frac{a+1 \pm \sqrt{(a-1)^2 - 4/b}}{2}$$

The number of fixed points is 1, 2, or 3 according as $(a-1)^2 - 4/b < 0$, $(a-1)^2 - 4/b = 0$, or $(a-1)^2 - 4/b > 0$. That is, the number of fixed points is

$$1 \text{ if } b < 4/(a-1)^2, \quad 2 \text{ if } b = 4/(a-1)^2, \quad \text{and } 3 \text{ if } b > 4/(a-1)^2 \quad (7.19)$$

For these practice problems, take $a = 1/2$ and $b = 12$ in Eq. (7.17).

7.6.1. Show the Fitzhugh-Nagumo equations (7.17) have a single fixed point.

7.6.2. Show the nullclines divide the phase plane into four regions. Find the general direction (NE, SE, SW, NW) of the vector field in each region.

7.6.3. Use the result of Practice Problem 7.6.2 to sketch the solution curve for initial values starting along the x axis, for $0 < x < 1$.

Practice Problem Solutions

7.6.1. Because $4/(1-a)^2 = 16 > 12 = b$, for these parameter values the Fitzhugh-Nagumo equations (7.17) have one fixed point by the criterion Eq. (7.19).

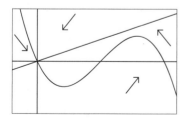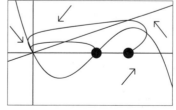

Figure 7.27. Left: vector field directions for Practice Problem 7.6.2. Right: some solution curves for Practice problem 7.6.3.

7.6.2. The y-nullcline, a straight line, divides the phase plane into two regions, above the line and below the line. Above the line, $x - by < 0$ so the vector field points downward. Below the line, $x - by > 0$ so the vector field points upward.

The x-nullcline, a cubic curve, divides the phase plane into two regions, above the curve and below the curve. Above the curve, $-x^3 + (a+1)x^2 - ax - y < 0$ so the vector field points to the left. Below the curve, $-x^3 + (a+1)x^2 - ax - y > 0$ so the vector field points to the right. Combining this information, we find the vector field directions shown on the left side of Fig. 7.27.

7.6.3. Starting along the x-axis for $0 < x < a$, solution curves move NW, crossing the y-nullcline pointing W. Starting along the x-axis for $a < x < 1$, solution curves move NE, crossing the x-nullcline pointing N, then move NW, crossing the y-nullcline pointing W. After crossing the y-nullcline, solution curves move SW, either converging to the fixed point $(0,0)$, or crossing the x-nullcline pointing S, then moving SE to the fixed point. Or it is possible that the solutions spiral in to the fixed point. See the right side of Fig. 7.27.

Exercises

7.6.1. Repeat Practice Problems 7.6.1, 7.6.2, and 7.6.3, modified where appropriate, for the Fitzhugh-Nagumo equations (7.17) with $a = 1/2$ and $b = 20$.

7.6.2. Write the x-nullcline equation for the Fitzhugh-Nagumo equations (7.17) as $y = -x(x - a)(x - 1)$. Does the basic shape of this nullcline (decreasing, then increasing, then decreasing again) as x ranges from 0 to 1 change if the nullcline equation is replaced by
(a) $y = -(x^2)(x - a)(x - 1)$,
(b) $y = -x(x^2 - a)(x - 1)$,
(c) $y = -x(x - a)(x^2 - 1)$, or
(d) $y = -x(x - a)(x - 1)^2$

7.6.3. (a) Plot the trajectory for Eq. (7.17) with $a = -0.15$, $b = 0.2$, $c = 0.5$, $z(t) = 0$, with $x(0) = 0.6$ and $y(0) = 0.4$. Take $0 \le t \le 100$. Describe the trajectory.
(b) Change a to $a = -0.35$. How does the trajectory change?

7.6.4. (a) Plot the trajectory for Eq. (7.17) with $a = -0.4$, $b = 0.2$, $c = 0.5$, $z(t) = 0$, with $x(0) = 0.1$ and $y(0) = 0.0$. Take $0 \le t \le 100$. Set MaxSteps to

2000. Describe the trajectory.

(b) Change $x(0)$ to 0.01. Describe the trajectory. Compare this with your answer for (a).

The remaining problems deal with the Fitzhugh-Nagumo equations (7.15). The Mathematica code to graph solutions of the Fitzhugh-Nagumo equations is in Sect. B.9.

7.6.5. For the Fitzhugh-Nagumo equations (7.15) with $a = 1$, $b = 0.05$, $c = 1$, $z(t) = 0$, and $x(0) = 0.6$ and $y(0) = 0.4$. Take $0 \le t \le 100$.
(a) Plot the population curves $x(t)$ and $y(t)$.
(b) Plot the trajectory $(x(t), y(t))$.
(c) Describe the long-term behavior of the system.

7.6.6. (a) Repeat Exercise 7.6.5 with b changed to 0.01.
(b) Set MaxSteps to 2000 and plot the population curves and trajectory for $0 \le t \le 200$.
(c) Now change b to -0.01 and repeat the steps of (b).

7.6.7. (a) Plot the population curves and the trajectory for Eq. (7.17) with $a = 1.0$, $b = -0.01$, $c = 2$, $z(t) = 0$, with $x(0) = 0.6$ and $y(0) = 0.4$. Set MaxSteps to 2000 and take $0 \le t \le 200$. Describe the population curves and the trajectory.
(b) Change c to 3 and repeat the steps of (a).

7.6.8. (a) Plot the population curves and the trajectory for Eq. (7.17) with $a = 0.95$, $b = -0.01$, $c = 3$, $z(t) = 0$, with $x(0) = 0.6$ and $y(0) = 0.4$. Set MaxSteps to 5000 and take $0 \le t \le 200$. Describe the population curves and the trajectory.
(b) Now set $z(t) = 1$. Plot and describe the population curves and the trajectory. Compare these plots to those of (a).
(c) Now set $z(t) = 0$ for $0 \le t < 10$, $z(t) = 1$ for $10 \le t < 100$, and then $z(t) = 0$ for $100 \le t \le 200$. Compare these plots to those of (a) and (b).

7.6.9. (a) Plot the population curves and the trajectory for Eq. (7.17) with $a = 0.85$, $b = 0.21$, $c = 3$, $z(t) = 0$, with $x(0) = 0.6$ and $y(0) = 0.4$. Set MaxSteps to 5000 and take $0 \le t \le 200$. Describe the population curves and the trajectory.
(b) Repeat (a) with $x(0) = 1.036$ and $y(0) = -0.6754$.
(c) Repeat (b) for $0 \le t \le 400$. Compare the trajectories of (b) and (c). Interpret the plots.
(d) Change b from 0.21 to 0.22. Does this modify your interpretation of (c)?

7.6.10. (a) Plot the population curves and the trajectory for Eq. (7.17) with $a = 1.1$, $b = -0.1$, $c = 1$, $z(t) = 0$, with $x(0) = 0.6$ and $y(0) = 0.4$. Set MaxSteps to 5000 and take $0 \leq t \leq 200$. Describe the population curves and the trajectory.
(b) Change a to 1.2 and interpret the plots.
(c) Explore some parameters. Can you find a long-term behavior other than convergence to a fixed point or convergence to a periodic trajectory?

Chapter 8 Linear systems and stability

A linear differential equation in the plane is a system of the form

$$dx/dt = ax + by \qquad dy/dt = cx + dy \qquad (8.1)$$

where a, b, c, and d are constants. That is, $f(x,y) = ax + by$ and $g(x,y) = cx + dy$. The origin is a fixed point for all linear systems, the only fixed point for most. In Sect. 8.2 we determine the stability of the origin by solving the differential equations and analyzing the behavior of the solutions. Most models of biological systems are nonlinear, and solving nonlinear differential equations can be much, much more difficult than is solving linear differential equations. So in Sects. 8.3–8.6 we'll develop another method, computing the eigenvalues of a matrix associated with a linear differential equation and determining the stability of the origin by the eigenvalues. If the last sentence contains unfamiliar words, don't worry. We'll describe matrices in Sect. 8.3, show how to compute eigenvalues and eigenvectors (after we say what they are) in Sect. 8.4, and in Sect. 8.5 show how the eigenvalues determine the stability of the origin. In the final section, Sect. 8.6, we present a shortcut that is especially useful for families of differential equations that depend on parameters.

We need a bit of time to develop these ideas, but they are worth the effort because of how very simple they make the problem of determining stability of fixed points.

8.1 SUPERPOSITION OF SOLUTIONS

Have you ever stood on a low bridge over a small, still pond and dropped two pebbles a couple of feet apart into the water? Dropping one pebble produces a sequence of concentric circular waves moving outward from where the pebble hit the water. Two pebbles produce two sequences of concentric circular waves which, when they meet, pass right through one another like a pair of ghosts in a Victorian novel.

Or you're listening to music (speakers, not earbuds) when the person sitting next to you asks you a question. You can hear the question and still hear the music. The sound waves pass right through one another.

These are examples of superposition. Here's why it works for solutions of linear differential equations. We'll do the calculation for 2-dimensional equations, but this generalizes in the obvious way to any dimension.

Suppose $(x_1(t), y_1(t))$ and $(x_2(t), y_2(t))$ are solutions of Eq. (8.1). Then for any constants k_1 and k_2,

$$(x_3(t), y_3(t)) = k_1(x_1(t), y_1(t)) + k_2(x_2(t), y_2(t))$$
$$= (k_1 x_1(t) + k_2 x_2(t), k_1 y_1(t) + k_2 y_2(t))$$

called a *linear combination* of $(x_1(t), y_1(t))$ and $(x_2(t), y_2(t))$, also is a solution of Eq. (8.1). This is called *linear superposition*; the proof of its validity is a simple calculation: (x_3', y_3')

$$= ((k_1 x_1 + k_2 x_2)', (k_1 y_1 + k_2 y_2)') = (k_1 x_1' + k_2 x_2', k_1 y_1' + k_2 y_2')$$
$$= (k_1(a x_1 + b y_1) + k_2(a x_2 + b y_2), k_1(c x_1 + d y_1) + k_2(c x_2 + d y_2))$$
$$= (a(k_1 x_1 + k_2 x_2) + b(k_1 y_1 + k_2 y_2), c(k_1 x_1 + k_2 x_2) + d(k_1 y_1 + k_2 y_2))$$
$$= (a x_3 + b y_3, c x_3 + d y_3)$$

Perhaps this argument has too many constants and variables. An example may clarify the calculation.

Example 8.1.1. *An example of linear superposition.* First, observe that both

$$(x, y) = (\cos(t), \sin(t)) \quad \text{and} \quad (x, y) = (-\sin(t), \cos(t))$$

are solutions of

$$x' = -y \quad y' = x \tag{8.2}$$

Then

$$(x, y) = 2(\cos(t), \sin(t)) - 3(-\sin(t), \cos(t))$$
$$= (2\cos(t) + 3\sin(t), 2\sin(t) - 3\cos(t))$$

also is a solution of Eq. (8.2):

$$x' = -2\sin(t) + 3\cos(t) = -(2\sin(t) - 3\cos(t)) = -y$$
$$y' = 2\cos(t) - 3(-\sin(t)) = 2\cos(t) + 3\sin(t) = x$$

That is, this linear combination of solutions of Eq. (8.2) again is a solution of Eq. (8.2). $\square$

In general, superposition fails for nonlinear differential equations. When it can be applied, superposition is a great simplifying force, suggesting that all solutions are linear combinations of a small number of simple solutions. On the other hand, this reinforces the reductionist approach to science: break a problem into small bits, solve each bit, and paste those solutions together to find a solution of the original problem. Sometimes this works just fine, sometimes not so much. We must be careful to avoid overzealous application of reductionism, especially in the presence of nonlinearities.

8.2 TYPES OF FIXED POINTS

While some linear systems have whole lines of fixed points (recall Practice Problem Solution 7.5.3, for example), most have a single fixed point and the asymptotic behavior of the system falls into one of a small number of categories. In this section we give examples of each category. In some particularly simple cases we can explain the asymptotic behavior now. The others will be analyzed in Sect. 8.5.

Example 8.2.1. *An asymptotically stable proper node.* The upper left graph of Fig. 8.1 is the vector field and some trajectories for the system

$$x' = -x \qquad y' = -y$$

Certainly, the only fixed point is the origin. Because this system is *uncoupled*, that is, x' depends on only x and y' depends on only y, it is easy to solve: apply the method of separation of variables (Sect. 4.6) to x and y individually. For example,

$$dx/dt = -x \quad \text{so} \quad dx/x = -t \quad \text{and} \quad \ln|x| = -t + c$$

Exponentiate to solve for x and substitute in $t = 0$ to evaluate the integration constant c. This gives

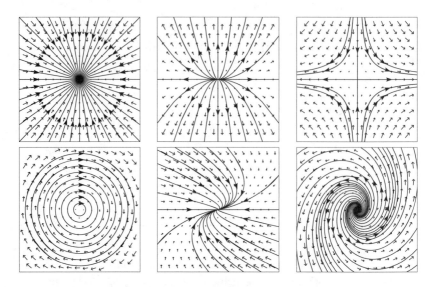

Figure 8.1. Left to right, top to bottom: vector fields and some sample trajectories for the linear systems of Examples 8.2.1 through 8.2.6.

$$x(t) = x(0)e^{-t}$$

Then the solutions of the system are

$$x(t) = x(0)e^{-t} \quad \text{and} \quad y(t) = y(0)e^{-t}$$

and so long as $x(0) \neq 0$,

$$y(t) = \frac{y(0)}{x(0)}x(t)$$

That is, the trajectories lie on lines $y = mx$, where the slope $m = y(0)/x(0)$, and with $(x(t), y(t)) \to (0,0)$ as $t \to \infty$. If $x(0) = 0$, the trajectory stays along the y-axis. $\square$

This solution has several properties common to many systems. The origin is an *asymptotically stable* fixed point because $(x(t), y(t)) \to (0,0)$ as $t \to \infty$. The origin is a *node* because the trajectories approach the origin and do not spiral around it as in Example 8.2.6. The origin is a *proper* node because trajectories approach the origin along all directions.

Example 8.2.2. *An unstable improper node.* The upper middle graph of Fig. 8.1 is the vector field and some trajectories for the system

$$x' = x \qquad y' = 2y$$

Here the solutions are $x(t) = x(0)e^t$ and $y(t) = y(0)e^{2t} = y(0)(e^t)^2$. Consequently, so long as $x(0) \neq 0$,

$$y(t) = \frac{y(0)}{x(0)^2}x(t)^2$$

That is, the trajectories lie along the family of parabolas, $y = mx^2$. Because $\sqrt{x(t)^2 + y(t)^2} \to \infty$ as $t \to \infty$, the origin is an *unstable* fixed point. This fixed point is an *improper node* because all but two of the trajectories, those with $(x(0), y(0)) = (0, k)$ for $k > 0$ and for $k < 0$, are tangent at the origin. $\square$

Example 8.2.3. *A (necessarily unstable) saddle point.* The upper right graph of Fig. 8.1 is the vector field and some trajectories for the system

$$x' = x \qquad y' = -y$$

The solutions are

$$x(t) = x(0)e^t \quad \text{and} \quad y(t) = y(0)e^{-t}$$

Combining these, we see that if $x(0) \neq 0$,

$$y(t) = y(0)(e^t)^{-1} = \frac{y(0)x(0)}{x(t)}$$

That is, the trajectories lie on hyperbolas $y = m/x$, or along the y-axis if $x(0) = 0$. The trajectories lying on the y-axis approach the origin as $t \to \infty$. For all other trajectories, as $t \to \infty$, $y(t) \to 0$ and $x(t) \to \pm\infty$. These trajectories move away from the origin, so the origin is *unstable*. Trajectories from initial points $(x(0), y(0))$ with $|y(0)| \gg |x(0)|$ move first toward the origin, then away, suggesting the path of a marble rolling on a saddle. For this reason, the origin is called a *saddle point*. $\square$

Example 8.2.4. *A center.* The lower left graph of Fig. 8.1 is the vector field and some trajectories for the system

$$x' = y \qquad y' = -x$$

This system is coupled, that is, x' depends on y and y' depends on x, so we cannot use separation of variables. However, this is a particularly simple coupled system, and another observation will lead us to the solution. Note

$$x'' = (x')' = y' = -x \quad \text{and} \quad y'' = (y')' = (-x)' = -y$$

The functions $\cos(t)$ and $\sin(t)$ satisfy these equations. Because this system is linear, the principle of linear superposition (Sect. 8.1) can be applied and so every combination of solutions is a solution; we'll take

$$x(t) = x(0)\cos(t) \quad \text{and} \quad y(t) = y(0)\sin(t)$$

For this system, the origin is called a *center*, a sensible name, because trajectories circle around the origin. Trajectories do not converge to the origin, but also they do not diverge to infinity. We'll call this a *weakly stable* fixed point. □

Example 8.2.5. *An asymptotically stable improper node.* The lower middle graph of Fig. 8.1 is the vector field and some solution curves for the system

$$x' = -x + y \qquad y' = -y$$

This is a bit trickier. The y' equation we can solve: $y(t) = y(0)e^{-t}$. What about the x' equation? Knowing y, it becomes

$$x' = -x + y(0)e^{-t}$$

Suppose we try

$$x(t) = x(0)e^{-t} + y(0)e^{-t}$$

Then

$$x'(t) = -x(0)e^{-t} - y(0)e^{-t} = -x(t)$$

Upon differentiating x, we lost the extra $y(0)e^{-t}$ term. What could we add to x that upon differentiating returns both $-x$ and $y(0)e^{-t}$? What about adding $ty(0)e^{-t}$? Then

$$x' = (x(0)e^{-t} + ty(0)e^{-t})' = -x(0)e^{-t} - ty(0)e^{-t} + y(0)e^{-t} = -x(t) + y(t)$$

We see the origin is a node, improper because all the trajectories are tangent to a single line (here it's the x-axis) at the origin. As $t \to \infty$, all trajectories $(x(t), y(t)) \to (0,0)$, so the origin is an asymptotically stable fixed point. □

If you think this technique is a bit *ad hoc*, we present a more algorithmic approach, the method of integrating factors, in Sect. A.1. Because we're interested in the long-term behavior near the origin, which we'll see how to determine in Sect. 8.5, we won't pursue exact solutions here. However, in Practice Problem 8.2.4 we'll present a minor variation of this example.

Example 8.2.6. *An unstable spiral.* The lower right graph of Fig. 8.1 is the vector field and some solution curves for the system

$$x' = x - 2y \qquad y' = 3x + y$$

This is trickier still, requiring an excursion into exponentials of complex numbers. But here's a glimpse. If $x > 0$ isn't too large and $y > 0$ isn't too small, then as y increases, x decreases, the beginning of a counterclockwise turn around the origin. Because the eigenvalue approach we'll see in Sect. 8.5

is straightforward, after some background is established, we won't pursue this intuition. In Sect. 8.5 we'll see that the origin is a *spiral*, an unstable fixed point in this case because as $t \to \infty$, the trajectory $(x(t), y(t))$ spirals out toward infinity. $\square$

Practice Problems

Describe the trajectories of these systems of differential equations.

8.2.1. $x' = 2x$ $\quad$ $y' = 3y$

8.2.2. $x' = -x$ $\quad$ $y' = 2y$

8.2.3. $x' = 2y$ $\quad$ $y' = -3x$

8.2.4. $x' = 2x$ $\quad$ $y' = 2y - x$

Practice Problem Solutions

8.2.1. From $x' = 2x$ we find $x(t) = x(0)e^{2t}$ and from $y' = 3y$ we find $y(t) = y(0)e^{3t}$. So long as $x(0) \neq 0$, solving the $x(t)$ equation for e^t gives $(x(t)/x(0))^{1/2} = e^t$. Substituting into the $y(t)$ equation gives

$$y(t) = \frac{y(0)}{x(0)^{3/2}}x(t)^{3/2}$$

These are semi-cubical parabolas. If $x(0) = 0$, then all $x(t) = 0$. The origin is an unstable fixed point.

8.2.2. From $x' = -x$ we find $x(t) = x(0)e^{-t}$ and from $y' = 2y$ we find $y(t) = y(0)e^{2t}$. So long as $x(0) \neq 0$, solving the $x(t)$ equation for e^t gives $(x(0)/x(t)) = e^t$. Substituting into the $y(t)$ equation gives

$$y(t) = \frac{y(0)x(0)^2}{x(t)^2}$$

and of course if $x(0) = 0$, then $x(t) = 0$ for all t. The origin is a saddle point, an unstable fixed point.

8.2.3. We'll Follow Example 8.2.4 and find

$$x'' = (x')' = (2y)' = 2(-3x) = -6x \quad \text{and} \quad y'' = (y')' = (-3x)' = -6y$$

In Sect. 8.5 we find the most general solution of this system. For the moment, suppose the solution has the form

$$x(t) = x(0)\cos(\sqrt{6}t) \quad \text{and} \quad y(t) = y(0)\sin(\sqrt{6}t)$$

The $\sqrt{6}$ appears because two differentiations, each with an application of the chain rule, produce a factor of 6. We see the trajectories are ellipses. The origin is a weakly stable fixed point.

8.2.4. From $x' = 2x$ we get $x(t) = x(0)e^{2t}$. Then $y' = -x + 2y$ becomes $y' = -x(0)e^{2t} + 2y$, and guided by Example 8.2.5 we'll guess

$$y(t) = -tx(0)e^{2t} + y(0)e^{2t}$$

Now let's check:

$$y' = (-tx(0)e^{2t} + y(0)e^{2t})' = -2tx(0)e^{2t} - x(0)e^{2t} + 2y(0)e^{2t}$$
$$-x + 2y = -x(0)e^{2t} + 2(-tx(0)e^{2t} + y(0)e^{2t})$$

Looks good. The origin is an unstable fixed point.

While this approach won't work in every situation, it will for every system of the form

$$x' = ax + by, \ y' = ay \quad \text{or} \quad x' = ax, \ y' = ay + bx$$

We'll see how to solve the more general case in Sect. A.1.

Exercises

For Probs. 8.2.1–8.2.9 find the solutions and determine the type (saddle, center, etc.) and stability of the fixed point at the origin.

8.2.1. (a) $x' = -2x$, $y' = -3y$ (b) $x' = 2x$, $y' = 5y$

8.2.2. (a) $x' = -3x$, $y' = 3y$ (b) $x' = -3x$, $y' = 4y$

8.2.3. (a) $x' = 2x$, $y' = 5y$ (b) $x' = -x/2$, $y' = y/4$

8.2.4. (a) $x' = -2x$, $y' = 5y$ (b) $x' = x/3, y' = 2y$

8.2.5. (a) $x' = 2x$, $y' = 3y$, $z' = 4z$ (b) $x' = -2x$, $y' = -3y$, $z' = -4z$

8.2.6. (a) $x' = 2x$, $y' = 3y$, $z' = -4z$ (b) $x' = 2x$, $y' = -3y$, $z' = -4z$

8.2.7. (a) $x' = -x$, $y' = 2y$, $z' = -2z$ (b) $x' = 2x$, $y' = -2y, z' = 2z$

8.2.8. (a) $x' = -2x + y$, $y' = -2y$ (b) $x' = 2x - 3y$, $y' = 2y$

8.2.9. (a) $x' = 2x$, $y' = 2y + 5x$ (b) $x' = 2x - y + 3z$, $y' = 2y$, $z' = 2z$

8.2.10. Find the solutions of $x' = kx$, $y' = ky$. For all constants $k \neq 0$. Does the type of fixed point at the origin depend on the sign of k?

8.3 THE MATRIX FORMALISM

With some linear algebra (see Sect. A.7) the system (8.1) can be written as the product of a matrix and a vector

$$\begin{bmatrix} dx/dt \\ dy/dt \end{bmatrix} = \begin{bmatrix} a & b \\ c & d \end{bmatrix} \begin{bmatrix} x \\ y \end{bmatrix} \tag{8.3}$$

where a matrix M and a vector $\vec{v}$,

$$M = \begin{bmatrix} a & b \\ c & d \end{bmatrix} \quad \text{and} \quad \vec{v} = \begin{bmatrix} x \\ y \end{bmatrix}$$

are multiplied by

$$\begin{bmatrix} a & b \\ c & d \end{bmatrix} \begin{bmatrix} x \\ y \end{bmatrix} = \begin{bmatrix} ax + by \\ cx + dy \end{bmatrix}$$

More compactly, Eq. (8.3) can be written as

$$\frac{d\vec{v}}{dt} = M\vec{v} \tag{8.4}$$

The matrix M is called a 2×2 matrix because it has 2 rows and 2 columns. The vector $\vec{v}$ is a 2×1 matrix. In order to be able to multiply matrices AB, the number of columns of A must equal the number of rows of B. For example, (8.5) illustrates how to multiply 2×2 matrices.

$$\begin{bmatrix} a & b \\ c & d \end{bmatrix} \begin{bmatrix} w & x \\ y & z \end{bmatrix} = \begin{bmatrix} aw + by & ax + bz \\ cw + dy & cx + dz \end{bmatrix} \tag{8.5}$$

What happens if these matrices are multiplied in the other order? Try an example, rather than a general calculation. The order doesn't matter when multiplying numbers, but order does matter for some operations, "putting on your socks" and "putting on your shoes," for example. The situation is a bit more complicated for matrix multiplication: for some pairs of matrices, the order matters, for some it doesn't.

Other matrix arithmetic

- To add matrices of the same size, add the corresponding entries.

$$\begin{bmatrix} 1 & 2 \\ 3 & -1 \end{bmatrix} + \begin{bmatrix} 3 & 2 \\ -3 & -4 \end{bmatrix} = \begin{bmatrix} 1+3 & 2+2 \\ 3-3 & -1-4 \end{bmatrix} = \begin{bmatrix} 4 & 4 \\ 0 & -5 \end{bmatrix}$$

- To multiply a matrix by a number (scalar multiplication), multiply every matrix entry by that number.

$$3\begin{bmatrix} 2 & -1 \\ -3 & 5 \end{bmatrix} = \begin{bmatrix} 3 \cdot 2 & 3 \cdot (-1) \\ 3 \cdot (-3) & 3 \cdot 5 \end{bmatrix} = \begin{bmatrix} 6 & -3 \\ -9 & 15 \end{bmatrix}$$

- To subtract matrices, multiply a matrix by -1 then add the matrices, or equivalently subtract corresponding matrix entries

- Matrix division is trickier, and not always possible. We describe how to do this in Sect. 8.4.

If real numbers are the only arithmetic you know, the arithmetic of matrices may seem strange. If you continue your studies of math, you'll find arithmetics more surprising still.

Exercises

For these problems use these matrices.

$$A = \begin{bmatrix} 1 & 2 \\ -1 & 3 \end{bmatrix} \quad B = \begin{bmatrix} -2 & 1 \\ 3 & 2 \end{bmatrix} \quad C = \begin{bmatrix} 2 & 3 \end{bmatrix} \quad D = \begin{bmatrix} 2 \\ 3 \end{bmatrix}$$

$$E = \begin{bmatrix} 1 & 2 & 1 \\ -1 & 3 & 0 \\ 2 & 1 & 2 \end{bmatrix} \quad F = \begin{bmatrix} -2 & 1 & -1 \\ 2 & 3 & 2 \\ 3 & 0 & -2 \end{bmatrix} \quad G = \begin{bmatrix} 2 & 3 & -2 \end{bmatrix} \quad H = \begin{bmatrix} 2 \\ 3 \\ -1 \end{bmatrix}$$

8.3.1. Compute these matrix products, or explain why the product cannot be computed.
(a) AB (b) BA (c) AD (d) AC (e) CB (f) DB (g) CD (h) DC

8.3.2. Compute these matrix products, or explain why the product cannot be computed.
(a) EF (b) FE (c) EE (d) EH (e) FG (f) GH (g) HG (h) BE

8.3.3. Calculate $A + B$, $B + A$, $E + F$, and $F + E$. While matrix multiplication depends on the order of the matrices, does matrix addition depend on the order of the matrices?

8.3.4. Let I denote the 2×2 matrix DC, and let J denote the 3×3 matrix HG. Compute $A + I$ and $E + J$.

8.3.5. Compute $(AB)I$ and $A(BI)$. Compute $(EF)J$ and $E(FJ)$. Although it is not commutative, matrix multiplication is associative.

8.4 EIGENVALUES AND EIGENVECTORS

Sometimes a technique that solves a 1-dimensional problem can be adapted to solve at least some cases of the corresponding 2-dimensional problem. Likely higher dimensions introduce complications not present in lower dimensions, but still, expanding the simpler model is a good start. That's just what we'll do now.

Recall the differential equation $dx/dt = \lambda x$ has solution $x(t) = x(0)e^{\lambda t}$. We hypothesize that Eq. (8.4) has solution

$$\vec{v}(t) = \begin{bmatrix} x(t) \\ y(t) \end{bmatrix} = \begin{bmatrix} x(0)e^{\lambda t} \\ y(0)e^{\lambda t} \end{bmatrix} = \begin{bmatrix} x(0) \\ y(0) \end{bmatrix} e^{\lambda t}$$

How can we find λ?

With this hypothesis, $d\vec{v}/dt = \lambda \vec{v}$, so Eq. (8.4) becomes

$$M\vec{v} = \lambda \vec{v} \tag{8.6}$$

We say $\vec{v}$ is an *eigenvector* of M with *eigenvalue* λ, and Eq. (8.6) is called the *eigenvector equation*.

If $\vec{v}$ is the *zero vector*, that is, both entries of $\vec{v}$ are 0, then Eq. (8.6) holds for all matrices M and all constants λ. Certainly, this is not interesting, so we require that $\vec{v}$ not be the zero vector.

To find the λ for which there is an eigenvector, note that Eq. (8.6) can be written as

$$M\vec{v} - \lambda\vec{v} = \vec{0} = \begin{bmatrix} 0 \\ 0 \end{bmatrix} \tag{8.7}$$

We would like to factor out the $\vec{v}$, but M is a matrix and λ is a number. However, we can write $\lambda\vec{v}$ as a matrix times $\vec{v}$ in this way

$$\lambda\vec{v} = \lambda(I\vec{v}) = (\lambda I)\vec{v}$$

where I is the *identity matrix*,

$$I = \begin{bmatrix} 1 & 0 \\ 0 & 1 \end{bmatrix} \quad \text{and} \quad \lambda I = \begin{bmatrix} \lambda & 0 \\ 0 & \lambda \end{bmatrix}$$

The matrix I is called the identity matrix because

$$MI = IM = M$$

Use Eq. (8.5) to verify this.

Now Eq. (8.7) can be written as

$$(M - \lambda I)\vec{v} = \vec{0} \qquad (8.8)$$

The final step we need is that the *inverse* M^{-1} of the matrix

$$M = \begin{bmatrix} a & b \\ c & d \end{bmatrix}$$

is

$$M^{-1} = \frac{1}{\det(M)} \begin{bmatrix} d & -b \\ -c & a \end{bmatrix} \qquad (8.9)$$

where $\det(M) = ad - bc$, the *determinant* of M. Certainly, M^{-1} exists if and only if $\det(M) \neq 0$. The matrix M^{-1} is called the inverse of M because

$$MM^{-1} = M^{-1}M = I$$

For example, take

$$M = \begin{bmatrix} 1 & 2 \\ 3 & 4 \end{bmatrix}$$

Then $\det(M) = 1 \cdot 4 - 3 \cdot 2 = -2$ and

$$M^{-1} = \frac{1}{-2} \begin{bmatrix} 4 & -2 \\ -3 & 1 \end{bmatrix} = \begin{bmatrix} -2 & 1 \\ 3/2 & -1/2 \end{bmatrix}$$

Now if $\det(M - \lambda I) \neq 0$, then multiply both sides of Eq. (8.8) by $(M - \lambda I)^{-1}$. On the left side this gives

$$(M - \lambda I)^{-1}((M - \lambda I)\vec{v}) = ((M - \lambda I)^{-1}(M - \lambda I))\vec{v} = I\vec{v} = \vec{v}$$

and on the right side

$$(M - \lambda I)^{-1}\vec{0} = \vec{0}$$

That is, if $\det(M - \lambda I) \neq 0$, then $\vec{v} = \vec{0}$ is the only solution of Eq. (8.8). So in order for us to have a nonzero eigenvector (remember that we placed this requirement on eigenvectors), the eigenvalue λ must satisfy the *characteristic equation*

$$\det(M - \lambda I) = 0 \qquad (8.10)$$

Because it will simplify the computation of eigenvectors, we'll mention one more result, a consequence of Eq. (8.6).

Proposition 8.4.1. If $\vec{v}$ is an eigenvector of a matrix M, then for any non-zero constant k, $k\vec{v}$ is an eigenvector of M.

Proof. We know that $M\vec{v} = \lambda\vec{v}$; we want to show that $M(k\vec{v}) = \lambda k\vec{v}$. This is a straightforward calculation once we write M and $\vec{v}$ in components.

$$M(k\vec{v}) = \begin{bmatrix} a & b \\ c & d \end{bmatrix}\left(k\begin{bmatrix} v_1 \\ v_2 \end{bmatrix}\right) = \begin{bmatrix} a & b \\ c & d \end{bmatrix}\begin{bmatrix} kv_1 \\ kv_2 \end{bmatrix} = \begin{bmatrix} akv_1 + bkv_2 \\ ckv_1 + dkv_2 \end{bmatrix}$$

$$= k\begin{bmatrix} av_1 + bv_2 \\ cv_1 + dv_2 \end{bmatrix} = k\begin{bmatrix} a & b \\ c & d \end{bmatrix}\begin{bmatrix} v_1 \\ v_2 \end{bmatrix} = kM\vec{v} = k\lambda\vec{v}$$

Any (non-zero) multiple of an eigenvector is an eigenvector; an eigenvector determines a line of eigenvectors. The line passes through the origin, the only point of the line which is not an eigenvector. □

We have assembled a fair number of ideas; time for an example.

Example 8.4.1. *Solving a linear system of differential equations by finding eigenvalues and eigenvectors.* Consider Eq. (8.4) with

$$M = \begin{bmatrix} 1 & 1 \\ 4 & 1 \end{bmatrix}$$

The eigenvalues are the solutions of

$$0 = \det\begin{bmatrix} 1-\lambda & 1 \\ 4 & 1-\lambda \end{bmatrix} = (1-\lambda)^2 - 4 = \lambda^2 - 2\lambda - 3 = (\lambda - 3)(\lambda + 1)$$

so the eigenvalues are 3 and -1.

Now we find the eigenvectors for these eigenvalues. The equation

$$\begin{bmatrix} 1 & 1 \\ 4 & 1 \end{bmatrix}\begin{bmatrix} x \\ y \end{bmatrix} = 3\begin{bmatrix} x \\ y \end{bmatrix}$$

becomes the *component equations* of the eigenvector equation

$$\begin{array}{rl} x + y &= 3x \\ 4x + y &= 3y \end{array} \quad \text{that is,} \quad \begin{array}{rl} -2x + y &= 0 \\ 4x - 2y &= 0 \end{array}$$

It's clear that these two equations on the right are equivalent, and are satisfied by all numbers x and y related by $y = 2x$. Prop. 8.4.1 guarantees that something like this must happen. Either one component equation will be a multiple of the other, or one component equation will be tautologous, for example, $x = x$, an equation that contains no information about the variables. This is because if neither component equation is tautologous nor is one a multiple of the other, then the two component equations would determine only one value of x and one of y. That is, there would be only one eigenvector. This contradicts Prop. 8.4.1. Consequently, when finding an eigenvector for a 2×2 matrix M, if one component equation of $M\vec{v} = \lambda\vec{v}$ isn't tautologous, then that equation suffices to determine an eigenvector. In other words, we needn't check that the other component equation is a multiple of this one or is tautologous. As soon as we find one component equation that is not tautologous, that's all we need to determine an eigenvector. For higher-dimensional systems, in Chapter 13 we'll see that it's a bit more complicated. But for 2×2 matrices, one non-tautologous equation is all we need. In Sect. B.10 we'll do this by Mathematica.

To make it easier to compare our eigenvectors, if possible we'll solve for y in terms of x and take $x = 1$. When would this not be possible? If one of the component equations is $x = 0$. Otherwise, we can set $x = 1$.

By a similar argument we see that the first component equation of $M\vec{v} = -1\vec{v}$ is $x + y = -x$. This is non-tautologous, so it's all we need to find an eigenvector for $\lambda = -1$. Check that

$$\begin{bmatrix} x \\ y \end{bmatrix} = \begin{bmatrix} 1 \\ 2 \end{bmatrix} e^{3t} \quad \text{and} \quad \begin{bmatrix} x \\ y \end{bmatrix} = \begin{bmatrix} 1 \\ -2 \end{bmatrix} e^{-t}$$

are solutions of Eq. (8.4). The most general solution is a linear combination of these

$$\begin{bmatrix} x \\ y \end{bmatrix} = A \begin{bmatrix} 1 \\ 2 \end{bmatrix} e^{3t} + B \begin{bmatrix} 1 \\ -2 \end{bmatrix} e^{-t}$$

for constants A and B. That this is a solution follows from the superposition principle of linear systems, Sect. 8.1. $\square$

The case of a zero eigenvalue, $\lambda = 0$, we avoid because we are concerned only with isolated fixed points.

Proposition 8.4.2. If $\lambda = 0$ is an eigenvalue of M, then the line determined by an eigenvector $\vec{v}_0$ of $\lambda = 0$ consists of fixed points of Eq. (8.4).

Proof. At any point $k\vec{v}_0$, we have

$$\left. \frac{d\vec{v}}{dt} \right|_{k\vec{v}_0} = M(k\vec{v}_0) = kM\vec{v}_0 = k\lambda\vec{v}_0 = k0\vec{v}_0 = \vec{0}$$

That is, at every point on the line $k\vec{v}_0$, $d\vec{v}/dt = \vec{0}$, so every point on this line is a fixed point. $\square$

Because the characteristic equation is quadratic, eigenvalues can be complex. In Sect. 8.5 we present a way to deal with complex eigenvalues when they arise in the solutions of real differential equations. If the differential equations model the growth of two interacting populations, the interpretation of complex eigenvalues and eigenvectors might not be so clear. We'll see how to do this.

A final comment, a shortcut really. A matrix is called *upper triangular* if all the entries below the diagonal are 0, *lower triangular* if all the entries above the the diagonal are 0, and *triangular* if it is either upper triangular or lower triangular. Examples:

$$\begin{bmatrix} 1 & 2 \\ 0 & 4 \end{bmatrix} \text{ is upper triangular,} \qquad \begin{bmatrix} 1 & 0 \\ 3 & 4 \end{bmatrix} \text{ is lower triangular}$$

We mention these because the eigenvalues of a triangular matrix are its diagonal entries. For example,

$$0 = \det \begin{bmatrix} 1-\lambda & 2 \\ 0 & 4-\lambda \end{bmatrix} = (1-\lambda) \cdot (4-\lambda) - 0 \cdot 2 = (1-\lambda) \cdot (4-\lambda)$$

So the eigenvalues are 1 and 4. This is a useful trick.

Practice Problem

8.4.1. Find the eigenvalues and eigenvectors of these matrices

$$\text{(a)} \begin{bmatrix} 1 & 1 \\ 0 & 2 \end{bmatrix} \qquad \text{(b)} \begin{bmatrix} 2 & 3 \\ 4 & 0 \end{bmatrix} \qquad \text{(c)} \begin{bmatrix} 1 & 1 \\ -1 & 1 \end{bmatrix}$$

Practice Problem Solution

8.4.1. (a) The characteristic equation, Eq. (8.10), is

$$0 = (1-\lambda) \cdot (2-\lambda) - 0 \cdot 2$$

so the eigenvalues are $1, 2$.

For $\lambda = 1$, the first component equation is $1 \cdot x + 1 \cdot y = 1 \cdot x$. This is not tautologous so we needn't bother with the second component equation. We find $y = 0$ and all eigenvectors have the form $\begin{bmatrix} x \\ 0 \end{bmatrix}$; we'll take $\begin{bmatrix} 1 \\ 0 \end{bmatrix}$.

For $\lambda = 2$, the first component equation, $1 \cdot x + 1 \cdot y = 2 \cdot x$, is non-tautologous, so suffices to determine the eigenvectors. This gives $y = x$. Taking $x = 1$ we get $\begin{bmatrix} 1 \\ 1 \end{bmatrix}$.

(b) The characteristic equation is

$$0 = (2 - \lambda) \cdot (0 - \lambda) - 4 \cdot 3 = -\lambda^2 + 2\lambda - 12$$

so the eigenvalues are $1 \pm \sqrt{13}$.

For $\lambda = 1 + \sqrt{13}$, the first component equation is $2 \cdot x + 3 \cdot y = (1 + \sqrt{13}) \cdot x$. This equation is not tautologous, so it suffices and gives $y = ((\sqrt{13} - 1)/3)x$.

Then taking $x = 1$, we obtain $\begin{bmatrix} 1 \\ (\sqrt{13} - 1)/3 \end{bmatrix}$. By a similar argument, we see an

eigenvector for $\lambda = 1 - \sqrt{13}$ is $\begin{bmatrix} 1 \\ -(1 + \sqrt{13})/3 \end{bmatrix}$.

(c) The characteristic equation is

$$0 = (1 - \lambda) \cdot (1 - \lambda) - (-1) \cdot 1 = \lambda^2 - 2\lambda + 2$$

so the eigenvalues are $1 \pm i$.

For $\lambda = 1 + i$, the first component equation is $1 \cdot x + 1 \cdot y = (1 + i) \cdot x$. This is non-tautologous, so it suffices and gives $y = ix$. Taking $x = 1$, we obtain $\begin{bmatrix} 1 \\ i \end{bmatrix}$.

Similarly, we see an eigenvector for $\lambda = 1 - i$ is $\begin{bmatrix} 1 \\ -i \end{bmatrix}$.

Exercises

8.4.1. Which of these equations are tautologous?
(a) $x + y - x = y$, (b) $x + y = x - y$
(c) $x(x + 1) + y = (x + y)(x - y) + x + y(y + 1)$

8.4.2. Find the eigenvalues and eigenvectors of these matrices.

(a) $\begin{bmatrix} -1 & 0 \\ 3 & 2 \end{bmatrix}$ (b) $\begin{bmatrix} 0 & 1 \\ 3 & 4 \end{bmatrix}$ (c) $\begin{bmatrix} -2 & 5 \\ -1 & 0 \end{bmatrix}$ (d) $\begin{bmatrix} 1 & 2 \\ 3 & 0 \end{bmatrix}$

(e) $\begin{bmatrix} 1 & 2 \\ 3 & -1 \end{bmatrix}$ (f) $\begin{bmatrix} 1 & 2 \\ 3 & 1 \end{bmatrix}$ (g) $\begin{bmatrix} 1 & 2 \\ 3 & 2 \end{bmatrix}$ (h) $\begin{bmatrix} 1 & 2 \\ 3 & 3 \end{bmatrix}$

(i) $\begin{bmatrix} 0 & -1 \\ 3 & 1 \end{bmatrix}$ (j) $\begin{bmatrix} 0 & -1 \\ 3 & -1 \end{bmatrix}$ (k) $\begin{bmatrix} -1 & -1 \\ 3 & -1 \end{bmatrix}$ (l) $\begin{bmatrix} -1 & -2 \\ 3 & -1 \end{bmatrix}$

8.5 EIGENVALUES AT FIXED POINTS

Now we'll compare the stability of the fixed point at the origin for Examples 8.2.1–8.2.6 with the eigenvalues of the matrices. These comparisons will be our key for relating eigenvalues to dynamics near a fixed point.

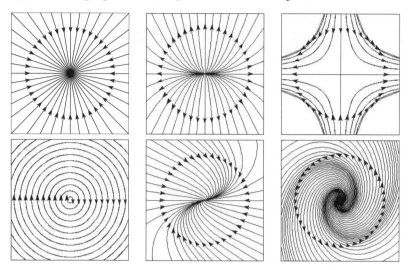

Figure 8.2. Left to right, top to bottom. Some solution curves for Cases 1 through 6. Solution curves with vector fields are plotted in Fig. 8.1.

Case 1. For the system

$$\begin{bmatrix} x' \\ y' \end{bmatrix} = \begin{bmatrix} -1 & 0 \\ 0 & -1 \end{bmatrix} \begin{bmatrix} x \\ y \end{bmatrix}$$

the matrix is triangular so its eigenvalues are -1 and -1. When we try to find eigenvectors, both the component equations are tautologous, $x = x$ and $y = y$. What could this mean? It means that every vector is an eigenvector. We'll take $\begin{bmatrix} 1 \\ 0 \end{bmatrix}$ and $\begin{bmatrix} 0 \\ 1 \end{bmatrix}$ and the general solution is

$$\begin{bmatrix} x \\ y \end{bmatrix} = A \begin{bmatrix} 1 \\ 0 \end{bmatrix} e^{-t} + B \begin{bmatrix} 0 \\ 1 \end{bmatrix} e^{-t} \quad \text{Note that} \quad \begin{bmatrix} x(0) \\ y(0) \end{bmatrix} = \begin{bmatrix} A \\ B \end{bmatrix}$$

So as $t \to \infty$, we see

$$\begin{bmatrix} x \\ y \end{bmatrix} \to \begin{bmatrix} 0 \\ 0 \end{bmatrix}$$

so the origin is asymptotically stable.

Case 2. For the system

$$\begin{bmatrix} x' \\ y' \end{bmatrix} = \begin{bmatrix} 1 & 0 \\ 0 & 2 \end{bmatrix} \begin{bmatrix} x \\ y \end{bmatrix}$$

the matrix is triangular, so its eigenvalues are 1 and 2. For $\lambda = 1$ the component equations are $x = x$ and $2y = y$. The first is tautologous, the second gives $y = 0$. Then $\begin{bmatrix} x \\ 0 \end{bmatrix}$ is an eigenvector. We'll take $x = 1$. For $\lambda = 2$ the component equations are $x = 2x$ and $2y = 2y$. The first gives $x = 0$, so $\begin{bmatrix} 0 \\ y \end{bmatrix}$ is an eigenvector. Take $y = 1$. Then the general solution is

$$\begin{bmatrix} x \\ y \end{bmatrix} = A \begin{bmatrix} 1 \\ 0 \end{bmatrix} e^t + B \begin{bmatrix} 0 \\ 1 \end{bmatrix} e^{2t} \quad \text{and again} \quad \begin{bmatrix} x(0) \\ y(0) \end{bmatrix} = \begin{bmatrix} A \\ B \end{bmatrix}$$

As $t \to \infty$, trajectories escape to infinity, so the origin is unstable.

Case 3. For the system

$$\begin{bmatrix} x' \\ y' \end{bmatrix} = \begin{bmatrix} 1 & 0 \\ 0 & -1 \end{bmatrix} \begin{bmatrix} x \\ y \end{bmatrix}$$

the matrix is triangular, so its eigenvalues are 1 and -1. For $\lambda = 1$ the component equations are $x = x$ and $-y = y$. The second equation gives $y = 0$. Then $\begin{bmatrix} x \\ 0 \end{bmatrix}$ is an eigenvector and we take $x = 1$. For $\lambda = -1$ the component equations are $x = -x$ and $-y = -y$. The first equation gives $x = 0$, so $\begin{bmatrix} 0 \\ y \end{bmatrix}$ is an eigenvector and we'll take $y = 1$. The general solution is

$$\begin{bmatrix} x \\ y \end{bmatrix} = A \begin{bmatrix} 1 \\ 0 \end{bmatrix} e^t + B \begin{bmatrix} 0 \\ 1 \end{bmatrix} e^{-t} \quad \text{One more time} \quad \begin{bmatrix} x(0) \\ y(0) \end{bmatrix} = \begin{bmatrix} A \\ B \end{bmatrix}$$

All solutions with $A \neq 0$ escape to infinity as $t \to \infty$, so the origin is unstable.

Cases 4 and 6 involve complex eigenvalues and eigenvectors, so let's establish some general results before looking at these particular cases. First, because the entries of the matrix M are real, the characteristic equation is a quadratic with real coefficients. Consequently, if the roots are complex, they are conjugates: $\lambda_{\pm} = \alpha \pm i\beta$. Moreover, if $\vec{v}$ is an eigenvector for the complex eigenvalue λ, then the conjugate of $\vec{v}$, by which we mean the vector formed by taking the conjugate of each element of $\vec{v}$, is an eigenvector for the conjugate of λ. The conjugate provides no information about the solutions not already available in a complex eigenvalue and eigenvector, so we'll always use λ_+. Moreover, provided the first component of an eigenvector is not 0, we'll choose the eigenvector with the first component equal to 1. Other choices are possible, but always making the same choices will facilitate comparing your solutions to ours. Once complex eigenvalues and eigenvectors are encountered, algebraic comparisons become more tedious.

Two more preliminary steps are needed before we can get to the examples. If you're wondering what to do with $e^{\lambda t}$ for complex eigenvalues λ, recall *Euler's formula*. We'll give a proof in Sect. A.5.

$$e^{i\theta} = \cos(\theta) + i\sin(\theta) \tag{8.11}$$

Combined with the familiar $e^{a+b} = e^a e^b$, we have an expression for $e^{\lambda t}$ for complex $\lambda = \alpha + i\beta$:

$$e^{(\alpha+i\beta)t} = e^{\alpha t}e^{i\beta t} = e^{\alpha t}(\cos(\beta t) + i\sin(\beta t)) \tag{8.12}$$

Finally, if x and y represent populations, what sense can we make of

$$\begin{bmatrix} x \\ y \end{bmatrix} = \vec{v}e^{\lambda t}$$

with complex λ and $\vec{v}$? Write $x = u + iv$ and $y = w + iz$. Then we can split the sides of

$$\begin{bmatrix} x \\ y \end{bmatrix}' = \begin{bmatrix} a & b \\ c & d \end{bmatrix}\begin{bmatrix} x \\ y \end{bmatrix} \tag{8.13}$$

into real and imaginary parts:

$$\begin{bmatrix} x \\ y \end{bmatrix}' = \begin{bmatrix} x' \\ y' \end{bmatrix} = \begin{bmatrix} u' + iv' \\ w' + iz' \end{bmatrix} = \begin{bmatrix} u' \\ w' \end{bmatrix} + i\begin{bmatrix} v' \\ z' \end{bmatrix}$$

and

$$\begin{bmatrix} a & b \\ c & d \end{bmatrix}\begin{bmatrix} u+iv \\ w+iz \end{bmatrix} = \begin{bmatrix} a & b \\ c & d \end{bmatrix}\left(\begin{bmatrix} u \\ w \end{bmatrix} + i\begin{bmatrix} v \\ z \end{bmatrix}\right) = \begin{bmatrix} a & b \\ c & d \end{bmatrix}\begin{bmatrix} u \\ w \end{bmatrix} + i\begin{bmatrix} a & b \\ c & d \end{bmatrix}\begin{bmatrix} v \\ z \end{bmatrix}$$

Then we can rewrite Eq. (8.13) as

$$\begin{bmatrix} u' \\ w' \end{bmatrix} + i\begin{bmatrix} v' \\ z' \end{bmatrix} = \begin{bmatrix} a & b \\ c & d \end{bmatrix}\begin{bmatrix} u \\ w \end{bmatrix} + i\begin{bmatrix} a & b \\ c & d \end{bmatrix}\begin{bmatrix} v \\ z \end{bmatrix}$$

An equality of complex quantities implies their real parts are equal and their imaginary parts are equal, so

$$\begin{bmatrix} u' \\ w' \end{bmatrix} = \begin{bmatrix} a & b \\ c & d \end{bmatrix}\begin{bmatrix} u \\ w \end{bmatrix} \qquad \begin{bmatrix} v' \\ z' \end{bmatrix} = \begin{bmatrix} a & b \\ c & d \end{bmatrix}\begin{bmatrix} v \\ z \end{bmatrix}$$

That is, we have shown

Proposition 8.5.1. If λ is a complex eigenvalue of M and $\vec{v}$ an eigenvector for λ, then both the real part and the imaginary part of $\vec{v}e^{\lambda t}$ are solutions of $\vec{u}' = M\vec{u}$.

Case 4. For the system

$$\begin{bmatrix} x' \\ y' \end{bmatrix} = \begin{bmatrix} 0 & 1 \\ -1 & 0 \end{bmatrix} \begin{bmatrix} x \\ y \end{bmatrix}$$

the characteristic equation is $\lambda^2 + 1 = 0$ and so the eigenvalues are i and $-i$. We'll take $\lambda = i$. The first component equation is $y = ix$. This is not tautologous, so it's all we need to find an eigenvector. Taking $x = 1$ we have $\begin{bmatrix} 1 \\ i \end{bmatrix}$. The complex solution is

$$\begin{bmatrix} 1 \\ i \end{bmatrix} e^{it} = \begin{bmatrix} 1 \\ i \end{bmatrix} (\cos(t) + i\sin(t)) = \begin{bmatrix} \cos(t) + i\sin(t) \\ i\cos(t) - \sin(t) \end{bmatrix} = \begin{bmatrix} \cos(t) \\ -\sin(t) \end{bmatrix} + i \begin{bmatrix} \sin(t) \\ \cos(t) \end{bmatrix}$$

The general solution is constructed from the real and imaginary parts

$$\begin{bmatrix} x \\ y \end{bmatrix} = A \begin{bmatrix} \cos(t) \\ -\sin(t) \end{bmatrix} + B \begin{bmatrix} \sin(t) \\ \cos(t) \end{bmatrix} \quad \text{So} \quad \begin{bmatrix} x(0) \\ y(0) \end{bmatrix} = \begin{bmatrix} A \\ B \end{bmatrix}$$

We see solutions are *closed curves*: the curve forms a loop. These trajectories neither escape to infinity nor converge to the origin. We say the origin is weakly stable, stable (because solutions do not escape to infinity) but not asymptotically stable (because solutions do not converge to the origin). In Case 6 we do an example with a complex eigenvalue having a non-zero real part. There we'll see that the origin is either unstable or asymptotically stable, depending on the sign of the real part of the eigenalues.

Case 5. The system

$$\begin{bmatrix} x' \\ y' \end{bmatrix} = \begin{bmatrix} -1 & 1 \\ 0 & -1 \end{bmatrix} \begin{bmatrix} x \\ y \end{bmatrix}$$

has eigenvalues -1 and -1 because the matrix is triangular. The component equations are $-x + y = -x$ and $-y = -y$. Consequently, $y = 0$. Unlike Case 1 in which both component equations are tautologous and so there are eigenvectors that are not all parallel to one another, here every eigenvector is a multiple of $\begin{bmatrix} 1 \\ 0 \end{bmatrix}$. Then $\begin{bmatrix} x \\ y \end{bmatrix} = \begin{bmatrix} 1 \\ 0 \end{bmatrix} e^{-t}$ is a solution, the only solution of the form eigenvector $e^{\text{eigenvalue } t}$. How do we find another?

We'll adapt the approach of Ex. 8.2.5 to this case. Take

$$\vec{v} = \begin{bmatrix} a_1 \\ a_2 \end{bmatrix} te^{-t} + \begin{bmatrix} b_1 \\ b_2 \end{bmatrix} e^{-t} = \begin{bmatrix} (a_1 t + b_1)e^{-t} \\ (a_2 t + b_2)e^{-t} \end{bmatrix}$$

and solve

$$\frac{d\vec{v}}{dt} = \begin{bmatrix} -1 & 1 \\ 0 & -1 \end{bmatrix} \vec{v} \tag{8.14}$$

How are we to do this? By differentiating the expression for $\vec{v}$ we have

$$\frac{d\vec{v}}{dt} = \begin{bmatrix} (a_1 - b_1)e^{-t} - a_1 te^{-t} \\ (a_2 - b_2)e^{-t} - a_2 te^{-t} \end{bmatrix}$$

On the other hand,

$$\begin{bmatrix} -1 & 1 \\ 0 & -1 \end{bmatrix} \vec{v} = \begin{bmatrix} (-b_1 + b_2)e^{-t} + (-a_1 + a_2)te^{-t} \\ -b_2 e^{-t} - a_2 te^{-t} \end{bmatrix}$$

Equate the corresponding terms of these vectors. Then the coefficients of e^{-t} must be equal, and the coefficients of te^{-t} must be equal. That is,

$$a_1 - b_1 = -b_1 + b_2 \qquad\qquad -a_1 = -a_1 + a_2 \quad \text{from the first row}$$
$$a_2 - b_2 = -b_2 \qquad\qquad -a_2 = -a_2 \quad \text{from the second row}$$

Both the upper right and lower left equations give $a_2 = 0$. The upper left equation gives $a_1 = b_2$, and the lower right equation is a tautology. Then b_1 and b_2 are undetermined by Eq. (8.14), $a_1 = b_2$ and $a_2 = 0$. Taking $b_1 = b_2 = 1$, we obtain $\vec{v} = \begin{bmatrix} t+1 \\ 1 \end{bmatrix} e^{-t}$. The general solution is

$$\begin{bmatrix} x \\ y \end{bmatrix} = A \begin{bmatrix} 1 \\ 0 \end{bmatrix} e^{-t} + B \begin{bmatrix} t+1 \\ 1 \end{bmatrix} e^{-t} \quad \text{So} \quad \begin{bmatrix} x(0) \\ y(0) \end{bmatrix} = \begin{bmatrix} A+B \\ B \end{bmatrix}$$

That is,

$$\begin{bmatrix} x \\ y \end{bmatrix} = (x(0) - y(0)) \begin{bmatrix} 1 \\ 0 \end{bmatrix} e^{-t} + y(0) \begin{bmatrix} t+1 \\ 1 \end{bmatrix} e^{-t}$$

Recall that for all $n > 0$ $\lim_{t\to\infty} t^n e^{-t} = 0$. If you don't recall this, apply l'Hôpital's rule to t^n/e^t. Then we see the origin is asymptotically stable.

Distinguishing Case 1 from Case 5, the two possibilities when the eigenvalues are equal, isn't difficult. All the necessary information is in the explanations of these cases. If this isn't clear, more detail is presented in Sect. A.7.

Case 6. The system

$$\begin{bmatrix} x' \\ y' \end{bmatrix} = \begin{bmatrix} 1 & -2 \\ 3 & 1 \end{bmatrix} \begin{bmatrix} x \\ y \end{bmatrix}$$

has characteristic equation $0 = (1 - \lambda)^2 - (-2) \cdot 3 = \lambda^2 - 2\lambda + 7$ and so the eigenvalues are $1 + i\sqrt{6}$ and $1 - i\sqrt{6}$. As we mentioned, we need only one complex eigenvalue, so we'll take $1 + i\sqrt{6}$. The first component equation is $x - 2y = (1 + i\sqrt{6})x$. This isn't tautologous, so it's all we need to find eigenvectors. The first component equation gives $y = -i(\sqrt{6}/2)x$. As usual, take $x = 1$ and find the eigenvector $\begin{bmatrix} 1 \\ -i\sqrt{6}/2 \end{bmatrix}$. The complex solution is

$$\begin{bmatrix} 1 \\ -i\sqrt{6}/2 \end{bmatrix} e^{(1+i\sqrt{6})t} = \begin{bmatrix} 1 \\ -i\sqrt{6}/2 \end{bmatrix} e^t (\cos(\sqrt{6}t) + i\sin(\sqrt{6}t))$$

$$= \begin{bmatrix} \cos(\sqrt{6}t) + i\sin(\sqrt{6}t) \\ (-i\sqrt{6}/2)\cos(\sqrt{6}t) + (\sqrt{6}/2)\sin(\sqrt{6}t) \end{bmatrix} e^t$$

$$= \begin{bmatrix} \cos(\sqrt{6}t) \\ (\sqrt{6}/2)\sin(\sqrt{6}t) \end{bmatrix} e^t + i \begin{bmatrix} \sin(\sqrt{6}t) \\ (-\sqrt{6}/2)\cos(\sqrt{6}t) \end{bmatrix} e^t$$

Applying Prop. 8.5.1, we see that the general solution is

$$\begin{bmatrix} x \\ y \end{bmatrix} = A \begin{bmatrix} \cos(\sqrt{6}t) \\ (\sqrt{6}/2)\sin(\sqrt{6}t) \end{bmatrix} e^t + B \begin{bmatrix} \sin(\sqrt{6}t) \\ (-\sqrt{6}/2)\cos(\sqrt{6}t) \end{bmatrix} e^t$$

Then

$$\begin{bmatrix} x(0) \\ y(0) \end{bmatrix} = \begin{bmatrix} A \\ -B\sqrt{6}/2 \end{bmatrix} \quad \text{that is} \quad \begin{bmatrix} A \\ B \end{bmatrix} = \begin{bmatrix} x(0) \\ -\sqrt{6}y(0)/3 \end{bmatrix}$$

Recognizing the roles of eigenvalues in determining the stability of solutions, these six cases suggest this result, which can be proved with a bit more linear algebra. For the reason given in Prop. 8.4.2 we shall suppose that $\lambda_1 \neq 0$ and $\lambda_2 \neq 0$. Then

- If $\text{Re}(\lambda_1)$ and $\text{Re}(\lambda_2)$ are negative, the origin is asymptotically stable.
- If at least one of $\text{Re}(\lambda_1)$ and $\text{Re}(\lambda_2)$ is positive, the origin is unstable.
- If $\text{Re}(\lambda_1) = \text{Re}(\lambda_2) = 0$, the origin is weakly stable.

In addition, the eigenvalues can determine the type of fixed point at the origin.

- If (real) $\lambda_1 > 0 > \lambda_2$, the origin is a saddle point.
- If (real) $\lambda_1, \lambda_2 > 0$, the origin is an unstable node.
- If (real) $\lambda_1, \lambda_2 < 0$, the origin is an asymptotically stable node.
- If λ_1 and λ_2 are complex with positive real part, the origin is an unstable spiral.
- If λ_1 and λ_2 are complex with negative real part, the origin is an asymptotically stable spiral.
- If λ_1 and λ_2 are imaginary, the origin is a center, weakly stable.

In Ch. 9 we'll see that almost all of these results can be applied to nonlinear differential equations as well, once we figure out which matrix to use. But the proof of this is a bit subtle. The main point to take from this section is that eigenvalues determine the type and stability of the fixed point of linear systems, and with a bit more work, they determine the type and stability of most fixed points of nonlinear differential equations.

In Sect. A.12.2 we'll see how to apply the techniques of Sect. 16.2 to derive these solutions.

Practice Problem

8.5.1. Find the general solution of $\vec{v}' = M\vec{v}$ for

(a) $M = \begin{bmatrix} 1 & 2 \\ 0 & 2 \end{bmatrix}$ (b) $M = \begin{bmatrix} 3 & 1 \\ -1 & 1 \end{bmatrix}$ (c) $M = \begin{bmatrix} 1 & -1 \\ 2 & 1 \end{bmatrix}$

Practice Problem Solution

8.5.1. (a) The matrix is triangular so the eigenvalues are 1 and 2. The first component equation for $\lambda = 1$ is $x + 2y = x$ so $y = 0$. The first component equation for $\lambda = 2$ is $x + 2y = 2x$ so $y = x/2$. We see that the eigenvectors are $\begin{bmatrix} 1 \\ 0 \end{bmatrix}$ and $\begin{bmatrix} 1 \\ 1/2 \end{bmatrix}$. Following the example of Case 2, the general solution is

$$\begin{bmatrix} x \\ y \end{bmatrix} = A \begin{bmatrix} 1 \\ 0 \end{bmatrix} e^t + B \begin{bmatrix} 1 \\ 1/2 \end{bmatrix} e^{2t}$$

(b) The characteristic equation is $0 = (3-\lambda)(1-\lambda) - (-1)\cdot 1 = \lambda^2 - 4\lambda + 4 = (\lambda-2)^2$, so $\lambda = 2$ is the only eigenvalue. The first component equation is $3x + y = 2x$, the eigenvector is $\begin{bmatrix} 1 \\ -1 \end{bmatrix}$, and the solution is $\begin{bmatrix} x \\ y \end{bmatrix} = \begin{bmatrix} 1 \\ -1 \end{bmatrix} e^{2t}$. Because all eigenvectors are multiples of this vector, we are in the situation described in Case 5. Then to find a second solution, write

$$\vec{v} = \begin{bmatrix} a_1 \\ a_2 \end{bmatrix} te^{2t} + \begin{bmatrix} b_1 \\ b_2 \end{bmatrix} e^{2t} = \begin{bmatrix} (a_1 t + b_1)e^{2t} \\ (a_2 t + b_2)e^{2t} \end{bmatrix}$$

and solve

$$\frac{d\vec{v}}{dt} = \begin{bmatrix} 3 & 1 \\ -1 & 1 \end{bmatrix} \vec{v} \tag{8.15}$$

First, differentiating the expression for $\vec{v}$ we obtain

$$\frac{d\vec{v}}{dt} = \begin{bmatrix} (a_1 + 2b_1)e^{2t} + 2a_1 te^{2t} \\ (a_2 + 2b_2)e^{2t} + 2a_2 te^{2t} \end{bmatrix}$$

Next,

$$\begin{bmatrix} 3 & 1 \\ -1 & 1 \end{bmatrix} \vec{v} = \begin{bmatrix} (3b_1 + b_2)e^{2t} + (3a_1 + a_2)te^{2t} \\ (-b_1 + b_2)e^{2t} + (-a_1 + a_2)te^{2t} \end{bmatrix}$$

Equating coefficients of e^{2t} and of te^{2t} for the terms of the vectors in Eq. (8.15) gives the equations

$$a_1 + 2b_1 = 3b_1 + b_2 \qquad\qquad 2a_1 = 3a_1 + a_2$$
$$a_2 + 2b_2 = -b_1 + b_2 \qquad\qquad 2a_2 = -a_1 + a_2$$

The two equations on the right are equivalent, both giving the relation $a_2 = -a_1$. With this substitution, the two equations on the left are equivalent, both giving the relation $b_1 = a_1 - b_2$. That is, a_1 and b_2 are not determined by Eq. (8.15), but a_2 and b_1 are. Taking $a_1 = b_2 = 1$ gives $a_2 = -1$ and $b_1 = 0$ and so

$$\vec{v} = \begin{bmatrix} t \\ -t+1 \end{bmatrix} e^{2t}$$

and the general solution is

$$\begin{bmatrix} x \\ y \end{bmatrix} = A \begin{bmatrix} 1 \\ -1 \end{bmatrix} e^{2t} + B \begin{bmatrix} t \\ -t+1 \end{bmatrix} e^{2t}$$

(c) The characteristic equation is $0 = (1 - \lambda)(1 - \lambda) - 2(-1) = \lambda^2 - 2\lambda + 3$, so the eigenvalues are $1 \pm i\sqrt{2}$. Taking $1 + i\sqrt{2}$, the first component equation is $x - y = (1 + i\sqrt{2})x$. This gives $y = -i\sqrt{2}x$ and we find the eigenvector $\begin{bmatrix} 1 \\ -i\sqrt{2} \end{bmatrix}$.

We follow the method of Case 6. The complex solution is

$$\begin{bmatrix} 1 \\ -i\sqrt{2} \end{bmatrix} e^{(1+i\sqrt{2})t} = \begin{bmatrix} 1 \\ -i\sqrt{2} \end{bmatrix} e^t (\cos(\sqrt{2}t) + i\sin(\sqrt{2}t))$$

$$= \begin{bmatrix} \cos(\sqrt{2}t) + i\sin(\sqrt{2}t) \\ -i\sqrt{2}\cos(\sqrt{2}t) + \sqrt{2}\sin(\sqrt{2}t) \end{bmatrix} e^t$$

$$= \begin{bmatrix} \cos(\sqrt{2}t) \\ \sqrt{2}\sin(\sqrt{2}t) \end{bmatrix} e^t + i \begin{bmatrix} \sin(\sqrt{2}t) \\ -\sqrt{2}\cos(\sqrt{2}t) \end{bmatrix} e^t$$

Then by Prop. 8.5.1 we see the general solution is

$$\begin{bmatrix} x \\ y \end{bmatrix} = A \begin{bmatrix} \cos(\sqrt{2}t) \\ \sqrt{2}\sin(\sqrt{2}t) \end{bmatrix} e^t + B \begin{bmatrix} \sin(\sqrt{2}t) \\ -\sqrt{2}\cos(\sqrt{2}t) \end{bmatrix} e^t$$

Exercises

Find the general solution of $d\vec{v}/dt = M\vec{v}$, where M is

8.5.1. (a) $\begin{bmatrix} 1 & 3 \\ 3 & 1 \end{bmatrix}$ (b) $\begin{bmatrix} 1 & 3 \\ -3 & 1 \end{bmatrix}$ (c) $\begin{bmatrix} 2 & 0 \\ -1 & 2 \end{bmatrix}$ (d) $\begin{bmatrix} -2 & -1 \\ 3 & 0 \end{bmatrix}$

8.5.2. (a) $\begin{bmatrix} 1 & 2 \\ 0 & 1 \end{bmatrix}$ (b) $\begin{bmatrix} 1 & -1 \\ 3 & 1 \end{bmatrix}$ (c) $\begin{bmatrix} 1 & -1 \\ 3 & 0 \end{bmatrix}$ (d) $\begin{bmatrix} 0 & -1 \\ 3 & 2 \end{bmatrix}$

8.5.3. (a) $\begin{bmatrix} 1 & 4 \\ -1 & -3 \end{bmatrix}$ (b) $\begin{bmatrix} -3 & 1 \\ -1 & -1 \end{bmatrix}$ (c) $\begin{bmatrix} 1 & 2 \\ 3 & 1 \end{bmatrix}$ (d) $\begin{bmatrix} -1 & 2 \\ 3 & 0 \end{bmatrix}$

8.6 THE TRACE-DETERMINANT PLANE

In Sect. 8.5 we've seen that the eigenvalues of the matrix M determine the type and stability of the fixed point at the origin. Of course, the eigenvalues are determined by the coefficients a, b, c, and d of M. Can we find the stability and the type (spiral, center, node, saddle) of the fixed point without computing the eigenvalues? In particular, can the stability and type be read from some functions of the matrix in Eq. (8.3)? The eigenvalues are the solutions of

$$0 = (a - \lambda)(d - \lambda) - cb = \lambda^2 - (a+d)\lambda + ad - bc$$

We know $ad - bc = \det(M)$, the determinant of M. The *trace* of M is defined as $\mathrm{tr}(M) = a + d$, so the eigenvalue equation can be rewritten as

$$0 = \lambda^2 - \mathrm{tr}\lambda + \det$$

where the M in $\mathrm{tr}(M)$ and $\det(M)$ is omitted, being understood. The eigenvalues are given by

$$\lambda_\pm = \frac{\mathrm{tr} \pm \sqrt{\mathrm{tr}^2 - 4\det}}{2} \tag{8.16}$$

Rather than constructing a 4-dimensional plot using a, b, c, and d as coordinates, all the relevant properties of the eigenvalues are set by two coordinates: trace and determinant. A map of stability and type can be plotted in the trace-determinant (tr-det) plane.

To understand the relation between trace, determinant, and fixed point type, we'll use these formulas, easily derived by algebraic simplification

$$\lambda_+ + \lambda_- = \mathrm{tr} \tag{8.17}$$

$$\lambda_+ \cdot \lambda_- = \det \tag{8.18}$$

Because we shall be referring to it often, write

$$\Delta = \mathrm{tr}^2 - 4\det$$

The eigenvalues are complex or real according as $\Delta < 0$ and $\Delta \geq 0$, so the curve $\Delta = 0$ divides the tr-det plane into two regions, $\Delta < 0$ and $\Delta > 0$. See Fig. 8.3.

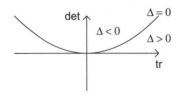

Figure 8.3. The tr-det plane, divided by $\Delta = 0$.

Let's consider stability and type together. We'll use Eqs. (8.16), (8.17), and (8.18).

- If det < 0, then Eq. (8.16) shows that both eigenvalues are real. Eq. (8.18) shows the product of the eigenvalues is negative, so one must be positive and one must be negative, so the fixed point is a saddle point, necessarily unstable.

- If tr > 0, det > 0, and $\Delta \geq 0$, the eigenvalues are real and Eq. (8.17) shows that at least one is positive. Then by Eq. (8.18), det > 0 means both eigenvalues are positive, so the fixed point is an unstable node.
- If tr < 0, det > 0, and $\Delta \geq 0$, the eigenvalues are real and Eq. (8.17) shows that at least one is negative. Then det > 0 and Eq. (8.18) show the product is positive, so both eigenvalues must be negative. That is, the fixed point is an asymptotically stable node.
- If tr > 0 and $\Delta < 0$, the eigenvalues are complex with positive real part (for complex eigenvalues, tr is twice the real part), so the fixed point is an unstable spiral.
- If tr < 0 and $\Delta < 0$, the eigenvalues are complex with negative real part, so the fixed point is an asymptotically stable spiral.
- If tr = 0 and $\Delta < 0$, the eigenvalues are complex with zero real part, so the fixed point is a center, weakly stable.

These observations are gathered in Fig. 8.4.

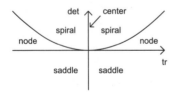 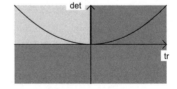

Figure 8.4. Left: fixed point type. Right: stability–asymptotically stable (light region), unstable (dark region) and stable (thick line).

There's a subtle issue about the tr-det plane: each point of the plane corresponds to a *lot* of matrices. Here's why. Given a value tr for trace and a value det for the determinant, a 2×2 matrix can be written as

$$\begin{bmatrix} a & b \\ c & d \end{bmatrix} = \begin{bmatrix} a & b \\ (\det - a(\mathrm{tr} - a))/b & \mathrm{tr} - a \end{bmatrix}$$

so long as $b \neq 0$. There's a simpler formula when $b = 0$. So a 2-dimensional collection of matrices, here the a-b plane, corresponds to a given value of tr and det.

For later reference, we note

the fixed point is asymptotically stable if $\mathrm{tr} < 0$ and $\det > 0$.   (8.19)

In Sect. 9.2 we'll need to pay special attention to the *boundaries* in the tr-det plane. These are

- $\det = 0$, which separates saddle points from nodes. From Eq. (8.18) we see that this boundary is where at least one eigenvalue is 0.
- $\mathrm{tr} = 0, \det > 0$, which separares asymptotically stable spirals from unstable spirals. From Eq. (8.16) we see that this boundary is where the eigenvalues are imaginary.
- $\det = \mathrm{tr}^2/4$, which separates nodes from spirals. From Eq. (8.16) we see that this boundary is where both eigenvalues are equal, and $\lambda_\pm = \mathrm{tr}/2$.

In this section we do not consider the case $\det = 0$, because on this line in the tr-det boundary at least one of the eigenvalues is 0 and by Prop. 8.4.2 we see that 0 eigenvalues are excluded when we insist on isolated fixed points.

Now the trace and determinant contain no information not in the eigenvalues, so why bother with the tr-det plane analysis? There are at least two reasons. First, often locating a point in the tr-det plane is easier than computing the eigenvalues. For example, if $\det < 0$, an easy calculation, then we know the fixed point is a saddle point. If $\det > 0$ we must do a bit more work, specifically, determine if the point $(\mathrm{tr}, \det)$ lies above or below the parabola $\det = \mathrm{tr}^2/4$. Often this involves solving a quadratic equation, about the same work as computing the eigenvalues.

Second, sometimes one or more of the entries of the matrix M can depend on parameters–reproduction rate, death rate, or competition–that can vary with circumstance. These parameters can show up inside the square root of the eigenvalue formula, in which case some care can be required to sort out how the type of fixed point depends on the parameters. Almost always, it's easier to plot the path through the tr-det plane defined by varying the parameters. For over a decade I taught the course from which these notes grew. The first exam always included a parameterized differential equation problem. Despite my comments that these are better approached by the tr-det plane, about a quarter of the class (of between 30 and 45 students) computed eigenvalues and tried to sort out the parameter ranges for each type of fixed point. In all those years, three students solved the problem using eigenvalues. The tr-det plane is a useful technique.

Practice Problems

Classify the fixed point type of the origin for the system $d\vec{v}/dt = M\vec{v}$ for

8.6.1. $M(a) = \begin{bmatrix} a & 1 \\ 1 & 2 \end{bmatrix}$ with all a, $-\infty < a < \infty$. Plot the path of this matrix in the tr-det plane as a function of a.

8.6.2. $M(b) = \begin{bmatrix} 1 & b \\ 1 & 2 \end{bmatrix}$ with all b, $-\infty < b < \infty$. Plot the path of this matrix in the tr-det plane as a function of b.

Practice Problem Solutions

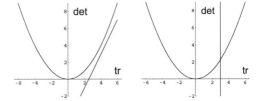

8.6.1. For $M(a)$ observe $\mathrm{tr} = 2 + a$ and $\det = 2a - 1$. Eliminating a from these equations, we find $\det = 2(\mathrm{tr} - 2) - 1 = 2\mathrm{tr} - 5$

Figure 8.5. The trace-determinant plot for Practice Problems 8.6.1 (left) and 8.6.2 (right).

This is a straight line, lying entirely below the $\det = \mathrm{tr}/4$ parabola. To see that the line doesn't cross the parabola, solve $\mathrm{tr}^2/4 = 2\mathrm{tr} - 5$ for tr, obtaining $\mathrm{tr} = 4 \pm 2i$. If the line did intersect the parabola this equation would have two real solutions (the line intersects the parabola at two points) or one real solution (the line intersects the parabola at one point); complex solutions mean the line misses the parabola. From the map of fixed point types in the tr-det plane (left side of Fig. 8.4), we see the origin is a saddle for $\det < 0$, that is, for $a < 1/2$, and an unstable node for $\det > 0$, that is, for $a > 1/2$.

8.6.2. For $M(b)$ observe $\mathrm{tr} = 3$ and $\det = 2 - b$, a vertical line in the tr-det plane. From the map of fixed point types in the tr-det plane (left side of Fig. 8.4), we see the origin is a saddle point for $\det < 0$ (so for $2 < b$), an unstable node for $0 < \det < 9/4$ (so $-1/4 < b < 2$), and an unstable spiral for $9/4 < \det$

(so $b < -1/4$). The value det $= 9/4$ comes from the intersection of the parabola det $=$ tr$^2/4$ and the line tr $= 3$.

Exercises

Classify the fixed point type of the origin for the system $d\vec{v}/dt = M\vec{v}$, for all a, $-\infty < a < \infty$. Plot the path of this matrix in the tr-det plane as a function of a.

8.6.1. (a) $\begin{bmatrix} 1 & 1 \\ a & 2 \end{bmatrix}$ (b) $\begin{bmatrix} 1 & 1 \\ 1 & a \end{bmatrix}$ (c) $\begin{bmatrix} a & -1 \\ 1 & -1 \end{bmatrix}$ (d) $\begin{bmatrix} a & a \\ 1 & 2 \end{bmatrix}$

8.6.2. (a) $\begin{bmatrix} a & a \\ 0 & 2 \end{bmatrix}$ (b) $\begin{bmatrix} a & 1 \\ 1 & a \end{bmatrix}$ (c) $\begin{bmatrix} a & 1 \\ -1 & a \end{bmatrix}$ (d) $\begin{bmatrix} a & 1 \\ a & -1 \end{bmatrix}$

8.6.3. (a) $\begin{bmatrix} a & 1 \\ a & 0 \end{bmatrix}$ (b) $\begin{bmatrix} 1 & a \\ a & 2 \end{bmatrix}$ (c) $\begin{bmatrix} a & 1 \\ -1 & 0 \end{bmatrix}$ (d) $\begin{bmatrix} a & -1+a^2 \\ 1 & 0 \end{bmatrix}$

Chapter 9   Nonlinear systems and stability

Nonlinear differential equations can have several fixed points and can exhibit behaviors more complicated than those of linear differential equations. While in most cases the stability of the fixed points can be determined by a method similar to that we used for linear systems, the limit cycles we can see for nonlinear differential equations cannot occur for linear differential equations. For these we need to develop some new techniques.

9.1   PARTIAL DERIVATIVES

A nonlinear differential equation in the plane is

$$\begin{bmatrix} dx/dt \\ dy/dt \end{bmatrix} = \begin{bmatrix} g(x,y) \\ h(x,y) \end{bmatrix} = \vec{F}(x,y) \tag{9.1}$$

As with linear differential equations, the fixed points are the solutions (x_*, y_*) of the fixed point equations

$$g(x_*, y_*) = h(x_*, y_*) = 0. \tag{9.2}$$

How do we test fixed point stability?

For a 1-dimensional system $dx/dt = f(x)$, we know the stability of a fixed point x_* is determined by $f'(x_*)$. So we must ask what could we mean by the derivative of $\vec{F}(x,y)$. We'll need two steps to do this.

First, recall that for a function $f(x)$ the derivative is defined by

$$\frac{df}{dx} = \lim_{\Delta x \to 0} \frac{f(x + \Delta x) - f(x)}{\Delta x}$$

For a function $g(x,y)$ of two variables, we can define two obvious derivatives, one holding y constant and varying x, the other holding x constant and varying y. These are

$$\frac{\partial g}{\partial x} = \lim_{\Delta x \to 0} \frac{g(x + \Delta x, y) - g(x,y)}{\Delta x}, \quad \frac{\partial g}{\partial y} = \lim_{\Delta y \to 0} \frac{g(x, y + \Delta y) - g(x,y)}{\Delta y}$$

Computationally this is easy: apply the familiar differentiation rules to one variable holding the other constant.

Example 9.1.1. *A sum of single-variable terms.* For $g(x,y) = x^2 + y^3$ we compute $\partial g/\partial x = 2x$ and $\partial g/\partial y = 3y^2$. $\square$

Example 9.1.2. *A product of single-variable terms.* For $g(x,y) = x^2 y^3$ we compute $\partial g/\partial x = 2xy^3$ (treat y^3 as a constant) and $\partial g/\partial y = x^2 3y^2$ (treat x^2 as a constant). $\square$

Example 9.1.3. *A chain rule calculation.* For $g(x,y) = \sin(xy^2)$,

$$\frac{\partial g}{\partial x} = \cos(xy^2)\frac{\partial xy^2}{\partial x} = \cos(xy^2)y^2, \quad \frac{\partial g}{\partial y} = \cos(xy^2)\frac{\partial xy^2}{\partial y} = \cos(xy^2)2xy$$

For more complicated functions some care is needed to keep straight the chain rule factors. $\square$

To illustrate the validity of this process, we use the definition of the partial derivative to compute $\partial g/\partial x$ for Example 9.1.2.

$$\frac{\partial g}{\partial x} = \lim_{\Delta x \to 0} \frac{(x + \Delta x)^2 y^3 - x^2 y^3}{\Delta x} = \lim_{\Delta x \to 0} \frac{2xy^3 \Delta x + y^3 (\Delta x)^2}{\Delta x} = 2xy^3$$

Example 9.1.3 illustrates one form of the chain rule for partial derivatives:

$$\frac{\partial}{\partial x} f(h(x,y)) = f'(h(x,y)) \cdot \frac{\partial h}{\partial x} \tag{9.3}$$

Another form applies if the variables x and y are functions of another variable, say $x = x(t)$ and $y = y(t)$. Then

$$\frac{dg}{dt} = \frac{\partial g}{\partial x}\frac{dx}{dt} + \frac{\partial g}{\partial y}\frac{dy}{dt} \tag{9.4}$$

For instance, if $g(x,y) = x^2 + xy^3$, $x(t) = \cos(t)$, and $y(t) = \sin(t)$, then

$$\frac{dg}{dt} = (2x + y^3)(-\sin(t)) + (3xy^2)(\cos(t))$$

$$= (2\cos(t) + \sin^3(t))(-\sin(t)) + (3\cos(t)\sin^2(t))(\cos(t))$$

The second step is the *derivative matrix*. For a function $\vec{F}: \mathbb{R}^2 \to \mathbb{R}^2$, written as $\vec{F} = \begin{bmatrix} g(x,y) \\ h(x,y) \end{bmatrix}$, the derivative of $\vec{F}(x,y)$ is defined by

$$D\vec{F} = \begin{bmatrix} \partial g/\partial x & \partial g/\partial y \\ \partial h/\partial x & \partial h/\partial y \end{bmatrix} \tag{9.5}$$

Here we'll get practice with computations. In Sect. 9.2 we'll apply eigenvalue analysis to the derivative matrix (9.5) in order to determine the stability of fixed points of nonlinear differential equations.

Practice Problems

9.1.1. Compute $\partial g/\partial x$ and $\partial g/\partial y$ for $g(x,y) = xy\sin(x^2 + y^3)$.

9.1.2. Compute $\partial g/\partial x$ and $\partial g/\partial y$ for $g(x,y) = \ln(1 + x^2 + \cos(xy))$.

9.1.3. Find $D\vec{F}$ for $\vec{F} = \begin{bmatrix} x + \sin(xy) \\ e^{x^2 y^3} \end{bmatrix}$.

9.1.4. Compute $\partial g/\partial s$ for $g(x,y) = x^2 + y^3 + xy$, $x(s,t) = st$, and $y(s,t) = \sin(s+t)$.

Practice Problem Solutions

9.1.1. Use the product rule and the chain rule

$$\frac{\partial g}{\partial x} = y\sin(x^2 + y^3) + xy\cos(x^2 + y^3)2x$$

$$\frac{\partial g}{\partial y} = x\sin(x^2 + y^3) + xy\cos(x^2 + y^3)3y^2$$

9.1.2. Use the product rule and the chain rule

$$\frac{\partial g}{\partial x} = \frac{2x - \sin(xy)y}{1 + x^2 + \cos(xy)} \quad \text{and} \quad \frac{\partial g}{\partial y} = \frac{-\sin(xy)x}{1 + x^2 + \cos(xy)}$$

9.1.3. Apply Eq. (9.5), with $g(x,y) = x + \sin(xy)$ and $h(x,y) = e^{x^2y^3}$.

$$D\vec{F} = \begin{bmatrix} 1 + \cos(xy)y & \cos(xy)x \\ e^{x^2y^3}2xy^3 & e^{x^2y^3}x^23y^2 \end{bmatrix}$$

9.1.4. Modify Eq. (9.4) for x and y functions of s and t:

$$\frac{\partial g}{\partial s} = \frac{\partial g}{\partial x}\frac{\partial x}{\partial s} + \frac{\partial g}{\partial y}\frac{\partial y}{\partial s} \qquad (9.6)$$

For these g, x, and y, we obtain

$$\frac{\partial g}{\partial s} = (2x + y)t + (3y^2 + x)\cos(s + t)$$

$$= (2st + \sin(s + t))t + (3\sin(s + t)^2 + st)\cos(s + t)$$

Exercises

In Exercises 9.1.1–9.1.4 compute $\dfrac{\partial g}{\partial x}$ and $\dfrac{\partial g}{\partial y}$; in 9.1.5 and 9.1.6 compute $\dfrac{\partial g}{\partial t}$; in 9.1.7 compute $\dfrac{\partial g}{\partial s}$. In 9.1.8, 9.1.9, and 9.1.10, describe the set of all (x,y) for which $\dfrac{\partial f}{\partial x} = 0$. Here the word "describe" means "write the equation for and give the name of" the set.

9.1.1. $g(x,y) = \tan(xy^2 + x^2y^3)$

9.1.2. $g(x,y) = \sin(x^2y^2)\cos(x^2 + y^2)$

9.1.3. $g(x,y) = e^{(x-y)/(1+x^2+y^2)}$

9.1.4. $g(x,y) = e^{(x^2-y^2)/(1+x^2+y^2)}$

9.1.5. $g(x,y) = \ln(1 + x^2 + y^4)$, $x(s,t) = t^2 + s^3$, and $y(s,t) = t/s$

9.1.6. $g(x,y) = (x^2 + y^3)^9$, $x(s,t) = \sin(s + t)$, and $y(s,t) = \cos(st)$

9.1.7. $g(x,y) = (x^2 + y^3)^9$, $x(s,t) = \sin(s^2t^2)$, and $y(s,t) = \cos(s^2 + t^2)$

9.1.8. $f(x,y) = e^{x^2-yx}$

9.1.9. $f(x,y) = \cos(x^2 + y^2)$

9.1.10. $f(x,y) = \ln(5x + x^3/3 + xy^2)$

9.2 THE HARTMAN-GROBMAN THEOREM

For two-variable differential equations (9.1), we'll apply the eigenvalue analysis of linear systems to the eigenvalues of $D\vec{F}$ at every fixed point (x_*,y_*). As we'll see in the *Hartman-Grobman theorem*, Thm. 9.2.1, most of the results for linear systems apply in this setting. For instance, we can use the tr-det plane as we did in Sect. 8.6, so long as the trace and the determinant do not lie on any of the three boundaries (i) $\det = 0$, (ii) $\text{tr} = 0$ and $\det > 0$, and (iii) $\det = \text{tr}^2/4$. For linear systems, trace and determinant on boundary (i) correspond to non-isolated fixed points, on on (ii) to centers, and on (iii) to repeated (necessarily real) eigenvalues and so to proper or improper nodes. For nonlinear differential equations none of these correspondences need to hold at tr-det boundary points. At points off these boundaries, nonlinear differential equations exhibit the same type of dynamics (saddle, node, spiral) as do linear systems with the same trace and determinant.

Theorem 9.2.1. For every fixed point (x_*,y_*) of Eq. (9.1), suppose the eigenvalues of $D\vec{F}(x_*,y_*)$ are λ_1 and λ_2. Then the fixed point (x_*,y_*) is

(1)   a saddle point if λ_1 and λ_2 are real with opposite signs,

(2)   an unstable node if λ_1 and λ_2 are real and positive,

(3)   an asymptotically stable node if λ_1 and λ_2 are real and negative,

(4)   an asymptotically stable spiral if λ_1 and λ_2 are complex with negative real parts, and

(5)   an unstable spiral if λ_1 and λ_2 are complex with positive real parts.

Sketch of the proof. In first semester calculus we learn that for nearby points x and x_*,

$$f(x) \approx f(x_*) + f'(x_*)(x - x_*)$$

In Sect. 10.8 we'll see that this is the first step in a sequence of increasingly accurate approximations of $f(x)$ made by the addition of higher derivatives of f.

A similar approximation holds for 2-dimensional vector functions of two variables. For points (x,y) very near to (x_*,y_*)

$$\vec{F}(x,y) \approx \vec{F}(x_*,y_*) + D\vec{F}(x_*,y_*) \begin{bmatrix} x - x_* \\ y - y_* \end{bmatrix}$$

So if (x_*,y_*) is a fixed point of $\vec{F}$ we have

$$\vec{F}(x,y) \approx \begin{bmatrix} x_* \\ y_* \end{bmatrix} + D\vec{F}(x_*,y_*) \begin{bmatrix} x - x_* \\ y - y_* \end{bmatrix}$$

Then Eq. (9.1) gives

$$\begin{bmatrix} x' \\ y' \end{bmatrix} \approx \begin{bmatrix} x_* \\ y_* \end{bmatrix} + D\vec{F}(x_*, y_*) \begin{bmatrix} x - x_* \\ y - y_* \end{bmatrix}$$

That is, the matrix $D\vec{F}(x_*, y_*)$ governs the approximate path of a trajectory near the fixed point.

From our Chapter 8 study of linear systems, we expect that the eigenvalues of $D\vec{F}(x_*, y_*)$ determine the stability of this system, at least if small variations of the matrix don't alter the relevant character of the eigenvalues. (This is where the "approximate" of "approximate path of a trajectory" comes in.) The character of the eigenvalues in cases (1)–(5) of the theorem is preserved under small perturbations of the matrix. However, for eigenvalues corresponding to any of the tr-det boundaries, an arbitrarily small perturbation of the matrix can move (tr, det) off that boundary. That is, for eigenvalues of matrices $D\vec{F}(x_*, y_*)$ on a tr-det boundary, these eigenvalues tell us nothing about the dynamics of trajectories near the fixed point (x_*, y_*). $\square$

A less sketchy argument is found in Sect. A.9. The theorem also gives a local map between trajectories of the nonlinear and linear systems; we'll mention this in Sect. A.9. Time for some examples.

Example 9.2.1. *Stability of the fixed points of nonlinear differential equations.* Consider the system

$$dx/dt = 2x - x^2 - xy \qquad dy/dt = 3y - y^2 - 2xy$$

The fixed points are the solutions of

$$0 = 2x - x^2 - xy = x(2 - x - y) \quad \text{and} \quad 0 = 3y - y^2 - 2xy = y(3 - y - 2x)$$

There are two choices for the first equation and two choices for the second, so four combinations: $(x_*, y_*) = (0,0), (2,0), (0,3),$ and $(1,1)$. Now

$$D\vec{F}(x,y) = \begin{bmatrix} 2 - 2x - y & -x \\ -2y & 3 - 2y - 2x \end{bmatrix}$$

Substitution of the fixed point coordinates gives

$$D\vec{F}(0,0) = \begin{bmatrix} 2 & 0 \\ 0 & 3 \end{bmatrix} \qquad\qquad D\vec{F}(2,0) = \begin{bmatrix} -2 & -2 \\ 0 & -1 \end{bmatrix}$$

$$D\vec{F}(0,3) = \begin{bmatrix} -1 & 0 \\ -6 & -3 \end{bmatrix} \qquad\qquad D\vec{F}(1,1) = \begin{bmatrix} -1 & -1 \\ -2 & -1 \end{bmatrix}$$

The eigenvalues at $(0,0)$ are 2 and 3, so $(0,0)$ is an unstable node. The eigenvalues at $(2,0)$ are -2 and -1, so $(2,0)$ is an asymptotically stable node. The eigenvalues at $(0,3)$ are -1 and -6, so $(0,3)$ is an asymptotically stable node. The eigenvalues at $(1,1)$ are $1 \pm \sqrt{2}$, so $(1,1)$ is a saddle point, unstable of course. □

Example 9.2.2. *The derivative of a linear system.* Here we'll show that the approach of this section generalizes the method of Sect. 8.5 to determine the stability of the origin (the only fixed point) of linear systems. First note

$$\begin{bmatrix} x' \\ y' \end{bmatrix} = \begin{bmatrix} a & b \\ c & d \end{bmatrix} \begin{bmatrix} x \\ y \end{bmatrix} = \begin{bmatrix} ax + by \\ cx + dy \end{bmatrix} = \vec{F}(x,y)$$

Now we compute

$$D\vec{F} = \begin{bmatrix} \dfrac{\partial}{\partial x}(ax+by) & \dfrac{\partial}{\partial y}(ax+by) \\ \dfrac{\partial}{\partial x}(cx+dy) & \dfrac{\partial}{\partial y}(cx+dy) \end{bmatrix} = \begin{bmatrix} a & b \\ c & d \end{bmatrix}$$

So we see the derivative of a linear system is the coefficient matrix of the system. Consequently, the eigenvalues of the derivative matrix are the eigenvalues of the coefficient matrix. □

Example 9.2.3. *The stability of fixed points of the Fitzhugh-Nagumo equations.* Recall the Fitzhugh-Nagumo equations (7.15)

$$x' = c(y + x - x^3/3 + z) \qquad y' = -(x + a + by)/c$$

We'll take $z = 0$. Then the derivative matrix is

$$D\vec{F} = \begin{bmatrix} c(1 - x^2) & c \\ -1/c & -b/c \end{bmatrix}$$

Recall that the x- and y-nullclines given in Eq. (7.16) are

$$y = -x + x^3/3 \qquad \text{and} \qquad y = (-x + a)/b$$

Exact expressions for these fixed points are messy, so we'll compute the eigenvalues of the derivative matrix at the fixed points of the a- and b-values of the three examples of Fig. 7.22. For all, take $c = 1$.

$a = b$	fixed point(s)
1	$(1.4423, -0.4423)$
-3.221954	$(-1.1447, 0.6447)$, $(2.2894, 1.7106)$
-2	$(-1.6312, 0.1844)$, $(-0.7669, 0.6166)$, $(2.3981, 2.1991)$

- Then at the fixed point $(1.4423, -0.4423)$ the eigenvalues of the derivative matrix are $-1.0401 \pm 0.9992i$ so the fixed point is an asymptotically stable spiral.
- At the fixed point $(-1.1447, 0.6447)$ the eigenvalues are 2.91158 and 0.00004 so the fixed point is an unstable node.
- At the fixed point $(2.2894, 1.7106)$ the eigenvalues are -4.10487 and 3.08547 so the fixed point is a saddle point.
- At the fixed point $(-1.6312, 0.1844)$ the eigenvalues are 1.70269 and -1.3635 so the fixed point is a saddle point.
- At the fixed point $(-0.7669, 0.6166)$ the eigenvalues are $1.20593 \pm 0.607829i$ so the fixed point is an unstable spiral.
- At the fixed point $(2.3981, 2.1991)$ the eigenvalues are -4.5995 and 1.8485 so the fixed point is a saddle point.

A sketch of the nullclines for each case may show something interesting about how trajectories near fixed points interact with one another. □

Practice Problems

This section has a lot of practice problems because the techniques introduced here are among the most important we'll see in the book. Many repetitions insure that the methods find comfortable homes in your minds.

For Practice Problems 9.2.1–9.2.4,

(a) graph and write the equations of the nullclines,
(b) find the fixed points,
(c) find the eigenvalues of the derivative at each fixed point, and
(d) determine the stability and type of each fixed point.

9.2.1. $x' = x + y^2 \qquad y' = x + y$

9.2.2. $x' = 1 - xy \qquad y' = x - y^3$

9.2.3. $x' = y - x^2 \qquad y' = y - 2 + x^2$

9.2.4. $x' = x + x^2 + y^2 \qquad y' = y - xy$

For the population models of Practice Problems 9.2.5 and 9.2.6,

(a) sketch the nulclines and find the fixed points,
(b) find the eigenvalues of the derivative at each fixed point,
(c) determine the stability and type of each fixed point,
(d) determine the limiting behavior of x and y as $t \to \infty$, and
(e) interpret (d) in terms of the populations.

9.2.5. $dx/dt = x(1.5 - x - 0.5y)$ $dy/dt = y(2 - y - 0.75x)$

9.2.6. $dx/dt = x(1.5 - x - 0.5y)$ $dy/dt = y(2 - 0.5y - 1.5x)$

For the systems in Practice Problems 9.2.7–9.2.10,
(a) find the fixed points and
(b) determine their type and stability.

9.2.7. $x' = x^3 - x - y$ $y' = 3x - y$

9.2.8. $x' = x(4 - x - y)$ $y' = y(9 - x^2 - y^2)$

9.2.9. $x' = -2x + x^2 + y^2$ $y' = -2y + x^2 + y^2$

9.2.10. $x' = y - x^3$ $y' = x - y^3$

9.2.11. Consider the system

$$x' = k \cdot y \qquad y' = -y - \sin(x)$$

where k is a constant.
(a) Assume $-3\pi/2 \le x \le 3\pi/2$. Show the fixed points of the system are $(0,0)$ and $(\pm\pi,0)$.
(b) Find the eigenvalues of the derivative at $(0,0)$ and at $(\pm\pi,0)$.
(c) For what range of k-values is $(0,0)$ asymptotically stable? For what range of k-values are $(\pm\pi,0)$ asymptotically stable?

Practice Problem Solutions

Recall that we represent the x-nullcline by thick, solid gray curves or lines, and the y-nullcline by thick dashed gray curves or lines.

9.2.1. (a) The x-nullcline is $x = -y^2$; the y-nullcline is $x = -y$. See the top left graph of Fig. 9.1.
(b) The fixed points are the intersections of the nullclines: $(0,0)$ and $(-1,1)$.
(c) First note

$$D\vec{F} = \begin{bmatrix} 1 & 2y \\ 1 & 1 \end{bmatrix} \quad \text{so} \quad D\vec{F}(0,0) = \begin{bmatrix} 1 & 0 \\ 1 & 1 \end{bmatrix} \quad \text{and} \quad D\vec{F}(-1,1) = \begin{bmatrix} 1 & 2 \\ 1 & 1 \end{bmatrix}$$

The eigenvalues at $(0,0)$ are 1 and 1. The eigenvalues at $(-1,1)$ are $1 \pm \sqrt{2}$.
(d) We see that $(0,0)$ is an unstable node because both eigenvalues are positive;

the fixed point $(-1,1)$ is a (necessarily unstable) saddle point because one eigenvalue is positive and one is negative.

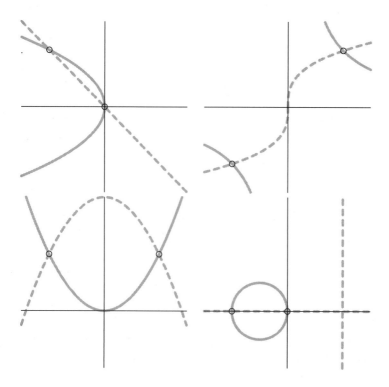

Figure 9.1. Nullclines and fixed points for Practice problems 9.2.1–9.2.4.

9.2.2. (a) The x-nullcline is $x = 1/y$; the y-nullcline is $x = y^3$. See the top right graph of Fig. 9.1.

(b) The fixed points are the intersections of the nullclines: $(1,1)$ and $(-1,-1)$.

(c) First note

$$D\vec{F} = \begin{bmatrix} -y & -x \\ 1 & -3y^2 \end{bmatrix} \text{ so } D\vec{F}(1,1) = \begin{bmatrix} -1 & -1 \\ 1 & -3 \end{bmatrix}, D\vec{F}(-1,-1) = \begin{bmatrix} 1 & 1 \\ 1 & -3 \end{bmatrix}$$

The eigenvalues at $(1,1)$ are -2 and -2. The eigenvalues at $(-1,-1)$ are $-1 \pm \sqrt{5}$.

(d) The fixed point $(1,1)$ is an asymptotically stable node because both eigenvalues are negative; the fixed point $(-1,-1)$ is a saddle point because one eigenvalue is positive and one is negative.

9.2.3. (a) The x-nullcline is $y = x^2$; the y-nullcline is $y = 2 - x^2$. See the bottom left graph of Fig. 9.1.

(b) The fixed points are the intersections of the nullclines: $(-1, 1)$ and $(1, 1)$.

(c) First note

$$D\vec{F} = \begin{bmatrix} -2x & 1 \\ 2x & 1 \end{bmatrix} \text{ so } D\vec{F}(-1,1) = \begin{bmatrix} 2 & 1 \\ -2 & 1 \end{bmatrix}, \ D\vec{F}(1,1) = \begin{bmatrix} -2 & 1 \\ 2 & 1 \end{bmatrix}$$

The eigenvalues at $(-1, 1)$ are $(3 \pm i\sqrt{7})/2$. The eigenvalues at $(1, 1)$ are $(-1 \pm \sqrt{17})/2$.

(d) The fixed point $(-1, 1)$ is an unstable spiral because the eigenvalues are complex with positive real parts; the fixed point $(1, 1)$ is a saddle point because one eigenvalue is negative and the other is positive.

9.2.4. (a) The x-nullcline is $(x + 1/2)^2 + y^2 = 1/4$ (complete the square); the y-nullcline is $y = 0$ and $x = 1$. See the bottom right graph of Fig. 9.1.

(b) The fixed points are the intersections of the nullclines: $(-1, 0)$ and $(0, 0)$.

(c) First note

$$D\vec{F} = \begin{bmatrix} 1 + 2x & 2y \\ -y & 1 - x \end{bmatrix} \text{ so } D\vec{F}(-1,0) = \begin{bmatrix} -1 & 0 \\ 0 & 2 \end{bmatrix}, \ D\vec{F}(0,0) = \begin{bmatrix} 1 & 0 \\ 0 & 1 \end{bmatrix}$$

The eigenvalues at $(-1, 0)$ are -1 and 2. The eigenvalues at $(0, 0)$ are 1 and 1.

(d) The fixed point $(-1, 0)$ is a saddle point (unstable) because one eigenvalue is negative and one is positive; the fixed point $(0, 0)$ is an unstable node because both eigenvalues are positive.

9.2.5. (a) The x-nullcline is $x = 0$ and $y = 3 - 2x$; the y-nullcline is $y = 0$ and $y = 2 - (3/4)x$. See the left graph of Fig. 9.2. The fixed points are the intersections of the nullclines:

$$(0, 0) \qquad (3/2, 0) \qquad (0, 2) \qquad (4/5, 7/5)$$

(b) First note

$$D\vec{F} = \begin{bmatrix} 3/2 - 2x - y/2 & -x/2 \\ -3y/4 & 2 - 3x/4 - 2y \end{bmatrix}$$

Then

$$D\vec{F}(0,0) = \begin{bmatrix} 3/2 & 0 \\ 0 & 2 \end{bmatrix} \qquad D\vec{F}(3/2,0) = \begin{bmatrix} -3/2 & -3/4 \\ 0 & 7/8 \end{bmatrix}$$

$$D\vec{F}(0,2) = \begin{bmatrix} 1/2 & 0 \\ -3/2 & -2 \end{bmatrix} \qquad D\vec{F}(4/5,7/5) = \begin{bmatrix} -4/5 & -2/5 \\ -21/20 & -7/5 \end{bmatrix}$$

The eigenvalues are

$3/2, 2$ for $(0,0)$, $-3/2, 7/8$ for $(3/2,0)$, $1/2, -2$ for $(0,2)$, and

$$\frac{-11 \pm \sqrt{51}}{10} \approx -1.814, -0.386 \text{ for } (4/5, 7/5)$$

(c) The fixed point $(0,0)$ is an unstable node because both eigenvalues are positive. The fixed points $(3/2,0)$ and $(0,2)$ are saddle points because for both one eigenvalue is positive and one is negative. The fixed point $(4/5, 7/5)$ is an asymptotically stable node because both eigenvalues are negative.

(d) So long as both x and y are non-zero, $(x,y) \to (4/5, 7/5)$. If $y = 0$, then $x \to 3/2$. If $x = 0$, then $y \to 2$.

(e) Populations consisting of x alone converge to $x = 3/2$. Populations consisting of y alone converge to $y = 2$. Populations with non-zero x and y converge to $(4/5, 7/5)$.

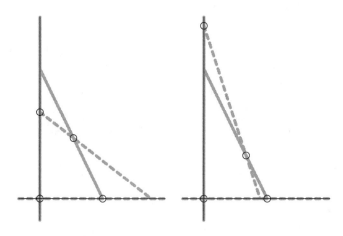

Figure 9.2. Nullclines and fixed points for Practice Problems 9.2.5 and 9.2.6.

9.2.6. (a) The x-nullcline is $x = 0$ and $y = 3 - 2x$; the y-nullcline is $y = 0$ and $y = 4 - 3x$. See the right graph of Fig. 9.2. The fixed points are the intersections of the nullclines:

$$(0,0) \qquad (3/2,0) \qquad (0,4) \qquad (1,1)$$

(b) First note

$$D\vec{F} = \begin{bmatrix} 3/2 - 2x - y/2 & -x/2 \\ -3y/2 & 2 - 3x/2 - y \end{bmatrix}$$

Then

$$DF(0,0) = \begin{bmatrix} 3/2 & 0 \\ 0 & 2 \end{bmatrix} \qquad DF(3/2,0) = \begin{bmatrix} -3/2 & -3/4 \\ 0 & -1/4 \end{bmatrix}$$

$$DF(0,4) = \begin{bmatrix} -1/2 & 0 \\ -6 & -2 \end{bmatrix} \qquad DF(1,1) = \begin{bmatrix} -1 & -1/2 \\ -3/2 & -1/2 \end{bmatrix}$$

The eigenvalues are

$3/2, 2$ for $(0,0)$, $-3/2, -1/4$ for $(3/2,0)$, $-1/2, -2$ for $(0,4)$, and

$$\frac{-3 \pm \sqrt{13}}{4} \approx -1.651,\ 0.151 \text{ for } (1,1)$$

(c) The fixed point $(0,0)$ is an unstable node because both eigenvalues are positive. The fixed points $(3/2,0)$ and $(0,4)$ are asymptotically stable nodes because for both fixed points both eigenvalues are negative. The fixed point $(1,1)$ is a saddle point because one eigenvalue is positive and the other is negative.

(d) For (x,y) near enough to $(3/2,0)$, $(x,y) \to (3/2,0)$. For (x,y) near enough to $(0,4)$, $(x,y) \to (0,4)$.

(e) Unless it starts exactly at $(1,1)$ (or $(0,0)$, of course), eventually one species becomes extinct, or remains extinct if the starting point is on an axis.

9.2.7. (a) The x-nullcline is $x^3 - x - y = 0$, that is, $y = x^3 - x$. The y-nullcline is $3x - y = 0$, that is, $y = 3x$. The x-coordinates of the fixed points are the solutions of $x^3 - x = 3x$, that is, $x^3 - 4x = 0$, so $x = 0, \pm 2$. Then the fixed points are $(-2,-6)$, $(0,0)$, and $(2,6)$.

(b) First compute the derivative

$$DF(x,y) = \begin{bmatrix} -1 + 3x^2 & -1 \\ 3 & -1 \end{bmatrix}$$

At the fixed points we have

$$DF(0,0) = \begin{bmatrix} -1 & -1 \\ 3 & -1 \end{bmatrix} \quad \text{and} \quad DF(\pm 2, \pm 6) = \begin{bmatrix} 11 & -1 \\ 3 & -1 \end{bmatrix}$$

The eigenvalues are $-1 \pm i\sqrt{3}$ for $(0,0)$ and $5 \pm \sqrt{33}$ for both $(2,6)$ and $(-2,-6)$. So the fixed point at the origin is an asymptotically stable spiral (complex eigenvalues with negative real parts) and the fixed points $(\pm 2, \pm 6)$ are saddle points (one positive eigenvalue, one negative eigenvalue).

9.2.8. (a) The x-nullcline is the lines $x = 0$ and $y = -x + 4$. The y-nullcline is the line $y = 0$ and the circle $x^2 + y^2 = 9$. The fixed points are the intersections

of the x-nullcline and the y-nullcline. That is, $(0,0)$, $(2+1/\sqrt{2},2-1/\sqrt{2})$, and
$(2-1/\sqrt{2},2+1/\sqrt{2})$.
(b) First, compute the derivative

$$D\vec{F}(x,y) = \begin{bmatrix} 4-2x-y & -x \\ -2xy & 9-x^2-3y^2 \end{bmatrix}$$

At the fixed points we have

$$D\vec{F}(0,0) = \begin{bmatrix} 4 & 0 \\ 0 & 9 \end{bmatrix}$$

$$D\vec{F}(2+1/\sqrt{2},2-1/\sqrt{2}) = \begin{bmatrix} -2-1/\sqrt{2} & -2-1/\sqrt{2} \\ -7 & -9+4\sqrt{2} \end{bmatrix}$$

$$D\vec{F}(2-1/\sqrt{2},2+1/\sqrt{2}) = \begin{bmatrix} -2+1/\sqrt{2} & -2+1/\sqrt{2} \\ -7 & -9-4\sqrt{2} \end{bmatrix}$$

The eigenvalues are 4 and 9 for $(0,0)$, so this fixed point is an unstable node.
The eigenvalues are approximately -7.39 and 1.34 for $(2+1/\sqrt{2},2-1/\sqrt{2})$,
so this fixed point is a saddle point. The eigenvalues are approximately -15.30
and -0.65 for $(2-1/\sqrt{2},2+1/\sqrt{2})$, so this fixed point is an asymptotically
stable node. The $\lambda \approx -15.30$ has an interesting interpretation. Trajectories near
the line passing through this fixed point and in the direction $\langle 1,10.8361 \rangle$ (an
eigenvector for this eigenvalue) will converge toward the fixed point about 23.5
times more rapidly than will trajectories near the line through the fixed point and
in the direction $\langle 1,-0.499645 \rangle$ (an eigenvector for the eigenvalue $\lambda \approx -0.65$).
This difference of speeds suggests that any system modeled by this differential
equation has two substantially different time scales.

9.2.9. (a) Completing the square, we see the x-nullcline is $(x-1)^2+y^2=1$ and
the y-nullcline is $x^2+(y-1)^2=1$. The fixed points are the intersections of the
x- and y-nullclines; that is, the points $(0,0)$ and $(1,1)$.
(b) First compute the derivative

$$D\vec{F}(x,y) = \begin{bmatrix} -2+2x & 2y \\ 2x & -2+2y \end{bmatrix}$$

At the fixed points we have

$$D\vec{F}(0,0) = \begin{bmatrix} -2 & 0 \\ 0 & -2 \end{bmatrix} \quad \text{and} \quad D\vec{F}(1,1) = \begin{bmatrix} 0 & 2 \\ 2 & 0 \end{bmatrix}$$

The eigenvalues are -2 and -2 for $(0,0)$ so the fixed point at the origin is an asymptotically stable node. The eigenvalues are ± 2 for $(1,1)$ so this fixed point is a saddle point.

9.2.10. (a) The x-nullcline is $y = x^3$; the y-nullcline is $x = y^3$. The fixed points are the intersections of the x- and y-nullclines; that is, the points $(-1,-1)$, $(0,0)$, and $(1,1)$.
(b) First compute the derivative

$$D\vec{F}(x,y) = \begin{bmatrix} -3x^2 & 1 \\ 1 & -3y^2 \end{bmatrix}$$

At the fixed points we have

$$D\vec{F}(0,0) = \begin{bmatrix} 0 & 1 \\ 1 & 0 \end{bmatrix} \quad \text{and} \quad D\vec{F}(\pm 1, \pm 1) = \begin{bmatrix} -3 & 1 \\ 1 & -3 \end{bmatrix}$$

The eigenvalues are ± 1 for $(0,0)$ and $-2, -4$ for $(\pm 1, \pm 1)$. So $(0,0)$ is a saddle point and $(-1,-1)$ and $(1,1)$ are asymptotically stable nodes.

9.2.11. (a) The x-nullcline is $y = 0$; the y-nullcline is $y = \sin(x)$. In the range $-3\pi/2 \leq x \leq 3\pi/2$, the nullclines intersect at the points $(0,0)$ and $(\pm\pi, 0)$.
(b) First compute the derivative

$$D\vec{F}(x,y) = \begin{bmatrix} 0 & k \\ -\cos(x) & -1 \end{bmatrix}$$

At the fixed points we have

$$D\vec{F}(0,0) = \begin{bmatrix} 0 & k \\ -1 & -1 \end{bmatrix} \quad \text{and} \quad D\vec{F}(\pm\pi, 0) = \begin{bmatrix} 0 & k \\ 1 & -1 \end{bmatrix}$$

The eigenvalues are $(-1 \pm \sqrt{1-4k})/2$ for $(0,0)$ and $(-1 \pm \sqrt{1+4k})/2$ for $(\pm\pi, 0)$. So all three fixed points are unstable.
(c) The fixed point $(0,0)$ is asymptotically stable if the eignvalues are complex (because they have real parts, $-1/2$), or real and negative. That is,

- *complex* if $1 - 4k < 0$, that is, $1/4 < k$, and
- *real and negative* if $0 < 1 - 4k$ and $\sqrt{1-4k} < 1$, that is, $0 < k < 1/4$. Combining these, the fixed point $(0,0)$ is asymptotically stable for $0 < k$.
  The fixed points $(\pm\pi, 0)$ are asymptotically stable if the eigenvalues are complex (because they have negative real parts, $-1/2$), or real and negative. That is,
- *complex* if $1 + 4k < 0$, that is, $k < -1/4$, and

- *real and negative* if $0 < 1+4k$ and $\sqrt{1+4k} < 1$, that is, $-1/4 < k < 0$.
  Combining these, the fixed points $(\pm\pi, 0)$ are asymptotically stable for $k < 0$.

Exercises

In these problems
(a) sketch the nullclines,
(b) find the fixed points graphically and numerically,
(c) evaluate the derivative at each fixed point,
(d) find the eigenvalues of these derivatives, and
(e) determine the stability and type of each fixed point to which the Hartman-Grobman theorem can be applied.

9.2.1. $x' = y - x^2 \qquad y' = x - y^3$

9.2.2. $x' = xy \qquad y' = 1 - x^2 - y^2$

9.2.3. $x' = y + 1 + x^2 \qquad y' = x + 1 + y^2$

9.2.4. $x' = x^2 - y^2 \qquad y' = 1 - y^2$

9.2.5. $x' = x^2 + y^2 - 4 \qquad y' = y - 1$

9.2.6. $x' = (x-1)^2 + y^2 - 1 \qquad y' = x^2 + (y-1)^2 - 1$

9.2.7. $x' = y + x^2 - 1 \qquad y' = y - x^2 + 1$

9.2.8. $x' = y - \cos(x) \qquad y' = y - \sin(x) \qquad 0 \le x, y \le \pi$

9.2.9. $x' = y - e^x \qquad y' = y - x - 2$

9.2.10. $x' = y - x^4 \qquad y' = y - x^2$

9.3 THE PENDULUM AS GUIDE

So the Hartman-Grobman theorem tells us how to determine the stability of fixed points, except when the eigenvalues give a trace and determinant on one of the three boundaries of regions in the tr-det plane described in Sect. 8.6. Recall these boundaries are

- det $= 0$; at least one eigenvalue is 0.
- tr $= 0$, det > 0; the eigenvalues are imaginary.
- det $=$ tr$^2/4$; the eigenvalues are equal.

In this section we'll present a physical illustration of a method, developed in Sects. 9.4 and 9.5, to determine the stability in some of these boundary cases. In Sect. A.9 we'll use this construction to sketch a proof of part of the Hartman-Grobman theorem for systems of differential equations that are not too nonlinear.

We'll start with a frictionless pendulum, a system of nonlinear differential equations with periodic trajectories, and study its total (= kinetic + potential) energy. In Example 9.4.4 we'll add in friction and see what changes.

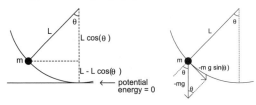

Figure 9.3. Computing the energy (left) and the forces (right) on the pendulum.

Imagine a pendulum of length L and of mass m. Denote by θ the angle of the pendulum from downward, and scale the potential energy so it is 0 at the lowest point of the pendulum's swing. See the left side of Fig. 9.3. The potential energy U of the pendulum is mgh, where g is the gravitational acceleration and h is the height above the zero point of the potential energy, that is,

$$U = mgh = mgL(1 - \cos(\theta)) \qquad (9.7)$$

Recall that the kinetic energy is $T = (1/2)mv^2$. For motion along a circle of radius L, the distance traveled as the angle (measured in radians) swings through θ is the fraction of the circumference $L2\pi$ determined by an arc subtended by an angle θ from the center, that is, $L\theta$. So because L is constant

$$v = (L\theta)' = L\theta'$$

and the kinetic energy of the pendulum is

$$T = \frac{1}{2}mv^2 = \frac{1}{2}mL^2(\theta')^2 \qquad (9.8)$$

To avoid notational confusion later, write $x = \theta$ and $y = \theta'$. Combining Eqs. (9.7) and (9.8), we see the total energy $V(x,y)$ of the pendulum is

$$V(x,y) = mgL(1 - \cos(x)) + \frac{1}{2}mL^2y^2 \qquad (9.9)$$

What equations describe the motion of a pendulum? Recall Newton's second law of motion

$$\vec{F} = \frac{d\vec{p}}{dt} = \frac{d(m\vec{v})}{dt} = m\vec{v}'$$

where $\vec{F}$ denotes the force applied to the object, $\vec{p} = m\vec{v}$ the momentum, and $\vec{v}' = \vec{a}$ is the acceleration. The gravitational force acting on the pendulum is $-mg$, the minus because the gravitational force is directed downward. Some of this force, the component in the direction of the pendulum line, maintains the tension on the line and does not cause the pendulum to move. The force component tangent to the circular path of the pendulum, and consequently perpendicular to the line, accelerates the pendulum. As seen on the right side of Fig. 9.3, this component is $-mg\sin(\theta)$. Then for this pendulum, Newton's second law is

$$-mg\sin(\theta) = mv' = m(L\theta')' = mL\theta''$$

Canceling m and recalling $x = \theta$ and $y = \theta'$, this becomes

$$-g\sin(x) = Ly'$$

This is a single equation. How do we get a system of two equations? Recall $y = \theta' = x'$. Then we have

$$x' = y \qquad y' = -\frac{g}{L}\sin(x) \qquad\qquad (9.10)$$

From this, we see that the state of the pendulum, and consequently its future motion, is determined by its position θ and speed θ', so the phase plane has coordinates $x = \theta$ and $y = \theta'$. In Fig. 9.4 we illustrate the vector field (x', y') and a trajectory in the phase plane. Say the pendulum starts from the point A. This indicates a positive initial displacement ($\theta > 0$) and zero initial speed ($\theta' = 0$).

The pendulum is pulled some distance to the right and released from rest.

As it begins to fall toward the lowest point of its swing, the pendulum accelerates to the left. That is, θ decreases toward 0 and θ' becomes ever more negative. At the point B the pendulum weight is directly under the pivot point.

Because it has some leftward speed, the pendulum continues swinging past B, but now gravity slows the motion of the pendulum. The pendulum slows (θ' goes to 0),

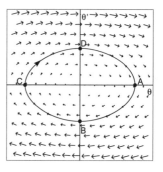

Figure 9.4. A trajectory and the vector field for the pendulum.

while the pendulum moves ever leftward. At the phase plane point C, the pendulum is as far left as it can go, and is at rest.

The pendulum begins to move to the right, with increasing rightward velocity ($\theta' > 0$), until point D. Again, the pendulum is vertical, but now moving to the right. From this point, gravity slows the pendulum until it stops momentarily at its right-most point. This occurs at phase plane point A, and the process repeats.

This picture and the accompanying description suggest that the pendulum has periodic trajectories. Can the total energy of the pendulum support this?

Suppose $(x(t), y(t))$ is a trajectory of the pendulum, a solution curve for Eq. (9.10). How does $V(x,y)$ change as t increases? First, apply the several variable chain rule, Eq. (9.4),

$$\frac{d}{dt}V(x(t),y(t)) = \frac{\partial V}{\partial x}\cdot\frac{dx}{dt} + \frac{\partial V}{\partial y}\cdot\frac{dy}{dt}$$
$$= mgL\sin(x)\cdot x' + mL^2 y\cdot y'$$
$$= mgL\sin(x)\cdot y + mL^2 y\cdot\left(-\frac{g}{L}\sin(x)\right) = 0$$

where the penultimate equality is the result of the differential equation (9.10). We have shown that the total energy remains unchanged along every pendulum trajectory. From this, how do we deduce that the trajectories are periodic?

Suppose $(x(t_0),y(t_0))$ crosses the positive θ-axis at the point $(\theta_0,0)$. The argument accompanying Fig. 9.4 shows the trajectory will cross the positive θ-axis at some later time $t_1 > t_0$. By Eq. (9.9) we see

$$V(\theta_0,0) = mgL(1 - \cos(\theta_0)) \quad \text{and} \quad V(\theta_1,0) = mgL(1 - \cos(\theta_1))$$

We assume the pendulum does not swing around in a complete circle, so $-\pi < \theta < \pi$. Then $\cos(\theta_0) = \cos(\theta_1)$ implies $\theta_0 = \theta_1$ because θ_0 and θ_1 lie on the positive θ-axis, so $0 < \theta_0, \theta_1 < \pi$, and consequently the trajectory is periodic because successive crossings of the positive x-axis occur at the same point.

This section was just an example to prepare our intuition. No need for practice problems or exercises here. We'll have plenty of those in the next section.

9.4 LIAPUNOV FUNCTIONS

For a general nonlinear differential equation

$$x' = f(x,y) \qquad y' = g(x,y) \tag{9.11}$$

can we find a function $V(x,y)$ that acts like the total energy of the pendulum, and from properties of this function can we deduce the stability of fixed points at which the derivative has trace and determinant on one of the boundaries of the tr-det plane? Such a function is called a *Liapunov* (or *Lyapunov* or *Ljapunov*) *function*. It's the same Laipunov, the Russian mathematician and physicist Aleksandr Liapunov, who invented Liapunov exponents.

We'll develop some background concepts, state Liapunov's theorem, and illustrate its use with some examples. In Sect. A.8 we'll sketch a proof of one case of the theorem, and in Sect. A.9 use Liapunov's theorem to prove the Hartman-Grobman theorem, at least if the differential equation is not too nonlinear.

Suppose $(0,0)$ is a fixed point of the differential equation, the only fixed point in the disc $D = \{(x,y) : x^2 + y^2 < r^2\}$, for some $r > 0$. In Sect. 9.5 we'll see how to deal with fixed points not at the origin.

A function $V : D \to \mathbb{R}$ is called

- *positive definite* if $V(0,0) = 0$ and $V(x,y) > 0$ for all $(x,y) \neq (0,0)$,
- *negative definite* if $V(0,0) = 0$ and $V(x,y) < 0$ for all $(x,y) \neq (0,0)$,
- *positive semidefinite* if $V(x,y) \geq 0$ for all $(x,y) \in D$, and
- *negative semidefinite* if $V(x,y) \leq 0$ for all $(x,y) \in D$.

In a moment we'll see examples of such functions, and develop some strategies to find them.

Applying the chain rule, Eq. (9.6), and the system Eq. (9.11) we see

$$V' = \frac{dV}{dt} = \frac{\partial V}{\partial x}\frac{dx}{dt} + \frac{\partial V}{\partial y}\frac{dy}{dt} = \frac{\partial V}{\partial x}f(x,y) + \frac{\partial V}{\partial y}g(x,y) \qquad (9.12)$$

Taking $V =$ potential energy $+$ kinetic energy for the frictionless pendulum, in Sect. 9.3 we saw that along every phase plane trajectory $V' = 0$, which implied the orbits are periodic. In Example 9.4.4 we'll see that if friction is added to the pendulum, then $V' < 0$ except at those points where the direction of swing reverses. Even though the total energy is only negative semidefinite and not negative definite, because we know something about how the system works we can show that the total energy decreases to 0 and the pendulum settles down to the origin. This shows the origin is an asymptotically stable fixed point.

But also, the fact that V' is negative semidefinite opens the possibility to try a different choice for a Liapunov function, which we'll also do in Example 9.4.4. This emphasizes that a Liapunov function need not be related to any physical quantity. We just need to find a function V with the right definiteness for V and V'. These behaviors of V and V' were generalized by Liapunov into this theorem.

Theorem 9.4.1. *Liapunov's theorem.* Suppose we can find a function $V : D \to \mathbb{R}$ with continuous partial derivatives. Then

1. if V is positive definite and V' is negative definite, then $(0,0)$ is asymptotically stable,
2. if V is positive definite and V' is negative semidefinite, then $(0,0)$ is stable,
3. if V is positive definite and V' is positive definite, then $(0,0)$ is unstable, and
4. if V is negative definite and V' is negative definite, then $(0,0)$ is unstable.

Conditions (3) and (4) of the theorem can be weakened a bit and still guarantee that the origin is unstable:

3.' If V' is positive definite and every disc centered at $(0,0)$ contains at least one point at which V is positive, then $(0,0)$ is unstable.
4.' If V' is negative definite and every disc centered at $(0,0)$ contains at least one point at which V is negative, then $(0,0)$ is unstable.
   And we should mention that conditions (1) and (2) have mirror images:
1.' If V is negative definite and V' is positive definite, then $(0,0)$ is asymptotically stable.
2.' If V is negative definite and V' is positive semidefinite, then $(0,0)$ is stable.

As far as definiteness (positive and negative definite, positive and negative semidefinite) is concerned, V and V' have 16 combinations. We've mentioned 6. What do you think happens with the other 10?

The point that sometimes causes confusion is really the strength of Liapunov's theorem: the function V needn't have a connection with energy or for that matter with any of the quantities modeled by the differential equation. Any function V will do, so long as V and $V' = (\partial V / \partial x)f + (\partial V / \partial y)g$ have appropriate sorts of definiteness. Sometimes considerations of energy or other physical variables can guide the choice of V. But other times it's better to look at f, look at g, make a guess for V, and if the definitenesses don't work out, modify your guess and try again.

First exposure to Liapunov functions can be a bit puzzling, so we'll begin with an easy example.

Example 9.4.1. *A simple system of nonlinear differential equations.* Consider the system

$$x' = -y - x^3 \qquad y' = x - y^3$$

A plot of the nullclines reveals that the origin is the only fixed point. The derivative matrix is

$$DF(0,0) = \begin{bmatrix} 0 & -1 \\ 1 & 0 \end{bmatrix}$$

The eigenvalues of $DF(0,0)$ are i and $-i$, so the Hartman-Grobman theorem cannot be applied. Can we find a Liapunov function? Note the cubic terms of both x' and y' are negative, but the linear terms have opposite signs. We try to find a positive definite V for which the linear terms of x' and y' contribute terms of V' that cancel. This may seem pretty vague, but with practice we'll develop some intuition. The simplest positive definite function of two variables, often our first guess for V, is

$$V(x,y) = x^2 + y^2$$

Then

$$V' = \frac{\partial V}{\partial x} \cdot x' + \frac{\partial V}{\partial y} \cdot y' = 2x(-y-x^3) + 2y(x-y^3) = -2(x^4+y^4)$$

Certainly, this V' is negative definite, so by condition (1) of Liapunov's theorem, $(0,0)$ is asymptotically stable for this system. □

Finding a Liapunov function in Ex. 9.4.1 was straightforward. The most obvious positive definite function, $x^2 + y^2$, worked because the terms $2x(-y)$ and $2y(x)$ subtracted out. To emphasize this is not our usual luck when seeking Liapunov functions, next we consider a slightly less simple example.

Example 9.4.2. *A bit less simple system of nonlinear differential equations.* Now consider

$$x' = -2y + x^3 \qquad y' = x + y^3$$

Again, plotting the nullclines shows that the origin is the only fixed point. The eigenvalues of $DF(0,0)$ are $\pm i\sqrt{2}$, so again Hartman-Grobman can't be applied. Let's try $V(x,y) = x^2 + y^2$ again. Unless there's an obvious reason not to use this, it's always a good first guess. We have a positive definite V and for V' we find

$$V' = \frac{\partial V}{\partial x} \cdot x' + \frac{\partial V}{\partial y} \cdot y' = 2x(-2y+x^3) + 2y(x+y^3) = 2x^4 + 2y^4 - 2xy$$

Now we have a problem: $V'(x,0) = 2x^4$ is positive except at $x=0$ and $V'(x,x) = 4x^4 - 2x^2$ is negative for $0 < |x| < 1/\sqrt{2}$. So V' is neither positive definite nor negative definite, for all discs D about the origin.

The problem is that, unlike in Example 9.4.1, here the xy terms do not subtract out because the $(\partial V/\partial x) \cdot x'$ term contributes $-4xy$ and the $(\partial V/\partial y) \cdot$

y' term contributes $2xy$. The simplest way to deal with this is to change V to

$$V(x,y) = x^2 + 2y^2$$

This still is positive definite, and now

$$V' = 2x(-2y + x^3) + 4y(x + y^3) = 2x^4 + 4y^4$$

This V' is positive definite, so by condition (3) of Liapunov's theorem, the origin is unstable for this system. □

So far we've been able to recognize positive definite or negative definite functions easily: $ax^{2n} + by^{2n}$ is positive definite if the coefficients a and b are positive, and negative definite if a and b are negative. But some functions that include an xy term are positive definite and some are negative definite. Here's how to tell.

Proposition 9.4.1. The function $V(x,y) = ax^2 + bxy + cy^2$ is positive definite if and only if $a > 0$ and $4ac - b^2 > 0$, negative definite if and only if $a < 0$ and $4ac - b^2 > 0$.

We explore the proof of this proposition later in this section.

Example 9.4.3. *A system trickier still.* Now consider this system

$$x' = xy^2 - x^3 \qquad y' = x^2y - 2y^3$$

The x-nullcline is the solutions of $0 = xy^2 - x^3 = x(y^2 - x^2)$, that is, the lines $x = 0$ and $y = \pm x$. The y-nullcline is the solutions of $0 = y(x^2 - 2y^2)$, that is, the lines $y = 0$ and $y = \pm(1/\sqrt{2})x$. The x-nullcline intersects the y-nullcline only at the origin, so that's the only fixed point.

The derivative matrix is

$$D\vec{F} = \begin{bmatrix} y^2 - 3x^2 & 2xy \\ 2xy & x^2 - 6y^2 \end{bmatrix} \quad \text{and so} \quad D\vec{F}(0,0) = \begin{bmatrix} 0 & 0 \\ 0 & 0 \end{bmatrix}$$

Both eigenvalues of $D\vec{F}(0,0)$ are 0, so this matrix is on the $\det = 0$ boundary of the tr-det plane and thus Hartman-Grobman won't tell us anything. We'll try a Liapunov function.

As usual, start with the positive definite $V(x,y) = x^2 + y^2$. Then

$$V' = \frac{\partial V}{\partial x}x' + \frac{\partial V}{\partial y}y' = 2x(xy^2 - x^3) + 2y(x^2y - 2y^3) = -2x^4 + 4x^2y^2 - 4y^4$$

Here the trick of Example 9.4.2 won't work: because the $x^2 y^2$ term of $(\partial V/\partial x)x'$ has the same sign as the $x^2 y^2$ term of $(\partial V/\partial y)y'$, we can't make them subtract out without changing the sign of either $\partial V/\partial x$ or $\partial V/\partial y$. But this would destroy the positive definiteness of V. Maybe you can find a significantly different form of V that will work. For now, we'll take a different approach.

So we'd like to apply Prop. 9.4.1, but V' consists of quartic terms, while the proposition involves quadratic terms. But if we substitute $u = x^2$ and $v = y^2$ we find

$$V'(u,v) = -2u^2 + 4uv - 4v^2$$

and we can apply the proposition to this. Matching coefficients of this $V'(u,v)$ with those of $V(x,y)$ in the proposition, we find $a = -2$, $b = 4$, and $c = -4$. Then $a = -2 < 0$ and $4ac - b^2 = 16 > 0$. We conclude that V' is negative definite. Then by Liapunov's theorem, the origin is asymptotically stable because V is positive definite and V' is negative definite. $\square$

Now we return to the pendulum, but to a more realistic model, one that includes friction.

Example 9.4.4. *The damped pendulum.* This example is complicated enough that one of my students suggested dropping the letter "p" in the word "damped." But recall that complicated examples can teach subtle tricks. We should treasure these examples.

Friction reduces velocity at a rate proportional to the velocity, so the damped pendulum equations are

$$x' = y \qquad y' = -\frac{g}{L}\sin(x) - cy \qquad (9.13)$$

where $c > 0$ is the friction proportionality constant. Intuition, or experience, perhaps uncommon in an era having few grandfather clocks, suggests that the damped pendulum eventually stops swinging. That is, $(\theta, \theta') = (0,0)$ is an asymptotically stable fixed point. This follows from the Hartman-Grobman theorem because

$$D\vec{F}(x,y) = \begin{bmatrix} 0 & 1 \\ -(g/L)\cos(x) & -c \end{bmatrix}$$

and the eigenvalues of $D\vec{F}(0,0)$ are

$$\frac{-cL \pm \sqrt{c^2 L^2 - 4gL}}{2L}$$

If these eigenvalues are real, both are negative because $c^2 L^2 - 4gL < c^2 L^2$ and $\sqrt{c^2 L^2 - 4gL} < cL$. If the eigenvalues are complex, both have negative real parts.

In either case, the fixed point $(0,0)$ is asymptotically stable, as we expected.

Can we test the stability of the origin of the damped pendulum with a Liapunov function? By Liapunov's theorem (Thm. 9.4.1), we seek a positive definite Liapunov function V with V' negative definite. Let's try the total energy Eq. (9.9) of the pendulum. So long as we restrict $x = \theta$ to lie in the range $-\pi < x < \pi$ so the pendulum does not swing all the way round in a circle, the total energy $V(x,y)$ is positive definite. However, for V' we find

$$V' = \frac{\partial V}{\partial x} \cdot x' + \frac{\partial V}{\partial y} \cdot y'$$

$$= mgL\sin(x) \cdot y + mL^2 y \cdot \left(-\frac{g}{L}\sin(x) - cy \right) = -cmL^2 y^2$$

No matter how small the disc D centered at $(0,0)$, on that disc V' is negative semidefinite, because $V' = 0$ all along the line $0 = y = \theta'$. As we see in Fig. 9.4, trajectories cross this line at the points where the direction of swing reverses. However, so long as $x \neq 0$, points (x,y) where $y = 0$ are not fixed points. Consequently, along any trajectory, V' is negative except for isolated (non-fixed) points where it is 0, so the total energy V decreases to 0, the pendulum stops, and the origin is an asymptotically stable fixed point.

Can we find a better Liapunov function, one where V' is negative definite and so without knowledge of pendulum physics, we can conclude that the origin is asymptotically stable? For the moment, let's stick with quadratic V, but more general than $x^2 + y^2$.

First, simplify the damped pendulum equations by rescaling (take units so $g = L$ and $c = 1$) to absorb the constants:

$$x' = y \qquad y' = -\sin(x) - y$$

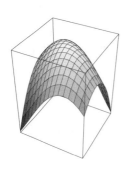

As a first guess for a quadratic V, we take $V(x,y) = x^2 + xy + y^2$, positive definite by Prop. 9.4.1 because here $a = b = c = 1$ so $a > 0$ and $4ac - b^2 = 3 > 0$. With this choice of V, we find

$$V' = (2x+y) \cdot y + (x+2y) \cdot (-\sin(x) - y)$$
$$= xy - y^2 - 2y\sin(x) - x\sin(x)$$

The plot of Fig. 9.5 shows V' is negative definite, at least on a small enough disc centered at the origin. The highest point of this graph is $V'(0,0) = 0$, so all other plotted points have $V'(x,y) < 0$ and we see that V' is negative definite. (Here we must restrict the domain of

Figure 9.5. Plot of V'.

V' to $-1 \leq x \leq 1, -1 \leq y \leq 1$, because V' does take on positive values for large enough x and y.) Then by Liapunov's theorem, we deduce, once again, that the origin is asymptotically stable for the damped pendulum. $\square$

Again, we can obtain the conclusion of this last example much more easily from the eigenvalues of the derivative matrix. Liapunov's theorem is more useful when at the fixed point the derivative matrix has trace and determinant on one of the three boundaries in the tr-det plane.

We'll make a few more comments about this example.

First, suppose you don't want to plot $V'(x,y)$. Is there a calculation we can use to tell if V' is positive definite or negative definite? There is. If $V'(0,0) = 0$ and $(0,0)$ is a local max of V', then V' is negative definite near $(0,0)$; if $V'(0,0) = 0$ and $(0,0)$ is a local min of V', then V' is positive definite near $(0,0)$. So we need a way to tell if $(0,0)$ is a local max or a local min for V'. To see how to do this, first recall the second-derivative test for local max and local min from single-variable calculus. We'll assume the second derivative of $f(x)$ is continuous. If $x = x_0$ is a local max or a local min, then $f'(x_0) = 0$. If $f''(x_0) > 0$, then near x_0 the graph of f is concave up and x_0 is a local min. If $f''(x_0) < 0$, then near x_0 the graph of f is concave down and x_0 is a local max. Here's how to extend this idea to the second-derivative test for two variables.

Second-derivative test Suppose $F(x,y)$ has continuous second partial derivatives, and suppose

$$\frac{\partial F}{\partial x}(x_0, y_0) = \frac{\partial F}{\partial y}(x_0, y_0) = 0$$

Also suppose

$$\left(\frac{\partial^2 F}{\partial x^2}(x_0, y_0)\right) \cdot \left(\frac{\partial^2 F}{\partial y^2}(x_0, y_0)\right) - \left(\frac{\partial^2 F}{\partial x \partial y}(x_0, y_0)\right)^2 > 0 \qquad (9.14)$$

- If $(\partial^2 F/\partial x^2)(x_0, y_0) > 0$, then (x_0, y_0) is a local min.
- If $(\partial^2 F/\partial x^2)(x_0, y_0) < 0$, then (x_0, y_0) is a local max.

We'll sketch a proof in Sect. A.6. Why use $(\partial^2 F/\partial x^2)(x_0, y_0)$ instead of $(\partial^2 F/\partial y^2)(x_0, y_0)$? The answer is that inequality (9.14) guarantees that $(\partial^2 F/\partial x^2)(x_0, y_0) > 0$ if and only if $(\partial^2 F/\partial y^2)(x_0, y_0) > 0$, so we could state the second-derivative test using either $\partial^2 F/\partial x^2$ or $\partial^2 F/\partial y^2$.

Note that Prop. 9.4.1 is a consequence of the second-derivative test:

$$\frac{\partial^2 F}{\partial x^2} = \frac{\partial}{\partial x}\frac{\partial}{\partial x}\left(ax^2 + bxy + cy^2\right) = \frac{\partial}{\partial x}\left(2ax + by\right) = 2a$$

and similarly $\partial^2 F/\partial x \partial y = b$ and $\partial^2 F/\partial y^2 = 2c$. Then the inequality (9.14) becomes $4ac - b^2 > 0$, exactly the condition of Prop. 9.4.1. Also, $\partial^2 F/\partial x^2 = 2a$ so the sign of $\partial^2 F/\partial x^2$ equals the sign of a.

And now we return to $V'(x,y) = xy - y^2 - 2y\sin(x) - x\sin(x)$ of Example 9.4.4. We'll compute the second partial derivatives

$$\frac{\partial^2 V'}{\partial x^2} = \frac{\partial}{\partial x}\frac{\partial}{\partial x}\left(xy - y^2 - 2y\sin(x) - x\sin(x)\right)$$
$$= \frac{\partial}{\partial x}\left(y - 2y\cos(x) - \sin(x) - x\cos(x)\right)$$
$$= 2y\sin(x) - \cos(x) - \cos(x) + x\sin(x)$$

$$\frac{\partial^2 V'}{\partial x \partial y} = \frac{\partial}{\partial x}\frac{\partial}{\partial y}\left(xy - y^2 - 2y\sin(x) - x\sin(x)\right)$$
$$= \frac{\partial}{\partial x}\left(x - 2y - 2\sin(x)\right) = 1 - 2\cos(x)$$

$$\frac{\partial^2 V'}{\partial y^2} = \frac{\partial}{\partial y}\frac{\partial}{\partial y}\left(xy - y^2 - 2y\sin(x) - x\sin(x)\right)$$
$$= \frac{\partial}{\partial y}\left(x - 2y - 2\sin(x)\right) = -2$$

At $(x,y) = (0,0)$ we have

$$\frac{\partial^2 V'}{\partial x^2}(0,0) = -2$$

and

$$\left(\frac{\partial^2 V'}{\partial x^2}(0,0)\right) \cdot \left(\frac{\partial^2 V'}{\partial y^2}(0,0)\right) - \left(\frac{\partial^2 V'}{\partial x \partial y}(0,0)\right)^2 = (-2)\cdot(-2) - (-1)^2 = 3$$

So the second-derivative test shows that V' is negative definite near $(0,0)$. Do you like this better than graphing $V'(x,y)$ for (x,y) near $(0,0)$?

Finally, near the start of Ex. 9.4.4, we considered the case that the eigenvalues of $D\vec{F}(0,0)$ could be real. Why mention this case at all? Shouldn't pendulum differential equations have complex eigenvalues, to account for the oscillatory motion of the pendulum? It turns out that for high enough friction, the pendulum drifts down to rest without overshooting the bottom of its arc. Imagine a grandfather clock in a tank of honey.

Practice Problems

9.4.1. For $\vec{F} = \begin{bmatrix} x^3 - y \\ x + y^3 \end{bmatrix} = \begin{bmatrix} g(x,y) \\ h(x,y) \end{bmatrix}$, verify that the eigenvalues of $D\vec{F}(0,0)$ are imaginary. Find a Liapunov function to determine the stability of the origin.

9.4.2. Determine the stability of the fixed point(s) for $g(x,y) = -x^3 + xy^2$ and $h(x,y) = 2x^2y - 5y^3$.

Practice Problem Solutions

9.4.1. Plotting the x-nullcline $(y = x^3)$ and the y-nullcline $(x = -y^3)$ shows that the origin is the only fixed point.

Next, compute $D\vec{F} = \begin{bmatrix} 3x^2 & -1 \\ 1 & 3y^2 \end{bmatrix}$, so the eigenvalues of $D\vec{F}(0,0)$ are $\pm i$ and we can't apply the Hartman-Grobman theorem. The symmetry, up to sign, of $\vec{F}$ suggests we might have some cancellations for a symmetric V, so we'll try $V(x,y) = x^2 + y^2$, certainly positive definite. Then

$$\frac{dV}{dt} = \frac{\partial V}{\partial x}g(x,y) + \frac{\partial V}{\partial y}h(x,y) = 2x(x^3 - y) + 2y(x + y^3) = 2x^4 + 2y^4$$

also positive definite, so by condition (3) of Liapunov's theorem the origin is an unstable fixed point.

9.4.2. We'll note that the x-nullcline is the lines $x = 0$ and $y = \pm x$; the y-nullcline is the line $y = 0$ and the lines $y = \pm\sqrt{2/5}x$. The x- and y-nullclines intersect only at the origin.

Next, compute $D\vec{F} = \begin{bmatrix} -3x^2 + y^2 & 2xy \\ 4xy & 2x^2 - 15y^2 \end{bmatrix}$, so the eigenvalues of $D\vec{F}(0,0)$ are 0 and 0, and once again the Hartman-Grobman theorem cannot be applied.

To apply Liapunov's theorem, start with the positive definite function $V(x,y) = x^2 + y^2$. Then

$$\frac{dV}{dt} = \frac{\partial V}{\partial x}(-x^3 + xy^2) + \frac{\partial V}{\partial y}(2x^2y - 5y^3)$$
$$= 2x(-x^3 + xy^2) + 2y(2x^2y - 5y^3) = -2x^4 + 6x^2y^2 - 10y^4$$

This may not look like a quadratic function, but it is quadratic in x^2 and y^2. That is, taking $u = x^2$ and $v = y^2$, we obtain

$$V'(u,v) = -2u^2 + 6uv - 10v^2$$

Then $V'(u,v)$ has $a = -2$, $b = 6$, and $c = -10$. With $a < 0$ and $4ac - b^2 = 44 > 0$ we see that Prop. 9.4.1 guarantees that $V'(u,v)$ is negative definite, so $V'(x,y)$ is negative definite. It follows by condition (1) of Liapunov's theorem that the origin is asymptotically stable.

Exercises

9.4.1. For $\vec{F} = \begin{bmatrix} -x^3 - y \\ x - y^3 \end{bmatrix}$, verify that the eigenvalues of $D\vec{F}(0,0)$ are imaginary. Find a Liapunov function to determine the stability of the origin.

9.4.2. For $\vec{F} = \begin{bmatrix} x^5 - y \\ x + y^5 \end{bmatrix}$, verify that the eigenvalues of $D\vec{F}(0,0)$ are imaginary. Find a Liapunov function to determine the stability of the origin.

9.4.3. Construct a differential equation $x' = f(x,y)$, $y' = g(x,y)$ with the origin being the only fixed point, both eigenvalues of $D\vec{F}(0,0)$ are 0, and the origin is unstable. Show your choice of f and g satisfy these three properties. Hint: look at Example 9.4.3.

9.4.4. Show how to convert every differential equation of the form $x'' = f(x,x')$ into a pair of first-order differential equations.

9.4.5. Convert the differential equation $x'' + 2x^2x' + x = 0$ into a pair of first-order equations. Show the origin is a fixed point and that at the origin the derivative matrix has imaginary eigenvalues. Show the origin is stable. Is the origin asymptotically stable?

For the functions $\vec{F}$ of Exercises 9.4.6–9.4.9, find the eigenvalues of $D\vec{F}(0,0)$. Find a Liapunov function to determine the stability of the origin.

9.4.6. $\vec{F} = \begin{bmatrix} -x^3 + xy^2 \\ -3x^2y - 2y^3 \end{bmatrix}$

9.4.7. $\vec{F} = \begin{bmatrix} x^3 - y^3 \\ 2xy^2 + 2x^2y + 2y^3 \end{bmatrix}$

9.4.8. $\vec{F} = \begin{bmatrix} -3x^3 + y^3 \\ -xy^2 \end{bmatrix}$

9.4.9. $\vec{F} = \begin{bmatrix} x^3 - y^2 \\ xy + x^2y + 3y^3 \end{bmatrix}$

9.4.10. For $\vec{F} = \begin{bmatrix} ay + bx(x^2 + y^2) \\ -ax + by(x^2 + y^2) \end{bmatrix}$, show the fixed point $(0,0)$ is asymptot-

ically stable if $b < 0$, and is unstable if $b > 0$.

9.5 FIXED POINTS NOT AT THE ORIGIN

The eigenvalue analysis of fixed point stability can be applied to any fixed point: just evaluate the derivative matrix at that point and find the eigenvalues. But our use of Liapunov functions are for a fixed point at the origin. Suppose for a fixed point not at the origin the derivative matrix has imaginary eigenvalues. What can we do? It's simple: change variables to move the fixed point to the origin. We'll sketch the general approach and then work through some examples. The process is straightforward, if we're careful with some details.

Suppose (x_0, y_0) is a fixed point of the differential equation

$$x' = f(x, y) \qquad y' = g(x, y)$$

We change variables to

$$u = x - x_0 \qquad v = y - y_0$$

That is, we translate the point (x_0, y_0) to the origin. Then the differential equation becomes

$$u' = x' = f(x, y) = f(u + x_0, v + y_0) \quad v' = y' = g(x, y) = g(u + x_0, v + y_0)$$

To show that this works, we'll compute the derivative matrix at the origin and show the eigenvalues agree with those of the original derivative matrix at (x_0, y_0). Then we'll apply this idea to Liapunov functions.

Example 9.5.1. *Eigenvalues untranslated and translated.* For this example we'll use the differential equation

$$x' = f(x, y) = y - x^2 \qquad y' = g(x, y) = x - y^2$$

The x-nullcline is $y = x^2$ and the y-nullcline is $x = y^2$ so the fixed points are $(0,0)$ and $(1,1)$. For this example we're interested in the second fixed point. The derivative matrix is

$$D\vec{F}(x, y) = \begin{bmatrix} -2x & 1 \\ 1 & -2y \end{bmatrix} \qquad \text{and so} \qquad D\vec{F}(1,1) = \begin{bmatrix} -2 & 1 \\ 1 & -2 \end{bmatrix}$$

The eigenvalues are $\lambda = -1$ and $\lambda = -3$. The fixed point is an asymptotically stable node. We've known how to do this since Sect. 9.2.

Now let's check what happens when we translate the fixed point to the origin. The new variables are

$$u = x - 1 \qquad v = y - 1$$

and the differential equation becomes

$$u' = x' = f(x,y) = f(u+1, v+1) = v+1 - (u+1)^2 = v - u^2 - 2u$$
$$v' = y' = g(x,y) = g(u+1, v+1) = u+1 - (v+1)^2 = u - v^2 - 2v$$

For this system the derivative matrix is

$$D\vec{F}(u,v) = \begin{bmatrix} -2u-2 & 1 \\ 1 & -2v-2 \end{bmatrix} \qquad \text{and so} \quad D\vec{F}(0,0) = \begin{bmatrix} -2 & 1 \\ 1 & -2 \end{bmatrix}$$

and again the eigenvalues are $\lambda = -1$ and $\lambda = -3$. This is how translating the fixed point to the origin works, and illustrates why we needn't do this to compute eigenvalues. $\square$

Example 9.5.2. *The Lotka-Volterra system, revisited.* From Sect. 7.3 recall the Lotka-Volterra equations

$$x' = -ax + bxy \qquad y' = cy - dxy$$

where a, b, c, and d are positive constants. We are interested in the fixed point $(c/d, a/b)$. There the derivative matrix eigenvalues are $\pm i\sqrt{ac}$.

Change variables to $u = x - c/d$ and $v = y - a/b$ so the differential equation becomes

$$u' = x' = -a(u + c/d) + b(u + c/d)(v + a/b) = (bc/d)v + buv$$
$$v' = y' = c(v + a/b) - d(u + c/d)(v + a/b) = -(ad/b)u - duv$$

It's easy to check, reinforcing Example 9.5.1, that the eigenvalues at $(0,0)$ are $\pm i\sqrt{ac}$, so Hartman-Grobman won't help.

As usual, we'll try $V(u,v) = u^2 + v^2$, positive definite, and so we want V' to be negative semidefinite.

$$V' = \frac{\partial V}{\partial u}u' + \frac{\partial V}{\partial v}v' = 2u((bc/d)v + buv) + 2v(-(ad/b)u - duv)$$
$$= 2uv((bc/d) - (ad/b) + bu - dv)$$

To show this V' is negative semidefinite does not seem so easy, so we'll try another approach.

Recall that the derivative of $\ln(x)$ is $1/x$. We can try to cancel factors from some of the terms by including logs in the Liapunov function. This will be easier to sort out if we do the initial calculation in terms of x and y, then convert to u and v. Take

$$V(x,y) = \alpha x + \beta \ln(x) + \gamma y + \delta \ln(y)$$

We'll define the constants α, β, γ, and δ in a moment. Then

$$V' = \frac{\partial V}{\partial x}(-ax + bxy) + \frac{\partial V}{\partial y}(cy - dxy)$$

$$= \left(\alpha + \frac{\beta}{x}\right)(-ax + bxy) + \left(\gamma + \frac{\delta}{y}\right)(cy - dxy)$$

$$= (\delta c - \beta a) - (\alpha a + \delta d)x + (\beta b + \gamma c)y + (\alpha b - \gamma d)xy$$

If we define $\alpha = -d$, $\beta = c$, $\gamma = -b$, and $\delta = a$, then we see that $V' = 0$.

Finally, we need to choose units so that $c/d < 1$ and $a/b < 1$. Then

$$V(u,v) = -d(u + c/d) + c\ln(u + c/d) - b(v + a/b) + a\ln(v + a/b)$$

To take x close to c/d and y close to a/b, we take $|u|$ and $|v|$ close to 0. Then $u + c/d$ and $v + a/b$ are positive and so the first and third terms of $V(u,v)$ are negative. Also, for small enough $|u|$ and $|v|$ we have $u + c/d < 1$ and $v + a/b < 1$, so the second and fourth terms of V are negative. That is, in some small disc centered at the origin, V is negative definite. And $V' = 0$ is positive semidefinite, maybe in a silly way, but still positive semidefinite. So by condition (2′) of Liapunov's theorem, the fixed point $(c/d, a/b)$ is stable.

Of course, $V' = 0$ also is negative semidefinite. We can use it in either way, but because V is negative definite, we'll take V' to be positive semidefinite. If you're worried that taking V' to be negative semidefinite could lead to a different conclusion about the stability of the fixed point, don't be. The combination of V negative definite and V' negative semidefinite gives no information about fixed point stability.

Could the fixed point be asymptotically stable? After all, asymptotically stable is just a kind of stable. Maybe a different Liapunov function could satisfy condition 1 or 1′ of the theorem. It turns out this isn't the case: the result that $V' = 0$ means V is constant along every trajectory, so trajectories can't spiral in to or out from the fixed point $(c/d, a/b)$. $\square$

Is finding the Liapunov function of Example 9.5.2 easier than the 4-dimensional plot of Sect. A.4? Neither seems particularly inviting. Think of this as a reminder that some problems just are difficult. Or maybe you can find a simpler proof.

Just because I've thought about these topics for decades doesn't mean that I've thought of everything. Fresh eyes sometimes see tricks old eyes have missed.

Practice Problems

9.5.1. Use a Liapunov function to verify that the fixed point $(1,1)$ of

$$x' = x^3 - 3x^2 + 4x - 2 \qquad y' = y^3 - 3y^2 + 5y - 3$$

is unstable. We can show stability with the derivative matrix. This and similar problems illustrate that Liapunov's method works for fixed points for which the derivative matrix trace and determinant do not lie on the boundaries in the tr-det plane.

9.5.2. Find the fixed points and determine their stability for the differential equation

$$x' = x^3 - 3x^2 + 3x + y \qquad y' = y^3 + 3y^2 + 3y - x + 2$$

Practice Problem Solutions

9.5.1. It's easy to check that for $(x,y) = (1,1)$ we have $x' = 0$ and $y' = 0$. The other solutions are complex. So change variables to $u = x-1$ and $v = y-1$. Then the differential equation becomes

$$u' = x' = (u+1)^3 - 3(u+1)^2 + 4(u+1) - 2 = u^3 + u$$
$$v' = y' = (v+1)^3 - 3(v+1)^2 + 5(v+1) - 3 = v^3 + 2v$$

As usual, we'll start with the (positive definite) Liapunov function $V(u,v) = u^2 + v^2$. Then

$$V' = 2u(u^3 + u) + 2v(v^3 + 2v) = 2u^4 + 2u^2 + 2v^4 + 4v^2$$

This is positive definite, so by condition (3) of Liapunov's theorem (9.4.1), $(u,v) = (0,0)$ is unstable. We conclude that $(x,y) = (u+1, v+1) = (1,1)$ is unstable.

9.5.2. To find the fixed points, sketch the nullclines or solve the nullcline equations. (Numerically is better: Mathematica does give one real solution and four pairs of complex solutions, but the real one is all we need. Do you care that one of the other solutions is $(x,y) = (1-(-1)^{1/8}, -1+(-1)^{3/8})$?) We find that the only fixed point is $(1,-1)$.

The derivative matrix is

$$DF = \begin{bmatrix} 3x^2 - 6x + 3 & 1 \\ -1 & 3y^2 + 6y + 3 \end{bmatrix} \text{ and so } DF(1,-1) = \begin{bmatrix} 0 & 1 \\ -1 & 0 \end{bmatrix}$$

The eigenvalues at the fixed point are $\lambda = \pm i$, so the Hartman-Grobman theorem won't help.

To apply Liapunov's theorem, shift the fixed point to the origin by changing variables: $u = x - 1, v = y + 1$. The differential equation becomes

$$u' = x' = (u+1)^3 - 3(u+1)^2 + 3(u+1) + (v-1) = u^3 + v$$
$$v' = y' = (v-1)^3 + 3(v-1)^2 + 3(v-1) - (u+1) + 2 = -u + v^3$$

Once again, we'll start with the (positive definite) Liapunov function $V(u,v) = u^2 + v^2$. Then

$$V' = 2u(u^3 + v) + 2v(-u + v^3) = 2u^4 + 2v^4$$

This is positive definite, so by condition (3) of Liapunov's theorem (9.4.1), $(u,v) = (0,0)$ is unstable. We conclude that $(x,y) = (u+1, v-1) = (1,-1)$ is unstable.

Exercises

In Exercises 9.5.1–9.5.7 use a Liapunov function to test the stability of the indicated fixed point. In addition, check if the Hartman-Grobman theorem can determine the stability.

9.5.1. $x' = x^3 - 3x^2 + 2x + y - 1 \quad y' = y^3 - 3y^2 + 2y - x + 1$ at $(1,1)$.

9.5.2. $x' = x^3 - 3x^2 + 4x + y - 2 \quad y' = -x + y + 1$ at $(1,0)$.

9.5.3. $x' = x^3 - 3x^2 + xy + 3x - y - 1 \quad y' = y^3 - x^2 + 2x - 1$ at $(1,0)$.

9.5.4. $x' = x^3 + x - 2y - 2 \quad y' = y^3 + 3y^2 + 4y + 2x + 2$ at $(0,-1)$.

9.5.5. $x' = -x^3 + 3x^2 - 3x + 3y + 4 \quad y' = -y^3 - 3y^2 - 3y - 3x + 2$ at $(1,-1)$.

9.5.6. $x' = x^3 + xy^2 + 3x^2 + 2xy + y^2 + 2y + 4x + 2$
$y' = y^3 + x^2y + 3y^2 + 2xy + x^2 + 2x + 4y + 2$ at $(-1,-1)$.

9.5.7. $x' = x^3 - 3x^2 + xy - y + 2x$
$y' = y^3 - 3y^2 - x^2 + 2x + 3y - 2$ at $(1,1)$.

In Exercises 9.5.8–9.5.10 for all positive a, determine the stability of the indicated fixed point.

9.5.8. $x' = x^3 + y - 1 \quad y' = (a/3)y^3 - ay^2 + ay - x - a/3$ at $(0,1)$.

9.5.9. $x' = ax^3 + xy^2 - 2xy + x \quad y' = y^3 - x^2y + x^2 - 3y^2 + 3y - 1$ at $(0,1)$.

9.5.10. $x' = -(1+a)(x+1)^3 + xy^2 + y^2 \quad y' = -(1+a)y^3 + y(1+x)^2$ at $(-1,0)$.

9.6 LIMIT CYCLES

In this and the next two sections we'll investigate limit cycles, periodic trajectories that in the phase plane appear as closed curves. In fact, limit cycles are closed curves that are *simple* because trajectories cannot cross themselves. They can be circles, ellipses, or lopsided ellipses, but not figure-eights.

That's the cycle part; what about the limit part? Trajectories near the limit cycle spiral toward it or away from it.

Limit cycles can be exhibited by nonlinear differential equations in the plane, but not by linear equations. In this section we'll investigate the simplest limit cycles, those that are circles. These we can treat by converting the differential equation into polar coordinates. In Sect. 9.7 we study limit cycles with more general shapes by illustrating the Poincaré-Bendixson theorem, whose proof we sketch in Sect. A.10. This beautiful theorem provides conditions under which closed orbits or limit cycles are guaranteed. Then in Sect. 9.8 we'll present some additional results about limit cycles and fixed points. Proofs mostly use techniques from vector calculus and so we'll save the proofs for Sect. 17.6 and Sect. A.21.

We'll begin with a simple example

$$dx/dt = x + y - x(x^2 + y^2) \qquad dy/dt = -x + y - y(x^2 + y^2) \qquad (9.15)$$

Certainly, $(x_*, y_*) = (0,0)$ is a fixed point of Eq. (9.15). In example 9.6.1 we show that the origin is the only fixed point of this system.

To test the stability of $(0,0)$, compute the derivative

$$D\vec{F}(x,y) = \begin{bmatrix} 1 - 3x^2 - y^2 & 1 - 2xy \\ -1 - 2xy & 1 - x^2 - 3y^2 \end{bmatrix} \text{ and so } D\vec{F}(0,0) = \begin{bmatrix} 1 & 1 \\ -1 & 1 \end{bmatrix}$$

The eigenvalues of $D\vec{F}(0,0)$ are $1 \pm i$, so the origin is unstable, in fact, an unstable spiral. We have encountered unstable spirals before. What's special about this one?

The presence of the terms $x^2 + y^2$ in Eq. (9.15) suggests converting to polar coordinates. Recall $r^2 = x^2 + y^2$, where r denotes the distance to the origin. Differentiating this gives

$$2r(dr/dt) = 2x(dx/dt) + 2y(dy/dt)$$

Substituting in dx/dt and dy/dt from Eq. (9.15) we find

$$2r(dr/dt) = 2x(x + y - x(x^2 + y^2)) + 2y(-x + y - y(x^2 + y^2))$$
$$= 2x^2 + 2y^2 - 2(x^2 + y^2)(x^2 + y^2)$$

and consequently,

$$dr/dt = r(1 - r^2)$$

From this we see that

$$dr/dt > 0 \text{ for } r < 1, \quad dr/dt = 0 \text{ for } r = 1, \quad dr/dt < 0 \text{ for } r > 1$$

Thus trajectories inside $r = 1$ move out to $r = 1$, trajectories outside $r = 1$ move in to $r = 1$, and trajectories on $r = 1$ stay on $r = 1$. So except for the fixed point at the origin, trajectories move toward the circle $r = 1$. We can visualize this by plotting dr/dt as a function of r, as in Fig. 9.6. At $r = b$, for example, r increases along trajectories, moving toward the circle $r = 1$. At $r = c$, r decreases along trajectories, again moving toward $r = 1$.

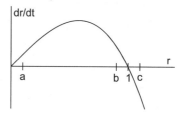

Figure 9.6. A plot of dr/dt vs r for Eq. (9.15).

While we're looking at this graph, note that it shows the origin is an unstable fixed point without calculating the eigenvalues of the derivative matrix. For trajectories near the origin, but not at the origin, the point $r = a$ in the plot of Fig. 9.6 for example, r increases along trajectories, so all trajectories near the origin move farther away from the origin. That is, the origin is unstable.

Note a source of possible confusion. *Circular limit cycles centered on the origin are determined by $dr/dt = 0$*. For the examples of Sect. 9.7, plotting the curve determined by $dr/dt = 0$ will *not* give the limit cycle. To plot these limit cycles, we'll need other tricks.

How do we know these trajectories travel all around the circle $r = 1$? Why couldn't they approach a point on that circle, or oscillate near an arc of the circle, ever more rapidly as they

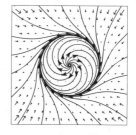

Figure 9.7. Trajectories and some vectors for Eq. (9.15).

approach the circle more closely? To answer this, in Example 9.6.2 we show $d\theta/dt = -1$, so all trajectories travel clockwise (recall positive angles are counterclockwise) at a constant rate. From this and $r' = 0$ for $r = 1$, we see $r = 1$ is a periodic trajectory. Nearby trajectories spiral out to, or in to, $r = 1$. Because nearby trajectories converge to this circle, it is called a *stable* limit cycle. See Fig. 9.7. Can this result be generalized? Can we detect limit cycles that are not circles? In the next section we shall see.

Example 9.6.1. *Why $(x_*, y_*) = (0,0)$ is the only fixed point of (9.15).* Multiply the x-nullcline equation $0 = x + y - x(x^2 + y^2)$ by x, multiply the y-nullcline equation $0 = -x + y - y(x^2 + y^2)$ by y, and add, obtaining

$$0 = (x^2 + y^2)(1 - (x^2 + y^2))$$

Then either $x^2 + y^2 = 0$, giving $(x,y) = (0,0)$, or $x^2 + y^2 = 1$. In the latter case, the x- and y-nullcline equations become $x + y - x = 0$ and $-x + y - y = 0$, giving $(x,y) = (0,0)$. But this contradicts $x^2 + y^2 = 1$, so the only fixed point is $(0,0)$. $\square$

Example 9.6.2. *Why $d\theta/dt = -1$ for all trajectories of (9.15), except of course for the fixed point.* From $\theta = \arctan(y/x)$ we see

$$\frac{d\theta}{dt} = \frac{1}{1 + (y/x)^2}\left(\frac{y}{x}\right)' = \frac{1}{1 + (y/x)^2}\frac{xy' - yx'}{x^2} = \frac{xy' - yx'}{x^2 + y^2} \tag{9.16}$$

Substituting in for x' and y' from Eq. (9.15) and simplifying, we obtain $xy' - yx' = -(x^2 + y^2)$ and so

$$\theta' = -1$$

Consequently, the trajectory at $r = 1$ travels at a constant rate, clockwise around the origin. $\square$

Practice Problems

9.6.1. For the system

$$x' = 2x + y + x(x^2 + y^2) - 3x\sqrt{x^2 + y^2}$$

$$y' = -x + 2y + y(x^2 + y^2) - 3y\sqrt{x^2 + y^2}$$

(a) show $dr/dt = 0$ on the circles $r = 1$ and $r = 2$,
(b) show $d\theta/dt$ is a non-zero constant on $r = 1$ and $r = 2$,
(c) deduce that $r = 1$ and $r = 2$ are closed trajectories, and
(d) show these are limit cycles and determine their stability.

9.6.2. Consider the system

$$x' = -4x + x(x^2 + y^2) - y(x^2 + y^2)$$
$$y' = -4y + y(x^2 + y^2) + x(x^2 + y^2)$$

(a) show $dr/dt = 0$ on the circle $r = 2$,
(b) show $d\theta/dt$ is a non-zero constant on $r = 2$,
(c) deduce that $r = 2$ is a closed trajectory, and
(d) show it is a limit cycle and determine its stability.

Practice Problem Solutions

9.6.1. Recalling $rr' = xx' + yy'$, we see
$rr' = xx' + yy' = 2r^2 + r^4 - 3r^3$ so we find
$r' = 2r + r^3 - 3r^2 = r(r-1)(r-2)$.
(a) That is, $r' = 0$ for $r = 0$ (the fixed point
at the origin) and $r = 1, 2$.

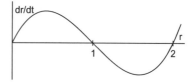

Figure 9.8. A plot of dr/dt.

(b) Next, because $xy' - yx' = -r^2$, from Eq. (9.16) we see $\theta' = -1$.
(c) Because θ' is a non-zero constant and $r' = 0$ on $r = 1$ and $r = 2$, we see these
are closed trajectories.
(d) From the graph of dr/dt as a function of r (Fig. 9.8), we see trajectories near
$r = 1$ spiral toward $r = 1$, making it a stable limit cycle, and trajectories near
$r = 2$ spiral away from it, making $r = 2$ an unstable limit cycle.

9.6.2. Substitute the expressions for x' and y' into $rr' = xx' + yy'$:

$$rr' = x(-4x + x(x^2 + y^2) - y(x^2 + y^2))$$
$$+ y(-4y + y(x^2 + y^2) + x(x^2 + y^2))$$
$$= -4(x^2 + y^2) + (x^2 + y^2)(x^2 + y^2) = -4r^2 + r^4$$

so $r' = -4r + r^3 = r(-4 + r^2)$.
(a) Then $r' = 0$ when $r = 0$ (the fixed point at the origin) and the circle $r = 2$.
Certainly, $r = -2$ is not a possibility.
(b) Next, because $xy' - yx' = (x^2 + y^2)^2$, from Eq. (9.16) we see $\theta' = r^2$ for all r,
so $\theta' = 4$ on the circle of radius 2.
(c) Consequently, $r = 2$ is a closed trajectory.
(d) Because $r' < 0$ for $0 < r < 2$ and $r' > 0$ for $r > 2$, we see that $r = 2$ is
an unstable limit cycle. Note that the fixed point at the origin is asymptotically
stable.

Exercises

9.6.1. Find all the circles where $r' = 0$, show $\theta' \neq 0$ for all points on these circles, show these circles are limit cycles, and determine their stability.

(a) $\quad x' = 3x - y + x(x^2 + y^2) - 4x\sqrt{x^2 + y^2}$

$\quad\quad y' = 3y + x + y(x^2 + y^2) - 4y\sqrt{x^2 + y^2}$

(b) $\quad x' = y + x(x^2 + y^2)^2 - 2x(x^2 + y^2)$

$\quad\quad y' = -x + y(x^2 + y^2)^2 - 2y(x^2 + y^2)$

(c) $\quad x' = 2x - 4y - x(x^2 + y^2)$

$\quad\quad y' = 2y + 4x - y(x^2 + y^2)$

(d) $\quad x' = 2x + y - 3x(x^2 + y^2) + x(x^2 + y^2)^2$

$\quad\quad y' = 2y - x - 3y(x^2 + y^2) + y(x^2 + y^2)^2$

(e) $\quad x' = 2x + y(x^2 + y^2) - 3x(x^2 + y^2)^{1/2} + x(x^2 + y^2)^{3/2}$

$\quad\quad y' = 2y - x(x^2 + y^2) - 3y(x^2 + y^2)^{1/2} + y(x^2 + y^2)^{3/2}$

9.6.2. Find all the circles where $r' = 0$, show $\theta' \neq 0$ for all points on these circles, show these circles are limit cycles, and determine their stability.

(a) $\quad x' = -x - y + x(x^2 + y^2)$

$\quad\quad y' = x - y + y(x^2 + y^2)$

(b) $\quad x' = -x - y + 2x(x^2 + y^2)^2$

$\quad\quad y' = x - y + 2y(x^2 + y^2)^2$

(c) $\quad x' = 6x - y - 5x(x^2 + y^2) + x(x^2 + y^2)^2$

$\quad\quad y' = x + 6y - 5y(x^2 + y^2) + y(x^2 + y^2)^2$

(d) $\quad x' = -x - y + 3x(x^2 + y^2)^{1/2} - 3x(x^2 + y^2) + x(x^2 + y^2)^{3/2}$

$\quad\quad y' = x - y + 3y(x^2 + y^2)^{1/2} - 3y(x^2 + y^2) + y(x^2 + y^2)^{3/2}$

(e) $x' = -4x + 6y + 8x(x^2 + y^2)^{1/2} - 5x(x^2 + y^2) + x(x^2 + y^2)^{3/2}$
$y' = -6x - 4y + 8y(x^2 + y^2)^{1/2} - 5y(x^2 + y^2) + y(x^2 + y^2)^{3/2}$

9.6.3. Based on what you've seen in Exercises 9.6.1 (c) and 9.6.2 (a) and (b), for the differential equation

$$x' = ax + by + cx(x^2 + y^2) \qquad y' = dx + ey + fy(x^2 + y^2)$$

(a) find relations between the constants a, b, c, d, e, and f that guarantee the system has a stable limit cycle, a circle on which $r' = 0$ and θ' is constant.
(b) Find relations between the constants that guarantee the differential equation has an unstable limit cycle.

9.6.4. Find relations between the constants a, b, c, d, e, and f that guarantee the differential equation

$$x' = ax + by + cx\sqrt{x^2 + y^2} + x(x^2 + y^2)$$
$$y' = dx + ey + fy\sqrt{x^2 + y^2} + y(x^2 + y^2)$$

(a) has no limit cycle,
(b) has exactly one circular limit cycle,
(c) has exactly two limit cycles.
(d) In cases (b) and (c), say whether θ' is positive or negative on each limit cycle.

9.6.5. Using the techniques developed in Sect. 9.5, (a) show that the differential equation

$$x' = -x^3 - xy^2 + 3x^2 + 2xy + y^2 - 3x - y$$
$$y' = -y^3 - x^2y + x^2 + 3y^2 + 2xy - 3x - 3y + 2$$

has a circular limit cycle centered at $(x,y) = (1,1)$. Compute θ' on this limit cycle.
(b) Show that the differential equation

$$x' = -x^3 - xy^2 + 3x^2 + y^2 - 2x + y$$
$$y' = -y^3 - x^2y + 2xy - x + 1$$

has a circular limit cycle centered at $(x,y) = (1,0)$. Compute θ' on this limit cycle.

9.7 THE POINCARÉ-BENDIXSON THEOREM

So far we have established the existence of limit cycles by showing that for some circles centered at the origin, $rr' = xx' + yy' = 0$ and $\theta' \neq 0$. In Exercise 9.6.5 we see how to shift the center of a circle to the origin, so the method of Sect. 9.6 can be adapted to show that circles not centered on the origin are limit cycles.

But this method will not work for limit cycles that are not circles. Because no matter how we might shift the origin, along all these curves the distance to the origin is not constant. So $r' \neq 0$.

One way to recognize this complication is that when converting to polar coordinates, the expression for r' is not a function of r alone, but has at least one term involving θ. For example,

$$x' = x + y + x^2 - x(x^2 + y^2) \qquad y' = -x + y - y(x^2 + y^2) \qquad (9.17)$$

Substitute these expressions for x' and y' in $rr' = xx' + yy'$, convert to polar coordinates, and divide by r. This gives

$$r' = r + r^2 \cos^3(\theta) - r^3 \qquad (9.18)$$

In these situations, we apply a beautiful result called the Poincaré-Bendixson theorem.

The central geometric idea is called a *trapping region*, a bounded area R in the plane. Here the term "bounded" means the region does not run away to infinity in any direction. In earlier versions of this course, some students suggested calling the region R a "hotel California region." (Disappointingly, none suggested names from songs by Radiohead, Kate Bush, Death Cab for Cutie, Tori Amos, or Lia Ices. I suggested they listen to better music.) We use a duller, but more obviously descriptive, name, "trapping region."

To show R is a trapping region for a differential equation $x' = f(x,y)$, $y' = g(x,y)$, it is enough to show that for every point (x,y) on the boundary curve or curves of R, the vector $\langle f(x,y), g(x,y) \rangle$ points into R. In a moment we'll illustrate this with an example. First, we state the theorem.

Theorem 9.7.1. *Poincaré-Bendixson theorem.* If R is a trapping region for a differential equation and R contains no fixed points for the differential equation, then R contains a closed trajectory, a periodic solution of the differential equation. Trajectories that enter R spiral in toward a closed trajectory in R.

The proof, sketched in Sect. A.10, is subtle, but the result is plausible. A trajectory that enters a trapping region and doesn't get stuck at a fixed point must accumulate somewhere. The only possibility we can imagine is that the trajectory converges to a limit cycle. Of course, a proof needs more than just saying we can't think of another possibility. But at least this theorem is believable. Let's apply it to the differential equation from the beginning of this section.

Example 9.7.1. *Finding a trapping region for the system (9.17).* We've seen that in polar coordinates the system (9.17) becomes (9.18),

$$r' = r + r^2 \cos^3(\theta) - r^3$$

To find a trapping region, we find upper and lower bounds for r', bounds not involving θ. Then we find a large circle on which the upper bound is negative, and a small circle on which the lower bound is positive. The region, called an *annulus*, will be a trapping region for the system.

To find the upper and lower bounds, first recall $-1 \le \cos^3(\theta) \le 1$, so multiplying through by r^2,

$$-r^2 \le r^2 \cos^3(\theta) \le r^2$$

Then adding $r - r^3$ to the inequality,

$$r - r^2 - r^3 \le r + r^2 \cos^3(\theta) - r^3 \le r + r^2 - r^3$$

That is,

$$r - r^2 - r^3 \le r' \le r + r^2 - r^3$$

So $ub(r) = r + r^2 - r^3$ is an upper bound for r' and $lb(r) = r - r^2 - r^3$ is a lower bound. These are promising, because for large r, r^3 dominates the upper bound and so for large enough r, the upper bound is negative, hence r' is negative. Similarly, for r near 0, r dominates the lower bound and so for small enough r, the lower bound is positive, hence r' is positive.

Specifically, $ub(2) = -2$, so $r' \le ub(2) = -2 < 0$ all around the circle of radius $r = 2$. Next, $lb(1/2) = 1/8$, so $r' \ge lb(1/2) = 1/8 > 0$ all around the circle of radius $r = 1/2$. Because $r' < 0$ on the circle of radius 2, trajectories cross that circle from the outside to the inside. Because $r' > 0$ on the circle of radius 1/2, trajectories cross that circle from the inside to the outside. Consequently, the annulus between $r = 1/2$ and $r = 2$ is a trapping region. The only fixed point of the system (9.17) is the origin, so the trapping region contains no fixed points. Then by the Poincaré-Bendixson theorem, the trapping region contains a closed trajectory. Trajectories crossing either boundary circle of the trapping region spiral toward this or some other closed trajectory, so the trapping region contains a limit cycle. □

The theorem tells us nothing about where the limit cycle lies in the trapping region, so if we are interested in a more accurate picture of the limit cycle, we want the smallest trapping region. To do this, we find where $ub(r) = 0$ and where $lb(r) = 0$:

$$0 = ub(r) = r(1 + r - r^2), \text{ so } r = 0, (1 \pm \sqrt{5})/2$$
$$0 = lb(r) = r(1 - r - r^2), \text{ so } r = 0, (-1 \pm \sqrt{5})/2$$

Then any circle of radius $r > (1 + \sqrt{5})/2$ can be the outer boundary of a trapping region, and any circle of radius $r < (-1 + \sqrt{5})/2$ can be the inner boundary of a trapping region.

Without obvious circular symmetry–differential equations with some terms other than $(x^2 + y^2)^n$–the existence of a trapping region can be harder to establish.

Example 9.7.2. *A trapping region for a chemical oscillator.* To illustrate some of the issues involved in finding a trapping region, we use an instance of a chemical oscillator described in Sect. 8.3 of [155]. The left side of Fig. 9.9 shows some solution curves for

$$\frac{dx}{dt} = 8 - x - \frac{4xy}{1 + x^2} \qquad \frac{dy}{dt} = bx\left(1 - \frac{y}{1 + x^2}\right) \qquad (9.19)$$

The x-nullcline is the curve $y = (8 - x)(1 + x^2)/(4x)$; the y-nullcline is the curve $y = 1 + x^2$ and the y-axis. These are pictured on the left side of Fig. 9.9. The x-nullcline does not intersect the y-axis, so the only fixed point is the intersection of the x-nullcline and $y = 1 + x^2$, that is, the point $(x_*, y_*) = (8/5, 89/25)$ The numerical approximations of the solutions plotted on the left side of Fig. 9.9 suggest that a limit cycle encloses this fixed point.

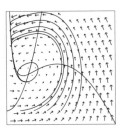 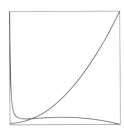

To deduce the existence of this limit cycle by the Poincaré-Bendixson theorem, we must find a trapping region that contains no fixed points. The vector field plot suggests that a large rectangle can serve as the outer boundary of a trapping region. The fixed point does lie within this rectangle, but if

Figure 9.9. Left: some solution curves and vectors for Eq. (9.19). Right: constructing a trapping region. The nullclines are shown.

the fixed point is an unstable spiral, then a small circle centered on the fixed point can be the inner boundary of the trapping region.

Showing the fixed point is an unstable spiral is relatively straightforward, if a bit tedious.

$$
D\vec{F} = \begin{bmatrix} \dfrac{\partial}{\partial x}\left(8 - x - \dfrac{4xy}{1+x^2}\right) & \dfrac{\partial}{\partial y}\left(8 - x - \dfrac{4xy}{1+x^2}\right) \\[2ex] \dfrac{\partial}{\partial x}\left(bx\left(1 - \dfrac{y}{1+x^2}\right)\right) & \dfrac{\partial}{\partial x}\left(bx\left(1 - \dfrac{y}{1+x^2}\right)\right) \end{bmatrix}
$$

$$
= \begin{bmatrix} \dfrac{-x^4 + 4x^2y - 2x^2 - 4y - 1}{(1+x^2)^2} & -\dfrac{4x}{1+x^2} \\[2ex] \dfrac{x^4 + x^2y + 2x^2 - y + 1}{(1+x^2)^2} & -\dfrac{x}{1+x^2} \end{bmatrix}
$$

So

$$
D\vec{F}(8/5, 89/25) = \begin{bmatrix} 67/89 & -160/89 \\ 128/89 & -40/89 \end{bmatrix}
$$

The eigenvalues of $D\vec{F}(8/5, 89/25)$ are $(27 \pm i\sqrt{70471})/178$. Because the eigenvalues are complex with positive real parts, the fixed point (x_*, y_*) is an unstable spiral.

A bit more work is needed to find a trapping region. First, note that $x' = 0$ all along the y-axis, so no trajectory crosses the y-axis. Next, $y' > 0$ all along the positive x-axis, so no trajectory crosses the positive x-axis from above to below. Third, along $x = 8$, $x' = 8 - 8 - 4 \cdot 8 \cdot y/(1 + 8^2) = -32y/65$, so for $y > 0$, every trajectory that crosses $x = 8$ crosses from right to left. Finally, for $y = 65, 0 \le x \le 8, y' \le 0$. Consequently, no trajectory crosses this line from below to above. Combining these observations, the rectangle shown on the right side of Fig. 9.9 (note the x- and y-axes are drawn to different scales, because $0 \le x \le 8$ and $0 \le y \le 65$) is the outer boundary of a trapping region: any trajectory that enters this region must stay in the region. Excising a small disc centered on the fixed point completes the construction of a trapping region that does not contain a fixed point. It follows from the Poincaré-Bendixson theorem that this region contains a limit cycle. $\square$

Example 9.7.3. *A limit cycle for the glycolysis model.* Glycolysis is a process by which cells obtain energy through breaking down sugars. The process, a metabolic pathway really, is very complicated: understanding all the steps took over a century. Despite this complexity, some of the dynamics of glycolysis is captured by E. E. Sel'kov's [147] model, expressed in Example 7.3.2 of [155] as

$$
x' = -x + ay + x^2 y = f(x, y) \qquad y' = b - ay - x^2 y = g(x, y) \qquad (9.20)
$$

where x denotes the concentration of ADP (adenosine diphosphate) and y the concentration of F6P (fructose 6-phosphate). Here a and b are positive constants, and x and y are (necessarily) non-negative concentrations.

The nullclines are the curves

$$y = \frac{x}{a+x^2} \quad \text{(x-nullcline)} \qquad y = \frac{b}{a+x^2} \quad \text{(y-nullcline)}$$

Equating the right-hand sides of the nullcline equations we find the fixed point has x-coordinate given by $x/(a+x^2) = b/(a+x^2)$, that is, $x = b$. Then the fixed point is $(b, b/(a+b^2))$. We'll find the range of a- and b-values where the fixed point is unstable, find a trapping region, and then deduce under what circumstances this glycolysis model exhibits a limit cycle.

The derivative matrix is

$$D\vec{F}(x,y) = \begin{bmatrix} -1+2xy & a+x^2 \\ -2xy & -(a+x^2) \end{bmatrix}$$

and so at the fixed point

$$D\vec{F}\left(b, \frac{b}{a+b^2}\right) = \begin{bmatrix} \dfrac{-a+b^2}{a+b^2} & a+b^2 \\ -\dfrac{2b^2}{a+b^2} & -(a+b^2) \end{bmatrix}$$

The eigenvalues are

$$\lambda_\pm = -\frac{a-b^2+(a+b^2)^2 \pm \sqrt{-4(a+b^2)^3+(a-b^2+(a+b^2)^2)^2}}{2(a^2+b^2)}$$

Ugh. Let's try the tr-det plane. From the derivative matrix at the fixed point we find

$$\text{tr} = -(a+b^2) + \frac{-a+b^2}{a+b^2} \quad \text{and} \quad \det = a+b^2$$

But even this is tricky: because of how the two parameters, a and b, appear in tr and in det, we cannot write an equation expressing det as a function of tr, or tr as a function of det. The closest we can come is

$$\text{tr} = \frac{\det - 2a}{\det}$$

We could draw a family of curves, one for each positive a, but that, too, would require some work to interpret. Inspecting the expressions for tr and det will show us the stability of the fixed point.

First, observe that $\det = a + b^2$ is positive, so the fixed point never is a saddle point. Recall that above the tr-axis in the tr-det plane, the fixed point is asymptotically stable if tr < 0 and unstable if tr > 0. The parameter values where the fixed point is stable are separated from those where the fixed point is unstable by the curve tr $= 0$. Keeping in mind that a and b are positive, tr $= 0$ is defined by

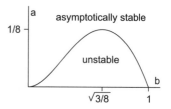

Figure 9.10. A plot of $a(b)$.

$$a = a(b) = \frac{-(1+2b^2) + \sqrt{1+8b^2}}{2}$$

In Fig. 9.10 we see a plot of $a(b)$ as a function of b. The requirement that $a(b) \geq 0$ implies $0 \leq b \leq 1$. Below this curve, the trace is positive and so the fixed point is unstable; above the curve the fixed point is asymptotically stable.

To find a trapping region, a good first step is to plot the nullclines. These divide the first quadrant into regions where the vector field points NE, SE, SW, and NW. In Fig. 9.11 we see plots of the x-nullcline (solid curve) and the y-nullcline (dashed curve) for the

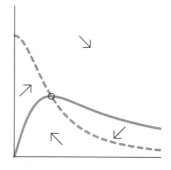

Figure 9.11. Nullclines of Eq. (9.20).

glycolysis model. Recall that in order for a vector field to have a closed trajectory, the trajectory must pass through regions with vectors in each of these four directions, in the order NE, SE, SW, and NW (starting anywhere in this sequence), or in the order NE, NW, SW, and SE. Of course, spirals show that this doesn't guarantee a closed trajectory.

To bound the trapping region, we can use parts of the x-axis (the vectors point NW), part of the y-axis (the vectors point NE, and level out to horizontal at the point $(b, b/a)$, where the y-nullcline crosses the y-axis), and also a vertical segment from the x-axis to the downward arc of the x-nullcline (the vectors point NW and SW). But how do we close up the boundary of the trapping region? We can top the region with the horizontal segment from $(0, b/a)$ to $(b, b/a)$ along which the vectors point SE. From Eq. (9.20) we see the slope of the vector field is

$$sl(x,y) = \frac{g(x,y)}{f(x,y)} = \frac{b - ay - x^2 y}{-x + ay + x^2 y} \tag{9.21}$$

At the point $(b, b/a)$, the slope is $sl(b, b/a) = -1$. This suggests closing the boundary of the trapping region with the line segment of slope -1 from the point $(b, b/a)$ till where it intersects the x-nullcline. (Finding the coordinates of this point involves solving a messy cubic, but happily, we don't need to know the coordinates.) In order for this segment to act as part of the boundary of a trapping region, along this segment the vector field vectors must point into the region. From Fig. 9.11 we see these vectors point SE, so in order to show that the vectors point into the trapping region, we must show that all along this segment the slope of the vectors is less than -1. Points along this segment have the form $(x, y) = (b + t, b/a - t)$ for $t > 0$. The algebra takes a few moments.

$$sl(b + t, b/a - t) < -1$$

$$\frac{-b^3 + (a - 2)b^2 t + (2a - 1)bt^2 + at(a + t^2)}{b^3 + (2 - a)b^2 t + (1 - 2a)bt^2 - at(1 + a + t^2)} < -1$$

This simplifies to $0 < at$, which is true.

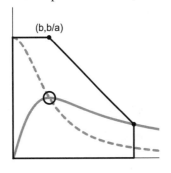

(b,b/a)

Figure 9.12. Trapping region of Eq. (9.20).

We can see that the five dark, heavy segments in Fig. 9.12 bound a trapping region for the glycolysis model of Eq. (9.20). (Do you see why the rectangular region formed by bumping out the diagonal segment to a corner that finished the rectangle isn't a trapping region?) However, we can't conclude that this region contains a closed trajectory, because the fixed point is in the region. For (a, b) above the curve in Fig. 9.10, the fixed point is asymptotically stable and trajectories that enter the trapping region converge to the fixed point. For (a, b) below the curve, the fixed point is unstable, a node or a spiral. So if we draw a small circle around this unstable fixed point, the region between this circle and the five dark, heavy segments in Fig. 9.12 form a trapping region without any fixed points. Then Poincaré-Bendixson implies this region contains a limit cycle: the ADP and F6P concentrations cycle repeatedly, until some system parameter changes. □

The Poincaré-Bendixson theorem has an implication for a type of behavior glimpsed by a few–Poincaré, Birkhoff, Hadamard, Cartwright, Littlewood, Lorenz–but not entering general scientific discourse until the 1970s. Specifically, differential equations in the plane cannot exhibit chaotic solutions. Fixed points, trajectories connecting fixed points, closed trajectories, limit cycles, and, of course, running away to infinity exhaust the list of behaviors for solutions of differential equations in the plane. In Chapter 13 we'll see that in 3 dimensions the world of differential equations is ever so much more interesting: the chaos we saw in Sect. 2.1 for the logistic map we'll find again in some differential equations with three or more variables. The inability to make long-term predictions for the logistic map appears again for many differential equations. We can find chaos in the motion of more than two planets under mutual gravitational attraction (Poincaré), in geodesics–shortest-distance paths–on surfaces that look like little bits of saddles sewn together (Birkhoff and Hadamard), in models of vacuum tubes (Google, if you don't know what these are) used in radar circuits (Cartwright and Littlewood), in models of the weather (Lorenz), and in many, many more examples.

For some of these, including weather and epidemic dynamics, we'd like to be able to make reliable long-term predictions. That we cannot sounds like bad news. The gold standard of a scientific theory is our ability to use it to make long-term predictions. If we cannot do this, is science an illusion? In the 1980s, incompleteness seemed to be the best we could hope for science.

But ... computers work, and so do weather satellites and GPS satellites, and solar panels, and malaria vaccines, and immunotherapy for some cancers, and on and on. Chaotic solutions to differential equations that model important natural processes forced us to rethink the goals of science. And in Sect. 13.9 we'll find some silver lining in this cloudy weather of unpredictability.

But the larger issue is that how the world works has nothing to do with our ease of understanding it. Humility is the first lesson every scientist should learn, and the last we should forget.

Practice Problems

9.7.1. Consider the system
$$x' = x + 2y - x(x^2 + y^2) \qquad y' = -x + y - y(x^2 + y^2)$$
(a) Show this system has a periodic trajectory. The only fixed point is the origin. You need not show this. Hint: recall $\cos(\theta)\sin(\theta) = (1/2)\sin(2\theta)$.
(b) Show the annulus $\sqrt{2}/2 < r < \sqrt{3/2}$ contains a limit cycle.

9.7.2. Consider the system
$$x' = x + (7/2)y - x(x^2 + y^2) \qquad y' = -2x + y - y(x^2 + y^2)$$
(a) Show this system has a periodic trajectory. The only fixed point is the origin. You need not show this. Same hint as Practice Problem 9.7.1.
(b) Show the annulus $1/2 \le r \le \sqrt{7}/4$ contains a limit cycle.

Practice Problem Solutions

9.7.1. (a) From $r^2 = x^2 + y^2$ we see

$$rr' = xx' + yy' = x(x + 2y - x(x^2 + y^2)) + y(-x + y - y(x^2 + y^2))$$

$$= x^2 + y^2 - (x^2 + y^2)^2 - xy = r^2 - r^4 - r^2 \cos(\theta)\sin(\theta) \qquad (9.22)$$

where the last line uses the polar coordinate identities $x = r\cos(\theta), y = r\sin(\theta)$. Divide Eq. (9.22) by r to obtain

$$r' = r(1 - \cos(\theta)\sin(\theta)) - r^3 \qquad (9.23)$$

Because $\cos(\theta)\sin(\theta) = (1/2)\sin(2\theta)$, we see

$$-1/2 \leq \cos(\theta)\sin(\theta) \leq 1/2$$

and so

$$1/2 \leq 1 + \cos(\theta)\sin(\theta) \leq 3/2$$

Combining this with Eq. (9.23), we see

$$r/2 - r^3 \leq r' \leq 3r/2 - r^3 \qquad (9.24)$$

so the upper and lower bounds on r' are $ub(r) = 3r/2 - r^3$ and $lb(r) = r/2 - r^3$. Because $ub(2) = -5$ and $lb(1/2) = 1/8$, we see the annulus between $r = 1/2$ and $r = 2$ is a trapping region that contains no fixed points, so by the Poincaré-Bendixson theorem must contain a limit cycle.

(b) Note the positive solution of $ub(r) = 0$ is $r = \sqrt{3/2}$ and the positive solution of $lb(r) = 0$ is $r = \sqrt{2}/2$. In fact, $ub(r) < 0$ for all $r > \sqrt{3/2}$, and $lb(r) > 0$ for all r, $0 < r < \sqrt{2}/2$. Thus the limit cycle must be contained in the annulus $\sqrt{2}/2 \leq r \leq \sqrt{3/2}$.

9.7.2. From $r^2 = x^2 + y^2$, $x = r\cos(\theta)$, and $y = r\sin(\theta)$ we see

$$rr' = xx' + yy' = x(x + (7/2)y - x(x^2 + y^2)) + y(-2x + y - y(x^2 + y^2))$$
$$= x^2 + y^2 - (x^2 + y^2)^2 + (3/2)xy = r^2 - r^4 + (3/2)r^2 \cos(\theta)\sin(\theta)$$

Divide by r to obtain

$$r' = r(1 + (3/2)\cos(\theta)\sin(\theta)) - r^3$$

From the previous problem, recall $-1/2 \leq \cos(\theta)\sin(\theta) \leq 1/2$. We see

$$1/4 \leq 1 + (3/2)\cos(\theta)\sin(\theta) \leq 7/4$$

Then

$$r/4 \leq r(1 + (3/2)\cos(\theta)\sin(\theta)) \leq 7r/4$$

That is,

$$lb(r) = r/4 - r^3 \le r' \le 7r/4 - r^3 = ub(r)$$

Now $ub(3/2) = -3/4$ and $lb(1/4) = 3/64$, so the annulus between $r = 1/4$ and $r = 3/2$ is a trapping region without fixed points. Then by the Poincaré-Bendixson theorem, this annulus contains a limit cycle.

(b) Because $ub(r) < 0$ for $r > \sqrt{7}/2$, and $lb(r) > 0$ for $0 < r < 1/2$, every annulus with inner radius $< 1/2$ and outer radius $> \sqrt{7}/2$ is a trapping region without fixed points, so contains a limit cycle.

Exercises

9.7.1. Show each of these differential equations has a limit cycle. For all, the origin is the only fixed point.

(a) $x' = x + \dfrac{y}{2} - x(x^2 + y^2)$ $y' = -2x + y - y(x^2 + y^2)$

(b) $x' = x + \dfrac{y}{4} - x(x^2 + y^2)$ $y' = -\dfrac{x}{2} + y - y(x^2 + y^2)$

(c) $x' = x + y + \dfrac{y^2}{2} - x(x^2 + y^2)$ $y' = -x + y - y(x^2 + y^2)$

(d) $x' = x + y + x^2 - x(x^2 + y^2)$ $y' = -x + y - y(x^2 + y^2)$

(e) $x' = x + y + \dfrac{x^2}{4} - x(x^2 + y^2)$ $y' = -x + y + \dfrac{y^2}{4} - y(x^2 + y^2)$

9.7.2. Show each of these differential equations has a limit cycle. For all, the origin is the only fixed point.

(a) $x' = x + \dfrac{y}{4} - x(x^2 + y^2)^2$ $y' = -\dfrac{x}{2} + y - y(x^2 + y^2)^2$

(b) $x' = x + \dfrac{y}{4} - x(x^2 + y^2)^3$ $y' = -\dfrac{x}{2} + y - y(x^2 + y^2)^3$

(c) $x' = x + \dfrac{3y}{2} - x(x^2 + y^2)$ $y' = -x + y - y(x^2 + y^2)$

(d) $x' = x + y + \dfrac{y^3}{2} - x(x^2 + y^2)$ $y' = -x + y - y(x^2 + y^2)$

(e) $x' = x + y + xy - x(x^2 + y^2)$ $y' = -x + y - \dfrac{x^2}{2} - y(x^2 + y^2)$

9.7.3. For this differential equation

$$x' = x + y - x^3 \qquad y' = -x + y - y^3$$

(a) plot the nullclines and show that the origin is the only fixed point.
(b) Show that the origin is an unstable spiral.
(c) Show that the square that bounds the region $-2 \le x \le 2$, $-2 \le y \le 2$ is a trapping region.
(d) Deduce that this differential equation has a limit cycle.

9.7.4. For this differential equation

$$x' = x + 2y - x^3 \qquad y' = -x + 2y - y^5$$

(a) plot the nullclines and show that the origin is the only fixed point.
(b) Show that the origin is an unstable spiral.
(c) Show that the square that bounds the region $-2 \le x \le 2$, $-2 \le y \le 2$ is a trapping region.
(d) Deduce that this differential equation has a limit cycle.

9.7.5. For this differential equation

$$x' = x \qquad y' = x - y - y^3$$

(a) plot the nullclines and show that the origin is the only fixed point.
(b) Show that the origin is a saddle point, necessarily unstable.
(c) Show there is no trapping region, consequently the Poincaré-Bendixson theorem cannot be applied, at least not in an obvious way, to this differential equation.

9.7.6. For this differential equation

$$x' = -x^3 + 12x^2 - 47x + y + 56 \qquad y' = -y^3 + 12y^2 - 47y - x + 64$$

(a) plot the nullclines and show that the point $(x,y) = (4,4)$ is the only fixed point.
(b) Show that the fixed point is an unstable spiral.

(c) Show that the square that bounds the region $2 \le x \le 6, 2 \le y \le 6$ is a trapping region.

(d) Deduce that this differential equation has a limit cycle.

(e) Can you solve this problem with the method of Sect. 9.5?

9.8   LIENARD'S AND BENDIXSON'S THEOREMS

In this section we present some additional results about stability of fixed points and the existence of limit cycles. We begin with some background about converting a second-order differential equation (a differential equation that involves a second derivative but no higher derivative) into a pair of first-order differential equations.

In Exercises 9.4.4 and 9.4.5 we saw the simplest conversion of

$$x'' + f(x)x' + g(x) = 0 \tag{9.25}$$

into a pair of first-order equations:

$$x' = y \qquad y' = x'' = -f(x)y - g(x) \tag{9.26}$$

There's another approach: Eq. (9.25) can be written as a *Lienard system*

$$x' = y - F(x) \qquad y' = -g(x) \tag{9.27}$$

where $F(x) = \int_0^x f(s)\, ds$. To see this, use Eq. (9.27) to compute x''

$$x'' = (y - F(x))' = y' - \frac{d}{dt}\int_0^x f(s)\, ds = -g(x) - f(x)x' \tag{9.28}$$

and we've obtained Eq. (9.25). Here the third equality comes from the second equation of Eq. (9.27), and from a combination of the fundamental theorem of calculus and the chain rule. To see this last point, observe

$$\frac{d}{dt}\int_0^x f(s)\, ds = f(x)\frac{dx}{dt} = f(x)x'$$

If this isn't clear, here's an example that may be more familiar, the type of problem everyone sees in first semester calculus

$$\frac{d}{dt}\int_0^{t^3} f(s)\, ds = f(t^3)\frac{d}{dt}t^3 = f(t^3)3t^2$$

For an illustration, we'll derive the Lienard form of the van der Pol equation

$$x'' + c(x^2 - 1)x' + x = 0$$

mentioned in Sect. 7.6. Putting this in the format of a Lienard system, Eq. (9.25), we see that $f(x) = c(x^2 - 1)$ and $g(x) = x$. Then

$$F(x) = \int_0^x f(s)\, ds = \int_0^x c(s^2 - 1)\, ds = c(x^3/3 - x)$$

Then the Lienard form of the van der Pol equation is

$$x' = y - F(x) = y - c(x^3/3 - x) \qquad y' = -g(x) = -x \qquad (9.29)$$

Why do we bother with Lienard systems? For some of these there is a simple test of the stability of a fixed point at the origin and also a simple test for the existence of a stable limit cycle.

Fixed points first. Let $G(x) = \int_0^x g(s)\, ds$.

Proposition 9.8.1. *Lienard's criterion for fixed points.* If for some $\delta > 0$ and for all x with $0 < |x| < \delta$, we have $G(x) > 0$ and $g(x)F(x) > 0$, then the origin is an asymptotically stable fixed point for Eq. (9.27). Under the same conditions on δ and x, if $G(x) > 0$ and $g(x)F(x) < 0$, then the origin is an unstable fixed point.

Proof. We'll use the Liapunov function $V(x, y) = y^2/2 + G(x)$. This is positive definite for $0 < |x| < \delta$ because $y^2/2 \geq 0$ and $G(x) \geq 0$, and because $y^2/2$ equals 0 only for $y = 0$ and $G(x)$ can equal 0 only for $x = 0$.

Next, observe that by the chain rule

$$\frac{dV}{dt} = \frac{\partial V}{\partial x}x' + \frac{\partial V}{\partial y}y' = g(y - F) + y(-g) = -gF$$

where the second equality comes from Eq. (9.27). We know that V is positive definite. If $gF > 0$, then dV/dt is negative definite and so the origin is asymptotically stable by Thm. 9.4.1. If $gF < 0$, then dV/dt is positive definite and the origin is unstable by Thm. 9.4.1, again. $\square$

Already we have two other methods for testing stability of fixed points, but we can use some help with limit cycles. As beautiful as it is, the Poincaré-Bendixson theorem can be tricky to use: finding a trapping region often is a challenge. Lienard's theorem provides another avenue to show a differential equation has a limit cycle, but unlike Poincaré-Bendixson, Lienard's criterion for limit cycles can be read easily from F and g. To understand Lienard's approach to limit cycles, we need some simple background on polynomials. A polynomial is called *even* if all the powers of x in the polynomial are even, and *odd* if all the powers of x are odd. For example, $x^2 + x^6 - x^8$ is even, $x - x^3 + x^5$ is odd, and $x + x^2 - x^4$ is neither even nor odd.

Theorem 9.8.1. *Lienard's criterion for limit cycles.* Suppose F and g are odd functions. Then the Lienard system (9.27) has a single limit cycle, and that limit cycle is stable, if

1. $xg(x) > 0$ for all $x \neq 0$,
2. $F(0) = 0$,
3. $F'(0) < 0$,
4. F has a single positive zero that is at $x = a$, and
5. for all $x \geq a$, $F(x)$ increases monotonically to ∞ as $x \to \infty$.

The proof is a bit involved, so we'll postpone it to Sect. A.21. Our theorem-to-example ratio is too high, so let's adjust that now.

Example 9.8.1. *A van der Pol limit cycle.* In the Lienard form, Eq. (9.29), of the van der Pol equation we have $F(x) = c(x^3/3 - x)$ and $g(x) = x$, so $G(x) = \int_0^x g(s)\, ds = x^2/2$. Then $G(x) > 0$ for all $|x| > 0$ and $g(x)F(x) = c(x^4/4 - x^2)$ is negative for $0 < |x| < 2$. (We know that for small $|x|$, $x^4 < x^2$ so we see that $g(x)F(x)$ is negative for small $|x|$. Next, observe that $g(x)F(x) = 0$ for $x = 0, \pm 2$.) Then by Lienard's criterion for fixed points (Prop. 9.8.1), the fixed point at the origin is unstable.

Now we'll apply Lienard's criterion for limit cycles (Theorem 9.8.1), ticking off the hypotheses of the theorem.

First, because they involve only odd powers of x, $F(x) = c(x^3/3 - x)$ and $g(x) = x$ are odd functions of x. Next, we check the numbered hypotheses:

1. $xg(x) = x^2 > 0$ for all $x \neq 0$.
2. $F(0) = c(0^3/3 - 0) = 0$.
3. $F'(x) = c(x^2 - 1)$, so $F'(0) = -c$, negative because $c > 0$.
4. By basic curve-sketching (or just plotting the graph of F), we see that the graph of F is decreasing for $0 \leq x < 1$, increasing for $x > 1$, and negative for $0 < x < \sqrt{3}$. The single positive zero of F occurs at $x = a = \sqrt{3}$.
5. For $x \geq \sqrt{3}$, F increases, ever more rapidly because $F''(x) = 2x > 0$, and so F increases monotonically to ∞ and $x \to \infty$.   Then by Lienard's criterion, the van der Pol equation has a unique limit cycle, and that limit cycle is stable. $\square$

Note the stability of the fixed point at the origin is easy to check using the method of Sect. 9.2:

$$D\vec{F} = \begin{bmatrix} -cx^2 + c & 1 \\ -1 & 0 \end{bmatrix} \quad \text{and so} \quad D\vec{F}(0,0) = \begin{bmatrix} c & 1 \\ -1 & 0 \end{bmatrix}$$

Then the eigenvalues of $D\vec{F}(0,0)$ are $\lambda = (c \pm \sqrt{c^2 - 4})/2$ and so the origin is an unstable spiral for $0 < c < 2$, an unstable node for $c > 2$. As expected, in addition to stability, the eigenvalues tell the type of the fixed point. Neither computing the eigenvalues nor checking the hypotheses of Prop. 9.8.1 is difficult.

For the van der Pol equation, finding a trapping region is a bit more work. For many of the examples of Sect. 9.7 this search was aided by the presence of $x^2 + y^2$ terms, suggesting conversion to polar coordinates. Without that simplification, the search can be tricky. So while Lienard's criterion gives no information about the size or location of the limit cycle, when it can be applied, usually it is straightforward.

A result in a different direction–conditions that preclude the existence of a periodic trajectory in a region–is Bendixson's criterion. Here we'll just state the result and give one example. We include Bendixson's criterion now to give a better sketch of the range of things we know about limit cycles. The proof is a straightforward application of Green's theorem, the topic of Sect. 17.4. So we'll prove Bendixson's criterion in Sect. 17.6, where we'll present some additional applications. First, we need one geometric notion, which we'll explore in more detail in Sect. 17.4. We say a region in the plane is *simply-connected* if it has no holes. The plane is simply-connected; the plane with a disc cut out is not simply-connected.

Theorem 9.8.2. *Bendixson's criterion.* Suppose f and g have continuous partial derivatives throughout a simply-connected region R. If $\partial f/\partial x + \partial g/\partial y$ has the same sign throughout R, then the differential equation $x' = f(x,y), y' = g(x,y)$ has no periodic trajectory that lies entirely in R.

Example 9.8.2. *A Fitzhugh-Nagumo limit cycle.* Recall the Fitzhugh-Nagumo equations (7.15) with $z = 0$,

$$x' = c\left(y + x - \frac{x^3}{3}\right) = f(x,y) \qquad y' = -\frac{x - a + by}{c} = g(x,y)$$

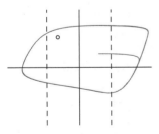

Figure 9.13. Regions for Example 9.8.2.

In Fig. 7.25 we saw that for $a = 0.8$, $b = 0.15$, and $c = 3.0$ the Fitzhugh-Nagumo equations have a limit cycle. In Fig. 9.13 we've indicated the fixed point by a small circle. We'll apply Bendixson's criterion to determine some bounds on the extent of the limit cycle. First, compute

$$\frac{\partial f}{\partial x} + \frac{\partial g}{\partial y} = c - cx^2 - \frac{b}{c}$$

Setting this expression equal to 0 we obtain $x = \pm\sqrt{1 - b/c^2}$. If $b/c^2 > 1$, then $\partial f/\partial x + \partial g/\partial y$ keeps the same sign throughout its domain and so Bendixson's criterion imposes no restriction on the extent of any limit cycles. But if $b/c^2 < 1$, the case in this example, we see that $x = \pm\sqrt{1 - b/c^2}$ divides the x-y plane into three simply-connected regions, $x < -\sqrt{1 - b/c^2}$ and $x > -\sqrt{1 - b/c^2}$ where $\partial f/\partial x + \partial g/\partial y$ is negative, and $-\sqrt{1 - b/c^2} < x < \sqrt{1 - b/c^2}$ where $\partial f/\partial x + \partial g/\partial y$ is positive. These regions are indicated by dashed vertical lines in Fig. 9.13. Bendixson's criterion implies that a limit cycle cannot lie entirely within one of these regions. Because we expect that a limit cycle will encompass the fixed point, indicated by the small circle, we might think we could find a limit cycle entirely within the middle regions but Bendixson shows this is impossible. In fact, we see that the limit cycle enters all three regions. $\square$

Bendixson's criterion doesn't give any information about whether a closed trajectory exists, but when it can be applied, Bendixson's criterion can say something about the size and location of any limit cycle.

Finally, we'll mention some results on counting limit cycles.

How many limit cycles can a 2-dimensional polynomial differential equation exhibit? A degree-n polynomial system has this form

$$x' = \sum_{i=0}^{n}\sum_{j=0}^{n-i} a_{ij}x^i y^j \qquad y' = \sum_{i=0}^{n}\sum_{j=0}^{n-i} b_{ij}x^i y^j$$

One result is

Theorem 9.8.3. *Dulac's theorem.* Every 2-dimensional polynomial differential equation has a finite number of limit cycles.

Fine, the number is finite, but what is it? (Henri Dulac's proof [32] contains an error. Correct proofs were given by Yu. Ilyashenko [67] and Jean Écalle [33].) More precisely, can we find the maximum number C_n of limit cycles for a degree-n polynomial system? If you think this should be easy to figure out, I'll mention that in 1900 David Hilbert posed 23 major mathematical problems at the second meeting of the International Congress of Mathematicians. The question we just asked is number 16 of Hilbert's list. Linear systems have no limit cycles, so $C_1 = 0$. S. Shi [149] and L. Chen and M. Wang [26] produced an $n = 2$ example with 4 limit cycles, but it isn't known if $C_2 = 4$. P. Yu and M. Han [181] produced an $n = 3$ example with 12 limit cycles, but it isn't known if $C_3 = 12$. Yes, this is a hard problem.

Practice Problems

9.8.1. Show that the system

$$x' = y - x(x^2 - 1) \qquad y' = -x$$

has a unique limit cycle, and that this limit cycle is not contained entirely in the region $-1/\sqrt{3} < x < 1/\sqrt{3}$.

9.8.2. Compare the limit cycle of the system

$$x' = y - x(x^2 - 1) \qquad y' = -x$$

with that of the system

$$x' = y - x(x^2 - 2) \qquad y' = -x$$

Practice Problem Solutions

9.8.1. In terms of Lienard's theorem, $F(x) = x(x^2 - 1)$ and $g(x) = x$, both are odd functions. We'll show that these functions satisfy the conditions of Lienard's theorem 9.8.1.

1. $xg(x) = x^2$ is positive for all $x \neq 0$,
2. $F(0) = 0(0^2 - 1) = 0$,
3. $F'(0) = -1 < 0$,
4. $F(x)$ has a single positive zero, at $x = 1$, and
5. for all $x \geq 1$, $F(x)$ increases to ∞ as $x \to \infty$.

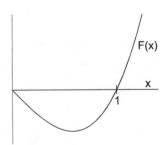

Figure 9.14. Plot of $F(x)$ for Practice Problem 9.8.1.

These last two points can be seen easily from the plot of F in Fig. 9.14. Then by Lienard's criterion this differential equation has a single limit cycle and that limit cycle is stable.

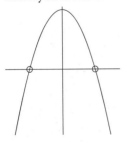

Figure 9.15. Plot of $\partial f/\partial x + \partial f/\partial y$ for Practice Problem 9.8.1.

Now we'll apply Bendixson's criterion 9.8.2. Here $f(x,y) = y - x(x^2 - 1)$ and $g(x,y) = -x$, so

$$\frac{\partial f}{\partial x} + \frac{\partial g}{\partial y} = -3x^2 + 1$$

We must identify regions on which this sum of partial derivatives does not change sign.

Fig. 9.15 shows the plot of $\partial f/\partial x + \partial g/\partial y$. The zeros of this function are $x =$

$\pm 1/\sqrt{3}$. In the interval $-1/\sqrt{3} < x < 1/\sqrt{3}$ we see that $\partial f/\partial x + \partial g/\partial y$ is positive. Because it does not change signs in this interval, Bendixson's criterion implies that the limit cycle of this equation cannot be contained entirely in this region.

9.8.2. In Fig. 9.16 we see the limit cycle for system (a)
$$x' = y - x(x^2 - 1), \quad y' = -x$$
and for system (b)
$$x' = y - x(x^2 - 2), \quad y' = -x$$

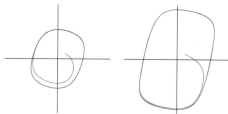

Both graphs have $-4 \le x \le 4$ and $-4 \le y \le 4$, so we see that limit cycle (b) is larger.

Figure 9.16. Limit cycles for (a) (left) and (b) (right) for Practice Problem 9.8.2.

One way to deduce this change of limit cycle extent without plotting the solution curves is to see how the application of Bendixson's criterion changes. We compute $\partial f/\partial x + \partial g/\partial y$ and obtain $-3x^2 + 1$ for (a), $-3x^2 + 2$ for (b). The zeros of these functions are $x = \pm 1/\sqrt{3} \approx \pm 0.57735$ for (a) and $x = \pm\sqrt{2/3} \approx \pm 0.816497$ for (b). Between these bounds the sum of the partial derivatives is negative, so Bendixson's criterion implies that the limit cycle cannot lie entirely between these bounds. A larger distance between the bounds is consistent with a larger limit cycle.

Exercises

9.8.1. Show that the system
$$x' = y - x(x^4 - 1) \qquad y' = -x^3$$
has a unique stable limit cycle, and that this limit cycle is not contained in the region $-1/5^{1/4} < x < 1/5^{1/4}$.

9.8.2. Show that the system
$$x' = y - x(x^2 - 1)^3 \qquad y' = -x$$
has a unique stable limit cycle, and that this limit cycle is not contained in the region $-1/\sqrt{7} < x < 1/\sqrt{7}$, or in the region $1/\sqrt{7} < x$.

9.8.3. Suppose the system
$$x' = y - F(x) \qquad y' = -g(x)$$

satisfies the conditions of Lienard's criterion 9.8.1, so has a stable limit cycle among its trajectories. So long as g is an odd function and $xg(x) > 0$ for all $x \neq 0$, the differential equation will have a limit cycle. The goal of this and the next problem is to gain some understanding of how the form of g influences the shape of the limit cycle. Take $F(x) = x(x^4 - 1)$. Plot the limit cycles and comment on their differences and similarities for

(a) $g(x) = x$,   (b) $g(x) = x^3$, and   (c) $g(x) = x^3 + x$.

9.8.4. Continuing the investigation of Exercise 9.8.3, take $F(x) = x^5 - 3x^3$ and comment on the shape of the limit cycle for

(a) $g(x) = x^3$,   (b) $g(x) = 2x^3$, and   (c) $g(x) = 5x^3$.

9.8.5. For continuously differentiable functions $A(y)$ and $B(x)$, show that if the differential equation

$$x' = x^3 - 3x + A(y) \qquad y' = y^3 + B(x)$$

has a limit cycle, that limit cycle must cross the circle $x^2 + y^2 = 1$.

9.8.6. For continuously differentiable functions $A(y)$ and $B(x)$, show that if the differential equation

$$x' = x^4 + A(y) \qquad y' = yx^3 - y^2 + B(x)$$

has a limit cycle, that limit cycle is not contained in quadrant 2 or in quadrant 4.

9.8.7. For continuously differentiable functions $A(y)$ and $B(x)$, show that if the differential equation

$$x' = x^2y + A(y) \qquad y' = xy^2 + B(x)$$

has a limit cycle, that limit cycle is not contained in any single quadrant.

9.8.8. Show that the differential equation

$$x' = x\sin\left(\frac{1}{x^2+y^2}\right) + y \qquad y' = y\sin\left(\frac{1}{x^2+y^2}\right) - x$$

defined on the xy-plane minus the origin, has infinitely many limit cycles. Why doesn't this contradict Dulac's theorem 9.8.3?

9.8.9. Show each of these differential equations has a unique stable limit cycle.

(a) $x' = y - (-x - x^3 + x^5)$ $y' = -x^3$

(b) $x' = y - (-2x - x^3 + x^5)$ $y' = -2x^3$

(c) Plot the limit cycles of (a) and (b) and describe their differences.

9.8.10. Find a polynomial differential equation with two limit cycles. Circles centered on the origin probably are the simplest to construct.

Chapter 10 Infinite series and power series

We begin with a slight modification of one of Zeno's paradoxes. Suppose you are standing 1 m from the door, and you really, really want to leave the room. Before you reach the door, first you must pass through the point half-way to the door, then through the point half-way between where you are and the door, and so on. First you must travel 1/2 m, then 1/4 m, then 1/8 m, and so on. At each observation, you have traveled half-way between the last observation and the door, so you can never get to the door. Can you find a resolution to this paradox? If not, you may be stuck in whatever room you're reading this. I hope you brought your dinner.

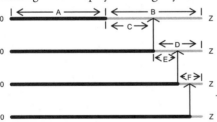

Figure 10.1. An illustration of Zeno's paradox

Fig. 10.1 illustrates several stages of Zeno's paradox. The starting point is labeled 0, the goal Z. In the first line you travel A, one-half of the distance from 0 to Z, leaving the distance B to travel. In the second line you travel C, one-half of B, leaving the distance D to trvael. In the third line you travel E, one-half of D, leaving the distance F to travel.

Two points are easy to see. First, after any finite number of steps of this form, you never quite reach Z, because with each step you travel half the distance remaining to Z. After successive steps the distance remaining to Z is

$$\frac{1}{2},\frac{1}{4},\frac{1}{8},\frac{1}{16},\frac{1}{32},\frac{1}{64},\frac{1}{128},\frac{1}{256},\frac{1}{512},\frac{1}{1024},\frac{1}{2048},\frac{1}{4096},\dots$$

This leads to the second point. For every positive distance δ, no matter how microscopically tiny, eventually your distance to Z will be less than δ. This is because for every $\delta > 0$, for large enough n

$$\frac{1}{2^n} < \delta$$

If you make infinitely many such steps, then in the limit you reach the point Z. This plausible conclusion is a consequence of an important property of the real numbers, mentioned in Sect. A.10. The property, completeness, means that there are no holes or gaps in the real numbers.

What we have done is argued that

$$\frac{1}{2}+\frac{1}{4}+\frac{1}{8}+\frac{1}{16}+\cdots=1$$

Can this calculation be generalized and made more rigorous?

Given a number r, denote by A_N the sum

$$1+r+r^2+\cdots+r^{N-1}$$

We find a simple expression for A_N, though it will take a few steps.

$$1+r+r^2+\cdots+r^{N-1}=A_N$$
$$r\cdot(1+r+r^2+\cdots+r^{N-1})=r\cdot A_N$$
$$r+r^2+r^3+\cdots+r^{N-1}+r^N=r\cdot A_N$$
$$1+r+r^2+r^3+\cdots+r^{N-1}+r^N=1+r\cdot A_N$$
$$(1+r+r^2+r^3+\cdots+r^{N-1})+r^N=1+r\cdot A_N$$
$$A_N+r^N=1+r\cdot A_N$$

Solving for A_N we find

$$A_N=\frac{1-r^N}{1-r}=\frac{1}{1-r}-\frac{r^N}{1-r} \tag{10.1}$$

If $|r| < 1$, the term $r^N/(1-r)$ goes to 0 as $N\to\infty$ and we see

$$A=\lim_{N\to\infty}A_N=\frac{1}{1-r}$$

That is, as long as $|r| < 1$, we can sum the *geometric series*

$$1 + r + r^2 + r^3 + r^4 + \cdots = \frac{1}{1-r} \tag{10.2}$$

To illustrate the range of behaviors series exhibit, we survey two more results. Consider these examples,

$$\sum_{n=1}^{\infty} \frac{1}{n} = \lim_{N \to \infty} \sum_{n=1}^{N} \frac{1}{n} \quad \text{and} \quad \sum_{n=1}^{\infty} \frac{1}{n^2} = \lim_{N \to \infty} \sum_{n=1}^{N} \frac{1}{n^2}$$

We shall see that the first series *diverges*, that is, the $N \to \infty$ limit does not exist, and the second *converges*, that is, the limit exists. Determining if a series converges or diverges can be challenging, but for these two examples the argument is simple, once we see a trick. The trick is to note that

$$\frac{1}{n} = f(n) \quad \text{and} \quad \frac{1}{n^2} = g(n) \quad \text{where} \quad f(x) = \frac{1}{x} \quad \text{and} \quad g(x) = \frac{1}{x^2}$$

then to observe that the area of the family of rectangles on the left side of Fig. 10.2 equals $\sum 1/n$, and the area on the right side equals $\sum 1/n^2$. From this we see

$$\sum_{n=1}^{\infty} \frac{1}{n} > \int_{1}^{\infty} \frac{1}{x} dx = \ln(x) \Big|_{1}^{\infty} = \infty$$

$$\sum_{n=1}^{\infty} \frac{1}{n^2} < \int_{1}^{\infty} \frac{1}{x^2} dx = -\frac{1}{x} \Big|_{1}^{\infty} = 1$$

This is an example of the Integral Test, the subject of Sect. 10.1.

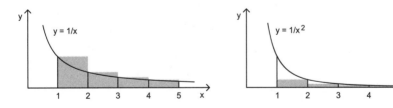

Figure 10.2. Comparing the graphs of $f(x) = 1/x$ and $g(x) = 1/x^2$ with families of rectangles

To determine if a series converges, we must learn a collection of tricks–a lot of tricks–so we may as well get started.

10.1 THE INTEGRAL TEST

The conclusions of the last section are obvious from geometry by comparing areas. How general is this approach? What conditions on the function f guarantee that the improper integral converges if and only if the series converges? We want the area under the curve to be bounded above and below by the appropriate collections of rectangles, so these conditions suffice:

(1) f is positive, (2) f is continuous, and (3) f is decreasing.

These guarantee that

$$f(n) = \max\{f(x) : n \leq x \leq n+1\}, \ f(n+1) = \min\{f(x) : n \leq x \leq n+1\}$$

Consequently, the graph of $y = f(x)$ lies below the rectangles of heights $f(1)$, $f(2), f(3), \dots$, and lies above the rectangles of heights $f(2), f(3), f(4), \dots$. This is the basis of the

Integral Test Suppose the function f satisfies conditions (1), (2), and (3) above. Then

- if $\int_1^\infty f(x)dx$ converges, then $\sum_{n=1}^\infty f(n)$ converges, and
- if $\int_1^\infty f(x)dx$ diverges, then $\sum_{n=1}^\infty f(n)$ diverges

This result has many extensions. For example, neither the series nor the integral need start with 1. Also, it is clear that the convergence of a series is not determined by its first 2 or 10 or 100 or 10^{10} terms, all of which have finite sums, so the function $f(x)$ need be only eventually decreasing.

As an application we find for which $p > 0$ does the *p-series* $\sum_{n=1}^\infty 1/n^p$ converge?

Example 10.1.1. *The convergence of a p-series.* To apply the Integral Test, we want a function $f(x)$ for which $f(n) = 1/n^p$, so take $f(x) = 1/x^p$. We check the conditions of the Integal Test: (1) holds because we'll need only $x \geq 0$ and so f is positive, (2) holds because rational functions are continuous on their domains and the domain of f is all real numbers except 0, and (3) holds because f is decreasing since $f'(x) = -px^{-p-1}$, negative because $p > 0$ and $x \geq 1$.

From our study of improper integrals, we know $\int_1^\infty 1/x^p dx$ converges for $p > 1$ and diverges for $p \leq 1$.

We required $p > 0$ in order to use the Integral Test, but we can treat the other values of p more directly.

For $p = 0$ the series is $\sum_{n=1}^\infty 1/n^0 = \sum_{n=1}^\infty 1$, which certainly diverges.

For $p < 0$, the nth term $1/n^p \to \infty$ as $n \to \infty$, so for these values of p the series certainly diverges. That is,

> The p-series converges for $p > 1$ and diverges for $p \le 1$.

In Sect. 10.2 we'll see this simple application of the Integral Test is quite useful. $\square$

Practice Problems

10.1.1. Does the series $\sum_{n=1}^{\infty} n/(n^2+1)$ converge or diverge?

10.1.2. Does the series $\sum_{n=1}^{\infty} n/(n^2+1)^2$ converge or diverge?

Practice Problem Solutions

10.1.1. The function $f(x) = x/(x^2+1)$ is positive, continuous, and decreasing. This last is because $f'(x) = (1-x^2)/(1+x^2)^2$ is negative for $x > 1$. By the substitution $u = x^2 + 1$ we see

$$\int_1^{\infty} \frac{x}{x^2+1}\,dx = \frac{1}{2}\int_2^{\infty} \frac{1}{u}\,du = \frac{1}{2}\ln(u)\Big|_2^{\infty}$$

This diverges, so the series diverges by the Integral Test.

10.1.2. The function $f(x) = x/(x^2+1)^2$ is positive, continuous, and decreasing. This last is because $f'(x) = (1-3x^2)/(1+x^2)^3$ is negative for $x > 1/\sqrt{3}$. By the substitution $u = x^2 + 1$ we see

$$\int_1^{\infty} \frac{x}{(x^2+1)^2}\,dx = \frac{1}{2}\int_2^{\infty} \frac{1}{u^2}\,du = \frac{1}{2}\frac{-1}{u}\Big|_2^{\infty} = \frac{1}{4}$$

This converges, so the series converges by the Integral Test.

Exercises

10.1.1. Do these series converge or diverge?

(a) $\sum_{n=1}^{\infty} \frac{n^2}{n^3+1}$ (b) $\sum_{n=2}^{\infty} n^2 e^{-n}$ (c) $\sum_{n=3}^{\infty} \frac{\ln(n)}{n}$ (d) $\sum_{n=2}^{\infty} \frac{n^3}{n^4+1}$

10.1.2. Do these series converge or diverge?

(a) $\sum_{n=3}^{\infty} \frac{\ln(n^2)}{n}$ (b) $\sum_{n=1}^{\infty} \frac{n}{(n^2+1)^3}$ (c) $\sum_{n=1}^{\infty} \frac{1+2n}{n^2+n}$ (d) $\sum_{n=1}^{\infty} \frac{e^n}{e^{2n}+1}$

10.1.3. Use an idea from the proof of the Integral Test to show

$$\ln(n+1) \leq 1 + \frac{1}{2} + \frac{1}{3} + \cdots + \frac{1}{n} \leq 1 + \ln(n)$$

10.1.4. Find lower and upper bounds for $\sum_{n=1}^{N} 1/n$ for $N = 10^3, 10^6, 10^9$, and 10^{12}.

10.1.5. (Harder.) Suppose a_n is a sequence of positive numbers decreasing to 0. Use the idea of the proof of the Integral Test to show that

$$\sum_{n=1}^{\infty} a_n \text{ converges if and only if } \sum_{n=1}^{\infty} 2^n a_{2^n} \text{ converges}$$

10.2 THE COMPARISON TEST

The Comparison Test is based on very simple geometry, so simple that a proof should be evident. (If you did Exercise 10.1.5, you've already discovered this.)

Comparison Test Suppose for all n, $0 < a_n \leq b_n$.

 1. If $\sum_{n=1}^{\infty} b_n$ converges, then $\sum_{n=1}^{\infty} a_n$ converges.
 2. If $\sum_{n=1}^{\infty} a_n$ diverges, then $\sum_{n=1}^{\infty} b_n$ diverges.

This proof is illustrated on the left side of Fig. 10.3, and the argument is clear. If the series of larger terms converges, then certainly the series of smaller terms converges. If the series of smaller terms diverges, then certainly the series of larger terms diverges. The right side of Fig. 10.3 emphasizes a point mentioned with the Integral Test, that in determining convergence by the relative values of the sequence terms, the first few, or first few million, terms are not important: only the eventual relationship between the a_n and b_n is relevant.

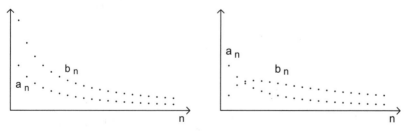

Figure 10.3. Comparison of the terms of $\sum_{n=1}^{\infty} a_n$ and $\sum_{n=1}^{\infty} b_n$.

For example, by the p-series convergence condition, $\sum_{n=1}^{\infty} 1/n^2$ converges. Now for all n,

$$n^2 + 1 > n^2 \quad \text{from which we see} \quad \frac{1}{n^2+1} < \frac{1}{n^2}$$

and so by the Comparison Test $\sum_{n=1}^{\infty} 1/(n^2+1)$ converges.

Similarly, we know the $p = 1$ p-series $\sum_{n=1}^{\infty} 1/n$ diverges. This is called the *harmonic series*. Then for all n,

$$n - 1 < n \quad \text{from which we see} \quad \frac{1}{n-1} > \frac{1}{n}$$

and so by the Comparison Test, $\sum_{n=1}^{\infty} 1/(n-1)$ diverges.

So far, so good, but what about $\sum_{n=1}^{\infty} 1/(n+1)$ and $\sum_{n=1}^{\infty} 1/(n^2-1)$? We can't apply the Comparison Test directly, because $1/(n+1) < 1/n$, and being less than a divergent series tells us nothing. Similarly, $1/(n^2-1) > 1/n^2$, and being greater than a convergent series tells us nothing. On the other hand, for large n, $1/(n+1)$ is pretty much identical to $1/n$, and $1/(n^2-1)$ is pretty much identical to $1/n^2$. So somehow the convergence behavior of $1/n$ and of $1/n^2$ should tell us about that of $1/(n+1)$ and of $1/(n^2-1)$, but how?

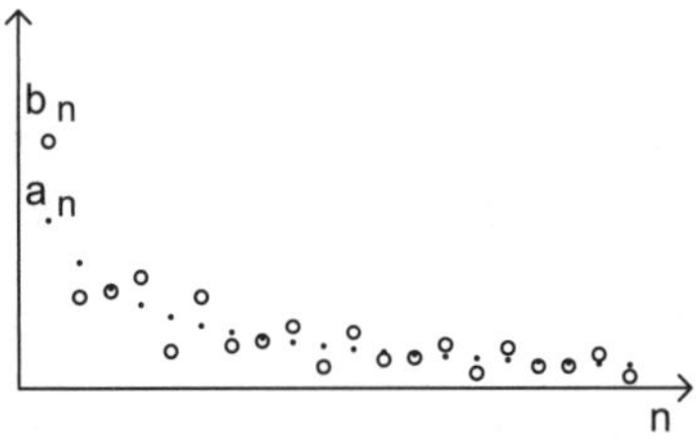

Figure 10.4. The a_n are represented by dots, the b_n by circles.

Or maybe, even worse, suppose the b_n oscillate around the a_n, sometimes above, sometimes below, but with ever decreasing amplitude, so it appears that the terms of both series are approaching the same behavior. See Fig. 10.4. Surely the convergence of one determines the convergence of the other.

In fact, this is the content of the Limit Comparison Test.

Limit Comparison Test Suppose $a_n > 0$ and $b_n > 0$ for all n, and $\lim_{n \to \infty} a_n/b_n = L$.

1. If $L < \infty$ and $\sum_{n=1}^{\infty} b_n$ converges, then the series $\sum_{n=1}^{\infty} a_n$ converges.
2. If $L > 0$ and $\sum_{n=1}^{\infty} b_n$ diverges, then the series $\sum_{n=1}^{\infty} a_n$ diverges.

For example,

$$\lim_{n \to \infty} \frac{1/(n^2-1)}{1/n^2} = \lim_{n \to \infty} \frac{n^2}{n^2-1} = \lim_{n \to \infty} \frac{1}{1-1/n^2} = 1$$

so $\sum_{n=1}^{\infty} 1/(n^2-1)$ converges by the Limit Comparison Test because $\sum_{n=1}^{\infty} 1/n^2$ converges.

Similarly,

$$\lim_{n \to \infty} \frac{1/(n+1)}{1/n} = \lim_{n \to \infty} \frac{n}{n+1} = \lim_{n \to \infty} \frac{1}{1+1/n} = 1$$

so $\sum_{n=1}^{\infty} 1/(n-1)$ diverges by the Limit Comparison Test because $\sum_{n=1}^{\infty} 1/n$ diverges.

Why is the Limit Comparison Test true? Here we cannot rely on transparent geometry, but must think a little. That $\lim_{n\to\infty} a_n/b_n = L$ means that for large enough n, a_n/b_n gets close to L and stays close to L. If L is positive, there are positive numbers α and β, with $0 < \alpha < L < \beta$ with

$$\alpha < \frac{a_n}{b_n} < \beta \quad \text{for all sufficiently large } n$$

That is,

$$\alpha b_n < a_n < \beta b_n \quad \text{for all sufficiently large } n \tag{10.3}$$

Now if $\sum_{n=1}^{\infty} b_n$ converges, so does $\sum_{n=1}^{\infty} \beta b_n$. From the right inequality of Eq. (10.3) and part 1 of the Comparison Test, we see that $\sum_{n=1}^{\infty} a_n$ converges.

If $\sum_{n=1}^{\infty} b_n$ diverges, so does $\sum_{n=1}^{\infty} \alpha b_n$. From the left inequality of Eq. (10.3) and part 2 of the Comparison Test, we see that $\sum_{n=1}^{\infty} a_n$ diverges.

Do you see how to extend the argument to show convergence even if $L = 0$ and divergence even if $L = \infty$?

Practice Problem

10.2.1. Do these series converge or diverge?

$$\text{(a)} \sum_{n=1}^{\infty} \frac{1}{n^2 + n} \qquad \text{(b)} \sum_{n=1}^{\infty} \frac{\sqrt{n^3 + 4}}{n^2 + 2n + 3} \qquad \text{(c)} \sum_{n=1}^{\infty} \sin\left(\frac{\pi}{n}\right)$$

Practice Problem Solution

10.2.1. (a) We could compare the series with $1/n$, which diverges, or with $1/n^2$, which converges. But with which one? We see

$$\frac{1}{n^2 + n} < \frac{1}{n} \quad \text{and} \quad \frac{1}{n^2 + n} < \frac{1}{n^2}$$

Having terms less than those of a diverging series tells us nothing, but having terms less than those of a converging series tells us that the series of (a) converges. (b) First we must find a series to use for comparison. For large n the numerator $\sqrt{n^3 + 4}$ is close to $n^{3/2}$, and the denominator $n^2 + 2n + 3$ is close to n^2. Then for large n the terms of the series are close to $n^{3/2}/n^2 = 1/n^{1/2}$. This is the series we use for comparison.

$$\frac{\sqrt{n^3+4}}{\frac{n^2+2n+3}{\frac{1}{n^{1/2}}}} = \frac{n^{3/2}\sqrt{1+\frac{4}{n^2}}}{n^2\left(1+\frac{2}{n}+\frac{3}{n^2}\right)}\cdot n^{1/2} = \frac{\sqrt{1+\frac{4}{n^2}}}{1+\frac{2}{n}+\frac{3}{n^2}}$$

As $n \to \infty$, this fraction goes to 1. Because $\sum_{n=1}^{\infty} 1/n^{1/2}$ is a p-series with $p < 1$, this series diverges, and so the series of (b) diverges by the Limit Comparison Test.

(c) Recall $\lim_{x\to 0} \sin(x)/x = 1$, so we see $\lim_{n\to\infty} \sin(\pi/n)/(\pi/n) = 1$. Because $\sum_{n=1}^{\infty} \pi/n$ is just π times the harmonic series, which diverges, the series of (c) diverges by the limit comparison theorem.

Exercises

Do these series converge or diverge?

10.2.1. (a) $\displaystyle\sum_{n=1}^{\infty} \frac{1}{3^n+2}$ (b) $\displaystyle\sum_{n=1}^{\infty} \frac{3^n+2}{5^n+4}$ (c) $\displaystyle\sum_{n=1}^{\infty} \frac{1}{1+2+\cdots+n}$

10.2.2. (a) $\displaystyle\sum_{n=2}^{\infty} \frac{1}{n-1}$ (b) $\displaystyle\sum_{n=2}^{\infty} \frac{1}{n^2-\sqrt{n}}$ (c) $\displaystyle\sum_{n=1}^{\infty} \frac{1}{3^n-2}$

10.2.3. (a) $\displaystyle\sum_{n=1}^{\infty} \frac{\sqrt{n^3+4}}{n^2+2n+3}$ (b) $\displaystyle\sum_{n=1}^{\infty} \sin^2\left(\frac{\pi}{n}\right)$ (c) $\displaystyle\sum_{n=1}^{\infty} \frac{1+\cos^2(n)}{n^3}$

10.2.4. Sometimes the Comparison Test can give information on the values to which convergent series converge. Find upper bounds for these sums.

(a) $\displaystyle\sum_{n=1}^{\infty} \frac{1}{2^n+(1/n)}$ (b) $\displaystyle\sum_{n=1}^{\infty} \frac{2}{5^n+(1/n^2)}$ (c) $\displaystyle\sum_{n=0}^{\infty} \frac{2^n}{5^n+1/(n+1)}$

10.2.5. Show $\displaystyle\frac{1}{3} < \sum_{n=1}^{\infty} \frac{1}{3^n+1} < \frac{1}{2}$.

10.3 ALTERNATING SERIES

So far we have considered mostly series of positive terms. If the signs of the terms alternate, then testing convergence is a much simpler matter. On the other hand, the limit to which some of these series converge depends delicately on the order in which the terms are added. How can this be? Let's see.

Suppose all a_n are positive. An *alternating series* is any series of the form

$$\sum_{n=1}^{\infty}(-1)^{n+1}a_n = a_1 - a_2 + \cdots \quad \text{or} \quad \sum_{n=1}^{\infty}(-1)^n a_n = -a_1 + a_2 - \cdots \quad (10.4)$$

Leibnitz found conditions under which alternating series converge. We sketch a proof at the end of this section.

Alternating Series Test An alternating series (10.4) converges if (1) for all n, $a_n \geq a_{n+1}$, and (2) $a_n \to 0$ as $n \to \infty$.

As we have seen, convergence is not determined by any finite collection of a_n, so the first condition of the Alternating Series Test need hold only for all $n \geq M$ for some M.

Example 10.3.1. *A simple alternating series.* Does the series

$$\sum_{n=1}^{\infty}(-1)^{n+1}\frac{\sqrt{n}+1}{\sqrt{n}+2}$$

converge or diverge? This is certainly an alternating series. To check that the terms are eventually decreasing, compute the derivative of $f(x) = (\sqrt{x}+1)/(\sqrt{x}+2)$, obtaining

$$f'(x) = \frac{2-\sqrt{x}}{2\sqrt{x}(x+2)^{3/2}}$$

negative for $x > 4$. Thus the terms of the alternating series are decreasing for $n > 4$. However, condition (2) of the Alternating Series Test is not satisfied, because

$$\lim_{n \to \infty}\frac{\sqrt{n}+1}{\sqrt{n}+2} = 1$$

Alone, not satisfying the hypotheses of the Alternating Series Test is not sufficient reason to conclude that the series does not converge. The test says that if its conditions are satisfied, then the series converges. It *DOES NOT* say that if the conditions are not satisfied, then the series does not converge. However, for this example we can conclude that the series diverges, because the way in which the second condition is not satisfied is a violation of the nth term test: if $a_n \nrightarrow 0$, then $\sum a_n$ does not converge. This is the simplest test, and not often useful, but don't forget it. □

For another example, the alternating harmonic series $\sum_{n=1}^{\infty}(-1)^{n+1}/n$ converges, because it satisfies the conditions of the Alternating Series Test. That the harmonic series diverges, but the alternating harmonic series converges, motivates these definitions:

The series $\sum_{n=1}^{\infty} a_n$ *converges absolutely* if the series $\sum_{n=1}^{\infty} |a_n|$ converges.

A series *converges conditionally* if it converges but does not converge absolutely.

Example 10.3.2. *Conditional or absolute convergence.* Does the series $\sum_{n=1}^{\infty} (-1)^n / \sqrt{n+1}$ converge absolutely, converge conditionally, or diverge?

Certainly, this is an alternating series, with $1/\sqrt{n+1}$ a decreasing function of n satisfying $\lim_{n\to\infty} 1/\sqrt{n+1} \to 0$. So we know the series converges by the Alternating Series Test. Does it converge absolutely? The absolute values of the terms are $1/\sqrt{n+1}$, looking much like $1/\sqrt{n}$. However, $1/\sqrt{n} > 1/\sqrt{n+1}$ and $\sum_{n=1}^{\infty} 1/\sqrt{n}$ is a p-series, divergent because $p = 1/2$, so the Comparison Test cannot be used. (Recall that being smaller than a divergent series tells us nothing.) This suggests using the Limit Comparison Test: $\lim_{n\to\infty} \sqrt{n+1}/\sqrt{n} = 1$. Consequently, $\sum_{n=1}^{\infty} 1/\sqrt{n+1}$ diverges. The series of this example converges but does not converge absolutely, so must converge conditionally. □

One relation between conditional and absolute convergence is more-or-less what we expect.

Absolute Convergence theorem. If $\sum_{n=1}^{\infty} a_n$ converges absolutely, then the series converges.

At the end of this section we sketch a proof of the Absolute Convergence theorem. On the other hand, the alternating harmonic series shows that a convergent series need not converge absolutely.

Absolutely convergent series have a property, of great use to us, that appears to be a subtle extension to infinite series of the commutative law of addition. For example, this theorem was proved by Bernhard Riemann in 1867.

Rearrangement theorem. If $\sum_{n=1}^{\infty} a_n$ converges absolutely, and $\sum_{n=1}^{\infty} b_n$ is ANY rearrangement of $\sum_{n=1}^{\infty} a_n$–that is, there is a 1-1 correspondence between $\{a_i\}$ and $\{b_j\}$–then $\sum_{n=1}^{\infty} b_n = \sum_{n=1}^{\infty} a_n$.

For finite series, indeed this is just the commutative law of addition. But in general, this law does not extend to infinite series. In fact, every infinite series that converges conditionally can be rearranged to converge (conditionally) to any number at all. Let's explore this.

The alternating harmonic series illustrates this point. The terms with odd denominators all are positive, and

$$\sum_{i=1}^{\infty} \frac{1}{2n-1}$$

diverges to $+\infty$. Fig. 10.5 shows this by a modification of the Integral Test. The shaded rectangles have areas that sum to $+\infty$, because they cover the area under the graph of $1/x$, which we know diverges to $+\infty$. Now the sum of the areas of these rectangles is

$$2 \cdot 1 + 2 \cdot \frac{1}{3} + 2 \cdot \frac{1}{5} + 2 \cdot \frac{1}{7} + \cdots$$

If $\sum_{n=1}^{\infty} 1/(2n-1)$ converged, then so would $2\sum_{n=1}^{\infty} 1/(2n-1) = \sum_{n=1}^{\infty} 2/(2n-1)$, but Fig. 10.5 shows this is not so.

A similar argument shows that

$$\sum_{n=1}^{\infty} -\frac{1}{2n}$$

diverges to $-\infty$.

Now pick any number at all, say $\pi/2$. Add the positive terms of the alternating harmonic series until the sum first exceeds $\pi/2$. Because the series of positive terms diverges, we know the sum of positive terms eventually will exceed $\pi/2$.

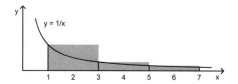

Figure 10.5. The sum of the odd terms of the alternating harmonic series diverges to ∞.

$$1 + \frac{1}{3} + \frac{1}{5} + \frac{1}{7} \approx 1.67619 > \frac{\pi}{2}$$

Now add the negative terms until the sum drops below $\pi/2$.

$$1 + \frac{1}{3} + \frac{1}{5} + \frac{1}{7} - \frac{1}{2} \approx 1.17619 < \frac{\pi}{2}$$

Now add the remaining positive terms until the sum exceeds $\pi/2$.

$$1 + \frac{1}{3} + \frac{1}{5} + \frac{1}{7} - \frac{1}{2} + \frac{1}{9} + \frac{1}{11} + \frac{1}{13} + \frac{1}{15} + \frac{1}{17} \approx 1.58062 > \frac{\pi}{2}$$

Now add the remaining negative terms until the sum drops below $\pi/2$.

$$1 + \frac{1}{3} + \frac{1}{5} + \frac{1}{7} - \frac{1}{2} + \frac{1}{9} + \frac{1}{11} + \frac{1}{13} + \frac{1}{15} + \frac{1}{17} - \frac{1}{4} \approx 1.33062 < \frac{\pi}{2}$$

Continue in this fashion. At each step, we have used finite collections of positive and of negative terms. Removing the first few (trillion) terms of a diverging series still leaves a diverging series, so always enough positive and negative terms remain to continue the sum to exceed $\pi/2$ and to drop below $\pi/2$. Moreover, as $n \to \infty$, both $1/(2n-1)$ and $-1/(2n)$ go to 0, so the amount of overshoot and of undershoot of $\pi/2$ goes to 0. That is, this recipe gives a rearrangement of the alternating harmonic series that converges to $\pi/2$.

But you may not see how this is a frightening result, so let's take a moment. Does the method we just sketched work for only the number $\pi/2$? Of course not. We can rearrange the alternating harmonic series to converge to $\sqrt{2}$; or to e; or to $10^{10^{10^{34}}}$ (Skewes' number); or to $-10^{10^{10^{34}}}$; or to your social security number; or to the year, month, day, and hour you were born; or to the year, month, day, and hour I'll die. If we convert letters, spaces, and punctuation into numbers, the alternating series can be rearranged to converge to the number that represents the text of your first term paper; or of the last letter I wrote to my mother; or of each story written by Haruki Murakami, or Alice Munro, or José Saramago. Everything, and I do mean *everything*, is encoded in the alternating harmonic series. But before you get excited and think you can decode the next winning Powerball numbers, I'll mention that although we know that number does come from some rearrangement of the alternating harmonic series, we do not know which rearrangement. The alternating harmonic series does not bestow prescience, but it can help grow our humility. Sometimes we too easily dismiss the scope of simple mathematical statements. "Can be rearranged to converge to any number" is such a statement. I hope the story of this paragraph helps you realize how very little we understand, even of something as simple as math.

If the commutative law of addition extended to allow arbitrary rearrangements of infinite series, all rearrangements of every convergent series would have the same sum. But we've just seen that they needn't.

So now maybe it seems strange that all rearrangements of an absolutely convergent series have the same sum, but this is the content of Riemann's Rearrangement theorem. The proof is a bit involved, so we omit it here.

Sketch of a proof of the Alternating Series Test. Recall S_n denotes the sum of the first n terms of the series. Because this is a finite sum, terms can be associated in any way

$$S_{2m} = (a_1 - a_2) + (a_3 - a_4) + (a_5 - a_6) + \cdots + (a_{2m-1} - a_{2m}) \tag{10.5}$$

$$= a_1 - (a_2 - a_3) - (a_4 - a_5) - \cdots - (a_{2m-2} - a_{2m-1}) - a_{2m} \tag{10.6}$$

Because all the a_i are positive, it follows from condition (1) of the Alternating Series Test that all the bracketed terms of Eq. (10.5) are non-negative, so the sequence $\{S_{2m}\}$ is non-decreasing. Similarly, each bracketed term $(a_{2i} - a_{2i+1})$ in Eq. (10.6) is non-negative, so all S_{2m} are bounded above by a_1. That is, the even partial sums constitute a non-decreasing sequence bounded above and so

$$\lim_{m \to \infty} S_{2m} = L \tag{10.7}$$

This plausible result is called the Monotone Convergence theorem. Next, because $S_{2m+1} = S_{2m} + a_{2m+1}$ and $a_{2m+1} \to 0$ as $m \to 0$, so

$$\lim_{m \to \infty} S_{2m+1} = \lim_{m \to \infty} (S_{2m} + a_{2m+1}) = \lim_{m \to \infty} S_{2m} = L \qquad (10.8)$$

Combining Eqs. (10.7) and (10.8), we see $\lim_{n \to \infty} S_n = L$. That is, $\sum_{n=1}^{\infty} a_n$ converges.

Sketch of a proof of the Absolute Convergence theorem. Suppose $\sum_{n=0}^{\infty} a_n$ converges absolutely. We show that this series converges. First, observe that for all n, $-|a_n| \le a_n \le |a_n|$, and so

$$0 \le a_n + |a_n| \le 2|a_n| \qquad (10.9)$$

Because $\sum_{n=0}^{\infty} a_n$ converges absolutely, $2 \sum_{n=0}^{\infty} |a_n| = \sum_{n=0}^{\infty} 2|a_n|$ converges. Then $\sum_{n=0}^{\infty} (a_n + |a_n|)$ converges by the Comparison theorem applied to Eq. (10.9). Because the two series on the left of 10.10 converge, so does their difference:

$$\sum_{n=0}^{\infty} (a_n + |a_n|) - \sum_{n=0}^{\infty} |a_n| = \sum_{n=0}^{\infty} (a_n + |a_n| - |a_n|) = \sum_{n=0}^{\infty} a_n \qquad (10.10)$$

That is, the series $\sum_{n=0}^{\infty} a_n$ converges.

Practice Problems

10.3.1. Do these series diverge, converge conditionally, or converge absolutely?

(a) $\sum_{n=1}^{\infty} \frac{(-1)^n n}{n^2 + 2}$ (b) $\sum_{n=1}^{\infty} \frac{(-1)^n n}{e^n}$ (c) $1 + \frac{1}{3} - \frac{1}{2} - \frac{1}{4} + \frac{1}{5} + \frac{1}{7} - \frac{1}{6} - \frac{1}{8} + \cdots$

10.3.2. For which p does the alternating p-series $1 - 1/2^p + 1/3^p - 1/4^p + \cdots$ diverge? For which p does it converge conditionally? For which p does it converge absolutely?

Practice Problem Solutions

10.3.1. (a) To determine absolute convergence, test the convergence of the series with terms $|a_n| = n/(n^2 + 2)$. The terms of this series look like $n/n^2 = 1/n$, which diverges. Because $n/(n^2 + 2) < 1/n$, we cannot apply the Comparison Test. Instead, apply the Limit Comparison Test.

$$\lim_{n \to \infty} \frac{n/(n^2 + 2)}{1/n} = \lim_{n \to \infty} \frac{1}{1 + 2/n^2} = 1$$

Then the series $\sum_{n=1}^{\infty} |a_n|$ diverges and consequently the series $\sum_{n=1}^{\infty} a_n$ does not converge absolutely.

To test conditional convergence, we apply the Alternating Series Test. Consider the function $f(x) = x/(x^2 + 2)$ and compute the derivative $f'(x) = (2 - x^2)/(x^2 + 2)^2$. Because $f'(x)$ is negative for $x > \sqrt{2}$, the terms $|a_n|$ eventually are decreasing. Also, $\lim_{x\to\infty} f(x) = 0$, so $\lim_{n\to\infty} a_n = 0$. Then by the Alternating Series Test, the series converges. Because it does not converge absolutely, the series converges conditionally.

(b) Take $f(x) = x/e^x$, so $a_n = (-1)^n f(n)$. Note that $f'(x) = (1-x)/e^x$, so $f(x)$ is decreasing for $x \geq 1$, and by l'Hôpital's rule, $\lim_{x\to\infty} x/e^x = 0$. Then by the Alternating Series Test, the series converges.

To determine if the convergence is conditional or absolute, test the series with terms $b_n = |a_n|$. Because e^x grows faster than any polynomial, we expect the series converges. To verify this, apply the Limit Comparison Test to b_n and $1/n^2$.

$$\lim_{n\to\infty} \frac{n/e^n}{1/n^2} = \lim_{n\to\infty} \frac{n^3}{e^n} = 0$$

To see the last equality, replace n by x and apply l'Hôpital's rule three times. Then $\sum b_n$ converges by the Limit Comparison Test, and so $\sum a_n$ converges absolutely.

(c) If the series converged absolutely, then by the Rearrangement theorem, every rearrangement would converge to the same limit. But an obvious rearrangement of this series is the alternating harmonic series, which does not converge absolutely. Consequently, the series of (c) does not converge absolutely. To see that the series converges conditionally, group the terms in pairs, $(1 + 1/3) - (1/2 + 1/4) + (1/5 + 1/7) - (1/6 + 1/8) + \cdots$, and apply the Alternating Series Test.

10.3.2. To determine absolute convergence, test the p-series $1 + 1/2^p + 1/3^p + \cdots$. Using the Integral Test, we know the p-series converges for $p > 1$ and diverges for $p \leq 1$. Consequently, the alternating p-series converges absolutely for $p > 1$.

For $0 < p \leq 1$, apply the Alternating Series Test. For $f(x) = 1/x^p$, we find $f'(x) = -p/x^{p+1}$ and so $f(x)$ is decreasing. Also, $\lim_{n\to\infty} 1/n^p = 0$ so the alternating p-series converges. Because the series does not converge absolutely in this range of p-values, the series converges conditionally.

For $p \leq 0$, the series diverges by the nth term test.

Exercises

Do these series diverge, converge conditionally, or converge absolutely?

10.3.1. (a) $1 - \dfrac{1}{3} + \dfrac{1}{9} - \dfrac{1}{27} + \cdots$ (b) $\displaystyle\sum_{n=1}^{\infty} (-1)^n \dfrac{n^2 - 2n + 3}{n^2 + 2n + 5}$

10.3.2. (a) $\displaystyle\sum_{n=2}^{\infty} (-1)^n \dfrac{1}{n^2 - 1}$ (b) $\displaystyle\sum_{n=1}^{\infty} (-1)^n \dfrac{1}{n + \sqrt{n}}$

10.3.3. (a) $\displaystyle\sum_{n=1}^{\infty} (-1)^n \dfrac{\sqrt{n+1}}{n^2}$ (b) $\displaystyle\sum_{n=2}^{\infty} (-1)^n \dfrac{\ln(n)}{n}$

10.3.4. (a) $\displaystyle\sum_{n=2}^{\infty} (-1)^n \dfrac{n}{\ln(n)}$ (b) $\displaystyle\sum_{n=1}^{\infty} (-1)^n \dfrac{n^2}{e^n}$

10.3.5. (a) $\displaystyle\sum_{n=1}^{\infty} (-1)^n \dfrac{\cos^2(n)}{n^2}$ (b) $\displaystyle\sum_{n=1}^{\infty} (-1)^n \dfrac{\sqrt{n^2 + 1}}{n^2}$

10.4 THE ROOT AND RATIO TESTS

Next, we discuss two tests, the Ratio Test and the Root Test, of great power for determining the convergence of numerical series, and also, as we shall see, for finding the radius of convergence of power series. Unlike the Integral Test and the Comparison Tests, which rely on identifying functions or other series, the Ratio and Root Tests are based on just the series being investigated.

The Ratio Test Suppose that $a_n > 0$ for large enough n and

$$\lim_{n \to \infty} \frac{a_{n+1}}{a_n} = L$$

Then the series $\sum_{n=1}^{\infty} a_n$ (1) converges if $L < 1$, (2) diverges if $L > 1$, and (3) can do anything if $L = 1$.

The idea of the proof is very simple. Suppose $L < 1$. Then for any M, $L < M < 1$, there is an N for which $n > N$ implies $a_{n+1}/a_n < M$, so $a_{n+1} < Ma_n$. Continue this line of argument to see

$$a_{n+2} < Ma_{n+1} < M^2 a_n, \ a_{n+3} < Ma_{n+2} < M^3 a_n, \ \ldots, \ a_{n+k} < M^k a_n$$

Then the series that begins with a_n converges by comparison with a geometric series, convergent because its ratio is $M < 1$. Because any finite collection of terms of a series does not determine its convergence, we can ignore the first N terms and deduce that $\sum_{n=1}^{\infty} a_n$ converges.

If $L > 1$, then for every M, $1 < M < L$, there is an N for which $n > N$ implies $a_{n+1}/a_n > M$, so $a_{n+1} > M a_n$. This leads to $a_{n+k} > M^k a_n$. The series diverges by the nth term test.

If $L = 1$, consider two series, the divergent harmonic series with $a_n = 1/n$, and the convergent $p = 2$ series with $a_n = 1/n^2$. For the first,

$$\lim_{n \to \infty} \frac{a_{n+1}}{a_n} = \lim_{n \to \infty} \frac{1/(n+1)}{1/n} = \lim_{n \to \infty} \frac{n}{n+1} = 1$$

and for the second,

$$\lim_{n \to \infty} \frac{a_{n+1}}{a_n} = \lim_{n \to \infty} \frac{1/(n+1)^2}{1/n^2} = \lim_{n \to \infty} \frac{n^2}{(n+1)^2} = 1$$

From these examples we see that $L = 1$ says nothing about the convergence of the series.

Example 10.4.1. *A Ratio Test example.* Does the series $\sum_{n=1}^{\infty} (n!)^2/(2n)!$ converge or diverge? Factorials usually suggest we apply the Ratio Test.

$$\lim_{n \to \infty} \frac{a_{n+1}}{a_n} = \lim_{n \to \infty} \frac{((n+1)!)^2/(2(n+1))!}{(n!)^2/(2n)!} = \lim_{n \to \infty} \frac{((n+1)!)^2}{(n!)^2} \frac{(2n)!}{(2n+2)!}$$

$$= \lim_{n \to \infty} \left(\frac{(n+1)!}{n!} \right)^2 \frac{(2n)(2n-1)\cdots 1}{(2n+2)(2n+1)(2n)(2n-1)\cdots 1}$$

$$= \lim_{n \to \infty} (n+1)^2 \frac{1}{(2n+2)(2n+1)} = \lim_{n \to \infty} \frac{n^2 + 2n + 1}{4n^2 + 6n + 2} = \frac{1}{4}$$

So the series converges by the Ratio Test. □

The Root Test Suppose that $a_n \geq 0$ for large enough n and

$$\lim_{n \to \infty} (a_n)^{1/n} = L$$

Then the series $\sum_{n=1}^{\infty} a_n$ (1) converges if $L < 1$, (2) diverges if $L > 1$, and (3) can do anything if $L = 1$.

The idea of the proof for the case $L < 1$ again involves bounding the series by a convergent geometric series. For if $L < 1$, then for any M, $L < M < 1$, there is an N for which $n > N$ implies $(a_n)^{1/n} < M$, so $a_n < M^n$. Then the series from $n > N$ is bounded about by a geometric series, convergent because its ratio is $M < 1$. Recall again that convergence is determined by the eventual behavior of the series terms, so we see that the series converges.

If $L > 1$, for every M, $1 < M < L$, there is an N for which $n > N$ implies $(a_n)^{1/n} > M$, so $a_n > M^n$. Because $M > 1$, the series diverges by the nth term test.

For the $L=1$ case, we use the same examples, $a_n = 1/n$ and $a_n = 1/n^2$, that we used for the Ratio Test. Note

$$\lim_{n\to\infty} (a_n)^{1/n} = \lim_{n\to\infty} \left(\frac{1}{n}\right)^{1/n} = \lim_{n\to\infty} \frac{1}{n^{1/n}} = 1$$

where the last equality follows by the computation of $\lim_{x\to\infty} x^{1/x}$. Take $y = x^{1/x}$, so $\ln(y) = \ln(x)/x$. As $x \to \infty$, apply l'Hôpital's rule to obtain $\lim_{x\to\infty} \ln(x)/x = \lim_{x\to\infty} 1/x = 0$. Thus $\lim_{x\to\infty} x^{1/x} = e^0 = 1$.

Next,

$$\lim_{n\to\infty} (a_n)^{1/n} = \lim_{n\to\infty} \left(\frac{1}{n^2}\right)^{1/n} = \lim_{n\to\infty} \frac{1}{(n^{1/n})^2} = 1$$

so $L=1$ allows no inference about the convergence of the series.

Example 10.4.2. *A Root Test example.* Does $\sum_{n=1}^{\infty} (n/(2n+1))^n$ converge or diverge? The presence of the exponent n usually suggests that we apply the Root Test.

$$\lim_{n\to\infty} (a_n)^{1/n} = \lim_{n\to\infty} \left(\left(\frac{n}{2n+1}\right)^n\right)^{1/n} = \lim_{n\to\infty} \frac{n}{2n+1} = \frac{1}{2}$$

So the series converges by the Root Test. □

Practice Problem

10.4.1. Do these series converge or diverge?

(a) $\sum_{n=1}^{\infty} \frac{3^n}{n^n}$ (b) $\sum_{n=1}^{\infty} \frac{n!}{e^n}$ (c) $\sum_{n=1}^{\infty} \frac{(3n)!}{n!(2n)!}$

Practice Problem Solution

10.4.1. (a) The exponent n suggests the Root Test, with $a_n = 3^n/n^n = (3/n)^n$. Then

$$\lim_{n\to\infty} (a_n)^{1/n} = \lim_{n\to\infty} ((3/n)^n)^{1/n} = \lim_{n\to\infty} 3/n = 0$$

and we see the series converges by the Root Test.

(b) The factorial suggests we try the Ratio Test, with $a_n = n!/e^n$. Then

$$\lim_{n\to\infty} \frac{a_{n+1}}{a_n} = \lim_{n\to\infty} \frac{(n+1)!/e^{n+1}}{n!/e^n} = \lim_{n\to\infty} \frac{(n+1)!}{n!} \cdot \frac{e^n}{e^{n+1}} = \lim_{n\to\infty} \frac{n+1}{e}$$

That is, the ratio $a_{n+1}/a_n \to \infty$ as $n \to \infty$ and we see the series diverges by the Ratio Test.

(c) Factorials suggest the Ratio Test. Grouping like factors together we find

$$\frac{a_{n+1}}{a_n} = \frac{(3n+3)!}{(3n)!} \cdot \frac{n!}{(n+1)!} \cdot \frac{(2n)!}{(2n+2)!}$$

Canceling common factors gives

$$\frac{a_{n+1}}{a_n} = \frac{(3n+3)(3n+2)(3n+1)}{(n+1)(2n+1)(2n+2)} = \frac{27n^3 + \text{lower order terms}}{4n^3 + \text{lower order terms}}$$

as $n \to \infty$. The cubic terms dominate, so $\lim_{n\to\infty} a_{n+1}/a_n = 27/4$ and the series diverges by the Ratio Test.

Exercises

Do these series converge or diverge? Hint: recall $\lim_{n\to\infty}\left(1+\frac{x}{n}\right)^n = e^x$

10.4.1. (a) $\displaystyle\sum_{n=1}^{\infty} \frac{n^2}{3^n}$ (b) $\displaystyle\sum_{n=2}^{\infty} \frac{n}{(\ln(n))^n}$ (c) $\displaystyle\sum_{n=1}^{\infty} \frac{(2n)!}{n!2^n}$

10.4.2. (a) $\displaystyle\sum_{n=1}^{\infty} \frac{2\cdot 4\cdots(2n)}{(2n)!}$ (b) $\displaystyle\sum_{n=1}^{\infty} \frac{n^n}{2^{n^2}}$ (c) $\displaystyle\sum_{n=1}^{\infty} \frac{2^n+n}{n^3+1}$

10.4.3. (a) $\displaystyle\sum_{n=1}^{\infty} \frac{4^n}{n!}$ (b) $\displaystyle\sum_{n=1}^{\infty} \frac{n!}{10^n}$ (c) $\displaystyle\sum_{n=1}^{\infty} \frac{2^{n+2}}{n3^n}$

10.4.4. (a) $\displaystyle\sum_{n=1}^{\infty} \frac{2^n n!}{(n+2)!}$ (b) $\displaystyle\sum_{n=1}^{\infty} \frac{n^2+3n+5}{2^{n+1}}$ (c) $\displaystyle\sum_{n=1}^{\infty} \frac{n^n}{n!}$

10.4.5. (a) $\displaystyle\sum_{n=1}^{\infty} \frac{10^n}{4^{2n+1}}$ (b) $\displaystyle\sum_{n=1}^{\infty} \left(1+\frac{-1}{n}\right)^{n^2}$ (c) $\displaystyle\sum_{n=1}^{\infty} \frac{n2^n}{n+3^n}$

10.5 NUMERICAL SERIES PRACTICE

Problems in Sects. 10.1, 10.2, 10.3, and 10.4 come with strong hints: the section title suggests which test, or at least which of two tests, to apply. This is useful when learning the mechanics of those tests, but not so much when approaching the problem "Does this series converge?" without any additional information. The practice problems and excersess presented here come with no hints. The first issue is to determine (or guess) which test to try. Here, as with all of science, and most of life, practice is the best guide.

Practice Problems

Do these series converge or diverge?

10.5.1. (a) $\displaystyle\sum_{n=1}^{\infty} \frac{n!}{n^n}$ (b) $\displaystyle\sum_{n=1}^{\infty} \frac{n^n}{3^{n^2}}$

10.5.2. (a) $\displaystyle\sum_{n=2}^{\infty} \frac{1}{n\ln(n)}$ (b) $\displaystyle\sum_{n=1}^{\infty} \frac{1+\cos(n)}{n^3}$

Practice Problem Solutions

10.5.1. (a) Here we have both a factorial and an exponent n. In case you were thinking that factorial always means Ratio Test and that exponent n always means Root Test, this problem shows that rule cannot be applied to all series. Our intuition suggests we won't have much luck with roots of factorials, but maybe we can handle ratios of exponents. So let's try the Ratio Test.

$$\lim_{n\to\infty} \frac{a_{n+1}}{a_n} = \lim_{n\to\infty} \frac{(n+1)!/(n+1)^{n+1}}{n!/n^n} = \lim_{n\to\infty} \frac{(n+1)!}{n!} \frac{n^n}{(n+1)^{n+1}}$$

$$= \lim_{n\to\infty} (n+1) \frac{n^n}{(n+1)^{n+1}} = \lim_{n\to\infty} \frac{n^n}{(n+1)^n} = \lim_{n\to\infty} \left(\frac{n}{n+1}\right)^n .$$

Recall that $\lim_{n\to\infty}(1+(1/n))^n = e$. We see $\lim_{n\to\infty} a_{n+1}/a_n = 1/e < 1$ and so the series converges by the Ratio Test.

(b) The presence of n in exponents suggests the Root Test. The trick is to recognize the denominator is $3^{n^2} = (3^n)^n$, so a_n can be rewritten as

$$a_n = \frac{n^n}{(3^n)^n} = \left(\frac{n}{3^n}\right)^n .$$

Then $(a_n)^{1/n} = n/3^n$. Replacing n with x and applying l'Hôpital's rule, we see $\lim_{n\to\infty}(a_n)^{1/n} = 0$ and so the series converges by the Root Test.

10.5.2. (a) The terms are smaller than $1/n$, so the Comparison Test won't work. Without an n exponent, how to apply the Root Test isn't so clear. The ratio of consecutive terms looks like it will have limit 1, so probably the Ratio Test won't work.

Can we apply the Integral Test? Take $f(x) = 1/(x\ln(x))$. For $x > 1$ we see f is positive and continuous; f is decreasing because

$$f'(x) = -\frac{1+\ln(x)}{x^2\ln^2(x)}$$

is negative for $x > 1$. The hypotheses of the Integral Test are satisfied, so we can apply it. Substitute $u = \ln(x)$ so $du = (1/x)dx$ and we see

$$\int \frac{1}{x\ln(x)}\,dx = \int \frac{1}{u}\,du = \ln(u) + C = \ln(\ln(x)) + C$$

As $x \to \infty$, $\ln(\ln(x)) \to \infty$ so the series diverges by the Integral Test.

(b) Because $-1 \le \cos(n) \le 1$ for all n, we see $0 \le 1 + \cos(n) \le 2$ and so

$$0 \le \frac{1 + \cos(n)}{n^3} \le \frac{2}{n^3}$$

We know $\sum_{n=1}^{\infty} 1/n^3$ converges, so the series of (b) converges by comparison to a convergent series.

Exercises

Do these series converge or diverge?

10.5.1. (a) $\displaystyle\sum_{n=1}^{\infty} \frac{1}{3^n - 2}$ (b) $\displaystyle\sum_{n=1}^{\infty} \left(\frac{n+1}{2n}\right)^n$ (c) $\displaystyle\sum_{n=2}^{\infty} \frac{1}{n(\ln(n))^2}$

10.5.2. (a) $\displaystyle\sum_{n=1}^{\infty} \frac{1}{\sqrt{n}(2 + \sqrt{n})}$ (b) $\displaystyle\sum_{n=1}^{\infty} \sin\left(\frac{1}{n}\right)$ (c) $\displaystyle\sum_{n=1}^{\infty} n^2 e^{-n}$

10.5.3. (a) $\displaystyle\sum_{n=2}^{\infty} \frac{1}{\ln(n)}$ (b) $\displaystyle\sum_{n=1}^{\infty} \frac{1}{(n+1)(2n+3)}$ (c) $\displaystyle\sum_{n=1}^{\infty} \frac{\ln(n)}{n^3}$

10.5.4. (a) $\displaystyle\sum_{n=1}^{\infty} (-1)^n 2^{-1/n}$ (b) $\displaystyle\sum_{n=1}^{\infty} \frac{1}{n}\left(\frac{5}{4}\right)^n$ (c) $\displaystyle\sum_{n=1}^{\infty} \frac{n}{(\ln(n))^n}$

10.5.5. (a) $\displaystyle\sum_{n=1}^{\infty} n\left(\frac{4}{5}\right)^n$ (b) $\displaystyle\sum_{n=1}^{\infty} \frac{10^n}{n!}$ (c) $\displaystyle\sum_{n=1}^{\infty} (-1)^n \ln\left(1 + \frac{1}{n}\right)$

10.6 POWER SERIES

At the start of this chapter we showed that for any number r,

$$1 + r + r^2 + r^3 + \cdots + r^{n-1} = \frac{1 - r^n}{1 - r}$$

and that if $|r| < 1$, then

$$\sum_{n=0}^{\infty} r^n = \frac{1}{1 - r}$$

Replace r with x and observe that we have written the rational function $1/(1-x)$ as an infinite series:

$$\frac{1}{1-x} = 1 + x + x^2 + x^3 + x^4 + \cdots \qquad (10.11)$$

and this expansion is valid for all $|x| < 1$. Because the terms of this series involve non-negative integer powers of x, this is called a *power series*.

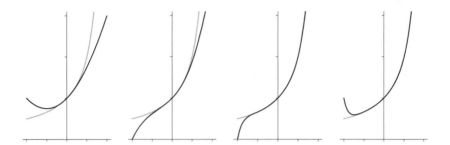

Figure 10.6. The graph of $1/(1-x)$ (in gray) plotted with the graphs of $p_n(x)$ for $n = 2, 3, 9,$ and 10 (in black).

How closely do the polynomials, obtained by truncating the series of Eq. (10.11), match the graphs of $1/(1-x)$? In Fig. 10.6 we plot the graph of $1/(1-x)$ and the graphs of $p_n(x) = 1 + x + x^2 + \cdots + x^n$, for $n = 2, 3, 9,$ and 10. Not surprisingly, for larger n, $p_n(x)$ is a closer match for $1/(1-x)$. Note that all $p_n(x)$ match reasonably well, and in the same sense, on the right side of the y-axis. However, on the left side for even n the graphs of $p_n(x)$ overestimates the graph of $1/(1-x)$, and for odd n the graphs of $p_n(x)$ underestimates the graph of $1/(1-x)$. This is a consequence of evenness or oddness of x^n, the dominant term of $p_n(x)$ for large $|x|$.

The $p_n(x)$ are polynomials, and so defined for all x. The domain of $1/(1-x)$ is all real numbers except $x = 1$, so certainly the polynomials cannot match $1/(1-x)$ very near $x = 1$. Perhaps surprisingly, the polynomials do not match well near $x = -1$. We return to this issue in Sect. 10.7, on the radius and interval of convergence.

In this section we find power series for several more functions by modifying Eq. (10.11). Try to find the series before consulting the answers.

Practice Problems

10.6.1. Find power series expansions for these functions. For what ranges of x-values are these expansions valid?

(a) $\dfrac{1}{1+x}$ (b) $\dfrac{1}{1-3x}$ (c) $\dfrac{1}{1-x^2}$ (d) $\dfrac{1}{2-x}$ (e) $\dfrac{x}{1+x}$

10.6.2. Write this series $x^2 - x^3 + x^4 - x^5 + x^6 - x^7 + \cdots$ as a ratio of polynomials.

Practice Problem Solutions

10.6.1. (a) Replace x in Eq. (10.11) with $-x$, to obtain

$$\frac{1}{1+x} = \frac{1}{1-(-x)} = 1 + (-x) + (-x)^2 + (-x)^3 + (-x)^4 - \cdots$$
$$= 1 - x + x^2 - x^3 + x^4 - \cdots \tag{10.12}$$

This series is a valid representation for $|-x| < 1$, i.e., for $|x| < 1$.

(b) Replace x in Eq. (10.11) with $3x$ to obtain

$$\frac{1}{1-3x} = 1 + (3x) + (3x)^2 + (3x)^3 + (3x)^4 + \cdots$$
$$= 1 + 3x + 3^2 x^2 + 3^3 x^3 + 3^4 x^4 + \cdots \tag{10.13}$$

This series is a valid representation for $|3x| < 1$, i.e., for $|x| < 1/3$.

(c) Replace x in Eq. (10.11) with x^2 to obtain

$$\frac{1}{1-x^2} = 1 + x^2 + (x^2)^2 + (x^2)^3 + (x^2)^4 + \cdots$$
$$= 1 + x^2 + x^4 + x^6 + x^8 + \cdots \tag{10.14}$$

This series is a valid representation for $|x^2| < 1$, i.e., for $|x| < 1$.

(d) This one is a bit trickier. In the power series for $1/(1-x)$, the 1 in the denominator is crucial. So to handle $1/(2-x)$ first factor out the 2.

$$\frac{1}{2-x} = \frac{1}{2(1-x/2)} = \frac{1}{2} \cdot \frac{1}{1-x/2}$$

Now adapt the result of Eq (10.13), with $x/2$ in place of $3x$:

$$\frac{1}{2} \cdot \frac{1}{1-x/2} = \frac{1}{2}\left(1 + \left(\frac{x}{2}\right) + \left(\frac{x}{2}\right)^2 + \left(\frac{x}{2}\right)^3 + \cdots\right)$$
$$= \frac{1}{2}\left(1 + \frac{1}{2}x + \frac{1}{2^2}x^2 + \frac{1}{2^3}x^3 + \cdots\right) \tag{10.15}$$

This series is a valid representation for $|x/2| < 1$, i.e., for $|x| < 2$.

(e) This one is simple: multiply the power series of Eq. (10.12) by x

$$\frac{x}{1+x} = x \cdot (1 - x + x^2 - x^3 + x^4 + \cdots)$$
$$= x - x^2 + x^3 - x^4 + x^5 - \cdots \tag{10.16}$$

This series has the same range of validity as that of $1/(1-x)$: $|x| < 1$.

10.6.2. Most of our series examples start with 1, so let's factor out x^2:

$$x^2 - x^3 + x^4 - x^5 + x^6 - x^7 + \cdots = x^2\left(1 - x + x^2 - x^3 + x^4 - x^5 + \cdots\right)$$

The bracketed terms are the power series for Practice Problem 10.6.1 (a), so

$$x^2 - x^3 + x^4 - x^5 + x^6 - x^7 + \cdots = \frac{x^2}{1+x}$$

Exercises

10.6.1. Find power series for these functions. On what intervals are these series valid? Hint: in Sect. 10.8 we'll see that to differentiate a power series, we differentiate each term.

(a) $\dfrac{1}{x^2 - 1}$ (b) $\dfrac{1}{x^2 - 4}$ (c) $\dfrac{2x}{x^2 - 4}$ (d) $\dfrac{-2x}{(x^2 - 4)^2}$

10.6.2. Find power series for these functions. On what intervals are these series valid?

(a) $\dfrac{1}{1 - x^3}$ (b) $\dfrac{3x^2}{(1 - x^3)^2}$ (c) $\dfrac{1}{8 - x^3}$ (d) $\dfrac{x^2}{(8 - x^3)^2}$

10.6.3. In Sect. 10.8 we'll find series not in terms of x^n, but in terms of $(ax + b)^n$. This problem gives practice with these series. Find series for these functions. On what intervals are these series valid? Hint: remember how to complete the square.

(a) $\dfrac{1}{x^2 + 2x}$ (b) $\dfrac{1}{x^2 + 2x - 3}$ (c) $\dfrac{1}{x^2 + 5x + 6}$ (d) $\dfrac{2x + 5}{(x^2 + 5x + 6)^2}$

10.6.4. Write these series as ratios of polynomials.
(a) $1 + 2x + 4x^2 + 8x^3 + 16x^4 + \cdots$ (b) $x + x^2 + x^3 + x^4 + x^5 + \cdots$
(c) $2 + x^2 + x^3 + x^4 + 2x^6 + x^8 + x^9 + x^{10} + 2x^{12} + x^{14} + x^{15} + x^{16} + 2x^{18} + \cdots$

10.6.5. Write these series as ratios of polynomials.
(a) $1 + 2x + 3x^2 + 4x^3 + 5x^4 + \cdots$. Hint: think of differentiation.
(b) $2 + 6x + 12x^2 + 20x^3 + \cdots$
(c) $x + x^2/2 + x^3/3 + x^4/4 + x^5/5 + \cdots$

10.7 RADIUS AND INTERVAL OF CONVERGENCE

The most general form of power series is

$$\sum_{n=1}^{\infty} a_n(x - b)^n \qquad (10.17)$$

Substituting in any number for x, the power series becomes a numerical series and so we can ask if that numerical series converges or diverges. The set of all x for which the power series (10.17) converges is called the *interval of convergence* of the power series. The possibile forms of this interval are limited; our study of the convergence of geometric series illustrates the main idea.

Radius of Convergence theorem Convergence of the power series (10.17) occurs in one of three ways:

1. the series converges for all x,
2. there is a number $R > 0$ for which the series converges absolutely if $|x - b| < R$ and diverges if $|x - b| > R$, and
3. the series converges only for $x = b$.

The number R of case (2) is called the *radius of convergence* of the power series. Unspecified by the theorem, the interval of convergence in (2) must take one of four forms:

$$(b - R, b + R), \quad [b - R, b + R), \quad (b - R, b + R], \quad \text{or} \quad [b - R, b + R]$$

Usually the radius of convergence can be determined by the Root Test or the Ratio Test. To find if the endpoints $b - R$ and $b + R$ belong to the interval of convergence, substitute $x = b - R$ and $x = b + R$ into the series (10.17) and apply the tests for convergence of numerical series. For any number x with $|x - b| < R$, the series (10.17) converges absolutely and so the Rearrangement theorem from Sect. 10.3 can be applied.

Example 10.7.1. *An infinite interval of convergence.* To find the radius and interval of convergence of the series

$$e^x = 1 + x + \frac{x^2}{2!} + \frac{x^3}{3!} + \frac{x^4}{4!} + \cdots$$

note that here $a_n = 1/n!$ and $b = 0$. Because of the factorial in a_n, let's apply the Ratio Test. Note that the terms of the power series are $a_n x^n$. Because the terms can be negative for some x, we use absolute values in the Ratio Test.

$$\lim_{n \to \infty} \left| \frac{a_{n+1} x^{n+1}}{a_n x^n} \right| = \lim_{n \to \infty} \left| \frac{x^{n+1}/(n+1)!}{x^n/n!} \right| = \lim_{n \to \infty} \left| x \frac{n!}{(n+1)!} \right|$$

$$= \lim_{n \to \infty} \left| \frac{x}{n+1} \right|$$

For every x, this limit is 0 and so the series converges by the Ratio Test. In this case, we say the radius of convergence is ∞ and the interval of convergence is $(-\infty, \infty)$. $\square$

Example 10.7.2. *A finite interval of convergence.* To find the radius and interval of convergence of the series

$$\sum_{n=1}^{\infty} \frac{n}{2^n}(x-3)^n$$

apply the Ratio Test.

$$\lim_{n\to\infty}\left|\frac{a_{n+1}x^{n+1}}{a_n x^n}\right| = \lim_{n\to\infty}\left|\frac{(n+1)(x-3)^{n+1}/2^{n+1}}{n(x-3)^n/2^n}\right|$$

$$= \lim_{n\to\infty}\left|\frac{n+1}{n}(x-3)\frac{2^n}{2^{n+1}}\right| = \left|\frac{x-3}{2}\right|$$

By the Ratio Test, the series converges for $|(x-3)/2| < 1$, that is, for $|x-3| < 2$, so the radius of convergence is 2. Rewrite the last inequality as $-2 < x-3 < 2$, that is, for $1 < x < 5$. To find the interval of convergence, we must test the convergence of the series evaluated at the endpoints of this interval.

For $x = 1$ the series becomes $\sum_{n=1}^{\infty}(n(-2)^n)/(2^n) = \sum_{n=1}^{\infty} n(-1)^n$, which diverges by the nth term test.

For $x = 5$ the series becomes $\sum_{n=1}^{\infty}(n2^n)/(2^n) = \sum_{n=1}^{\infty} n$, which diverges by the nth term test.

Then the interval of convergence for this series is $(1, 5)$. □

Example 10.7.3. *Another finite interval of convergence.* To find the radius and interval of convergence of the series

$$\sum_{n=1}^{\infty} \frac{(x+1)^n}{n}$$

the simple form of the fractions suggests using the Ratio Test.

$$\lim_{n\to\infty}\left|\frac{a_{n+1}x^{n+1}}{a_n x^n}\right| = \lim_{n\to\infty}\left|\frac{(x+1)^{n+1}/n+1}{(x+1)^n/n}\right|$$

$$= \lim_{n\to\infty}\left|\frac{(x+1)^{n+1}}{(x+1)^n}\cdot\frac{n}{n+1}\right| = |x+1|$$

(We can obtain this result with the Root Test if we recall $\lim_{n\to\infty}(1/n)^{1/n} = 1$.) The series converges for $|x+1| < 1$ so the radius of convergence is 1. Rewrite the inequality as $-1 < x+1 < 1$ or $-2 < x < 0$.

At $x = -2$ the series becomes $\sum_{n=1}^{\infty}((-1)^n)/n$, which converges by the Alternating Series Test.

At $x = 0$ the series becomes $\sum_{n=1}^{\infty}(1^n)/n$, the harmonic series, known to diverge.

The interval of convergence is $[-2, 0)$. □

Example 10.7.4. *Convergence at a single point.* To find the radius and interval of convergence of

$$\sum_{n=1}^{\infty} n!(x-2)^n$$

the factorial suggests an application of the Ratio Test.

$$\lim_{n\to\infty}\left|\frac{a_{n+1}(x-2)^{n+1}}{a_n(x-2)^n}\right| = \lim_{n\to\infty}\left|\frac{(n+1)!(x-2)^{n+1}}{n!(x-2)^n}\right|$$

$$= \lim_{n\to\infty}\left|\frac{(n+1)!}{n!}(x-2)\right| = \lim_{n\to\infty}|(n+1)(x-2)|$$

For all $x \neq 2$, this limit is ∞. The radius of convergence is 0 and the interval of convergence is the single point [2]. $\square$

Power series are effective tools for many calculations, but it's easy to forget that these solutions may be valid only for a limited range of x-values. With the techniques of this section we can determine this range.

Practice Problem

10.7.1. Find the radius and interval of convergence of these series.

(a) $\sum_{n=1}^{\infty} \dfrac{x^n}{\sqrt{n}}$ (b) $\sum_{n=2}^{\infty} \dfrac{(-1)^n x^n}{2^n \ln(n)}$ (c) $\sum_{n=1}^{\infty} \dfrac{n! x^n}{(2n)!}$ (d) $\sum_{n=1}^{\infty} \dfrac{n^n x^n}{2^n}$

Practice Problem Solution

10.7.1. (a) Apply the Ratio Test:

$$\lim_{n\to\infty}\left|\frac{x^{n+1}/\sqrt{n+1}}{x^n/\sqrt{n}}\right| = \lim_{n\to\infty}|x|\sqrt{\frac{n}{n+1}} = |x|$$

The series converges for $|x| < 1$, that is, for $-1 < x < 1$, so the radius of convergence is 1.

The radius of convergence also can be found by the Root Test, if we note that $\lim_{n\to\infty}(1/\sqrt{n})^{1/n} = \lim_{n\to\infty}(1/n^{1/n})^{1/2} = 1^{1/2}$ since $\lim_{n\to\infty} n^{1/n} = 1$.

At the left endpoint the series becomes $\sum_{n=1}^{\infty}(-1)^n/\sqrt{n}$, which converges by the Alternating Series Test.

At the right endpoint the series becomes $\sum_{n=1}^{\infty}1/\sqrt{n}$, a p-series with $p = 1/2$, so diverges.

Then the interval of convergence is $[-1,1)$.

(b) Apply the Ratio Test:

$$\lim_{n\to\infty}\left|\frac{(-1)^{n+1}x^{n+1}/(2^{n+1}\ln(n+1))}{(-1)^{n}x^{n}/(2^{n}\ln(n))}\right|=\lim_{n\to\infty}|-x|\cdot\frac{2^{n}}{2^{n+1}}\cdot\frac{\ln(n)}{\ln(n+1)}$$

$$=\frac{|-x|}{2}$$

where $\lim_{n\to\infty}\ln(n)/\ln(n+1)=1$ is obtained by an application of l'Hôpital's rule to $\lim_{x\to\infty}\ln(x)/\ln(x+1)$. By the Ratio Test, the series converges for $|-x|/2<1$, that is, for $-2<x<2$, so the radius of convergence is 2.

The radius of convergence can be obtained also by the Root Test by noting $1/(2^{n}\ln(n)^{1/n})=1/(2\ln(n)^{1/n})$. To evaluate $\lim_{n\to\infty}(\ln(n))^{1/n}$, let $y(x)=(\ln(x))^{1/x}$, so $\ln(y)=\ln(\ln(x))/x$. Then by l'Hôpital's rule, $\displaystyle\lim_{x\to\infty}\frac{\ln(\ln(x))}{x}=$

$\displaystyle\lim_{x\to\infty}\frac{1}{\ln(x)}\frac{1}{x}=0$, so $\lim_{x\to\infty}y=e^{0}=1$.

At the left endpoint the series becomes

$$\sum_{n=2}^{\infty}\frac{(-1)^{n}(-2)^{n}}{2^{n}\ln(n)}=\sum_{n=2}^{\infty}\frac{2^{n}}{2^{n}\ln(n)}=\sum_{n=2}^{\infty}\frac{1}{\ln(n)}$$

Now for $n\geq2$, $\ln(n)<n$, so $1/\ln(n)>1/n$ and the series diverges by comparison with the harmonic series.

At the right endpoint the series becomes $\displaystyle\sum_{n=2}^{\infty}\frac{(-1)^{n}2^{n}}{2^{n}\ln(n)}=\sum_{n=2}^{\infty}\frac{(-1)^{n}}{\ln(n)}$, which

converges by the Alternating Series Test.

The interval of convergence is $(-2,2]$.

(c) Factorials suggest the Ratio Test:

$$\lim_{n\to\infty}\frac{(n+1)!x^{n+1}/(2n+2)!}{n!x^{n}/(2n)!}=\lim_{n\to\infty}|x|\frac{(n+1)!}{n!}\cdot\frac{(2n)!}{(2n+2)!}$$

$$=\lim_{n\to\infty}|x|\frac{n+1}{(2n+2)(2n+1)}=0$$

for all x. This power series converges for all x, so the radius of convergence is ∞.

(d) Exponents suggest the Root Test: $\displaystyle\lim_{n\to\infty}\left(\frac{n^{n}|x|^{n}}{2^{n}}\right)^{1/n}=\lim_{n\to\infty}|x|\frac{n}{2}=\infty$ for all $x\neq0$. The radius of convergence is 0; the interval of convergence is [0].

Exercises

Find the radius and interval of convergence of these series.

10.7.1. (a) $\displaystyle\sum_{n=1}^{\infty}\frac{(x-1)^{n}}{n5^{n}}$ (b) $\displaystyle\sum_{n=1}^{\infty}\frac{(3x+1)^{n}}{n^{2}}$ (c) $\displaystyle\sum_{n=1}^{\infty}\frac{x^{n}}{n^{n}}$

10.7.2. (a) $\displaystyle\sum_{n=1}^{\infty} \frac{n^2 x^n}{2^n}$ (b) $\displaystyle\sum_{n=1}^{\infty} \frac{(2x-1)^{2n+1}}{\sqrt{n}}$ (c) $\displaystyle\sum_{n=1}^{\infty} \frac{2^n}{\sqrt{n}}(x-1)^n$

10.7.3. (a) $\displaystyle\sum_{n=1}^{\infty} \frac{3^{n^2}}{e^n}x^n$ (b) $\displaystyle\sum_{n=2}^{\infty} \frac{1}{(\ln(n))^n}(x-1)^n$ (c) $\displaystyle\sum_{n=1}^{\infty} \frac{\ln(n)}{2^n}(x+1)^n$

10.7.4. (a) $\displaystyle\sum_{n=1}^{\infty} (-1)^n \frac{3^n}{4^n}(x-2)^n$ (b) $\displaystyle\sum_{n=1}^{\infty} \frac{x^n}{n^3}$ (c) $\displaystyle\sum_{n=1}^{\infty} \frac{n^2 x^n}{3^n(n+1)}$

10.7.5. (a) $\displaystyle\sum_{n=1}^{\infty} \frac{x^n}{\sqrt{n^2+1}}$ (b) $\displaystyle\sum_{n=1}^{\infty} \frac{(-1)^n x^n}{\sqrt{n^2+1}}$ (c) $\displaystyle\sum_{n=1}^{\infty} \frac{(x-1)^n}{n\sqrt{n}}$

10.8 TAYLOR'S THEOREM, SERIES MANIPULATION

Recall that in Sect. 10.6 we interpreted the geometric series

$$1 + x + x^2 + x^3 + \cdots = \frac{1}{1-x} \quad \text{for} \quad |x| < 1$$

as a power series representation for the function $f(x) = 1/(1-x)$.

We'll see this is an example of *Taylor's theorem*. Denote by $f^{(n)}(a)$ the nth derivative of $f(x)$, evaluated at $x = a$.

Taylor's theorem Suppose $f(x)$ has derivatives of all orders at $x = a$. The *Taylor series* of $f(x)$ expanded about $x = a$ is

$$f(x) = \sum_{n=0}^{\infty} \frac{f^{(n)}(a)}{n!}(x-a)^n \tag{10.18}$$

Being power series, Taylor series have radii and intervals of convergence as defined in Sect. 10.7.

For instance, we'll find the Taylor series of $f(x) = \sin(x)$, expanded about $x = 0$. First, compute derivatives and look for a pattern.

$$
\begin{aligned}
f'(x) &= \cos(x) & f'(0) &= 1 \\
f''(x) &= -\sin(x) & f''(0) &= 0 \\
f'''(x) &= -\cos(x) & f'''(0) &= -1 \\
f^{(4)}(x) &= \sin(x) & f^{(4)}(0) &= 0 \\
f^{(5)}(x) &= \cos(x) & f^{(5)}(0) &= 1
\end{aligned}
$$

So we see

$$\sin(x) = x - \frac{x^3}{3!} + \frac{x^5}{5!} - \frac{x^7}{7!} + \frac{x^9}{9!} - \cdots$$

Applying the methods of Sect. 10.7, we see the Taylor series for $\sin(x)$ converges for all x.

A similar, but easier, argument shows that

$$e^x = 1 + x + \frac{x^2}{2!} + \frac{x^3}{3!} + \frac{x^4}{4!} + \cdots$$

and this series converges for all x.

Truncating the Taylor series at an exponent n gives the *Taylor polynomial* of order n. Fig. 10.7 shows the graph of $f(x) = \sin(x)$, together with the Taylor polynomials of orders 1, 3, 5, 7, 9, and 11. Note that higher-order Taylor polynomials match $f(x)$ over a larger interval.

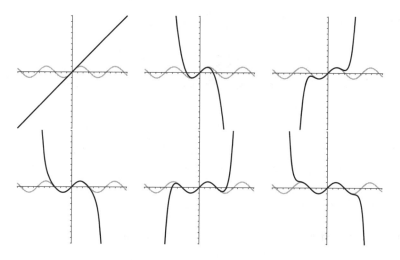

Figure 10.7. Taylor polynomials for $\sin(x)$ through order 11.

Given a function f represented by a power series,

$$f(x) = \sum_{n=0}^{\infty} b_n(x-a)^n \tag{10.19}$$

Then f' and $\int f(x)\, dx$ have the power series

$$f'(x) = \sum_{n=1}^{\infty} b_n n(x-a)^{n-1} \qquad \int f(x)\, dx = \sum_{n=0}^{\infty} b_n \frac{(x-a)^{n+1}}{n+1} + C \tag{10.20}$$

These series converge on the interior of the interval of convergence of (10.19), where (c,d) is the interior of $[c,d]$, $[c,d)$, $(c,d]$, and (c,d).

Example 10.8.1. *Derivative of a power series.* To find $f'(x)$ for

$$f(x) = x - \frac{x^3}{3!} + \frac{x^5}{5!} - \frac{x^7}{7!} + \frac{x^9}{9!} - \cdots$$

differentiate term-by-term

$$f'(x) = 1 - \frac{3x^2}{3!} + \frac{5x^4}{5!} - \frac{7x^6}{7!} + \frac{9x^8}{9!} - \cdots = 1 - \frac{x^2}{2!} + \frac{x^4}{4!} - \frac{x^6}{6!} + \frac{x^8}{8!} - \cdots$$

Recognizing that $f(x) = \sin(x)$, we have just found the power series for $f'(x) = \cos(x)$. $\square$

Now given a power series $f(x) = \sum_{n=0}^{\infty} b_n x^n$ and a function $g(x)$, we can substitute into the series for $f(x)$ obtaining an expression for the composition

$$f(g(x)) = \sum_{n=0}^{\infty} b_n (g(x))^n \qquad (10.21)$$

If $g(x)$ is a polynomial, then Eq. (10.21) is a series expansion for the composition. For example, from the series

$$\cos(x) = 1 - \frac{x^2}{2!} + \frac{x^4}{4!} - \frac{x^6}{6!} + \frac{x^8}{8!} - \cdots$$

we find the series for $\cos(x^2)$

$$\cos(x^2) = 1 - \frac{(x^2)^2}{2!} + \frac{(x^2)^4}{4!} - \frac{(x^2)^6}{6!} + \frac{(x^2)^8}{8!} - \cdots$$

Occasionally, the composition gives a Taylor series expansion even when g is not a polynomial.

Example 10.8.2. *A simple composition example.* To find the series, expanded about $x = 0$, for $\cos(\sqrt{x})$ we could use Eq. 10.18, taking many derivatives and using the chain rule. But there's a simpler approach: because the series for $\cos(u)$ contains only even powers of u, the substitution $u = \sqrt{x}$ gives only integer powers of x.

$$\cos(\sqrt{x}) = 1 - \frac{(\sqrt{x})^2}{2!} + \frac{(\sqrt{x})^4}{4!} - \frac{(\sqrt{x})^6}{6!} + \frac{(\sqrt{x})^8}{8!} - \cdots$$

$$= 1 - \frac{x}{2!} + \frac{x^2}{4!} - \frac{x^3}{6!} + \frac{x^4}{8!} - \cdots$$

If you have some extra time, or don't quite believe this, compute the first few terms of the Taylor series for $\cos(\sqrt{x})$ by evaluating the derivatives $(\cos(\sqrt{x}))^{(n)}$ at $x = 0$. $\square$

Example 10.8.3. *Integral of a power series.*
The integral $\int \sin(x^2)\, dx$ is not so easy to evaluate. I have no idea how to find an antiderivative for $\sin(x^2)$. Integration by parts does not help. If you wish, try the Mathematica command

    Integrate[Sin[x∧2],x]

Probably you won't be enlightened by the result. I wasn't.

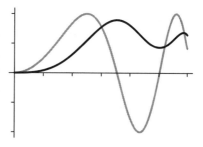

Figure 10.8. The graphs of $\sin(x^2)$ (gray) and order 39 Taylor polynomial (black) of its integral.

But suppose we ask a related question, to find a power series expansion for the integral. This we can do pretty easily:

$$\sin(u) = u - \frac{u^3}{3!} + \frac{u^5}{5!} - \frac{u^7}{7!} + \frac{u^9}{9!} - \frac{u^{11}}{11!} + \cdots$$

so $$\sin(x^2) = x^2 - \frac{x^6}{3!} + \frac{x^{10}}{5!} - \frac{x^{14}}{7!} + \frac{x^{18}}{9!} - \frac{x^{22}}{11!} + \cdots$$

Then applying Eq. (10.20) we find

$$\int \sin(x^2)\, dx = \frac{x^3}{3} - \frac{x^7}{7\cdot 3!} + \frac{x^{11}}{11\cdot 5!} - \frac{x^{15}}{15\cdot 7!} + \frac{x^{19}}{19\cdot 9!} - \frac{x^{23}}{23\cdot 11!} + \cdots$$

In Fig. 10.8 we see a plot of $\sin(x^2)$ and the order 39 Taylor polynomial of $\int \sin(x^2)\, dx$. Recalling the geometry of curve-sketching, we see that up to about $x = 3$ the graph of $\sin(x^2)$ is positive where the graph of the Taylor polynomial is increasing, and negative where the polynomial graph is decreasing. Adding more polynomial terms would give correct behavior over a wider range of x-values. $\square$

Example 10.8.4. *Another power series integral.* To find the Taylor series of $\arctan(x)$, expanded around $x = 0$, no obvious trick suggests itself, so let's start differentiating.

$$\frac{d}{dx}\arctan(x) = \frac{1}{1+x^2}$$

Do we really want to compute the higher derivatives of the arctan? Maybe we can find another way. We know the series for $1/(1+x^2)$:

$$\frac{1}{1+x^2} = 1 - x^2 + x^4 - x^6 + x^8 - x^{10} + x^{12} - \cdots$$

so the series for $\arctan(x)$ is

$$\arctan(x) = \int \frac{1}{1+x^2}\,dx$$

$$= \int (1 - x^2 + x^4 - x^6 + x^8 - x^{10} + x^{12} - \cdots)\,dx$$

$$= x - \frac{x^3}{3} + \frac{x^5}{5} - \frac{x^7}{7} + \frac{x^9}{9} - \frac{x^{11}}{11} + \frac{x^{13}}{13} - \cdots$$

Much easier than computing all the derivatives of $1/(1+x^2)$. $\square$

Example 10.8.5. *A simple product.* Find a power series for $x^3 e^x$ without computing any derivatives. This is easy. Just multiply each term of the Taylor series for e^x by x^3:

$$x^3 e^x = x^3 \cdot \left(1 + x + \frac{x^2}{2!} + \frac{x^3}{3!} + \frac{x^4}{4!} + \cdots\right)$$

$$= x^3 + x^4 + \frac{x^5}{2!} + \frac{x^6}{3!} + \frac{x^7}{4!} + \cdots$$

Not that we have anything against differentiation, but sometimes quicker approaches are better. $\square$

What if we want to find the series for $(x + x^2)e^x$ without computing any derivatives? Find the series for xe^x and $x^2 e^x$, then add together the coefficients of like powers of x.

$$xe^x = x + x^2 + \frac{x^3}{2!} + \frac{x^4}{3!} + \frac{x^5}{4!} + + \frac{x^6}{5!} + \cdots$$

$$x^2 e^x = x^2 + x^3 + \frac{x^4}{2!} + \frac{x^5}{3!} + \frac{x^6}{4!} + \frac{x^7}{5!} + \cdots$$

Then by the Rearrangement theorem, $(x + x^2)e^x$

$$= x + 2x^2 + \left(1 + \frac{1}{2!}\right)x^3 + \left(\frac{1}{2!} + \frac{1}{3!}\right)x^4 + \left(\frac{1}{3!} + \frac{1}{4!}\right)x^5 + \cdots$$

and for $n > 2$ the general term is

$$\left(\frac{1}{(n-2)!} + \frac{1}{(n-1)!}\right)x^n = \left(\frac{n-1}{(n-1)!} + \frac{1}{(n-1)!}\right)x^n = \frac{n}{(n-1)!}x^n$$

Although the bookkeeping is more tedious, this is the general idea for multiplying power series.

Example 10.8.6. *A complicated product.* To find the power series for $e^x/(1-x)$, write the series for $1/(1-x)$ and for e^x, and for each $n \geq 0$ find what product of a term from each series gives kx^n.

$$\frac{1}{1-x} = 1 + x + x^2 + x^3 + x^4 + x^5 + x^6 + \cdots \qquad (10.22)$$

$$e^x = \frac{1}{0!} + \frac{x}{1!} + \frac{x^2}{2!} + \frac{x^3}{3!} + \frac{x^4}{4!} + \frac{x^5}{5!} + \frac{x^6}{6!} + \cdots \qquad (10.23)$$

constant   multiply the 1 from (10.22) by the 1/0! from (10.23)

x multiply the 1 from (10.22) by the $x/1!$ from (10.23), and
           multiply the x from (10.22) by the 1 from (10.23):
           $(1/0! + 1/1!)x$

x^2 multiply the 1 from (10.22) by the $x^2/2!$ from (10.23),
           multiply the x from (10.22) by the $x/1!$ from (10.23), and
           multiply the x^2 from (10.22) by the 1/0! from (10.23):
           $(1/0! + 1/1! + 1/2!)x^2$

           $\cdots$

x^n $(1/0! + 1/1! + 1/2! + \cdots + 1/n!)x^n$

           $\cdots$

The first few terms are

$$\frac{e^x}{1-x} = 1 + 2x + \frac{5x^2}{2} + \frac{8x^3}{3} + \frac{65x^4}{24} + \frac{163x^5}{60} + \cdots$$

Is this faster than computing the derivatives and finding the Taylor coefficients directly? Try a few and decide for yourself. $\square$

Practice Problems

10.8.1. Find the $a = 0$ Taylor series for $f(x) = 1/(1+x)^3$.

10.8.2. Use Taylor series, not l'Hôpital's rule, to evaluate $\displaystyle\lim_{x \to 0} \frac{\sin(x) - x}{x^3}$.

10.8.3. Find the terms through order 4 of the $a = 0$ Taylor series for $f(x) = e^{x+x^2}$.

10.8.4. Find the $a = 0$ Taylor series for $\displaystyle\int \frac{1}{1-x^5}\, dx$.

Practice Problem Solutions

10.8.1. For $f(x) = 1/(1+x)^3$ we compute some derivatives and look for a pattern

$$f'(x) = \frac{-3}{(1+x)^4}, \quad f''(x) = \frac{3 \cdot 4}{(1+x)^5}, \quad f'''(x) = \frac{-3 \cdot 4 \cdot 5}{(1+x)^6}$$

and so on. Evaluating the function and its derivatives at $x = 0$ we find

$$f(0) = 1, \quad f'(0) = -3, \quad f''(0) = 3 \cdot 4, \quad f'''(0) = -3 \cdot 4 \cdot 5$$

and so on. The coefficients of the Taylor series, $f^{(n)}(0)/n!$, are

$$f(0) = 1, \quad \frac{f'(0)}{1!} = -3, \quad \frac{f''(0)}{2!} = \frac{3 \cdot 4}{2}, \quad \frac{f'''(0)}{3!} = \frac{-4 \cdot 5}{2}$$

We need a few more terms to find the pattern.

$$\frac{f^{(4)}(0)}{4!} = \frac{5 \cdot 6}{2}, \quad \frac{f^{(5)}(0)}{5!} = \frac{-6 \cdot 7}{2}$$

Now we see it:

$$\frac{f^{(n)}(0)}{n!} = \frac{(-1)^n (n+1)(n+2)}{2}$$

10.8.2. Substitute in the Taylor series for $\sin(x)$ to obtain

$$\sin(x) - x = \left(x - \frac{x^3}{3!} + \frac{x^5}{5!} - \frac{x^7}{7!} + \cdots \right) - x = -\frac{x^3}{3!} + \frac{x^5}{5!} - \frac{x^7}{7!} + \cdots$$

and so

$$\frac{\sin(x) - x}{x^3} = -\frac{1}{3!} + \frac{x^2}{5!} - \frac{x^4}{7!} + \cdots$$

As $x \to 0$, only the $-1/3!$ remains.

10.8.3. Substitute $u = x + x^2$ in the series

$$e^u = 1 + u + \frac{u^2}{2!} + \frac{u^3}{3!} + \frac{u^4}{4!} + \cdots$$

to obtain

$$e^{x+x^2} = 1 + (x + x^2) + \frac{(x+x^2)^2}{2!} + \frac{(x+x^2)^3}{3!} + \frac{(x+x^2)^4}{4!} + \cdots$$

$$= 1 + (x + x^2) + \frac{x^2 + 2x^3 + x^4}{2!} + \frac{x^3 + 3x^4 + 3x^5 + x^6}{3!} +$$

$$\frac{x^4 + 4x^5 + 6x^6 + 4x^7 + x^8}{4!} + \cdots$$

Now group together like powers of x.

$$e^{x+x^2} = 1 + x + \left(1 + \frac{1}{2!} \right) x^2 + \left(1 + \frac{1}{3!} \right) x^3 + \left(\frac{1}{2!} + \frac{3}{3!} + \frac{1}{4!} \right) x^4 + \cdots$$

10.8.4. The series for $1/(1 - x^5)$ can be found by substitution of $u = x^5$ into the series for $1/(1 - u)$.

$$\frac{1}{1 - x^5} = 1 + x^5 + (x^5)^2 + (x^5)^3 + (x^5)^4 + \cdots = 1 + x^5 + x^{10} + x^{15} + x^{20} + \cdots$$

Then the integral $\int 1/(1-x^5)\,dx$ has the series obtained by integrating this:

$$\int \frac{1}{1-x^5}\,dx = x + \frac{x^6}{6} + \frac{x^{11}}{11} + \frac{x^{16}}{16} + \frac{x^{21}}{21} + \cdots.$$

Exercises

10.8.1. Find the $a=1$ Taylor series for $f(x) = 1 + x - x^3$.

10.8.2. Without computing any derivatives, find the $a=0$ Taylor series for $f(x) = \sin^2(x)$. Hint: trigonometric identities.

10.8.3. Find the $a=0$ Taylor series for $f(x) = \ln(1+x)$. Find the radius of convergence of this series.

10.8.4. Without computing any derivatives, find the $a=0$ Taylor series for $f(x) = \cos(\sqrt{x^2+1})$ Leave the answer in terms of $(1+x^2)^n$.

10.8.5. Without using l'Hôpital's rule, evaluate $\lim\limits_{x\to 0} \dfrac{\tan(x) - \sin(x)}{x^3}$.

10.8.6. For what values of k is the limit $\lim\limits_{x\to 0} \dfrac{1 - kx^2 - \cos(3x)}{x^4}$ finite?

10.8.7. Without computing any derivatives, find the $a=0$ Taylor series for $f(x) = x \cdot \cos(x^2)$.

10.8.8. Without computing any derivatives, find the $a=0$ Taylor series for $f(x) = 1/(1-x)^3$.

10.8.9. Without computing any derivatives, find the $a=0$ Taylor series for $f(x) = (e^x - 1)/x$.

10.8.10. (a) Without computing any derivatives, find the first three terms of the $a=0$ Taylor series for $\sin(\ln(1+x))$.
(b) Now find the first three terms by an application of Eq. (10.18).

10.9 POWER SERIES SOLUTIONS

Recall a non-autonomous differential equation has the form $x' = f(x,t)$. That is, the time derivative of x is an explicit function of both x and t. One approach to solving these equations is to take a power series expansion $x(t) = \sum_{n=0}^{\infty} a_n x^n$, from this find expansions for dx/dt and for $f(x,t)$, and equate the coefficients of like powers of t in these expansions to find relations between the coefficients a_n. Sometimes these relations can be solved explicitly to obtain a closed form expression for the a_n. If we are very lucky, the coefficients can be recognized as those of the Taylor series of a familiar function. Of course, there are far more differential equations than familiar functions, so don't be too optimistic about carrying out this last step.

This sounds complicated, and the coefficient bookkeeping may be tedious, but the general approach is fairly straightforward. Easiest is to consider examples.

Example 10.9.1. *A series we recognize.* Solve $x' = 2tx$ subject to the condition $x(0) = 1$.   Write $x(t) = a_0 + a_1 t + a_2 t^2 + a_3 t^3 + \cdots$. The condition $x(0) = 1$ means $a_0 = 1$. Next,

$$x' = a_1 + 2a_2 t + 3a_3 t^2 + \cdots \quad \text{and} \quad 2tx = 2a_0 t + 2a_1 t^2 + 2a_2 t^3 + \cdots$$

Now match the coefficients of corresponding powers of t in x' and in $2tx$ to find relations between the coefficients. This is easily seen in a table.

	x'	$2tx$	
t^0	a_1	0	so $a_1 = 0$
t^1	$2a_2$	$2a_0$	so $a_2 = a_0 = 1$
t^2	$3a_3$	$2a_1$	so $a_3 = 2a_1/3 = 0$
t^3	$4a_4$	$2a_2$	so $a_4 = 2a/4 = 2/4 = 1/2$
t^4	$5a_5$	$2a_3$	so $a_5 = 2a_3/5 = 0$
t^5	$6a_6$	$2a_4$	so $a_6 = 2a_4/6 = 2^2/(6 \cdot 4) = 1/3!$
...	...	...	...

The pattern is clear: $a_{2n+1} = 0$ and $a_{2n} = 1/n!$. This gives

$$x(t) = 1 + t^2 + \frac{t^4}{2!} + \frac{t^6}{3!} + \frac{t^8}{4!} + \cdots$$
$$= 1 + t^2 + \frac{1}{2!}(t^2)^2 + \frac{1}{3!}(t^2)^3 + \frac{1}{4!}(t^2)^4 + \cdots = e^{t^2}$$

Checking this is correct is easy: $x(0) = e^0 = 1$, and $x'(t) = e^{t^2} 2t$ and $2tx = 2te^{t^2}$, so $x' = 2tx$. $\square$

Of course, the differential equation $dx/dt = 2tx$ can be solved by separation of variables:

$$\int \frac{dx}{x} = \int 2t dt \quad \text{so} \quad \ln(x) = t^2 + c \quad \text{and we have} \quad x(t) = x(0)e^{t^2}$$

agreeing, happily, with our power series calculation.

If the right-hand side cannot be written as a product of a function of x and a function of t, then separation of variables cannot be applied and we must revert to other approaches, power series, for example.

Example 10.9.2. *A series we don't recognize.* Solve $dx/dt = xt + t^2$ subject to $x(0) = 1$.   Note that separation of variables cannot be applied, so we start with the power series. From $x(t) = a_0 + a_1 t + a_2 t^2 + a_3 t^3 + \cdots$ we deduce $x' = a_1 + 2a_2 t + 3a_3 t^2 + 4a_4 t^3 + \cdots$ and

$$xt + t^2 = \left(a_0 t + a_1 t^2 + a_2 t^2 + a_3 t^3 + a_4 t^4 + \cdots\right) + t^2$$
$$= a_0 t + (a_1 + 1)t^2 + a_2 t^2 + a_3 t^3 + a_4 t^4 + \cdots$$

Here again we've applied the Rearrangement theorem, because power series converge absolutely in the interior of their interval of convergence. Equating coefficients of like powers of t in these series expansions gives

	x'	$xt + t^2$	
t^0	a_1	0	so $a_1 = 0$
t^1	$2a_2$	a_0	so $a_2 = a_0/2 = 1/2$
t^2	$3a_3$	$a_1 + 1$	so $a_3 = (a_1 + 1)/3 = 1/3$
t^3	$4a_4$	a_2	so $a_4 = a_2/4 = 1/(4 \cdot 2)$
t^4	$5a_5$	a_3	so $a_5 = a_3/5 = 1/(5 \cdot 3)$
t^5	$6a_6$	a_4	so $a_6 = a_4/6 = 1/(6 \cdot 4 \cdot 2)$
t^6	$7a_7$	a_5	so $a_7 = a_5/7 = 1/(7 \cdot 5 \cdot 3)$
$\cdots$	$\cdots$	$\cdots$	$\cdots$

and so on. We see different patterns of the even and odd coefficients:

$$a_{2n} = \frac{1}{2n(2n-2)\cdots 2} = \frac{1}{2^n n!}$$

$$a_{2n+1} = \frac{1}{(2n+1)(2n-1)\cdots 3} = \frac{2n(2n-2)\cdots 2}{(2n+1)!} = \frac{2^n n!}{(2n+1)!}$$

We (or at least, I) don't recognize these coefficients as those of the Taylor series of any familiar function, but that's fine. Only a small collection of functions have names, and there are infinitely many functions. For that matter, even when we know a name for a function, unless it's a ratio of polynomials, how do you think we compute its value at a particular number? How do we find $\sin(0.7236115987)$? With an adequate collection of terms from its Taylor series. So don't be disappointed if a solution of a differential equation is just formulas for the coefficients. $\square$

Non-autonomous differential equations can be solved in other ways, for example, the method of integrating factors of Sect. A.1 (we try integrating factors on $x' = xt + t^2$ in Example A.1.3) and the iterated integrals in the proof of the existence and uniqueness theorem of Sect. A.2. But the power series approach can be adapted to experimental situations, where the data are sufficient to estimate the first few coefficients in a power series representation, but not enough to guess closed form solutions.

Finally, in addition to sums of $b_n t^n$, reasonably well-behaved functions of t can be expressed through Fourier series, sums of $a_n \cos(nt) + b_n \sin(nt)$, and wavelets, where the sum is of wavelet basis functions, special kinds of functions that are non-zero over only a finite interval. This margin being too mean for my purposes, I shall not record these here. (Sorry, a math history joke. But the point is that now math is so immense that no book can contain all relevant topics.)

Practice Problems

10.9.1. Solve $x'(t) = t^2 x(t)$, subject to $x(0) = 2$.

10.9.2. Solve $x'(t) = x(t) + t^3$, subject to $x(0) = 1$.

Practice Problem Solutions

10.9.1. From the power series $x(t) = a_0 + a_1 t + a_2 t^2 + a_3 t^3 + \cdots$ we obtain $x'(t) = a_1 + 2a_2 t + 3a_3 t^2 + \cdots$ and $t^2 x(t) = a_0 t^2 + a_1 t^3 + a_2 t^4 + \cdots$. The condition $x(0) = 2$ gives $a_0 = 2$. Equating coefficients of like powers of t in x' and $t^2 x$, we obtain

	x'	$t^2 x$	
t^0	a_1	0	so $a_1 = 0$
t^1	$2a_2$	0	so $a_2 = 0$
t^2	$3a_3$	$a_0 = 2$	so $a_3 = a_0/3 = 2/3$
t^3	$4a_4$	a_1	so $a_4 = a_1/4 = 0$
t^4	$5a_5$	a_2	so $a_5 = a_2/5 = 0$
t^5	$6a_6$	a_3	so $a_6 = a_3/6 = 2/(6 \cdot 3)$
t^6	$7a_7$	a_4	so $a_7 = a_4/7 = 0$
$\cdots$	$\cdots$	$\cdots$	$\cdots$

Then, we find

$$a_{3n} = \frac{2}{3n \cdot 3(n-1) \cdots 6 \cdot 3} = \frac{2}{3^n n!} \quad \text{and} \quad a_{3n+1} = a_{3n+2} = 0$$

This gives

$$x(t) = 2 + 2\frac{t^3}{3} + 2\frac{(t^3/3)^2}{2!} + 2\frac{(t^3/3)^3}{3!} + \cdots = 2e^{t^3/3}$$

The check is straightforward: $x(0) = 2e^0 = 2$ and $x'(t) = 2e^{t^3/3}t^2 = x(t)t^2$. Here we are lucky. Don't expect the power series solution to be a recognizable function. Really.

Because the right-hand side is the product of a function of x and a function of t, we can solve this, too, by separation of variables:

$$\int \frac{dx}{x} = \int t^2 dt \quad \text{so} \quad \ln(x) = \frac{t^3}{3} + C \quad \text{and} \quad x = x(0)e^{t^3/3}$$

Don't expect this to work all the time, either.

10.9.2. From the power series $x(t) = a_0 + a_1 t + a_2 t^2 + a_3 t^3 + a_4 t^4 + \cdots$ we find $x' = a_1 + 2a_2 t + 3a_3 t^2 + 4a_4 t^3 + \cdots$ and $x + t^3 = a_0 + a_1 t + a_2 t^2 + (a_3 + 1)t^3 + a_4 t^4 + \cdots$. Equating coefficients of like powers of t in these series expansions gives

	x'	$x + t^3$	
t^0	a_1	a_0	so $a_1 = a_0 = 1$
t^1	$2a_2$	a_1	so $a_2 = a_1/2 = 1/2$
t^2	$3a_3$	a_2	so $a_3 = a_2/3 = 1/(3 \cdot 2)$
t^3	$4a_4$	$a_3 + 1$	so $a_4 = (a_3 + 1)/4 = 7/4!$
t^4	$5a_5$	a_4	so $a_5 = a_4/5 = 7/5!$
t^5	$6a_6$	a_5	so $a_6 = a_5/6 = 7/6!$
$\cdots$	$\cdots$	$\cdots$	$\cdots$

and so on. Then we see

$$x(t) = 1 + t + \frac{t^2}{2!} + \frac{t^3}{3!} + \frac{7t^4}{4!} + \frac{7t^5}{5!} + \cdots$$

$$= \left(7 + 7t + \frac{7t^2}{2!} + \frac{7t^3}{3!} + \frac{7t^4}{4!} + \cdots\right) - 6\left(1 + t + \frac{t^2}{2!} + \frac{t^3}{3!}\right)$$

$$= 7e^t - 6 - 6t - 3t^2 - t^3$$

Checking is straightforward: $x(0) = 7e^0 - 6 = 1$ and $x'(t) = 7e^t - 6 - 6t - 3t^2 = x(t) + t^3$. Again we have found a formula for the solution. In case you think this always happens, remember Example 10.9.2.

Exercises

10.9.1. Solve $x' = x + t$, subject to $x(0) = 1$.

10.9.2. Solve $x' = x + t^2$, subject to $x(0) = 1$.

10.9.3. Solve $x' = x + t^4$, subject to $x(0) = 1$.

10.9.4. Solve $x' = 2x + t$, subject to $x(0) = 1$.

10.9.5. Solve $x' = x + 2t$, subject to $x(0) = 1$.

10.9.6. Solve $x' = x + 2t$, subject to $x(0) = 2$.

10.9.7. Solve $x' = x + t - t^2$, subject to $x(0) = 1$.

10.9.8. Solve $x' = x + t - t^2$, subject to $x(0) = 2$.

10.9.9. Solve $x' = x + 2t - t^2$, subject to $x(0) = 1$.

10.9.10. Solve $x' = x + t + 1$, subject to $x(0) = 1$.

Chapter 11   Some probability

Most of the models we have seen are *deterministic*: if we know the initial conditions exactly, then we can describe the future behavior exactly. This is true of chaotic systems as well. Their familiar lack of long-term predictability is a consequence of sensitivity to initial conditions and inevitable uncertainty in measuring any physical conditions.

In preparation for our Chapter 14 study of systems whose dynamics involve randomness, here we present some basic probability. We begin by sketching some simple probability theory, first for discrete variables (think flipping coins or tossing dice), then we'll move on to continuous variables (think enzyme concentration or intervals between consecutive heartbeats), some rules for calculating probabilities and properties of distributions, and five important probability distributions: binomial, Poisson, normal, Cauchy, and Lévy. We'll fold in some biomedical applications, including Simpson's paradox, a sobering warning about the need for care when interpreting statistics. We conclude this chapter with brief sketches of classical and Bayesian hypothesis tests.

11.1 DISCRETE VARIABLES

Dice provide a straightforward introduction to the basic concepts of probability, interpreted as relative frequency. For a fair die, each number, 1 through 6, should come up about equally often. In 60 tosses of a die, we should get the number 4 about 10 times, but we could see it 12 or 15 times, or 6 times, or even no times. In 600 tosses we should see the number 4 about 100 times, but we could get different outcomes. The result of a fair die toss we'll denote by X, an example of a random variable, a variable whose outcome depends on random processes. (If this seems too informal, the mathematical definition of *random variable* is a measurable function defined on a probability space. We won't pursue this, but excellent sources are [27, 37, 129].) Then we write

$$P(X = 1) = \frac{1}{6}, \ P(X = 2) = \frac{1}{6}, \ \ldots, \ P(X = 6) = \frac{1}{6}$$

As an illustration of a slightly more complicated problem, consider an experiment that consists of tossing two dice, one painted red, the other green. The *sample space* is the set of all possible outcomes of the experiment. In this case, the sample space consists of pairs of numbers (red die number, green die number), hence of 36 pairs: $\{(1,1), (1,2), \ldots, (6,6)\}$. A *simple event* is an outcome of an instance of the experiment, for example, $(2,3)$. An *event* is a combination of simple events, for instance, the event "the sum of the numbers showing on the two dice is ≥ 10" consists of the simple events $(6,4)$, $(5,5)$, $(4,6)$, $(6,5)$, $(5,6)$, and $(6,6)$.

We use the *relative frequency* interpretation of probabilty. The probability $P(G)$ of observing an event G is the $N \to \infty$ limit of $G(N)/N$, where $G(N)$ denotes the number of times the event G is observed in N repetitions of the experiment. This is the *Law of Large Numbers* (LLN), explored in Sect. A.11.

For the experiment consisting of tossing a fair die, each simple event has probability $1/6$; for tossing two fair dice, each simple event has probability $1/36$; and so on.

We'll answer these questions when tossing two dice.

1. What is the probability of getting 7?
2. What is the probability of getting 6 or 7?
3. What is the probability of getting 2, 3, or 4?

Still assuming the dice are fair, to find the probability of getting a certain outcome, count the number of simple events giving that outcome and multiply by $1/36$, the probability of each simple event.

For question 1, a 7 can be obtained from exactly these simple events:

$$(\text{red die, green die}) = (1,6), (2,5), (3,4), (4,3), (5,2), (6,1)$$

That is, 6 simple events give an outcome of 7, so the probability of getting 7 is $6/36 = 1/6$. Now letting the variable X denote the outcome of tossing two dice, we write $P(X = 7) = 1/6$.

Another way to view this is to plot the possible outcomes, 2 through 12, along the x-axis, and above each outcome list the combinations giving that outcome. See Fig 11.1. Those combinations that give a 7 are shaded in Fig. 11.1. The probability of getting 7 is just the number of shaded boxes

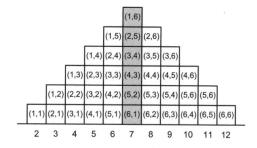

Figure 11.1. Dice tosses by outcome.

divided by the total number of boxes; equivalently, *the fraction of the area occupied by those outcomes corresponding to 7.*

Continuing in this way, we see $P(6 \leq X \leq 7)$ (question 2) and $P(X \leq 4)$ (question 3) are the fractions of the area shaded in the left and right sides of Fig. 11.2.

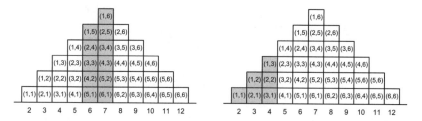

Figure 11.2. Shading the combinations for questions 2 and 3.

The probabilities can be found by calculating what portion of the total area is shaded. We find

$$P(6 \leq X \leq 7) = 11/36 \quad \text{and} \quad P(2 \leq X \leq 4) = 1/6$$

Practice Problems

11.1.1. When 5 fair coins are tossed, what is the probability of obtaining exactly 3 heads?

11.1.2. Suppose N fair coins are tossed. How large must N be so that the probability of getting exactly 5 heads is less than 0.1?

11.1.3. A fair die is rolled and then a fair coin is tossed the number of times that comes up on the die. What is the probability that no toss of the coin yields heads?

11.1.4. Suppose a box contains 6 red dice and 5 green dice. If two dice are drawn from the box, what is the probability that one is red and one is green? Note here the dice are just objects to be drawn, not to be tossed.

Practice Problem Solutions

11.1.1. Tossing 5 coins there are $2^5 = 32$ possible outcomes, from $HHHHH$ to $TTTTT$. The number of ways to get 3 heads in 5 tosses is

$$\binom{5}{3} = \frac{5!}{3!(5-3)!} = \frac{5 \cdot 4 \cdot 3 \cdot 2}{3 \cdot 2 \cdot 2} = 10$$

Then the probability of obtaining 3 heads in 5 tosses is $10/32 \approx 0.3223$.

11.1.2. Write $P(N, 5)$ for the probability of obtaining exactly 5 heads in N tosses. Because the coins are fair, the probability of any particular combination of N tosses is $(1/2)^N$, so all we need to determine is the number of combinations of N coins that gives exactly 5 heads. This is

$$\binom{N}{5} = \frac{N!}{5!(N-5)!} \quad \text{and so} \quad P(N, 5) = \frac{N!}{5!(N-5)!} \left(\frac{1}{2}\right)^N$$

Here are the first few values

$$P(5,5) \approx 0.03125 \quad P(6,5) \approx 0.09375 \quad P(7,5) \approx 0.16406$$
$$P(8,5) \approx 0.21875 \quad P(9,5) \approx 0.24609 \quad P(10,5) \approx 0.24609$$
$$P(11,5) \approx 0.22559 \quad P(12,5) \approx 0.19336 \quad P(13,5) \approx 0.15710$$
$$P(14,5) \approx 0.12219 \quad P(15,5) \approx 0.09164 \quad P(16,5) \approx 0.06665$$

and for $N > 16$ $P(N, 5)$ continues to decrease. So we see $P(N, 5) < 0.1$ for $N = 5$ or 6 and $N \geq 15$.

11.1.3. We'll find a more organized way to approach this problem in Sect. 11.3, but for now we'll reason directly. We'll toss the coin one of 1 through 6 times. For each number of tosses, find the probability of getting no heads (that is, all tails), then add these up, weighted by the probability, 1/6, of making that number of tosses.

Tossing a fair coin N times, the probability of getting N tails is $1/2^N$. Then the probability we want is

$$\frac{1}{6}\frac{1}{2}+\frac{1}{6}\frac{1}{2^2}+\frac{1}{6}\frac{1}{2^3}+\frac{1}{6}\frac{1}{2^4}+\frac{1}{6}\frac{1}{2^5}+\frac{1}{6}\frac{1}{2^6}$$

$$=\frac{1}{6}\frac{1}{2}\left(1+\frac{1}{2}++\frac{1}{2^2}+\frac{1}{2^3}+\frac{1}{2^4}+\frac{1}{2^5}\right)$$

$$=\frac{1}{6}\frac{1}{2}\left(\frac{1-1/2^6}{1-1/2}\right)=\frac{21}{128}\approx 0.16406$$

where we have used the formula $1+r+r^2+\cdots+r^N = (1-r^{N+1})/(1-r)$ for the sum of a finite geometric series.

11.1.4. Suppose the first die drawn is red. There are 6 out of 11 ways to get a red die, so the probability of the first die being red is 6/11. If the first chosen die is red, 10 dice remain, 5 red and 5 green, so the probability that the second chosen die is green is 5/10. Then the probability that the first die is red and the second is green is $(6/11)\cdot(5/10)$.

The other way we can get one die of each color is with the first die green, probability $= 5/11$, and the second red, probability $= 6/10$. Combining these, we see the probability of drawing one red die and one green die is

$$\frac{6}{11}\cdot\frac{5}{10}+\frac{5}{11}\cdot\frac{6}{10}=\frac{6}{11}$$

Exercises

11.1.1. When 10 fair coins are tossed, what is the probability of obtaining exactly 3 or 4 heads?

11.1.2. When two fair dice are tossed, is it more likely the sum is an even number or an odd number?

11.1.3. When three fair dice are tossed,
(a) what is the probability that the numbers sum to four?
(b) what is the probability that the numbers sum to nine?

11.1.4. Suppose N fair coins are tossed. How large must N be so that the probability of getting exactly 4 heads is less than 0.1?

11.1.5. Suppose N fair coins are tossed. How large must N be so that the probability of getting exactly 3 or 4 heads is less than 0.2?

11.1.6. A fair die is rolled and then a fair coin is tossed the number of times that comes up on the die. What is the probability that exactly one toss of the coin yields heads?

11.1.7. A fair die is rolled and then a fair coin is tossed the number of times that comes up on the die. What is the probability that exactly two tosses of the coin yields heads?

11.1.8. Suppose a box contains 6 red dice and 5 green dice. If three dice are drawn from the box,
(a) what is the probability that one is red and two are green?
(b) what is the probability that one is green and two are red?

11.1.9. Suppose a box contains 6 red dice and 5 green dice. Suppose also that one die is drawn from the box, then it is returned to the box and a second die is drawn. What is the probability that one die is red and one is green? (This is called "sampling with replacement.")

11.1.10. A die is rolled twice. What is the probability that the second roll gives a higher number than the first?

11.2 CONTINUOUS VARIABLES

Not all experiments have only a finite collection of possible outcomes. Measurements of the intervals between heartbeats are continuously distributed, as are the opening and closing times of ion channels in heart cells, and the concentration of drugs in a patient's blood. Physical measurements cannot be done with infinite accuracy; perhaps we measure interbeat intervals to the nearest millisecond, or to the nearest microsecond. Any of these graphs would be columns of little boxes, similar to Fig. 11.1. When we measure time on smaller scales, the graphs look more and more like the area under a curve. This suggests a Riemann sum definition of the integral. The function f approximated by the heights of the graph is called a *probability density function* (pdf); the interpretation of the probability as an area can be read by

$$P(a \leq X \leq b) = \int_a^b f(x)\,dx \qquad (11.1)$$

Every probability density function $f(x)$ must satisfy these two conditions

$$f(x) \geq 0 \text{ for all } x \text{ in the domain } D \tag{11.2}$$

and

$$\int_D f(x)dx = 1 \tag{11.3}$$

Now if the domain D is the interval $a \leq x \leq b$, then $\int_D f(x)dx = \int_a^b f(x)dx$. If $D = (-\infty, \infty)$, then as we recall from Sect. 5.5,

$$\int_{-\infty}^{\infty} f(x)dx = \lim_{R \to -\infty} \int_R^0 f(x)dx + \lim_{S \to \infty} \int_0^S f(x)dx$$

For example, the function

$$f_0(x) = 6x(1-x) \text{ for } 0 \leq x \leq 1$$

pictured on the left of Fig 11.3, is a probability density function. The shaded region is the area representing $P(1/4 \leq X \leq 1/2)$. Here's the calculation:

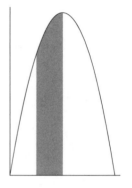

$$P(1/4 \leq X \leq 1/2) = \int_{1/4}^{1/2} f_0(x)dx$$

$$= (3x^2 - 2x^3)\Big|_{1/4}^{1/2} = \frac{11}{32} = 0.34375$$

Figure 11.3. The graph of f_0.

Why the factor of 6 in the definition of $f_0(x)$? It's to satisfy the condition of Eq. (11.3). Integrating $\int_0^1 x(1-x)dx = 1/6$, so multiplying by 6, we obtain a function that satisfies both requirements for a probability density function, Eq. (11.2) because $f_0(x) \geq 0$ for $0 \leq x \leq 1$, and (11.3) because of the factor of 6.

In fact, any function non-negative on its domain D (for example, $D = (-\infty, \infty)$, $D = [a, b]$, $D = [a, \infty)$, or $D = (-\infty, b]$) can be turned into a probability density function by dividing by $\int_D f(x)dx$. Here are some examples.

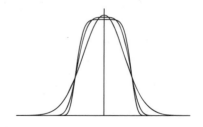

Figure 11.4. Graphs of $f_1, f_2,$ and f_3.

In Fig. 11.4 we see the graphs of three probability density functions, all symmetric across $x = 0$. The function that gives the graph with the highest maximum is

$$f_1(x) = k_1 e^{-x^2} = \frac{1}{\sqrt{\pi}} e^{-x^2}$$

The calculation that gives $k_1 = 1/\sqrt{\pi}$ is presented in Sect. A.16.

The functions that give the other two graphs are

$$f_2(x) = k_2 e^{-x^4} \qquad \text{where } k_2 = \left(\int_{-\infty}^{\infty} e^{-x^4} dx \right)^{-1} \approx \frac{1}{1.8128}$$

$$f_3(x) = k_3 e^{-x^6} \qquad \text{where } k_3 = \left(\int_{-\infty}^{\infty} e^{-x^6} dx \right)^{-1} \approx \frac{1}{1.85544}$$

The factors k_1, k_2, and k_3, and for that matter the factor of 6 in f_0, are called *normalization factors* for these probability distributions.

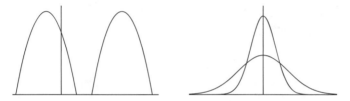

Figure 11.5. Left: $f_0(x + 3/4)$ and $f_0(x - 1/2)$. Right: $f_4(x)$ and $f_5(x)$.

Probability density functions can be shifted left or right by replacing x with $x \pm k$. The left graph of the first plot of Fig. 11.5 is a graph of $f_0(x + 3/4)$, the right is $f_0(x - 1/2)$. To see why replacing x by $x + 3/4$ shifts the graph to the left, note that in order for $f_0(x + 3/4)$ to be non-negative, we must have $0 \le x + 3/4 \le 1$, that is, $-3/4 \le x \le 1/4$. Because we will use x to represent all sorts of things, it may take negative values or be unbounded. So we need some flexibility with the domain of probability distribution functions.

In addition to shifting left or right, we also can make the distribution wider or narrower. For example, in the right image of Fig. 11.5 we see the graphs of

$$f_4(x) = k_4 e^{-2x^2} \qquad f_5(x) = k_5 e^{-0.5x^2}$$

where the normalization factors are $k_4 \approx 1/1.25331$ and $k_5 \approx 1/2.50663$. The taller, narrower graph is that of f_4. In general, narrower distributions mean the values of the random variable are more closely clustered about the middle of the distribution. Another way to say this is that with narrow distributions we are more confident about using the middle of the distribution to predict the next value of the random variable. This is just the first step in prediction, in using past behavior to estimate future behavior. A big song, many verses, not nearly finished yet.

Practice Problems

11.2.1. (a) Find the normalization factor for $f(x) = 1/x$ for $1 \leq x \leq 2$.
(b) For the probability distribution of (a), find the value of $x = a$ that gives $P(1 \leq X \leq a) = 1/2$.

11.2.2. (a) For each $p > 1$, find the normalization factor for $f(x) = x^{-p}, 1 \leq x < \infty$.
(b) For the probability distributions of (a), compute $P(1 \leq X \leq 2)$.

11.2.3. Here we explore the effect of the width of a distribution on probability calculations.
(a) Find $P(0 \leq X \leq 1)$ for $f_4(x) = \dfrac{1}{1.25331} e^{-2x^2}$.

(b) Find $P(0 \leq X \leq 1)$ for $f_5(x) = \dfrac{1}{2.50663} e^{-0.5x^2}$.

Practice Problem Solutions

11.2.1. (a) To find the normalization factor, integrate $f(x)$ over the indicated range, $1 \leq x \leq 2$.

$$\int_1^2 \frac{1}{x}\,dx = \ln(x)\Big|_1^2 = \ln(2) - \ln(1) = \ln(2)$$

So the probability density function is $\dfrac{1}{\ln(2)} \dfrac{1}{x}$.
(b) By Eq. (11.1) the goal $P(1 \leq X \leq a) = 1/2$ can be expressed as

$$\int_1^a \frac{1}{\ln(2)} \frac{1}{x}\,dx = \frac{1}{2} \qquad \text{so} \qquad \frac{1}{\ln(2)} \ln(x)\Big|_1^a = \frac{1}{2}$$

$$\text{and} \quad \frac{1}{\ln(2)} \ln(a) = \frac{1}{2} \qquad \text{Then} \quad \ln(a) = \frac{\ln(2)}{2} = \ln(2^{1/2})$$

and we see that $a = \sqrt{2}$.

11.2.2. (a) To find the normalization factor, compute the improper integral

$$\int_1^\infty x^{-p}\,dx = \frac{x^{-p+1}}{-p+1}\Big|_1^\infty = \frac{1}{p-1}$$

So the normalization factor is $p-1$.

(b) We'll use Eq. (11.1) to compute the probability

$$P(1 \le X \le 2) = \int_1^2 (p-1)x^{-p}dx = (p-1)\frac{x^{-p+1}}{1-p}\Big|_1^2 = 1 - 2^{-p+1}$$

11.2.3. (a) For f_4 we compute the probability by

$$P(0 \le X \le 1) = \int_0^1 f_4(x)\, dx = \int_0^1 \frac{1}{1.25331}e^{-2x^2}\, dx \approx 0.477251$$

(b) For f_5 we compute the probability by

$$P(0 \le X \le 1) = \int_0^1 f_4(x)\, dx = \int_0^1 \frac{1}{2.50663}e^{-0.5x^2}\, dx \approx 0.341345$$

We see that, not surprisingly, the wider the distribution, the smaller the probability of being in a given interval $a \le X \le b$.

Exercises

11.2.1. (a) Find the normalization factor for $1 + e^x$, $0 \le x \le 1$.
(b) For the probability distribution of (a) compute

$$P(0 \le X \le 1/2), \ P(1/2 \le X \le 3/4), \ \text{and} \ P(3/4 \le X \le 1)$$

11.2.2. (a) Show $\dfrac{\sin(x)}{2}$, $0 \le x \le \pi$, is a probability distribution function.
(b) Find an expression for $P(0 \le X \le a)$ as a function of a.

11.2.3. (a) Find the normalization factor for $1 - x^3$, $0 \le x \le 1$.
(b) For the probability distribution of (a), compute $P(0 \le X \le 1/2)$.

11.2.4. (a) Show $\frac{3}{2}(1 - x^2)$, $0 \le x \le 1$, is a probability density function.
(b) Find an expression for $P(0 \le X \le a)$ as a function of a.
(c) Find an expression for $P(b \le X \le 1)$ as a function of b.
(d) Find an expression for $P(a \le X \le b)$ as a function of a and b.

11.2.5. (a) Find the normalization factor for $\sqrt{1 - x^2}$, $0 \le x \le 1$.
(b) For the probability distribution of (a), compute $P(0 \le X \le 1/2)$.

11.2.6. (a) Show xe^x, $0 \le x \le 1$, is a probability density function.
(b) Find an expression for $P(0 \le X \le a)$ as a function of a.

11.2.7. (a) Find the normalization factor for $e^{-x}, 1 \leq x < \infty$.
(b) For the probability distribution of (a), for all $n \geq 1$ find an expression for $P(n \leq X \leq n+1)$.

11.2.8. (a) For all $k > 0$, find the normalization factor for $e^{-kx}, 1 \leq x < \infty$.
(b) For the probability distribution of (a) and for all $n \geq 1$, find an expression for $P(n \leq X \leq n+1)$.

11.2.9. (a) For $0 < p < 1$, find the normalization factor for $x^{-p}, 0 < x < 1$.
(b) For the probability distribution of (a) and for all $n \geq 1$, find an expression for $P(1/(n+1) \leq X \leq 1/n)$.

11.2.10. (a) Find the normalization factor for $x\sin(x), 0 \leq x \leq \pi$.
(b) For the probability distribution of (a), Find an expression for $P(0 \leq X \leq a)$ as a function of a.
(c) Find $P(0 \leq X \leq \pi/2)$.
(d) Find numerically the value x_0 of x that is the maximum for $x\sin(x)$. Carry four digits to the right of the decimal.
(e) Find numerically $P(0 \leq X \leq x_0)$. Carry four digits to the right of the decimal.

11.3 SOME COMBINATORIAL RULES

For concreteness, we present these concepts in the context of tossing dice, but the rules we discover can be applied in many other settings. Our goal is to develop general rules that extend and simplify the direct calculations presented in Sect. 11.1.

If a die is fair and the random variable X is the outcome of tossing the die, then $P(X = 1) = 1/6, P(X = 2) = 1/6, \ldots, P(X = 6) = 1/6$.

We'll start with a simple question. Say event E occurs when a fair die is tossed and we get a 1 or a 2. So E is $X = 1$ or $X = 2$. We can write this more compactly as $E = E_1 \cup E_2$ where E_1 is the simple event $X = 1$ and E_2 is the simple event $X = 2$. Our intuition tells us that when we toss a fair die, we should get a 1 or a 2 about a third of the time, that is, $P(E) = 1/3$. Can we calculate this using the decomposition of E as a union of simple events?

$$P(E) = P(E_1 \cup E_2) = P(E_1) + P(E_2) = \frac{1}{6} + \frac{1}{6} = \frac{2}{6}$$

as expected.

This is a special case of the *addition rule*. Suppose E_1 and E_2 are events. Then

$$P(E_1 \cup E_2) = P(E_1) + P(E_2) - P(E_1 \cap E_2) \tag{11.4}$$

The reason for the last term is that the sum $P(E_1) + P(E_2)$ counts twice all the simple events common to E_1 and E_2.

For example, suppose E_1 occurs when the toss of a die gives an even number, and E_2 occurs when the toss of a die gives a number larger than 3. Then $E_1 \cup E_2$ occurs when a die is tossed and we get 2, 4, 5, or 6, and $E_1 \cap E_2$ occurs when a die is tossed and we get a 4 or 6. Direct calculation shows that

$$P(E_1 \cup E_2) = \frac{4}{6} \text{ and } P(E_1 \cap E_2) = \frac{2}{6}$$

and applying the addition rule gives

$$P(E_1 \cup E_2) = P(E_1) + P(E_2) - P(E_1 \cap E_2) = \frac{1}{2} + \frac{1}{2} - \frac{2}{6} = \frac{4}{6}$$

Note the calculation $P(E_1) = 3/6$ is a special case of the addition rule, because the events $X = 2$, $X = 4$, and $X = 6$ are *mutually exclusive*, or *disjoint*: if a die toss gives a 2, it cannot simultaneously give a 4 or a 6. More generally, if E_1 and E_2 are mutually exclusive, then $P(E_1 \cap E_2) = 0$ and we see

$$P(E_1 \cup E_2) = P(E_1) + P(E_2) \qquad \text{if } E_1 \text{ and } E_2 \text{ are mutually exclusive} \qquad (11.5)$$

Events E_1 and E_2 are *independent* if the occurrence of one has no effect on the occurrence of the other. (In a moment we'll have another formulation of independence.)

The *multiplication rule* for independent events E_1 and E_2 is

$$P(E_1 \cap E_2) = P(E_1) \cdot P(E_2) \qquad \text{if } E_1 \text{ and } E_2 \text{ are independent} \qquad (11.6)$$

To construct more complicated examples we'll use two dice, one red, the other green. Say E_1 occurs when the red die is tossed and we get an even number, and E_2 occurs when the green die is tossed and we get a number greater than 3. Certainly, the outcome of tossing the red die has no effect on the outcome of tossing the green die. That is, E_1 and E_2 are independent. Then by the multiplication rule, Eq. (11.6),

$$P(E_1 \cap E_2) = P(E_1) \cdot P(E_2) = \frac{1}{2} \cdot \frac{1}{2} = \frac{1}{4}$$

As a check, can we calculate $P(E_1 \cap E_2)$ directly? First, observe that tossing both the red die and the green die gives 36 possible outcomes,

$$(R, G) = (1,1), (1,2), (1,3), \ldots, (5,6), (6,6)$$

Assuming fair dice, each is equally likely, so each has probability 1/36. These 36 events $(R, G) = (i, j)$ are mutually exclusive, so we can calculate $P(E_1 \cap E_2)$ by enumerating the possibilities,

$$(R, G) = (2,4), (2,5), (2,6), (4,4), (4,5), (4,6), (6,4), (6,5), (6,6)$$

and we see
$$P(E_1 \cap E_2) = \frac{9}{36} = \frac{1}{4}$$

We can see the necessity of independence by computing $P(E_1 \cap E_3)$, where E_3 occurs when the red die is tossed and we get a number greater than 3. (Note that E_2 is different from E_3.) Then $E_1 \cap E_3$ occurs when $R = 4$ or $R = 6$, so $P(E_1 \cap E_3) = 2/6$. Unthinking application of the multiplication rule gives $P(E_1) \cdot P(E_3) = (1/2) \cdot (1/2) \neq 1/3$.

Conditional probability is one approach to treating events that are not independent, and to giving another characterization of independent events. Write $P(E_1|E_2)$ for the probability the E_1 occurs, if we know that E_2 has occurred. This is called the *conditional probability* of E_1 given E_2. For example, suppose that when a die is tossed the event E_1 means an even number comes up, and the event E_2 means a number greater than 3 comes up. We are told that E_2 has occurred. Knowing that, what is the probability that E_1 has occurred? Two of the three outcomes in E_2 are even, so we expect $P(E_1|E_2) = 2/3$. That this is the case follows from the formula for conditional probabilities

$$P(E_1|E_2) = \frac{P(E_1 \cap E_2)}{P(E_2)} \tag{11.7}$$

Applying this to these events E_1 and E_2, we find

$$P(E_1|E_2) = \frac{P(E_1 \cap E_2)}{P(E_2)} = \frac{2/6}{1/2} = \frac{2}{3}$$

Conditional probability gives another formulation of independence of events. First, solve Eq. (11.7) for $P(E_1 \cap E_2)$:

$$P(E_1 \cap E_2) = P(E_1|E_2) \cdot P(E_2) \tag{11.8}$$

If the events E_1 and E_2 are independent, then E_2 has no influence on E_1, so $P(E_1|E_2) = P(E_1)$ and Eq. (11.8) gives us another derivation of the characterization of independence in terms of intersections $P(E_1 \cap E_2) = P(E_1) \cdot P(E_2)$, agreeing with our intuition.

Certainly Eq. (11.6) generalizes to larger collections of independent events. This multiplication rule sometimes is misapplied in settings where the independence of the events has not been established, or is false.

Now suppose the sample space consists of disjoint events $E_1, \ldots, E_n$. That is, the sample space is $E_i \cup \cdots \cup E_n$ and $E_i \cap E_j$ empty for all $i \neq j$. Under these conditions, we say $E_1, \ldots, E_n$ form a *partition* of the sample space. Then for any event A,

$$A = \bigcup_{i=1}^{n} (A \cap E_i)$$

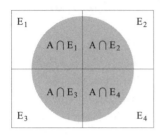

Figure 11.6. Partition of A.

Because the E_i are disjoint, the $A \cap E_i$ are disjoint. This is illustrated in Fig. 11.6 where A is represented by the gray disc. Applying Eq. (11.5),

$$P(A) = P\left(\bigcup_{i=1}^{n}(A \cap E_i)\right) = \sum_{i=1}^{n} P(A \cap E_i)$$

Finally, recalling that $P(A \cap E_i) = P(A|E_i) \cdot P(E_i)$ by Eq. (11.8), we find the *law of conditioned probabilities*:

$$P(A) = \sum_{i=1}^{n} P(A|E_i) \cdot P(E_i) \tag{11.9}$$

We'll illustrate the law of conditioned probabilities by calculating the probability of contracting the flu, if we know the probabilities of contraction depending on immune system status.

Example 11.3.1. *Probability of contracting the flu.* Suppose people with compromised immune systems have a 0.4 probability of contracting the flu in a given season, while people without compromised immune systems have a 0.2 probability. Further, suppose 30% of the population have compromised immune systems. Find the probability that a randomly selected person will contract the flu in this season. The problem can be described with these events.
E_1: this person has a compromised immune system,
E_2: this person does not have a compromised immune system, and
A: this person contracts the flu this season.
   Applying Eq. (11.9),

$$P(A) = P(A|E_1) \cdot P(E_1) + P(A|E_2) \cdot P(E_2) = 0.4 \cdot 0.3 + 0.2 \cdot 0.7 = 0.26$$

That the probability lies between 0.2 and 0.4 is clear. With this calculation we find where between these two values the probability lies. $\square$

We might imagine that $P(E_1|E_2)$ and $P(E_2|E_1)$ need not be related at all, but they are related in a very interesting fashion. Here's why: from Eq. (11.8) recall that

$$P(E_1 \cap E_2) = P(E_1|E_2) \cdot P(E_2)$$

Similarly,

$$P(E_2 \cap E_1) = P(E_2|E_1) \cdot P(E_1)$$

Of course, $P(E_1 \cap E_2) = P(E_2 \cap E_1)$, giving

$$P(E_1|E_2) \cdot P(E_2) = P(E_2|E_1) \cdot P(E_1)$$

Solving for $P(E_2|E_1)$ we obtain *Bayes' theorem*:

$$P(E_2|E_1) = \frac{P(E_1|E_2) \cdot P(E_2)}{P(E_1)} \tag{11.10}$$

To illustrate Bayes' theorem, we continue with the data of Ex. 11.3.1.

Example 11.3.2. *Probability of a compromised immune system.* Suppose a randomly selected person has the flu. What is the probability that this person has a compromised immune system? We need to find $P(E_1|A)$, so we'll apply Bayes' theorem.

$$P(E_1|A) = \frac{P(A|E_1) \cdot P(E_1)}{P(A)} = \frac{0.4 \cdot 0.3}{0.26} = 0.46$$

Because people with compromised immune systems have a higher probability of contracting the flu, intuition suggests that a person with the flu has an increased likelihood of having a compromised immune system. But intuition also tells us that this likelihood is tempered by the portion of the population with compromised immune systems. Bayes' theorem is our guide to account for the relative effects of these notions. ☐

Important to a correct understanding of Bayes' theorem is the recognition that $P(A|B) > 0$ does not imply that B in any way causes A. The events A and B may have a common cause, or their relation may be more complicated still.

We end this section with two measures important to interpreting the results of medical tests.

The *sensitivity* of a medical test is the probability that a diseased person will have a positive test result. The *specificity* of a test is the probability that a healthy person will have a negative test result. Certainly, these measures can be expressed as conditional probabilities, with these events represented by these symbols.

$+$	positive test result	$-$	negative test result
D	diseased	N	not diseased

Then the four conditional probabilities of the form P(test result|disease state) are

$P(+|D) = $ sensitivity	$P(+|N) = $ false positive
$P(-|D) = $ false negative	$P(-|N) = $ specificity (11.11)

In Practice Problem 11.3.3 we calculate these probabilities for a 1950 study of chest X-rays and tuberculosis.

Practice Problems

11.3.1. Find the probability that when two fair dice are tossed, the sum is ≥ 10, given that 5 appears on at least one die.

11.3.2. Suppose 4% of male CLL patients have an 11q deletion, and 1% of female CLL patients have an 11q deletion. Further, suppose 60% of all CLL patients are male. If a randomly selected blood sample from a CLL patient has a 11q deletion, what is the probability this patient is male? (These numbers are made up.)

11.3.3. Jacob Yerushalmy and coworkers [180] reviewed the X-ray diagnoses of 1820 patients, finding negative X-rays for 1739 patients without TB and 8 with TB; and positive X-rays for 51 patients without TB and 22 with TB. Find the sensitivity, specificity, probability of false positives, and probability of false negatives for chest X-rays as a TB diagnostic tool.

Practice Problem Solutions

11.3.1. Denote by A the event that the sum is ≥ 10, and by B the event that 5 occurs on at least one die. The problem is to find the conditional probability $P(A|B) = P(A \cap B)/P(B)$. So to keep track of the tosses of both dice, we can say one is red and the other green. Of the 36 possible outcomes of tossing two dice, six have a 5 on the red die and six have a 5 on the green die, but one of these twelve, 5 on both red and green, occurs in both lists of six. The event B occurs with these eleven tosses: (red, green) = (5, 1), (5, 2), (5, 3), (5, 4), (5, 5), (5, 6), (1, 5), (2, 5), (3, 5), (4, 5), (6, 5), so we see that $P(B) = 11/36$. The event $A \cap B$ occurs with these tosses: (5, 5), (5, 6), and (6, 5), so $P(A \cap B) = 3/36$. Then $P(A|B) = (3/36)/(11/26) = 3/11$.

11.3.2. Denote by Q the event that a CLL patient has an 11q deletion, by M the event that a CLL patient is male, and by F the event that a CLL patient is female. The problem is to compute the conditional probability $P(M|Q)$. We apply Bayes' theorem

$$P(M|Q) = \frac{P(Q|M) \cdot P(M)}{P(Q)}$$

We know $P(M) = 0.6$ and $P(Q|M) = 0.04$, but what is $P(Q)$? For this, we apply Eq. (11.9)

$$P(Q) = P(Q|M) \cdot P(M) + P(Q|F) \cdot P(F) = 0.04 \cdot 0.6 + 0.01 \cdot 0.4 = 0.028$$

Then $P(M|Q) = 0.04 \cdot 0.6 / 0.028 = 0.857$.

11.3.3. We compute the four conditional probabilities of Eq. (11.11) using Eq. (11.7). We estimate probabilities by relative frequencies, so for example

$$P(D) = \frac{\text{(number with TB)}}{\text{number surveyed}} = \frac{8 + 22}{1820} \approx 0.016$$

So we have

$$\text{sensitivity} = \frac{P(+ \cap D)}{P(D)} = \frac{22/1820}{(8 + 22)/1820} = \frac{22}{30} \approx 0.73$$

$$\text{false positive} = \frac{P(+ \cap N)}{P(N)} = \frac{51}{51 + 1739} = \frac{51}{1790} \approx 0.028$$

$$\text{false negative} = \frac{P(- \cap D)}{P(D)} = \frac{8}{8 + 22} = \frac{8}{30} \approx 0.27$$

$$\text{specificity} = \frac{P(- \cap N)}{P(N)} = \frac{1739}{1739 + 51} = \frac{1739}{1790} \approx 0.97$$

Exercises

11.3.1. In an experiment that consists of tossing two fair dice, suppose A is the event that the sum is even, and B is the event that the sum is ≤ 4. Are events A and B independent?

11.3.2. Find the probability that When two fair dice are tossed, the sum is ≥ 10, given that 5 appears on the first die.

11.3.3. (a) When two fair dice are tossed, what is the probability that the sum is even, given that the sum is greater than 6?
(b) When two fair dice are tossed, what is the probability that the sum is greater than 6, given that the sum is even?

11.3.4. Suppose 70% of CLL patients of age < 60 respond positively to rituximab, and 60% of CLL patients of age ≥ 60 respond to rituximab. Suppose further that 40% of all CLL patients have age < 60. A randomly selected patient responds positively to rituximab. What is the probability that this patient has age < 60? (Don't take these numbers seriously: they're made up.)

11.3.5. (a) Show that Eq. (11.8) can be generalized to more than two sets by proving
$$P(A_1 \cap A_2 \cap A_3) = P(A_1|A_2 \cap A_3) \cdot P(A_2|A_3) \cdot P(A_3)$$
(b) Suppose $P(A_1 \cap A_2 \cap A_3) = 1/40$ and $P(A_3) = 1/8$. Find a lower bound on $P(A_2|A_3)$.

11.3.6. Find the probability that when two fair dice are tossed, the sum is ≥ 10, given that 5 appears on at least one die.

11.3.7. Suppose E_1 and E_2 form a partition of the sample space, and suppose we know $P(E_1)$ and $P(E_2)$.
(a) For any event A if we know $P(A)$ and $P(A|E_1)$, can we calculate $P(A|E_2)$?
(b) If we know $P(A)$ and $P(A|E_1)$ can we calculate $P(E_2|A)$?

11.3.8. Suppose E_1, E_2, E_3, and E_4 form a partition of the sample space.
(a) If $P(A|E_1) = P(A|E_2) = P(A|E_3) = P(A|E_4) = 0.1$, find $P(A)$.
(b) If $P(A|E_1) = P(A|E_2) = P(A|E_3) = (1/2)P(A|E_4)$, find an expression for $P(A)$ in terms of $P(A|E_4)$ and $P(E_4)$.

11.3.9. Suppose 4% of the population has bird flu, and there is a proposed test for bird flu. Suppose 3% of those tested test positive, and of those who test positive, 70% have bird flu. What is the sensitivity of this test?

11.3.10. Suppose in a survey of 200,000 people, 300 are diseased. If a test has sensitivity 0.99 and specificity 0.98, find a relation between $P(-|D)$ and $P(+|N)$.

11.4 SIMPSON'S PARADOX

Now we'll see an important, and subtle, illustration of the relation between causality and correlation. The immediate lesson is that sometimes data can be partitioned so that the result of a test on each subset is the opposite of the result on the aggregate data. Further, in some situations the partitioned data give the correct guidance, while in others, the aggregate data should be followed. We'll give examples of Simpson's paradox in this section; in Sect. 11.5 we'll show why this should be called "Simpson's reversal" instead of "Simpson's paradox."

Example 11.4.1. *(Homer) Simpson's paradox.* We begin with an instance of a *real* Simpson's paradox. On a rainy day in Springfield, Lisa, Marge, Bart, and Homer play checkers. Denote by L_1 the number of games Lisa won and by L_2 the number of games Lisa played, and the corresponding notations for games played by Marge, Bart, and Homer.

$$\frac{L_1}{L_2} = \frac{0}{2} \qquad \frac{M_1}{M_2} = \frac{6}{8} \qquad \frac{B_1}{B_2} = \frac{2}{8} \qquad \frac{H_1}{H_2} = \frac{2}{2}$$

Partitioned by age, then the women do less well than the men:

$$\frac{L_1}{L_2} = \frac{0}{2} < \frac{2}{8} = \frac{B_1}{B_2} \qquad \frac{M_1}{M_2} = \frac{6}{8} \quad \frac{2}{2} = \frac{H_1}{H_2}$$

Yet the aggregate wins of the women and those of the men

$$\frac{L_1 + M_1}{L_2 + M_2} = \frac{0+6}{2+8} = \frac{6}{10} > \frac{4}{10} = \frac{2+2}{8+2} = \frac{B_1 + H_1}{B_2 + H_2}$$

show the women outplay the men. The aggregate data reverse the relation observed among the partitioned data. Which is right, the aggregate or the partitioned? In this context, what do we even mean by the word "right"?

Outside of the board games played by some admittedly very influential cartoon characters, why should we care? Let's see. □

Although milder versions had been noted about half a century before Edward Simpson's 1951 paper [150], since the mid 1970s [16] this effect has been called *Simpson's paradox*. Some history can be found in [116, 117, 119, 120]. (Don't worry about [117] being a technical report. You can download it from the UCLA website of the report's author, Judea Pearl.) A charming popular description is in Chapter 19, "Induction and probability," of Martin Gardner's *Time Travel and Other Mathematical Bewilderments* [44], a volume of annotated reprints of some of his "Mathematical Games" columns from *Scientific American*.

The basic construction is this: the relationship between $P(X)$ and $P(Y)$ reverses when conditioned on a third variable Z, regardless of the value of Z. That is, $P(X) < P(Y)$ despite the fact that $P(X|Z) > P(Y|Z)$ for all values of Z. In this form, Judea Pearl states that it should be called *Simpson's reversal*; the "paradox" reserved for situations where the reversal gives rise to disbelief. We'll see that the situation is more complex still, but this is best introduced by an example emphasizing the paradox. Choosing one from among several medical treatments for a serious disease will provide adequate emotional investment.

Example 11.4.2. *Kidney stones.* We'll use real data about two treatments for kidney stones [25, 70]. The treatments are open surgery (OS) and the less invasive percutaneous nephrolithotomy (PN). In this study, 350 patients were treated by OS and 350 by PN. Here are the success rates:

$$273/350 \approx 78 \quad \text{for OS} \qquad 289/350 \approx 82 \quad \text{for PN}$$

Here success means the patient remained free of kidney stones for at least three months after the procedure. The difference is modest, a 4% improvement in the success rate, but still, PN outperformed OS.

But now, suppose we partition the patients into those with small kidney stones (< 2 cm) and those with large. Then the data look a bit different.

	OS	PN
small stone	$81/87 \approx 93\%$	$234/270 \approx 87\%$
large stone	$192/263 \approx 73\%$	$55/80 \approx 69\%$

What do we see? When the data aren't partitioned by kidney stone size, PN has a higher success rate than OS. On the other hand, if the data are partitioned by size, then for both small and large stones, OS has a higher success rate than PN. If you *don't know* the size of the stone–for example, if the stone is of irregular shape and close to the size that distinguishes small from large–then you should make the choice of treatment based on the aggregate data and you'd choose PN. This despite the fact that for small stones and for large stones, OS has the better success rate. You see why this is called a paradox. If you've had kidney stones (I have), you'll understand why you'd want the treatment with the best success rate. □

An explanation is that there is a hidden variable, often called a *confounding variable* in statistics literature. Here the confounding variable is the severity of the case. In more severe cases physicians often choose the more reliable, but also riskier, treatment OS; in less severe cases the less risky treatment PN often was preferred. This accounts for the much larger denominators in the small stone and PN treatment, and the large stone and OS treatment categories.

Why not always use OS? It is a more major surgery with higher risks for old or infirm patients, and also OS uses more resources than PN, a concern in our world of limited resources. What about an otherwise healthy patient whose kidney stone is about 2 cm, the boundary between the large and small? In this case the right recommendation could be PN, which has better success for aggregate data.

But this isn't the end of the story. Other partitions of the group of patients could result in still different answers. How can we tell which partition, or if any partition at all, is appropriate for the problem we wish to address? Pearl [116, 117, 120] presents a graph theoretic approach to determine the causal relations between variables and thus which variables (which partitions) are relevant to the current decisions. Moreover, intuition about relevant causal relations must be deduced from stories about the data and cannot be found from statistics alone. Stories are important in our lives; that they have a useful role in the interpretation of tricky data feels appropriate. These are promising grounds for new applications of mathematics to medical decision-making.

In Sect. 11.5 we'll sketch Pearl's method for determining the appropriate partition. There are several steps, but the method removes the paradox from Simpson's paradox, and enables the correct interpretation of many experiments and observations.

Practice Problems

11.4.1. Suppose A has success rate $50/100$, B has $b/30$, C has $30/50$, and D has $d/200$. Find all (positive integer) values of b and d for which all three of these conditions hold:
(i) the success rate of A is greater than that of B,
(ii) the success rate of C is greater than that of D, and
(iii) the aggregate success of A and C is less than the aggregate success of B and D.

11.4.2. Suppose $A_1 = x > 0$, $A_2 = 1$, $B_1 = y > 0$, $B_2 = 1$, $C_1 = 1$, $C_2 = 2$, $D_1 = 2$, and $D_2 = 3$. Further, suppose

$$\frac{A_1}{A_2} < \frac{B_1}{B_2}, \quad \frac{C_1}{C_2} < \frac{D_1}{D_2} \quad \text{and} \quad \frac{A_1 + C_1}{A_2 + C_2} > \frac{B_1 + D_1}{B_2 + D_2}$$

Describe the region in the x, y plane where these three inequalities are satisfied.

Practice Problem Solutions

11.4.1. Conditions (i), (ii), and (iii) are

$$\frac{50}{100} > \frac{b}{30} \qquad \frac{30}{50} > \frac{d}{200} \qquad \frac{50+30}{100+50} < \frac{b+d}{30+200}$$

The first condition gives $b < 15$, the second $d < 120$, and the third $123 \le b+d$. (Recall b and d must be positive integers, so $122.667 < b+d$ means $123 \le b+d$.) These three conditions give the combinations

d	b	$b+d$
109	14	123
110	13,14	123,124
...	...	...
119	4,...,14	123,...,133

11.4.2. First note that with the given values, the second inequality is $1/2 < 2/3$, always true, so this inequality places no conditions on x and y. The first inequality becomes $y > x$ and the third becomes $y < (4/3)x - 2/3$. Note the lines $y = x$ and $y = (4/3)x - 2/3$ intersect at $x = y = 2$. The region lies above the line $y = x$ and below the line $y = (4/3)x - 2/3$.

Exercises

11.4.1. *An arithmetic paradox.* Suppose $x_1 = 1, x_2 = 1, x_3 = 2, x_4 = 2$ and $y_1 = 1, y_2 = 2, y_3 = 1, y_4 = 2$. Denote the average of the x_i by $\langle x_i \rangle = (x_1 + x_2 + x_3 + x_4)/4$ and similarly for the average of y_i.
(a) Show $\langle x_i + y_i \rangle = \langle x_i \rangle + \langle y_i \rangle$. We'll see this again in Prop. 11.6.1 (c).
(b) Show $\langle x_i \cdot y_i \rangle = \langle x_i \rangle \cdot \langle y_i \rangle$.
So where's the paradox, you wonder. How about this?
(c) For $x_1 = 1, x_2 = 1, x_3 = 2$ and $y_1 = 1, y_2 = 2, y_3 = 1$, does $\langle x_i \cdot y_i \rangle = \langle x_i \rangle \cdot \langle y_i \rangle$?
Okay, it's not a whole lot of a paradox, but you'd be surprised how many people think that because the average of the sum is the sum of the averages, that the average of the product must be the product of the averages. Sometimes it's true, sometimes not.

11.4.2. Continuing with the idea of Exercise 11.4.1 (c), here we'll compare the relative magnitudes of $\langle x_i \cdot y_i \rangle$ and $\langle x_i \rangle \cdot \langle y_i \rangle$.
(a) Take $x_1 = 2, x_2 = 3$ and $y_1 = 5, y_2 = 4$.
(b) Take $x_1 = 3, x_2 = 2$ and $y_1 = 5, y_2 = 4$.
(c) Why doesn't the reversal of (a) and (b) seem paradoxical?

11.4.3. Suppose A has success rate $10/30$, B has $b/60$, C has $20/40$, and D has $d/120$. Find all (positive integer) values of b and d for which all three of these conditions hold:
(i) the success rate of A is greater than that of B,
(ii) the success rate of C is greater than that of D, and
(iii) the aggregate success of A and C is less than the aggregate success of B and D.

11.4.4. Suppose A has success rate $60/100$, B has $b/60$, C has $c/10$, and D has $40/50$. Find all (positive integer) values of b and c for which all three of these conditions hold:
(i) the success rate of A is greater than that of B,
(ii) the success rate of C is greater than that of D, and

(iii) the aggregate success of A and C is less than the aggregate success of B and D.

Hint: don't forget that success rates cannot exceed 1.

11.4.5. Suppose A has success rate $60/90$, B has $b/60$, C has $c/20$, and D has $40/50$. Find all (positive integer) values of b and c for which all three of these conditions hold:

(i) the success rate of A is greater than that of B,

(ii) the success rate of C is greater than that of D, and

(iii) the aggregate success of A and C is less than the aggregate success of B and D.

Hint: don't forget that success rates cannot exceed 1.

11.4.6. Consider a drug trial of a population partitioned into urban and rural patients. Here are the data.

	Urban		Rural	
	Treated	Untreated	Treated	Untreated
Cured	1000	50	90	1000
Uncured	5000	950	10	1000

(a) For the aggregate population compute the conditional probabilities $P(C|T)$ and $P(C|U)$, where C denotes "cured," T denotes "treated," and U denotes "untreated."

(b) For the urban population compute $P(C|T)$ and $P(C|U)$.

(c) For the rural population compute $P(C|T)$ and $P(C|U)$.

(d) Compare the results of (a), (b), and (c).

11.4.7. Suppose $A_1 = 1$, $A_2 = x > 0$, $B_1 = 2$, $B_2 = y > 0$, $C_1 = 1$, $C_2 = 2$, $D_1 = 3$, and $D_2 = 4$. Further, suppose

$$\frac{A_1}{A_2} < \frac{B_1}{B_2}, \quad \frac{C_1}{C_2} < \frac{D_1}{D_2} \quad \text{and} \quad \frac{A_1 + C_1}{A_2 + C_2} > \frac{B_1 + D_1}{B_2 + D_2}$$

(a) Describe the region in the x,y plane where these three inequalities are satisfied.

(b) Describe the region in the x,y plane where the first two inequalities along with

$$\frac{A_1 + C_1}{A_2 + C_2} < \frac{B_1 + D_1}{B_2 + D_2}$$

are satisfied.

11.4.8. Suppose $A_1 = x > 0$, $A_2 = 1$, $B_1 = 2$, $B_2 = y > 0$, $C_1 = 1$, $C_2 = 3$, $D_1 = 1$, and $D_2 = 2$. Further, suppose

$$\frac{A_1}{A_2} < \frac{B_1}{B_2}, \quad \frac{C_1}{C_2} < \frac{D_1}{D_2} \quad \text{and} \quad \frac{A_1 + C_1}{A_2 + C_2} > \frac{B_1 + D_1}{B_2 + D_2}$$

Describe the region in the x, y plane where these three inequalities are satisfied.

11.4.9. For positive numbers A_1, A_2, B_1, B_2, C_1, C_2, D_1, and D_2, show that if

$$\frac{A_1}{A_2} < \frac{B_1}{B_2}, \quad \frac{C_1}{C_2} < \frac{D_1}{D_2}, \quad \text{and} \quad \frac{A_1 + C_1}{A_2 + C_2} > \frac{B_1 + D_1}{B_2 + D_2}$$

then

$$A_1 D_2 + C_1 B_2 > A_2 D_1 + C_2 B_1$$

11.4.10. This is a variation on Example 11.4.1. Find values of L_1, L_2, M_1, M_2, B_1, B_2, H_1, and H_2 so that partitioned by gender, the adults outplay the children

$$\frac{L_1}{L_2} < \frac{M_1}{M_1} \quad \text{and} \quad \frac{B_1}{B_2} < \frac{H_1}{H_2}$$

yet when we aggregate the wins of the children and the wins of the adults we have the reverse inequality

$$\frac{L_1 + B_1}{L_2 + B_2} > \frac{M_1 + H_1}{M_2 + H_2}$$

Note that your numbers must satisfy $L_2 + B_2 + M_2 + H_2 = 2(L_1 + B_1 + M_1 + H_1)$.

11.5 CAUSALITY CALCULUS

Because it is so useful to resolve Simpson's paradox by determining whether partitioned or aggregated data are appropriate, here we'll present the main points of Pearl's method. Judea Pearl is a UCLA computer scientist and statistician. His books *Causality* [116] and *The Book of Why* [120] are brilliant explications of how to analyze causality. Not surprisingly, he gives a thorough treatment of Simpson's paradox.

But before we get to the math, I must mention another point. The journalist Daniel Pearl, kidnapped and murdered in Afghanistan in 2002, was Judea Pearl's son. I have no children, so simply cannot imagine a parent's grief over losing a child to disease or accident. But to lose a child to murder must be a different level of heartbreak. I know no words to describe this category of despair, and I know lots of words. Of the many possible reactions to Daniel's murder, Judea, his wife Ruth, and other family and friends formed the Daniel Pearl Foundation, whose mission is to promote understanding across cultures. This is, in my opinion, the finest possible act that parents could do in the face of such a tragedy. The dedication of *Causality* is this:

> To Danny
> And the glowing audacity of goodness

Already I was fascinated by Judea Pearl's work, especially because it leads to the resolution of Simpson's paradox. Any tools that can help unpack the thicket of statistical inference are worth pursuing. That the architect of so many of these ideas also has a moving personal story is reason to say these few words about him, not just his work. Science has many brilliant people; a few are actual heros, those who make the world better through acts of personal bravery. Judea Pearl is a hero.

First, a comment on terminology. We mentioned that Pearl points out that the more correct term is *Simpson's reversal*, rather than Simpson's paradox. The "paradox" comes about when accepting all the inequalities of the reversal violates our intuition when we assume a causal dimension to conditional probabilities. We'll see that Pearl provides a way to determine which direction of the inequalities gives rise to the correct deductions, so we'll drop the term "paradox" in favor of "reversal."

Part of the issue is that building on a campaign of the influential statistician Karl Pearson, causality was not viewed to be a valid statistical concept for much of the 20th century. Statistics could measure correlation and contingency, but not causation. A good description is provided in the Epilogue of [116].

Simpson's reversal is three inequalities of this form

$$P(A|B) > P(A|\neg B), \tag{11.12}$$

$$P(A|B \cap X) < P(A|\neg B \cap X), \quad P(A|B \cap \neg X) < P(A|\neg B \cap \neg X) \tag{11.13}$$

or with all three inequalities reversed. Here A and B are particular events, X stands for any event, and $\neg B$ is "not B," the complement of the event B. We'll translate this formulation into words, using $X \cup \neg X$ as a partition of the data into two subsets. The occurrence of event B increases the probability of observing A in the whole data set (Eq. (11.12)), but decreases the probability of observing A in every subset (Eq. (11.13)). Often applications involve taking X and $\neg X$ to stand for a particular pair of complementary subsets, rather than all pairs of complementary subsets. While this reversal may seem impossible, we've seen examples already in Sect. 11.4.

Before we get to the resolution of Simpson's reversal, we'll begin with simple examples. We must illustrate how to construct causal diagrams, how to identify backdoor paths and check if they are blocked, and the rudiments of Pearl's *do*-calculus, leading to the Sure-Thing Principle.

Once we understand them, these techniques are straightforward. The hard work is building the model. That is, identifying the relevant factors and how they are related. Usually this includes the assumption that all factors not included have negligible effect on the causal relations to be studied.

We'll begin with a graphical representation of the model. A *causal diagram* is a graph whose vertices represent factors. We'll focus on *directed graphs*, that is, graphs whose edges are arrows. An arrow from A to B indicates a causal relation between A and B, that is, A is at least part of the cause of B. A *path* in the graph is any unbroken sequence of arrows in which the direction of the arrows need not be considered. In a *causal path* the directions of the arrows must be consistent. For example, $CDAB$ is a path from C to B, but it is not a causal path: CAB is a causal path from C to B.

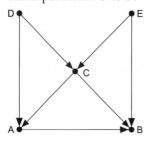

Figure 11.7. A causal diagram.

A path from X to Y is a *backdoor path* if at least one of the arrows along the path points toward X. That is, a backdoor path is not a causal path. For example, ACB and $ADCEB$ are backdoor paths from A to B. In ACB the arrow $C \to A$ points toward A; in $ADCEB$ the arrows $D \to A$ and $E \to C$ point toward A. A vertex X on a backdoor path is a *collider* if both of the path arrows that touch X point toward X, and is a *repeller* if both of the path arrows that touch X point away from X. On the backdoor path $ADCEB$ the vertex C is a collider because of the directions of the arrows $D \to C \leftarrow E$, and the vertices D and E are repellers because of the paths $A \leftarrow D \to C$ and $C \leftarrow E \to B$.

Pearl refers to these diagrams as *Directed Acyclic Graphs*, or DAGs. Here "directed" means that all of the edges are arrows; "acyclic" means that no causal path forms a loop. On pages 46 and 47 of [55] the authors argue that when time is considered, loops can be unfolded into acyclic graphs. We mention this in case you're thinking, "Wait a minute. Don't some biological systems have positive feedback loops, where each factor contributes to the growth of the others?" The answer is "Yes," but these loops can be removed with the explicit addition of time. We won't pursue this here, but in Sect. 13.8 of volume 2 we'll convert non-autonomous differential equations into autonomous differential equations by increasing the dimension of the system by one through the addition of time as a dependent variable.

Now we'll build an example and show how backdoor paths determine whether to use aggregate or partitioned data. The variable A means the treatment has been applied, $\neg A$ the treatment was not applied; B represents recovery, and $\neg B$ not recovery. The third variable X, a confounding variable, may mediate an indirect effect of A on B. The left table presents the results partitioned on X, while the right table presents the data with X and $\neg X$ combined. Here, for example, the recovery rate for X and A is $108/(108 + 72) = 0.6$.

	X		$\neg X$	
	A	$\neg A$	A	$\neg A$
B	108	42	12	54
$\neg B$	72	18	48	120
rec. rate	0.6	0.7	0.2	0.3

	combined	
	A	$\neg A$
B	120	96
$\neg B$	120	138
rec. rate	0.5	0.41

For the partitioned data, we see that in both groups X and $\neg X$, the recovery rate is greater for the untreated, $\neg A$, than for the treated, A. On the other hand, for the combined data the recovery rate for the treated group exceeds the recovery rate of the untreated.

Presented with a new patient, should we treat or not? The partitioned data suggest not to treat; the aggregate data suggest to treat. How can we tell which we should use? Dennis Lindley and Melvin Novick [85] showed that no statistical criterion always can determine the right choice, that we must look to the story, the narrative behind the data. The diagrams of Fig. 11.8 illustrate their argument.

Both of these graphs represent the data. Without knowing the underlying story we cannot distinguish which is correct. Suppose A is treatment for coronary artery calcification (CAC) and B is a reduction of the CAC rate.

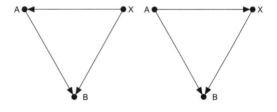

Figure 11.8. Two simple causal diagrams.

The arrow $A \to B$ represents the direct influence, good or bad, of the treatment A on the disease B. But there also may be indirect effects, X in this diagram.

If we take X to represent gender, say X for females and $\neg X$ for males, a drug treatment administered for CAC cannot possibly influence a patient's gender. So there can be no arrow $A \to X$. However, a patient's gender might influence whether the treatment is indicated, or if indicated, the expectation that the patient will take it. (My mother was much better about taking her medicines than was my father, who often complained that he'd "had enough of them durn pills.") So an arrow $X \to A$ is reasonable and the left diagram of Fig. 11.8 is appropriate if there are no other significant indirect influences. The partitioned data tell us that for both female patients and male patients the treatment is less effective than is doing nothing, while the aggregated data tell us that the treatment is more effective than is not treating. Here we see the problem: prescribe the drug only if we do not know the patient's gender. This is ridiculous. The partitioned data make sense and the treatment is not indicated.

But now suppose that instead of gender, X represents blood pressure. We can believe that the treatment influences blood pressure, so the diagram has an arrow $A \to X$. Now you might think that blood pressure could be a factor in whether the treatment is appropriate, and that certainly is true for some treatments. But adding an arrow $X \to A$ would produce a loop, so the diagram no longer would be a DAG. We'll ignore this second arrow and use the right diagram of Fig. 11.8. That is, the treatment has a direct effect on CAC, and an indirect effect on CAC through the blood pressure, regardless of the blood pressure. Then the aggregated data are appropriate and the treatment is recommended.

How do the graphs of Fig. 11.8 tell us when to use the aggregate data and when to use the partitioned? Pearl and other authors refer to "controlling for X" or "adjusting for X," but often we have no actual physical control over the factor X. Instead, we can condition on X. That is, partition the population into subsets called *strata* on which the value of X is constant. We use the partitioned data or the aggregate data according as the DAG indicates that we condition on X or do not condition on X.

To correctly assess the causal effect of A on B, we need to *block* all backdoor paths between A and B. To block a repeller, condition on it; to block a collider, do not condition on it or any of its descendants. Blocking a path eliminates all non-causal effects of A on B along that path. A path that is not blocked is *unblocked*. See [118], though note that instead of "repeller," Pearl uses the term "arrow-emitting variable."

In the left graph of Fig. 11.8, AXB is a backdoor path between A and B. On this path X is a repeller, so to block X we must condition on it and consequently use the partitioned data to determine the proper response.

In the right graph of Fig. 11.8, there is no backdoor path between A and B. So there is nothing to block and no need to condition on X. Then the aggregate data determine the proper response.

That was pretty easy, but then the graphs were quite simple. And of course the hard part of any of these problems is to build the model. We'll address one obvious point now. The graphs of Fig. 11.8 have only one vertex, X, in addition to A and B, so we have only one vertex to consider blocking. But what if the graph is more complicated? How can we tell which collections of vertices are sufficient to condition in order to find the causal effect of A on B? Is there only one such collection, or do some graphs have several?

Here's how we can tell. The method is outlined on page 43 of [55]. In order to eliminate all spurious (non-causal) paths from A to B, a set S of vertices is sufficient to control if

1. every vertex of S is not a descendant of either A or B,
2. every unblocked backdoor path from A to B contains a vertex from S, and
3. if every collider on a backdoor path from A to B either belongs to S or has a descendant in S, then S also must contain a noncollider along this path.

For instance, in the DAG of Fig. 11.7 we might think that controlling $S = \{C\}$ would suffice to block all spurious paths from A to B because except for the direct (causal) path $A \to B$, every path (ignoring loops, of course) from A to B passes through C. The first condition is clear: C is not a descendant of either A or B. For the second and third conditions, we need to list all the backdoor paths from A to B. These are

$$ACB, \; ADCB, \; ACEB, \text{ and } ADCEB$$

The second condition is satisfied because C, the only member of S, belongs to each of these paths. The third condition fails because C is a collider on the backdoor path $ADCEB$ and S contains no noncollider along this path because the only element of S is C.

But how, if we condition on C, that is, divide the population into classes on each of which the value of C is constant, could D influence E? This is an instance of *Berkson's paradox* [12]. Observations of a common consequence of independent causes can induce a dependence between those causes. For example, suppose Bart tosses a coin (event D) and Homer tosses a coin (event E). Each sees his own coin, but not the other's coin. Finally, suppose Lisa sees both coins and says that at least one of the coins is heads. This observation compromises the independence of D and E: if Homer sees that his coin has come up tails, then he knows Bart's coin came up heads. For complicated graphs, direct arguments of this type could be trying. How fortunate we have the three conditions enumerated above.

Back to Fig. 11.7. On the other hand, both $S = \{D, C\}$ and $S = \{C, E\}$ suffice to block all spurious paths from A to B. We'll check the first. Neither D nor C is a descendant of A or B. Every backdoor path contains $C \in S$. The only collider on the backdoor paths from A to B is $C \in S$, a collider on only $ADCEB$, and this path includes the noncollider $D \in S$. You can see why a similar argument shows that $S = \{C, E\}$ also is sufficient.

Of course, it is possible that a DAG has no sufficient set of vertices to block all spurious paths from A to B. Then redesign the experiment, if possible, to add the needed vertices or arrows, or estimate the magnitudes of the unblocked spurious effects and account for these in any deductions.

Much of the psychological problem that leads to the "paradox" of Simpson's reversal comes from the intuition that conditional probabilities $P(y|x)$ encode causal relations. Of course they need not. To avoid this problem, Pearl defines (Definition 3.2.1 and Sect. 3.4 of [116]) the *causal effect* of x on y, $P(y|do(x))$, taking into account the distributions of the variables related to x and y in the DAG that models the system. Despite the beauty of the *do*-calculus, details are a bit complicated, so we won't present them here. Rather, we recommend Pearl's books [116, 120], which are gems.

Using his notion of causal effect, Pearl proves the *Sure-Thing Principle*.

Theorem 11.5.1. An event A that increases the probability of an event B in each subpopulation also must increase the probability of B in the whole population, provided that A does not change the distribution of the subpopulations.

Causal analysis is a new addition to mathematical analysis. Already it has been useful in interpreting test results when the population can be partitioned in several ways. Eventually, appropriate attention to causal analysis may inform experimental design along more direct avenues. Much remains to do, but already we see that this is an exciting aspect of the relation between math and medicine.

Practice Problems

11.5.1. For the causal diagram on the right, list the backdoor paths between A and B. Find two sets S_1 and S_2, of vertices we can condition on in order to eliminate all spurious paths from A to B.

11.5.2. Suppose diet, pancreas function, and gender have direct effects on diabetes; exercise and pancreas function have direct effects on diet; and gender and exercise have direct effects on pancreas function. There are no other direct effects. Draw the causal diagram for this system. Identify a sufficient set S of factors to condition on to eliminate all spurious paths from diet to 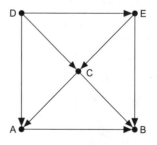 diabetes. This is an exercise in graph construction. I am not proposing this as a serious model of diabetes.

Practice Problem Solutions

11.5.1. Here are the backdoor paths from A to B.

  ACB, ACEB, ACDEB, ADCB, ADCEB, ADEB, ADECB

We've not included paths with (necessarily non-causal) loops, for example, *ACEDCB*. Among these paths, the only collider is C on *ADCEB*. Clearly, $S = \{C\}$ is not sufficient, because C does not lie on the path *ADEB*. On the other hand, $S_1 = \{C, D\}$ is sufficient. The vertex D lies on *ADEB*; the vertex C lies on all the other backdoor paths. For the only collider, C on *ADCEB*, this path contains a noncollider, $D \in S_1$. A similar argument shows that $S_2 = \{C, E\}$ is sufficient: E lies on the path *ADEB* that does not contain C, and E is a noncollider on *ADCEB*.

11.5.2. Each arrow in the DAG represents a direct effect proposed in the problem. The backdoor paths from diet to diabetes are listed below. Observe that pancreas function is on each path, but it is a collider on the last path and so $S = \{$pancreas function$\}$ doesn't satisfy the third condition for sufficiency: this path contains no noncollider in S. But both $S = \{$exercise, pancreas function$\}$ and $S = \{$pancreas function, gender$\}$ are sufficient sets.

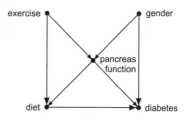

diet → pancreas function → diabetes

diet → exercise → pancreas function → diabetes

diet → pancreas function → gender → diabetes

diet → exercise → pancreas function → gender → diabetes

Exercises

The first two problems ask you to put some numbers on the statement of Berkson's paradox.

11.5.1. Suppose three fair coins are tossed independently of one another. Let X, Y, and Z denote the outcome of tossing the first, second, and third coins. Suppose we know that the number of heads showing on the three coins is at least 2. Compute $P(X = H | Y = H)$ and $P(X = H | ((Y = H) \cap (Z = H)))$. Interpret these results in terms of the effect of the given conditioning on the independence of these three coin tosses.

11.5.2. Suppose two fair dice are tossed independently of one another. Let X and Y denote the outcomes of the first and second dice. Suppose we know $X + Y \leq 7$. Compute $P(X \leq 2 | Y \geq 3)$. Compare this with the value of $P(X \leq 2 | Y \geq 3)$ if we do not know that $X + Y \leq 7$.

11.5.3. Suppose diet and exercise have direct effects on blood pressure; blood pressure, diet, and exercise have direct effects on glaucoma; and diet has a direct effect on exercise. There are no other direct effects. Draw the causal diagram for this system. Identify a sufficient set S of factors to condition on to eliminate all spurious paths from blood pressure to glaucoma, or show this diagram has no sufficient set. Again, this is not a real model, just an exercise for practice with the concepts.

11.5.4. Suppose genetics and neural architecture have direct effects on vision; genetics has a direct effect on neural architecture; vision has a direct effect on reading; and reading, vision, and neural architecture have direct effects on the onset of Alzheimer's disease. There are no other direct effects. Draw the causal diagram for this system. Identify a sufficient set S of factors to condition on to eliminate all spurious paths from reading to Alzheimer's, or show this diagram has no sufficient set. Once more, this is not a real model, just an exercise for practice with the concepts.

11.5.5–11.5.10. For each of the graphs in Fig. 11.9, list the (loopless) backdoor paths from A to B. Determine if there is a set of vertices we can condition on in order to eliminate all spurious paths from A to B.

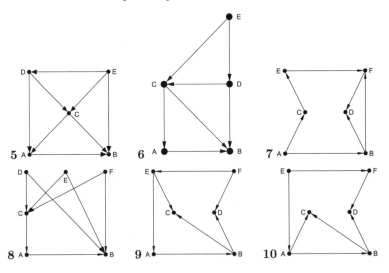

Figure 11.9. Causal diagrams for Exercises **11.5.5–11.5.10**.

11.6 EXPECTED VALUE AND VARIANCE

Returning to the dice toss example, the average, or *expected value*, of the experiment, over many repetitions, can be deduced from Fig. 11.1. The value 2 occurs in about 1/36 repetitions, the value 3 in about 2/36, . . . , the value 12 in about 1/36 repetitions. In general, if the possible outcomes of a measurement are $x_1, \ldots, x_n$, each occurring with probability $P(x_i)$, then the expected value of X is

$$\mathbb{E}(X) = x_1 \cdot P(x_1) + \cdots + x_n \cdot P(x_n) \qquad (11.14)$$

Consequently, the expected value of tossing two fair dice is

$$\mathbb{E}(X) = 2 \cdot \frac{1}{36} + 3 \cdot \frac{2}{36} + 4 \cdot \frac{3}{36} + \cdots + 12 \cdot \frac{1}{36} = 7$$

Note the expected value need never occur. For example, tossing three (fair) coins yields eight equally likely outcomes:

$$TTT, TTH, THT, THH, HTT, HTH, HHT, HHH$$

If X is the number of H, we have $P(X = 0) = 1/8$, $P(X = 1) = 3/8$, $P(X = 2) = 3/8$, and $P(X = 3) = 1/8$. Then the expected value is

$$\mathbb{E}(X) = 0 \cdot \frac{1}{8} + 1 \cdot \frac{3}{8} + 2 \cdot \frac{3}{8} + 3 \cdot \frac{1}{8} = \frac{3}{2}$$

and we never, ever, will get one and a half heads when we toss three coins.

For continuous distributions the analogous computation is

$$\mathbb{E}(X) = \int xf(x) \, dx \qquad (11.15)$$

where f is the probability density function and the limits of the integral are the bounds of the values of x. If $a \leq x \leq b$, then the integral is $\int_a^b$, if x can take any real value, the integral is $\int_{-\infty}^{\infty}$.

For $f(x) = 6x(1 - x)$, $0 \leq x \leq 1$, the expected value is

$$\mathbb{E}(X) = \int_0^1 xf(x)dx = \int_0^1 x6x(1 - x)dx = 2x^3 - \frac{3}{2}x^4 = \frac{1}{2}$$

Integrating the product of $f(x)$ and other powers of x can give useful information. We define the nth *moment* of x by

$$\mathbb{E}(X^n) = \int x^n f(x) \, dx \qquad (11.16)$$

for any integer $n > 0$. So the first moment is just the expected value. The second moment has a familiar interpretation. First, if $\mathbb{E}(X)$ is finite, then translate the variables by $\mathbb{E}(X)$. That is, consider the new variables $Y = X - \mathbb{E}(X)$. The second moment $\mathbb{E}(Y^2)$ is the *variance*, $\sigma^2(X)$, of X; the *standard deviation* of X is $\sigma(X) = \sqrt{\sigma^2(X)}$. This looks annoyingly obvious, but it's just a definition: the standard deviation is the square root of the variance. A simple formulation of $\sigma^2(X)$ is

$$\sigma^2(X) = \mathbb{E}(X^2) - \mathbb{E}(X)^2 \tag{11.17}$$

Here's why this is true:

$$\mathbb{E}((X - \mathbb{E}(X))^2) = \int_{-\infty}^{\infty} (x - \mathbb{E}(X))^2 f(x) \, dx$$

$$= \int_{-\infty}^{\infty} (x^2 - 2x\mathbb{E}(X) + \mathbb{E}(X)^2) f(x) \, dx$$

$$= \int_{-\infty}^{\infty} x^2 f(x) \, dx - 2\mathbb{E}(X) \int_{-\infty}^{\infty} xf(x) \, dx + \mathbb{E}(X)^2 \int_{-\infty}^{\infty} f(x) \, dx$$

$$= \mathbb{E}(X^2) - 2\mathbb{E}(X)\mathbb{E}(X) + \mathbb{E}(X)^2 = \mathbb{E}(X^2) - \mathbb{E}(X)^2$$

where we have used the fact that the constants $\mathbb{E}(X)$ and $\mathbb{E}(X^2)$ factor through integrals, and also Eq. (11.3) for the last term in the penultimate equality. Here are some useful properties of the expected value.

Proposition 11.6.1. For any constant A, (a) $\mathbb{E}(X + A) = \mathbb{E}(X) + A$, (b) $\mathbb{E}(A \cdot X) = A \cdot \mathbb{E}(X)$, and (c) $\mathbb{E}(X + Y) = \mathbb{E}(X) + \mathbb{E}(Y)$.

Before we jump into the proof, think of "expected value" as "average value" and each of the three pieces of the proposition should make sense. We'll work through a simple instance of (c) in Example 11.6.1.

Proof. We'll give the proof for discrete distributions. For continuous distributions the ideas are similar but the implementation requires considerably more sophisticated math: the branch of analysis called measure theory, usually not seen until near the end of an undergraduate mathematics program. The mantra of the early weeks of my graduate probability course was "A random variable is just a measurable function." Probability is an immense part of mathematics, far more than computing the odds in card games.

(a) Apply Eq. (11.14), the definition of expected value for discrete distributions,

$$\mathbb{E}(X) = \sum_i x_i P(X = x_i)$$

Now change variables, $Y = X + A$ and $y_i = x_i + A$. Then

$$\mathbb{E}(X + A) = \mathbb{E}(Y) = \sum_i y_i P(Y = y_i) = \sum_i (x_i + A)P(X + A = x_i + A)$$

$$= \sum_i (x_i + A)P(X = x_i) = \mathbb{E}(X) + A$$

In the last equality we used Eq. (11.14) and $\sum_i P(X = x_i) = 1$.

(b) Once more we'll change variables, this time to $Y = A \cdot X$ and $y_i = A \cdot x_i$.

$$\mathbb{E}(A \cdot X) = \mathbb{E}(Y) = \sum_i y_i P(Y = y_i) = \sum_i A \cdot x_i P(A \cdot X = A \cdot x_i)$$

$$= \sum_i A \cdot x_i P(X = x_i) = A \cdot \sum_i x_i P(X = x_i) = A \cdot \mathbb{E}(X)$$

(c) For this we need another construction, the *joint distribution*. For each possible value $x_i, i = 1, \ldots, m$ of X and $y_j, j = 1, \ldots, n$ of Y, $P(X = x_i, Y = y_j)$ is the probability of obtaining those values. It is defined by

$$P(X = x_i, Y = y_j) = P(X = x_i) \cdot P(Y = y_j | X = x_i) \qquad (11.18)$$

With the definition of conditional probability, Eq. (11.7), we obtain

$$P(X = x_i, Y = y_j) = P(X = x_i) \cdot \frac{P(\{X = x_i\} \cap \{Y = y_j\})}{P(X = x_i)}$$

$$= P(\{X = x_i\} \cap \{Y = y_j\}) \qquad (11.19)$$

In Example 11.6.2 we'll do a calculation, but for this proof we need only one property of the joint distribution:

$$P(X = x_i) = \sum_{j=1}^{n} P(X = x_i, Y = y_j) \quad \text{and}$$

$$P(Y = y_j) = \sum_{i=1}^{m} P(X = x_i, Y = y_j) \qquad (11.20)$$

When several variables are involved in the system, $P(X = x_i)$ and $P(Y = y_j)$ are called *marginal distributions*. Now we can prove property (c) of the expected value.

$$\mathbb{E}(X + Y) = \sum_i \sum_j (x_i + y_j) P(X = x_i, Y = y_j)$$

$$= \sum_i \sum_j x_i P(X = x_i, Y = y_j) + \sum_i \sum_j y_j P(X = x_i, Y = y_j)$$

$$= \sum_i x_i \sum_j P(X = x_i, Y = y_j) + \sum_j y_j \sum_i P(X = x_i, Y = y_j)$$

$$= \sum_i x_i P(X = x_i) + \sum_j y_j P(Y = y_j) \qquad \text{by Eq. (11.20)}$$

$$= \mathbb{E}(X) + \mathbb{E}(Y)$$

Here we have rearranged the order of the sums (sum over i then sum over j gives the same result as sum over j then sum over i, always valid for finite sums) and factored x_i through the sum over j and y_j through the sum over i, valid because i does not depend on j. Write out an example for $n = m = 2$ if these points are unclear. ☐

For independent events variances have a particularly nice behavior.

Proposition 11.6.2. If X and Y are independent, then

$$\sigma^2(X+Y) = \sigma^2(X) + \sigma^2(Y)$$

Proof. We'll apply Eq. (11.17) and Prop. 11.6.1 (b) and (c) to factor a multiplicative constant through an expected value, and to convert the expected value of a sum into the sum of expected values. That is,

$$
\begin{aligned}
\sigma^2(X+Y) &= \mathbb{E}((X+Y)^2) - (\mathbb{E}(X+Y))^2 \\
&= \mathbb{E}((X+Y)^2) - (\mathbb{E}(X) + \mathbb{E}(Y))^2 \\
&= \mathbb{E}(X^2 + 2XY + Y^2) - (\mathbb{E}(X)^2 + 2\mathbb{E}(X)\mathbb{E}(Y) + \mathbb{E}(Y)^2) \\
&= \mathbb{E}(X^2) + 2\mathbb{E}(XY) + \mathbb{E}(Y^2) - \mathbb{E}(X)^2 - 2\mathbb{E}(X)\mathbb{E}(Y) - \mathbb{E}(Y)^2 \\
&= \mathbb{E}(X^2) - \mathbb{E}(X)^2 + 2\mathbb{E}(XY) - 2\mathbb{E}(X)\mathbb{E}(Y) + \mathbb{E}(Y^2) - \mathbb{E}(Y)^2 \\
&= \sigma^2(X) + 2\Big(\mathbb{E}(XY) - \mathbb{E}(X)\mathbb{E}(Y)\Big) + \sigma^2(Y)
\end{aligned}
$$

For independent X and Y the bracketed terms sum to 0. Here's how to see this for discrete variables. The argument for continuous variables is similar.

$$
\begin{aligned}
\mathbb{E}(XY) &= \sum_i \sum_j x_i y_j P(x_i \cap y_j) = \sum_i \sum_j x_i y_j P(x_i) P(y_j) \\
&= \sum_i \sum_j x_i P(x_i) y_j P(y_j) = \sum_j \sum_i x_i P(x_i) y_j P(y_j) \\
&= \sum_j \Big(\sum_i x_i P(x_i)\Big) y_j P(y_j) = \Big(\sum_i x_i P(x_i)\Big)\Big(\sum_j y_j P(y_j)\Big) \\
&= \mathbb{E}(X)\mathbb{E}(Y)
\end{aligned}
$$

where the second equality comes from Eq. (11.6), the fourth from reverse the sums, and the sixth from factoring the constant $\sum_i x_i P(x_i)$ through the sum over j. ☐

Next we'll illustrate Prop. 11.6.1 and then give a simple example of a joint distribution.

Example 11.6.1. *Expected value of a sum.* Suppose X can take the values $x_1 = 1$, $x_2 = 2$, and $x_3 = 3$ with probabilities $1/2, 1/4$, and $1/4$. Suppose also that Y can take the values $y_1 = 4$ and $y_2 = 5$ with probabilities 1/3 and 2/3. Finally, suppose that X and Y are independent. Let's see what part (c) of Prop. 11.6.1 says in this case. Try not to be underwhelmed. Here the point is to show just how plausible the proposition is.

Once more we'll apply Eq. (11.14). First note that

$$\mathbb{E}(X) = 1 \cdot (1/2) + 2 \cdot (1/4) + 3 \cdot (1/4) = 7/4$$
$$\mathbb{E}(Y) = 4 \cdot (1/3) + 5 \cdot (2/3) = 14/3$$

so $\mathbb{E}(X) + \mathbb{E}(Y) = 7/4 + 14/3 = 77/12$.

Now we'll compute $\mathbb{E}(X + Y)$. This is straightforward because along with Eq. (11.19), by the multiplication rule of Eq. (11.6) the independence of X and Y means that

$$P(X = x_i, Y = y_j) = P(\{X = x_i\} \cap \{Y = y_j\})$$
$$= P(X = x_i) \cdot P(Y = y_j) \tag{11.21}$$

Then we have

$$\mathbb{E}(X + Y) = \sum_{i=1}^{3}\sum_{j=1}^{2}(x_i + y_j) \cdot P(X = x_i, Y = y_j)$$
$$= \sum_{i=1}^{3}\sum_{j=1}^{2}(x_i + y_j) \cdot P(X = x_i) \cdot P(Y = y_j)$$

$$= (1+4) \cdot (1/2) \cdot (1/3) + (1+5) \cdot (1/2) \cdot (2/3)$$
$$+ (2+4) \cdot (1/4) \cdot (1/3) + (2+5) \cdot (1/4) \cdot (2/3)$$
$$+ (3+4) \cdot (1/4) \cdot (1/3) + (3+5) \cdot (1/4) \cdot (2/3) = 77/12$$

This agrees with our calculation of $\mathbb{E}(X) + \mathbb{E}(Y)$. $\square$

Example 11.6.2. *Joint distribution.* Now we'll construct a joint distribution for two random variables that are not independent. Suppose two fair dice (one red, one green) are tossed, so there are 36 equally likely outcomes. Take the random variable X to be the sum of the numbers showing on the two dice, and Y to be the larger of the two numbers. Then X can take on the values 2 through 12, and Y can take on the values 1 through 6. Here is the table of values $P(X = x, Y = y)$ of the joint distribution.

	2	3	4	5	6	7	8	9	10	11	12
1	$\frac{1}{36}$	0	0	0	0	0	0	0	0	0	0
2	0	$\frac{2}{36}$	$\frac{1}{36}$	0	0	0	0	0	0	0	0
3	0	0	$\frac{2}{36}$	$\frac{2}{36}$	$\frac{1}{36}$	0	0	0	0	0	0
4	0	0	0	$\frac{2}{36}$	$\frac{2}{36}$	$\frac{2}{36}$	$\frac{1}{36}$	0	0	0	0
5	0	0	0	0	$\frac{2}{36}$	$\frac{2}{36}$	$\frac{2}{36}$	$\frac{2}{36}$	$\frac{1}{36}$	0	0
6	0	0	0	0	0	$\frac{2}{36}$	$\frac{2}{36}$	$\frac{2}{36}$	$\frac{2}{36}$	$\frac{2}{36}$	$\frac{1}{36}$

For example, the sum $X = 2$, maximum $Y = 1$ entry can occur in only one way, $(R, G) = (1,1)$, so $P(X = 2, Y = 1) = 1/36$. The $X = 3$, $Y = 2$ entry can occur in two ways, $(R, G) = (1,2)$ or $(2,1)$, so $P(X = 3, Y = 2) = 2/36$. All the other entries are computed similarly. Certainly the sum and the maximum value are not independent. If they were, then Eq. (11.21) would give

$$P(X = 3, Y = 2) = P(X = 3) \cdot P(Y = 2)$$
$$= P(\{(1,2), (2,1)\}) \cdot P(\{(1,2), (2,1), (2,2)\}) = \frac{2}{36} \cdot \frac{3}{36} = \frac{1}{216}$$

which is wrong by a lot. The joint distribution captures how these probabilities interact.

Note by Eq. (11.20) that summing the entries in each row gives the marginal distribution of Y:

$$P(Y = 1) = \frac{1}{36}, \ P(Y = 2) = \frac{3}{36}, \ P(Y = 3) = \frac{5}{36}, \ \cdots, \ P(Y = 6) = \frac{11}{36}$$

and summing the row entries gives the marginal distribution of X. □

Finally we're ready to think about higher moments. The third moment is used to define the *skewness* a measure of the asymmetry of the distribution about the mean. But instead of $\mathbb{E}((X - \mathbb{E}(X))^3)$, we'll divide by the standard deviation to adjust for the spread of the distribution. Specifically, skewness is

$$\text{sk}(X) = \mathbb{E}\left(\left(\frac{X - \mathbb{E}(X)}{\sigma(X)}\right)^3\right) \tag{11.22}$$

Symmetric distributions have $\text{sk}(X) = 0$. We'll see that the Poisson distribution is not symmetric. A lack of symmetry has implications for the length of one of the tails of the distribution, important for understanding how far from the mean events can stray on one side.

Using Prop. 11.6.1 we can find a simpler expression for skewness.

$$\mathrm{sk}(X) = \mathbb{E}\left(\left(\frac{X - \mathbb{E}(X)}{\sigma(X)}\right)^3\right)$$

$$= \frac{1}{\sigma(X)^3}\mathbb{E}\left(X^3 - 3X^2\mathbb{E}(X) + 3X\mathbb{E}(X)^2 - \mathbb{E}(X)^3\right)$$

$$= \frac{1}{\sigma(X)^3}\left(\mathbb{E}(X^3) - 3\mathbb{E}(X^2)\mathbb{E}(X) + 3\mathbb{E}(X)\mathbb{E}(X)^2 - \mathbb{E}(X)^3\right)$$

$$= \frac{1}{(\sigma(X)^2)^{3/2}}\left(\mathbb{E}(X^3) - 3\mathbb{E}(X^2)\mathbb{E}(X) + 2\mathbb{E}(X)^3\right) \tag{11.23}$$

where in the second equality we have used $\mathbb{E}(\mathbb{E}(X)^3) = \mathbb{E}(X)^3$, a special case of $\mathbb{E}(k) = k$ for every constant k. See Exercise 11.6.1.

Using Eq. (11.17) we see that an equivalent expression for skewness is

$$\mathrm{sk}(X) = \frac{\mathbb{E}(X^3) - 3\mathbb{E}(X)\sigma^2(X) - \mathbb{E}(X)^3}{\sigma(X)^3}$$

but we prefer Eq. (11.23).

The fourth moment is used to define the *kurtosis*, a measure of how sharp or flat the peak of the distribution is relative to the normal distribution.

$$\mathrm{kur}(X) = \mathbb{E}\left(\left(\frac{X - \mathbb{E}(X)}{\sigma(X)}\right)^4\right) \tag{11.24}$$

Computationally simpler is this equivalent expression for kurtosis

$$\mathrm{kur}(X) = \frac{\mathbb{E}(X^4) - 4\mathbb{E}(X^3)\mathbb{E}(X) + 6\mathbb{E}(X^2)\mathbb{E}(X)^2 - 3\mathbb{E}(X)^4}{\sigma(X)^4} \tag{11.25}$$

Exercise 11.6.2 asks you to derive (11.25) from the definition (11.24). As we'll see in Sect. 11.9, because the normal distribution has a kurtosis of 3, often what's calculated is the *excess kurtosis*, defined as $\mathrm{kur}(X) - 3$.

Skewness and kurtosis are used to test whether a distribution of experimental data is normal. While left-right symmetry about the mean is easy to see (though if it is off a bit, the skewness number helps us assess if the distribution departs significantly from symmetry), checking if the shape is close to that of a normal distribution is more difficult to do by eye. Kurtosis is a way to quantify how close to normal the shape of a distribution is. Kurtosis partitions distributions into three types.

- *Leptokurtic*: kurtosis exceeds 3, the distribution has a higher peak and fatter tails than the normal distribution.
- *Mesokurtic*: kurtosis equals 3, the distribution has the shape of a normal distribution.

- *Platykurtic*: kurtosis is less than 3, the distribution has a lower, broader peak and smaller tails than a normal distribution.

Time for some examples. We'll compute the mean, variance, skewness, and kurtosis for a discrete distribution and for a continuous distribution.

Example 11.6.3. *Die toss moments.* Here we'll study the distribution of tosses of a fair die. The values are $X = 1, 2, \ldots, 6$ and $p_i = P(X = i) = 1/6$ for all i. The expected value is

$$\mathbb{E}(X) = \sum_{i=1}^{6} i \cdot p_i = \left(\sum_{i=1}^{6} i \right) \cdot \frac{1}{6} = (1 + 2 + 3 + 4 + 5 + 6) \cdot \frac{1}{6} = \frac{7}{2} \quad (11.26)$$

For the variance we'll need the second moment

$$\mathbb{E}(X^2) = \sum_{i=1}^{6} i^2 \cdot p_i = \left(1^2 + 2^2 + 3^2 + 4^2 + 5^2 + 6^2 \right) \cdot \frac{1}{6} = \frac{91}{6}$$

Then by Eq. (11.17), the variance is given by

$$\sigma^2(X) = \mathbb{E}(X^2) - \mathbb{E}(X)^2 = \frac{91}{6} - \left(\frac{7}{2} \right)^2 = \frac{35}{12}$$

For the skewness we'll need the third moment

$$\mathbb{E}(X^3) = \sum_{i=1}^{6} i^3 \cdot p_i = \left(1^3 + 2^3 + 3^3 + 4^3 + 5^3 + 6^3 \right) \cdot \frac{1}{6} = \frac{441}{6}$$

We use Eq. (11.23) to compute the skewness

$$\begin{aligned} \mathrm{sk}(X) &= \frac{1}{\sigma(X)^3} \left(\mathbb{E}(X^3) - 3\mathbb{E}(X^2)\mathbb{E}(X) + 2\mathbb{E}(X)^3 \right) \\ &= \frac{1}{(35/12)^{3/2}} \left(\frac{441}{6} - 3\frac{91}{6}\frac{7}{2} + 2\left(\frac{7}{2} \right)^3 \right) = 0 \end{aligned}$$

This is not a surprise because the distribution is left-right symmetric.

Finally, to compute the kurtosis we need the fourth moment

$$\mathbb{E}(X^4) = \sum_{i=1}^{6} i^4 \cdot p_i = \left(1^4 + 2^4 + 3^4 + 4^4 + 5^4 + 6^4 \right) \cdot \frac{1}{6} = \frac{2275}{6}$$

and we'll use Eq. (11.25) to compute kurtosis

$$\begin{aligned} \mathrm{kur}(X) &= \frac{1}{\sigma(X)^4} \left(\mathbb{E}(X^4) - 4\mathbb{E}(X^3)\mathbb{E}(X) + 6\mathbb{E}(X^2)\mathbb{E}(X)^2 - 3\mathbb{E}(X)^4 \right) \\ &= \frac{1}{(35/12)^2} \left(\frac{2275}{6} - 4\frac{441}{6}\frac{7}{2} + 6\frac{91}{6}\left(\frac{7}{2} \right)^2 - 3\left(\frac{7}{2} \right)^4 \right) = \frac{303}{175} \end{aligned}$$

Because the kurtosis $303/175 \approx 1.73$ is less than 3, this distribution is platykurtic. $\Box$

Example 11.6.4. *Exponential moments.* Now we'll study the moments of the exponential distribution $f(x) = \lambda e^{-\lambda x}$ for $0 \leq x < \infty$ and positive λ.

The expected value is

$$\mathbb{E}(X) = \int_0^\infty x\lambda e^{-\lambda x}\,dx = \lim_{B\to\infty} \left(-\frac{(1+\lambda x)e^{-\lambda x}}{\lambda} \right) \Bigg|_0^B$$

$$= \frac{1}{\lambda} - \lim_{B\to\infty} \frac{1+\lambda B}{\lambda} e^{-\lambda B} = \frac{1}{\lambda}$$

where the last equality follows from the fact that as $B \to \infty$, $e^{-\lambda B}$ goes to 0 faster than B^n goes to ∞ for any positive n. (Apply l'Hôpital's rule.)

To find the variance we need the second moment

$$\mathbb{E}(X^2) = \int_0^\infty x^2\lambda e^{-\lambda x}\,dx = \frac{2}{\lambda^2} - \lim_{B\to\infty} \frac{2+\lambda B(2+\lambda B)}{\lambda^2} e^{-\lambda B} = \frac{2}{\lambda^2}$$

Then

$$\sigma^2(X) = \mathbb{E}(X^2) - \mathbb{E}(X)^2 = \frac{2}{\lambda^2} - \left(\frac{1}{\lambda}\right)^2 = \frac{1}{\lambda^2}$$

To find the skewness we need the third moment

$$\mathbb{E}(X^3) = \int_0^\infty x^3\lambda e^{-\lambda x}\,dx$$

$$= \frac{6}{\lambda^3} - \lim_{B\to\infty} \frac{2+\lambda B(2+\lambda B(3+\lambda B))}{\lambda^3} e^{-\lambda B} = \frac{6}{\lambda^3}$$

We use Eq. (11.23) to compute the skewness

$$\mathrm{sk}(X) = \frac{1}{\sigma(X)^3} \left(\mathbb{E}(X^3) - 3\mathbb{E}(X^2)\mathbb{E}(X) + 2\mathbb{E}(X)^3 \right)$$

$$= \frac{1}{(1/\lambda^2)^{3/2}} \left(\frac{6}{\lambda^3} - 3\frac{2}{\lambda^2}\frac{1}{\lambda} + 2\left(\frac{1}{\lambda}\right)^3 \right) = 2$$

On the right we see plots of the exponential distribution $f(x) = \lambda e^{-\lambda x}$ for $\lambda = 1/2, 1$, and 2. The lack of left-right symmetry is apparent. Maybe a bit surprising is that the skewness is independent of λ, the only parameter of the exponential distribution.

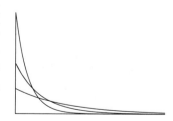

To find the kurtosis we need the fourth moment

$$\mathbb{E}(X^4) = \int_0^\infty x^3 \lambda e^{-\lambda x}\, dx$$

$$= \frac{24}{\lambda^4} - \lim_{B \to \infty} \frac{24 + \lambda B(24 + \lambda B(12 + \lambda B(4 + \lambda B)))}{\lambda^4} e^{-\lambda B} = \frac{24}{\lambda^4}$$

We'll use Eq. (11.25) to compute kurtosis

$$\text{kur}(X) = \frac{1}{\sigma(X)^4}\left(\mathbb{E}(X^4) - 4\mathbb{E}(X^3)\mathbb{E}(X) + 6\mathbb{E}(X^2)\mathbb{E}(X)^2 - 3\mathbb{E}(X)^4\right)$$

$$= \frac{1}{(1/\lambda^2)^2}\left(\frac{24}{\lambda^4} - 4\frac{6}{\lambda^3}\frac{1}{\lambda} + 6\frac{2}{\lambda^2}\left(\frac{1}{\lambda}\right)^2 - 3\left(\frac{1}{\lambda}\right)^4\right) = 9$$

Because the kurtosis is greater than 3, this distribution is leptokurtic. □

So why have we done all this? The calculations are easy, provided we can sum the series or evaluate the integral. Expected value and variance are important characteristics of distributions, and their interpretations are clear. One of the principal applications of skewness and kurtosis is to test if a distribution is normal.

Our examples are mathematical. We know all of the characteristics of the distributions studied, while in experimental situations we have only a sample, and almost always a sample with some noise stemming from measurement errors. The real question is: how do we test if a sample of data is consistent with the underlying data's being normally distributed? This belongs to the field of hypothesis testing in statistics.

Using sample data to make inferences about the the underlying distribution is another topic of hypothesis testing. This is the centerpiece of most introductory statistics courses; the subject has much subtlety and has filled many volumes. We'll say a bit about this in Sect. 11.11. How do we tell if the sample truly represents the characteristics of the population? How do we design the sample selection to respect this issue? Careers have been built on this. This is a central issue in modern polling theory. To emphasize the complexity of this topic, I'll mention that I'm writing this in early 2017 and so have little reason to place much trust in modern polling theory. Yes, this is a complicated problem.

Practice Problems

11.6.1. Suppose $P(X = 1) = P(X = 2) = 1/4$ and $P(X = 3) = P(X = 4) = P(X = 5) = P(X = 6) = 1/8$. Compute the expected value, variance, skewness, and kurtosis.

11.6.2. Verify that $f(x) = 6x(1-x)$, $0 \le x \le 1$, is a probability distribution. Compute the expected value, variance, skewness, and kurtosis for this distribution.

Practice Problem Solutions

11.6.1. Recall that in order to compute the mean, variance, skewness, and kurtosis, we need the first, second, third, and fourth moments, respectively. For this probability distribution

$$\mathbb{E}(X^n) = (1^n + 2^n) \cdot (1/4) + (3^n + 4^n + 5^n + 6^n) \cdot (1/8)$$

and so

$$\mathbb{E}(X) = 3, \quad \mathbb{E}(X^2) = 12, \quad \mathbb{E}(X^3) = 225/4, \quad \mathbb{E}(X^4) = 573/2$$

Then apply the formula for variance Eq. (11.17), the formula for skewness Eq. (11.23), and the formula for kurtosis Eq. (11.25). These give

$$\sigma^2 = \mathbb{E}(X^2) - \mathbb{E}(X)^2 = 12 - 3^2 = 3$$

$$sk(X) = \frac{1}{\sigma^3}\left(\mathbb{E}(X^3) - 3\mathbb{E}(X^2)\mathbb{E}(X) + 2\mathbb{E}(X)^3\right)$$

$$= \frac{1}{3^{3/2}}\left(\frac{225}{4} - 3\cdot 12 \cdot 3 + 2\cdot 3^3\right) = \frac{\sqrt{3}}{4}$$

and

$$kur(X) = \frac{1}{\sigma^4}\left(\mathbb{E}(X^4) - 4\mathbb{E}(X^3)\mathbb{E}(X) + 6\mathbb{E}(X^2)\mathbb{E}(X)^2 - 3\mathbb{E}(X)^4\right)$$

$$= \frac{1}{3^2}\left(\frac{573}{2} - 4\frac{225}{4}\cdot 3 + 6\cdot 12 \cdot 3^2 - 3\cdot 3^4\right) = \frac{11}{6}$$

11.6.2. For $0 \le x \le 1$, $f(x) = 6x(1-x)$ is non-negative. To show $f(x)$ is a probability distribution, all that remains is to note

$$\int_0^1 6x(1-x)\,dx = \left(3x^2 - 2x^3\right)\Big|_0^1 = 3 - 2 = 1$$

As usual, we'll need the first four moments.

$$\mathbb{E}(X) = \int_0^1 x6x(1-x)\,dx = \left(2x^3 - \frac{3}{2}x^4\right)\Big|_0^1 = \frac{1}{2}$$

$$\mathbb{E}(X^2) = \int_0^1 x^2 6x(1-x)\,dx = \left(\frac{15}{10}x^4 - \frac{6}{5}x^5\right)\Big|_0^1 = \frac{3}{10}$$

$$\mathbb{E}(X^3) = \int_0^1 x^3 6x(1-x)\, dx = \left(\frac{6}{5}x^5 - x^6\right)\Big|_0^1 = \frac{1}{5}$$

$$\mathbb{E}(X^4) = \int_0^1 x^4 6x(1-x)\, dx = \left(x^6 - \frac{6}{7}x^7\right)\Big|_0^1 = \frac{1}{7}$$

Then apply the formula for variance Eq. (11.17), the formula for skewness Eq. (11.23), and the formula for kurtosis Eq. (11.25). These give

$$\sigma^2 = \mathbb{E}(X^2) - \mathbb{E}(X)^2 = \frac{3}{10} - \left(\frac{1}{2}\right)^2 = \frac{1}{20}$$

$$\mathrm{sk}(X) = \frac{1}{\sigma^3}\left(\mathbb{E}(X^3) - 3\mathbb{E}(X^2)\mathbb{E}(X) + 2\mathbb{E}(X)^3\right)$$

$$= \frac{1}{(1/20)^{3/2}}\left(\frac{1}{5} - 3\frac{3}{10}\frac{1}{2} + 2\left(\frac{1}{2}\right)^3\right) = 0$$

$$\mathrm{kur}(X) = \frac{1}{\sigma^4}\left(\mathbb{E}(X^4) - 4\mathbb{E}(X^3)\mathbb{E}(X) + 6\mathbb{E}(X^2)\mathbb{E}(X)^2 - 3\mathbb{E}(X)^4\right)$$

$$= \frac{1}{(1/20)^2}\left(\frac{1}{7} - 4\frac{1}{5}\frac{1}{2} + 6\frac{3}{10}\left(\frac{1}{2}\right)^2 + 3\left(\frac{1}{2}\right)^4\right) = \frac{15}{7}$$

Exercises

11.6.1. Show that for any constant k, if $X = k$ for all X, then $\mathbb{E}(X) = k$. Give two arguments, one for discrete and one for continuous variables.

11.6.2. (a) Derive the expression (11.25) from the definition (11.24).
(b) Derive the equivalent expression

$$\mathrm{kur}(X) = (1/\sigma^4)(\mathbb{E}(X^4) - 4\mathbb{E}(X^3)\mathbb{E}(X) + 6\mathbb{E}(X)^2\sigma(X)^2 + 3\mathbb{E}(X)^4)$$

11.6.3. Suppose $P(X = 1) = P(X = 3) = 1/4$ and $P(X = 2) = P(X = 4) = P(X = 5) = P(X = 6) = 1/8$. Compute the expected value, variance, skewness, and kurtosis.

11.6.4. Suppose $P(X = 1) = P(X = 5) = 1/4$ and $P(X = 2) = P(X = 3) = P(X = 4) = P(X = 6) = 1/8$. Compute the expected value, variance, skewness, and kurtosis.

11.6.5. (a) Find the normalization factor of $x^2(1-x)$ for $0 \le x \le 1$.
(b) For the probability distribution of (a), find the expected value, variance, skewness, and kurtosis.

11.6.6. (a) Find the normalization factor of $x(1-x)^2$ for $0 \le x \le 1$.
(b) For the probability distribution of (a), find the expected value, variance, skewness, and kurtosis.

11.6.7. (a) Find the normalization factor of $x^2(1-x)^2$ for $0 \le x \le 1$.
(b) For the probability distribution of (a), find the expected value, variance, skewness, and kurtosis.

11.6.8. (a) Suppose $P(X=1)=p$ and $P(X=2)=1-p$. Compute the expected value, variance, skewness, and kurtosis.
(b) Plot the skewness and the kurtosis as a function of p for $0 < p < 1$.

11.6.9. (a) Show $f(x) = 2x$ for $0 \le x \le 1$ is a probability distribution.
(b) Find the expected value, variance, skewness, and kurtosis of this distribution.

11.6.10. (a) Show $f(x) = 3x^2$ for $0 \le x \le 1$ is a probability distribution.
(b) Find the expected value, variance, skewness, and kurtosis of this distribution.

11.7 THE BINOMIAL DISTRIBUTION

Before we discuss the normal distribution, we introduce the binomial, a discrete distribution that is used in models of genetic drift, among other things. In Sect. 11.8 we present an important limiting case, the Poisson distribution, whose first application was to model soldier deaths by horse kicks. Really.

Bernoulli trials are independent measurements with only two possible outcomes, A and B, whose probabilities $P(X=A)=p$ and $P(X=B)=1-p=q$ are unchanged as the trials continue. Repeated tosses of a coin are an example of Bernoulli trials.

Suppose we have a sequence of N independent trials. In this sequence A can occur $0,1,\ldots,N$ times. For any m, $0 \le m \le N$, what is the probability that A occurs exactly m times?

First of all, because the trials are independent, the probability of any sequence of N trials in which A occurs m times, and consequently B occurs $N-m$ times, is the product

$$p^m q^{N-m}$$

Recall Eq. (11.6): the probability of several independent events occurring is the product of the probabilities of the individual events. So now all we need to do is count the number of ways m occurrences of A can be arranged among N trials.

Remember the binomial expansions

$$(1+x)^2 = 1 + 2x + x^2$$
$$(1+x)^3 = 1 + 3x + 3x^2 + x^3$$
$$(1+x)^4 = 1 + 4x + 6x^2 + 4x^3 + x^4$$
$$(1+x)^5 = 1 + 5x + 10x^2 + 10x^3 + 5x^4 + x^5$$

$$\cdots$$

The coefficients are the familiar *binomial coefficients*

$$\binom{N}{m} = \frac{N!}{m!(N-m)!} \tag{11.27}$$

that is, the number of ways to arrange m items in a collection of size N. Then we see

$$P(A \text{ occurs } m \text{ times in } N \text{ trials}) = \frac{N!}{m!(N-m)!} p^m q^{N-m} \tag{11.28}$$

Here we've used Eq. (11.5) to justify adding the probabilities of each of the sequences with exactly m occurrences of A. We can do this because these sequences are mutually exclusive: if one sequence of N trials gives m As arranged $A \ldots AB \ldots B$, it does not also give m As arranged $BA \ldots AB \ldots B$. This second arrangement must occur in another sequence of N trials.

Proposition 11.7.1. The expected value of the number of times the value A occurs in N trials is

$$\mathbb{E}(X) = Np \tag{11.29}$$

Proof. The reason is a bit involved, but introduces some nice tricks. Recall the formula for the expected value of a discrete variable, Eq. (11.14), which we'll write more compactly as

$$\mathbb{E}(X) = \sum_{m=0}^{N} mP(A \text{ occurs exactly } m \text{ times in } N \text{ trials})$$

$$= \sum_{m=0}^{N} m \binom{N}{m} p^m q^{N-m} \tag{11.30}$$

Accounting for the extra factor of m will take some work. We'll start with the binomial expansion, illustrated above.

$$(1+x)^N = \sum_{m=0}^{N} \binom{N}{m} x^m \tag{11.31}$$

To introduce the factor of m, differentiate both sides with respect to x.

$$N(1+x)^{N-1} = \sum_{m=0}^{N} \binom{N}{m} mx^{m-1}$$

Now take $x = p/q$ and simplify. Recall $p + q = 1$.

$$N\left(1+\frac{p}{q}\right)^{N-1} = \sum_{m=0}^{N} \binom{N}{m} m\left(\frac{p}{q}\right)^{m-1}$$

$$N\left(\frac{q+p}{q}\right)^{N-1} = \sum_{m=0}^{N} \binom{N}{m} m\frac{p^{m-1}}{q^{m-1}}$$

$$N\left(\frac{1}{q}\right)^{N-1} \cdot pq^{N-1} = \left(\sum_{m=0}^{N} \binom{N}{m} m\frac{p^{m-1}}{q^{m-1}}\right) \cdot pq^{N-1}$$

$$Np = \sum_{m=0}^{N} \binom{N}{m} mp^{m}q^{N-m}$$

By Eq. (11.30), the last equality is Eq. (11.29). □

The trick "differentiate to produce a factor of m in the terms of the polynomial" can be adapted in the obvious way to compute higher moments.

Proposition 11.7.2. The variance of the number of times the value A occurs in N trials is

$$\sigma^2(X) = Np(1-p) \tag{11.32}$$

Proof. Recall Eq. (11.17) $\sigma^2(X) = \mathbb{E}(X^2) - \mathbb{E}(X)^2$, the definition of variance. We know

$$\mathbb{E}(X^2) = \sum_{m=0}^{N} m^2 \binom{N}{m} p^{m}q^{N-m}$$

so we need an additional factor of m^2 in the polynomial. Differentiating twice comes close, producing a factor of $m(m-1)$. With a little algebra, this will work. Again we start with the binomial expansion Eq. (11.31) and differentiate twice

$$N(N-1)(1+x)^{N-2} = \sum_{m=0}^{N} \binom{N}{m} m(m-1)x^{m-2}$$

$$= \sum_{m=0}^{N} \binom{N}{m} (m^2x^{m-2} - mx^{m-2})$$

Now substitute $x = p/q$ and simplify, and again recall that $p + q = 1$.

$$N(N-1)\left(1+\frac{p}{q}\right)^{N-2} = \sum_{m=0}^{N}\binom{N}{m}\left(m^2\left(\frac{p}{q}\right)^{m-2} - m\left(\frac{p}{q}\right)^{m-2}\right)$$

$$N(N-1)\left(\frac{q+p}{q}\right)^{N-2} = \sum_{m=0}^{N}\binom{N}{m}\left(m^2\frac{p^{m-2}}{q^{m-2}} - m\frac{p^{m-2}}{q^{m-2}}\right)$$

$$N(N-1)\frac{1}{q^{N-2}}\cdot p^2 q^{N-2} = \left(\sum_{m=0}^{N}\binom{N}{m}\left(m^2\frac{p^{m-2}}{q^{m-2}} - m\frac{p^{m-2}}{q^{m-2}}\right)\right)\cdot p^2 q^{N-2}$$

$$N(N-1)p^2 = \sum_{m=0}^{N}\binom{N}{m}m^2 p^m q^{N-m} - \sum_{m=0}^{N}\binom{N}{m}m p^m q^{N-m}$$

That is,

$$N(N-1)p^2 = \mathbb{E}(X^2) - \mathbb{E}(X) \tag{11.33}$$

Then

$$\sigma^2(X) = \mathbb{E}(X^2) - \mathbb{E}(X)^2 = \mathbb{E}(X^2) - \mathbb{E}(X) + \mathbb{E}(X) - \mathbb{E}(X)^2$$
$$= N(N-1)p^2 + \mathbb{E}(X) - \mathbb{E}(X)^2 = N(N-1)p^2 + Np - (Np)^2$$
$$= Np - Np^2$$

The third equality follows by Eq. (11.33), the fourth by Eq. (11.29). $\square$

We'll use these formulas for mean and variance when we study the Wright-Fisher model for genetic drift in Sect. 16.1.

To compute the skewness and the kurtosis of the binomial distribution, we'll need the third and fourth moments. Continue our approach for the first and second moments: for the third moment we'll differentiate both sides of

$$(1+x)^N = \sum_{m=0}^{N}\binom{N}{m}x^m$$

three times and substitute $x = p/q$, to obtain

$$N(N-1)(N-2)\left(1+\frac{p}{q}\right)^{N-3} = \sum_{m=0}^{N}\binom{N}{m}m(m-1)(m-2)\left(\frac{p}{q}\right)^{m-3}$$

Now simplify the left-hand side and multiply both sides by $p^3 q^{N-3}$,

$$N(N-1)(N-2)p^3 = \sum_{m=0}^{N} \binom{N}{m}(m^3 - 3m^2 + 2m)p^m q^{N-m}$$

$$= \sum_{m=0}^{N} m^3 \binom{N}{m}p^m q^{N-m} - 3\sum_{m=0}^{N} m^2 \binom{N}{m}p^m q^{N-m}$$

$$+ 2\sum_{m=0}^{N} m \binom{N}{m}p^m q^{N-m}$$

That is,

$$N(N-1)(N-2)p^3 = \mathbb{E}(X^3) - 3\mathbb{E}(X^2) + 2\mathbb{E}(X)$$

Solving for $\mathbb{E}(X^3)$ gives

$$\mathbb{E}(X^3) = N(N-1)(N-2)p^3 + 3\mathbb{E}(X^2) - 2\mathbb{E}(X)$$

Recall Eq. (11.33). We'd like to organize the last two terms of the right-hand side as some multiple of $\mathbb{E}(X^2) - \mathbb{E}(X)$. Then by Eqs. (11.33) and (11.29) we have

$$\mathbb{E}(X^3) = N(N-1)(N-2)p^3 + 3(\mathbb{E}(X^2) - \mathbb{E}(X)) + \mathbb{E}(X)$$
$$= N(N-1)(N-2)p^3 + 3(N(N-1)p^2) + Np \qquad (11.34)$$

With this expression and a bit of algebra we can compute the skewness of the binomial distribution.

Proposition 11.7.3. The skewness of the binomial distribution is

$$\frac{1-2p}{\sqrt{Np(1-p)}} \qquad (11.35)$$

Proof. Into the formula Eq. (11.23) for skewness substitute the expression Eq. (11.34) for $\mathbb{E}(X^3)$, the expression Eq. (11.32) for $\sigma^2(X)$, and the expression Eq. (11.29) for $\mathbb{E}(X)$. Simplify a whole lot (remember $\sigma^3(X) = \sigma^2(X)\sigma(X) = Np(1-p)\sqrt{Np(1-p)}$) and you'll get the expression Eq. (11.35) for skewness. $\square$

For later use, note that for each value of p, $0 < p < 1$, the $N \to \infty$ limit of the skewness is

$$\lim_{N\to\infty} \frac{1-2p}{\sqrt{Np(1-p)}} = 0 \qquad (11.36)$$

To compute the kurtosis of the binomial distribution we'll need the fourth moment. By now, the approach is clear: differentiate four times the binomial expansion for $(1+x)^N$ and solve for $\mathbb{E}(X^4)$.

$$\mathbb{E}(X^4) = N(N-1)(N-2)(N-3) + 6\mathbb{E}(X^3) - 11(\mathbb{E}(X^2) - \mathbb{E}(X)) + 5\mathbb{E}(X)$$

Substitute in Eq. (11.34) for $\mathbb{E}(X^3)$, Eq. (11.33) for $\mathbb{E}(X^2) - \mathbb{E}(X)$, and Eq. (11.29) for $\mathbb{E}(X)$ and then apply the expression Eq. (11.25) for kurtosis. After a lot of algebra we obtain

Proposition 11.7.4. The kurtosis of the binomial distribution is

$$3 - \frac{6}{N} + \frac{1}{Np(1-p)} \tag{11.37}$$

Also for later use, note that for each fixed value of p, $0 < p < 1$, the $N \to \infty$ limit of the kurtosis is

$$\lim_{N \to \infty} 3 - \frac{6}{N} + \frac{1}{Np(1-p)} = 3 \tag{11.38}$$

Because we'll see that the normal distribution has kurtosis $= 3$, this calculation presages the binomial approximation to the normal distribution of Example 11.9.2.

The binomial distribution has many applications. For instance, to predict the probabilities of combinations of wild type and mutant genes, to estimate diagnostic accuracies, to predict the number of times a disease will occur in a population, to model the encapsulation of target cells in heterogeneous mixtures. Once again, Google is your friend.

Practice Problems

11.7.1. Suppose 20% of the residents of a town have the flu. If a random sample of 50 residents is taken find
(a) the expected value of the number of that sample who have the flu,
(b) the variance of the sample number who have the flu, and
(c) the probability that fewer than 5 of the sampled people have the flu.

11.7.2. Given a sequence of Bernoulli trials X_i with success parameter p, denote by N the number of trials when the first success occurs. That is, $X_1 = \cdots = X_{N-1} = 0$ and $X_N = 1$.
(a) For each p show the pdf for N is $f(N) = (1-p)^{N-1}p$.
(b) Show $\mathbb{E}(N) = 1/p$. Hint: don't forget the trick to differentiate a geometric series.
(c) Show the variance of N is $\sigma^2(N) = (1-p)/p^2$.

Practice Problem Solutions

11.7.1. This is a binomial experiment with $p = 0.2$ and $N = 50$. Denote by X the number of sample members who have the flu.
(a) By Eq. (11.29), $\mathbb{E}(X) = Np = 50 \cdot 0.2 = 10$.
(b) By Eq. (11.32), $\sigma^2(X) = Np(1-p) = 50 \cdot 0.2 \cdot (1-0.2) = 8$.

(c) Because, for example, the events "finding 3 people in the sample with the flu" and "finding 4 people in the sample with the flu" are mutually exclusive, we can apply Eq. (11.5) to write $P(X < 5)$ as a sum of probabilities.

$$P(X < 5) = P(X = 0) + P(X = 1) + P(X = 2) + P(X = 3) + P(X = 4)$$

$$= \binom{50}{0} \cdot 0.2^0 \cdot 0.8^{50} + \binom{50}{1} \cdot 0.2^1 \cdot 0.8^{49} + \binom{50}{2} \cdot 0.2^2 \cdot 0.8^{48}$$

$$+ \binom{50}{3} \cdot 0.2^3 \cdot 0.8^{47} + \binom{50}{4} \cdot 0.2^4 \cdot 0.8^{46} \approx 0.018$$

11.7.2. (a) The probability density function is $f(n) = (1-p)^{n-1}p$, the probability of $n - 1$ failures and one success. Because the trials are independent, the probability of $n - 1$ failures followed by one success is just the product of the individual probabilities. Finally, we must show that $\sum_{N=1}^{\infty} f(N) = 1$. This is just summing a geometric series:

$$\sum_{N=1}^{\infty} (1-p)^{N-1}p = p \sum_{N=1}^{\infty} (1-p)^{N-1} = p \sum_{m=0}^{\infty} (1-p)^m = \frac{p}{1-(1-p)} = 1$$

(b) In the context of the random variable N, the expected value is

$$\mathbb{E}(N) = \sum_{n=1}^{\infty} nf(n) \tag{11.39}$$

Recall the proof of Prop. 11.7.1. We'll introduce the extra factor of n in the terms of $\mathbb{E}(N)$ by differentiating.

$$\mathbb{E}(N) = \sum_{n=1}^{\infty} n(1-p)^{n-1}p = p \sum_{n=1}^{\infty} n(1-p)^{n-1} = p \sum_{n=1}^{\infty} -\frac{d}{dp}(1-p)^n$$

$$= -p\frac{d}{dp} \sum_{n=1}^{\infty} (1-p)^n = -p\frac{d}{dp}\left(\frac{1-p}{1-(1-p)}\right) = -p\frac{d}{dp}\left(\frac{1}{p} - 1\right)$$

$$= -p\frac{-1}{p^2} = \frac{1}{p} \tag{11.40}$$

where to sum the series we've used the geometric series sum $x + x^2 + x^3 + \cdots = x/(1-x)$, valid because $x = 1 - p$ satisfies $|x| < 1$.

(c) To compute the variance of N we need the second moment, which we'll find by the second derivative.

$$\mathbb{E}(N^2) = \sum_{n=1}^{\infty} n^2 f(n) = \sum_{n=1}^{\infty} (n(n-1) + n) f(n)$$

$$= \sum_{n=1}^{\infty} n(n-1)(1-p)^{n-1} p + \sum_{n=1}^{\infty} nf(n)$$

$$= p(1-p) \sum_{n=1}^{\infty} n(n-1)(1-p)^{n-2} + \frac{1}{p}$$

where in the last equality we've applied Eqs. (11.39) and (11.40). Now recognize the terms in the sum as the second derivative of $(1-p)^n$. Then

$$\mathbb{E}(N^2) = p(1-p) \sum_{n=1}^{\infty} \frac{d^2}{dp^2} (1-p)^n + \frac{1}{p}$$

$$= p(1-p) \frac{d^2}{dp^2} \left(\frac{1}{p} - 1 \right) + \frac{1}{p} = p(1-p) \frac{2}{p^3} + \frac{1}{p} = \frac{2-p}{p^2}$$

where the second equality comes from summing the geometric series as in (b). Then by Eq. (11.17),

$$\sigma^2(N) = \mathbb{E}(N^2) - \mathbb{E}(N)^2 = \frac{2-p}{p^2} - \left(\frac{1}{p} \right)^2 = \frac{1-p}{p^2} \qquad (11.41)$$

Exercises

11.7.1. For a binomial distribution with parameters N and p, for fixed N, what value of p gives the largest variance? Why?

11.7.2. Is it more likely to get exactly one head in two tosses of a fair coin, exactly 2 heads in 4 tosses, or exactly 4 heads in 8 tosses?

11.7.3. Pneumococcal pneumonia has a 20% mortality rate. In a random sample of 10 pneumococcal pneumonia patients,
(a) what is the probability that only 1 will die from this disease,
(b) what is the probability that none will die from this disease, and
(c) what is the probability that 9 will die from this disease?

11.7.4. If a vaccine is 99% effective, in a random sample of 10 vaccinated people,
(a) what is the probability that none of the 10 have the disease for which the vaccine was designed, and
(b) what is the probability that 2 of the 10 have the disease?

11.7.5. Suppose 10% of a population has a disease detected by a complex blood test. Typically, blood from several people is mixed together and the mixture is tested. If any blood in the mixture is infected, the test is positive. Otherwise, the test is negative. If the test is positive, the people must be tested individually. What is the probability the mixture will test positive if the blood of N people is combined in the mixture to be tested? (I know of no such disease or test.)
(a) $N = 2$ (b) $N = 4$ (c) $N = 6$
(d) $N = 8$ (e) $N = 10$ (f) What conclusions can you draw?

11.7.6. Suppose the process for mass-producing a new flu vaccine delivers an adequate amount of medicine to the dose 98% of the time.
(a) In a run of 20 doses, what is the probability that at most 2 doses do not have an adequate amount of medicine?
(b) What is the probability that at least 3 doses do not have an adequate amount of medicine?

11.7.7. As in Practice Problem 11.7.2 let N denote the number of Bernoulli trials until the first success occurs, and let p be the probability of success. Show the skewness of N is $(2 - p)/\sqrt{1 - p}$. The kurtosis of N is $p^2/(1 - p)$, but we won't ask you to show this.

11.7.8. It is estimated that 1 in 1000 people in the Amazon are infected with a disease that may have hopped from capuchin monkeys.
(a) What is the expected number of people we need to sample before we encounter an infected individual?
(b) Suppose we sample 2000 people without encountering an infected person. Does this suggest there is no infection? Reason informally, standard deviations from the mean.

11.7.9. (a) For the first success random variable N from Practice Problem 11.7.2, find the value of the probability p for which the kurtosis equals 3.
(b) For the value of p found in (a), compute the skewness of N.
(c) Find the value of the probability p for which the kurtosis takes on a given positive value k.

11.7.10. If you feel that you must, derive the binomial kurtosis formula, Eq. (11.37). (If your teacher assigns this problem and gives no additional instruction, then you may turn in a blank page if you really do not feel that you must derive Eq. (11.37).)

11.8 THE POISSON DISTRIBUTION

Like the binomial distribution, the Poisson distribution is discrete. While the binomial distribution is characterized by two parameters, N and p, the Poisson distribution is determined by a single parameter λ and is defined by

$$P(X = n) = \frac{e^{-\lambda}\lambda^n}{n!} \qquad (11.42)$$

for $n = 0, 1, 2, \ldots$. We'll call this the λ-*Poisson distribution*. In order for this to define a distribution, we must check that $\sum_{n=0}^{\infty} P(X = n) = 1$. Recall the Taylor series $e^x = \sum_{n=0}^{\infty} x^n/n!$. Then

$$\sum_{n=0}^{\infty} P(X = n) = \sum_{n=0}^{\infty} \frac{e^{-\lambda}\lambda^n}{n!} = e^{-\lambda} \sum_{n=0}^{\infty} \frac{\lambda^n}{n!} = e^{-\lambda}e^{\lambda} = 1$$

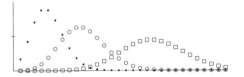

Figure 11.10. Three Poisson distribution plots.

In Fig. 11.10 we see points from three Poisson distributions. The dots represent points for $\lambda = 5$, the circles for $\lambda = 10$, and the squares for $\lambda = 20$. The mark on the vertical axis is at $P = 0.1$. In all graphs, $n = 1, \ldots, 30$, but of course we could take higher values of n. Note that as λ increases, the location of the distribution peak increases, as does its width. We'll see why in Props. 11.8.1 and 11.8.3. Note also that the distribution is asymmetric for small λ, and becomes more symmetric as λ increases. We'll see why when we compute the skewness of the Poisson distribution in Prop. 11.8.4.

This distribution was developed by the French mathematician Siméon-Denis Poisson and published in his 1837 book [128], though he didn't appear to grasp the importance of this discovery because only one page of this 428-page text is devoted to this distribution. More attention was given in 1898 by the Russian-German mathematician Ladislas von Bortkiewicz in his book [159], where he used the Poisson distribution to study a large number n of independent repetitions of Bernoulli trials with a very small success probability p. Specifically, he studied the number of Prussian cavalrymen (a large number) killed by horse kicks (unlikely events, independent of one another). Now the Poisson distribution is recognized as tremendously important, used for example to model the distribution of ER arrival times (large population, small probability any particular person makes an ER visit, typically one person's visit is independent of another's). We'll see some examples in a bit of detail. First, though, we'll explore some properties of the Poisson distribution, starting with the expected value and the variance.

Proposition 11.8.1. A λ-Poisson random variable X has expected value

$$\mathbb{E}(X) = \lambda \tag{11.43}$$

Proof. Use the expected value formula of Eq. (11.14). Then $\mathbb{E}(X) =$

$$\sum_{n=0}^{\infty} nP(X = n) = \sum_{n=1}^{\infty} n\frac{e^{-\lambda}\lambda^n}{n!} = e^{-\lambda}\sum_{n=1}^{\infty}\frac{n}{n!}\lambda^n = e^{-\lambda}\sum_{n=1}^{\infty}\frac{1}{(n-1)!}\lambda^n$$

$$= e^{-\lambda}\lambda\sum_{n=1}^{\infty}\frac{\lambda^{n-1}}{(n-1)!} = e^{-\lambda}\lambda\sum_{m=0}^{\infty}\frac{\lambda^m}{m!} = e^{-\lambda}\lambda e^{\lambda} = \lambda$$

The first equality, where the sum changes from initial index $n = 0$ to initial index $n = 1$, is because the $n = 0$ term contributes nothing to the sum. The third equality is simply cancellation a factor of n. The fifth equality is the change of the summation index from n to $m = n - 1$. $\square$

To compute the variance, skewness, and kurtosis, we'll need the second, third, and fourth moments. We'll compute these by the cancellation of factors from factorials. This is just algebra, but seeing the pattern can take some time, so we'll go carefully.

Proposition 11.8.2. The second, third, and fourth moments of the λ-Poisson random variable X are

$$\mathbb{E}(X^2) = \lambda^2 + \lambda, \ \mathbb{E}(X^3) = \lambda^3 + 3\lambda^2 + \lambda, \ \mathbb{E}(X^4) = \lambda^4 + 6\lambda^3 + 7\lambda^2 + \lambda$$

Proof. For each moment we need to express the coefficient of $e^{-\lambda}\lambda^n/n!$ as a sum of terms that will cancel factors of $n!$ in the denominator. The higher the moment, the more involved the algebra. First, $\mathbb{E}(X^2) =$

$$\sum_{n=0}^{\infty} n^2\frac{e^{-\lambda}\lambda^n}{n!} = e^{-\lambda}\sum_{n=1}^{\infty} n\frac{\lambda^n}{(n-1)!} = e^{-\lambda}\sum_{n=1}^{\infty}((n-1)+1)\frac{\lambda^n}{(n-1)!}$$

$$= e^{-\lambda}\left(\sum_{n=2}^{\infty}\frac{\lambda^n}{(n-2)!} + \sum_{n=1}^{\infty}\frac{\lambda^n}{(n-1)!}\right)$$

$$= e^{-\lambda}\left(\lambda^2\sum_{n=2}^{\infty}\frac{\lambda^{n-2}}{(n-2)!} + \lambda\sum_{n=1}^{\infty}\frac{\lambda^{n-1}}{(n-1)!}\right) = e^{-\lambda}\left(\lambda^2 e^{\lambda} + \lambda e^{\lambda}\right) = \lambda^2 + \lambda$$

where we've adjusted the summation indices to ignore terms that have a factor of 0.

On to the third moment.

$$\mathbb{E}(X^3) = \sum_{n=0}^{\infty} n^3 \frac{e^{-\lambda}\lambda^n}{n!} = e^{-\lambda} \sum_{n=1}^{\infty} n^2 \frac{\lambda^n}{(n-1)!}$$

$$= e^{\lambda} \sum_{n=1}^{\infty} ((n-1)+1)^2 \frac{\lambda^n}{(n-1)!}$$

$$= e^{\lambda} \sum_{n=1}^{\infty} ((n-1)^2 + 2(n-1) + 1) \frac{\lambda^n}{(n-1)!}$$

$$= e^{\lambda} \left(\sum_{n=2}^{\infty} (n-1)\frac{\lambda^n}{(n-2)!} + 2\sum_{n=2}^{\infty} \frac{\lambda^n}{(n-2)!} + \sum_{n=1}^{\infty} \frac{\lambda^n}{(n-1)!} \right)$$

$$= e^{\lambda} \left(\sum_{n=2}^{\infty} ((n-2)+1)\frac{\lambda^n}{(n-2)!} + 2\lambda^2\sum_{n=2}^{\infty} \frac{\lambda^{n-2}}{(n-2)!} + \lambda\sum_{n=1}^{\infty} \frac{\lambda^{n-1}}{(n-1)!} \right)$$

$$= e^{\lambda} \left(\lambda^3 \sum_{n=3}^{\infty} \frac{\lambda^{n-3}}{(n-3)!} + 3\lambda^2\sum_{n=2}^{\infty} \frac{\lambda^{n-2}}{(n-2)!} + \lambda\sum_{n=1}^{\infty} \frac{\lambda^{n-1}}{(n-1)!} \right)$$

$$= e^{\lambda} \left(\lambda^3 e^{\lambda} + 3\lambda^2 e^{\lambda} + \lambda e^{\lambda} \right) = \lambda^3 + 3\lambda^2 + \lambda$$

The argument for the fourth moment is similar. We just need to write n^4 as a sum of products of the factors n, $n-1$, $n-2$, and $n-3$. It's easy to verify that this works:

$$n^4 = n(n-1)(n-2)(n-3) + 6n(n-1)(n-2) + 7n(n-1) + n$$

Then we find $\mathbb{E}(X^4) =$

$$e^{-\lambda} \left(\sum_{n=4}^{\infty} \frac{\lambda^n}{(n-4)!} + 6\sum_{n=3}^{\infty} \frac{\lambda^n}{(n-3)!} + 7\sum_{n=2}^{\infty} \frac{\lambda^n}{(n-2)!} + \sum_{n=1}^{\infty} \frac{\lambda^n}{(n-1)!} \right)$$

$$= e^{-\lambda} \left(\lambda^4 \sum_{n=4}^{\infty} \frac{\lambda^{n-4}}{(n-4)!} + 6\lambda^3 \sum_{n=3}^{\infty} \frac{\lambda^{n-3}}{(n-3)!} + 7\lambda^2 \sum_{n=2}^{\infty} \frac{\lambda^{n-2}}{(n-2)!} \right.$$

$$\left. + \lambda\sum_{n=1}^{\infty} \frac{\lambda^{n-1}}{(n-1)!} \right)$$

$$= e^{-\lambda} \left(\lambda^4 e^{\lambda} + 6\lambda^3 e^{\lambda} + 7\lambda^2 e^{\lambda} + \lambda e^{\lambda} \right) = \lambda^4 + 6\lambda^3 + 7\lambda^2 + \lambda$$

This approach can be extended to compute any moments we want. $\square$

Now we can apply the expressions for these moments to compute variance, skewness, and kurtosis for Poisson distributions.

Proposition 11.8.3. The λ-Poisson random variable X has variance

$$\sigma^2(X) = \lambda \tag{11.44}$$

Proof. Use the variance formula of Eq. (11.17), $\sigma^2(X) = \mathbb{E}(X^2) - \mathbb{E}(X)^2$. Then substitute in the expressions for $\mathbb{E}(X)$ from Prop. 11.8.1 and for $\mathbb{E}(X^2)$ from Prop. 11.8.2. We find $\sigma^2(X) = (\lambda^2 + \lambda) - (\lambda)^2 = \lambda$ as claimed. $\square$

Proposition 11.8.4. The λ-Poisson random variable X has

$$\text{sk}(X) = \frac{1}{\sqrt{\lambda}} \qquad \text{kur}(X) = 3 + \frac{1}{\lambda} \tag{11.45}$$

Proof. We'll use the definitions of skewness Eq. (11.23) and of kurtosis Eq. (11.25), along with our calculations of the moments in Prop. 11.8.2. In fact, we've done all the hard work already when we computed the moments and variance:

$$\text{sk}(X) = \frac{1}{\sigma(X)^3}\left(\mathbb{E}(X^3) - 3\mathbb{E}(X^2)\mathbb{E}(X) + 2\mathbb{E}(X)^3\right)$$

$$= \frac{1}{\lambda^{3/2}}\left((\lambda^3 + 3\lambda^2 + \lambda) - 3(\lambda^2 + \lambda)\cdot\lambda + 2\lambda^3\right) = \frac{1}{\lambda^{3/2}}\lambda = \frac{1}{\sqrt{\lambda}}$$

where in the second line small brackets group the expressions for $\mathbb{E}(X^3)$ and $\mathbb{E}(X^2)$, and because $\sigma(X)^2 = \lambda$ by Prop. 11.8.3, $\sigma(X)^3 = \lambda^{3/2}$.

For the kurtosis we have

$$\text{kur}(X) = \frac{1}{\sigma(X)^4}\left(\mathbb{E}(X^4) - 4\mathbb{E}(X^3)\mathbb{E}(X) + 6\mathbb{E}(X^2)\mathbb{E}(X)^2 - 3\mathbb{E}(X)^4\right)$$

$$= \frac{1}{\lambda^2}\left((\lambda^4 + 6\lambda^3 + 7\lambda^2 + \lambda) - 4(\lambda^3 + 3\lambda^2 + \lambda)\cdot\lambda\right.$$

$$\left. + 6(\lambda^2 + \lambda)\cdot\lambda^2 - 3\lambda^4\right) = \frac{1}{\lambda^2}\left(3\lambda^2 + \lambda\right) = 3 + \frac{1}{\lambda}$$

Just simple algebra, after we've calculated the moments. $\square$

Next, we'll see a relation between the binomial distribution and the Poisson distribution that gives a more solid foundation for the claim that the distribution of many independent events, each of the same low probability, is Poisson.

Recall that binomial distributions are determined by two parameters, the number N of trials and the probability p of success of each trial. Suppose we have a sequence of binomial trials where $p_N = P(X = N)$ decreases as N increases, specifically so that Np_N is a constant, which we'll denote by λ. We'll see that

$$\lim_{N\to\infty}\binom{N}{k}p_N^k(1 - p_N)^{N-k} = \frac{e^{-\lambda}\lambda^k}{k!} \tag{11.46}$$

The left side is $\lim_{N\to\infty} P(X=k)$ for the binomial distribution with parameters N and p_N; the right is $P(X=k)$ for the Poisson distribution with parameter

$$\lambda = Np_N \tag{11.47}$$

This is the sense in which the Poisson distribution is the limit of binomial distributions. Let's see why Eq. (11.46) is true.

$$\binom{N}{k} p_N^k (1-p_N)^{N-k} = \frac{N!}{k!(N-k)!} \left(\frac{\lambda}{N}\right)^k \left(1-\frac{\lambda}{N}\right)^{N-k}$$

$$= \frac{\lambda^k}{k!} \left(1-\frac{\lambda}{N}\right)^N \frac{N!}{N^k(N-k)!} \left(1-\frac{\lambda}{N}\right)^{-k}$$

Now for any fixed k, $0 \le k \le N$ (the last inequality is no restriction, because for each k we let $N \to \infty$),

$$\frac{N!}{N^k(N-k)!} = \frac{N(N-1)\cdots(N-k+1)}{N^k}$$

$$= \frac{N}{N}\frac{N-1}{N}\frac{N-2}{N}\cdots\frac{N-(k-1)}{N}$$

$$= 1\cdot\left(1-\frac{1}{N}\right)\left(1-\frac{2}{N}\right)\cdots\left(1-\frac{k-1}{N}\right)$$

Then for each fixed k,

$$\lim_{N\to\infty} \frac{N!}{N^k(N-k)!} = \lim_{N\to\infty} \left(1-\frac{1}{N}\right)\left(1-\frac{2}{N}\right)\cdots\left(1-\frac{k-1}{N}\right) = 1$$

Recall that λ is constant. Then for each fixed k,

$$\lim_{N\to\infty} \left(1-\frac{\lambda}{N}\right)^{-k} = 1$$

And finally, recall the limit definition of the exponential

$$\lim_{N\to\infty} \left(1-\frac{\lambda}{N}\right)^N = e^{-\lambda}$$

Putting these pieces together we find

$$\lim_{N\to\infty} \binom{N}{k} p_N^k (1-p_N)^{N-k} = \frac{\lambda^k}{k!} e^{-\lambda}$$

which is what we set out to show.

The requirement $Np_N = \lambda$ is quite rigid and consequently unlikely to hold in any physical setting. However, Eq. (11.46) remains true when $Np_N = \lambda$ is replaced by $Np_N = \lambda_N$, provided that $\lim_{N\to\infty} \lambda_N = \lambda$. This does appear to be true in many circumstances.

Before we look at examples, we'll calculate the expected time between Poisson-distributed events, and explore the clustering of event times.

First, we must recognize that the Poisson distribution is discrete, k successes in N trials, but occurrence times are continuous variables. To accommodate continuous variables, we move from Poisson distributions to Poisson processes. So in fact to model ER arrival times we need Poisson processes, not distributions. Although they can occur in space as well as in time, initially we'll characterize Poisson processes in time:

1. events occur one at a time,
2. the numbers of events in disjoint time intervals are independent, and
3. the number of events on any time interval is Poisson distributed with parameter λ ($= \mathbb{E}$(number) by Prop. 11.8.1) proportional to the length of the interval.

Another reading of condition 3 is that λ is the expected number of events that occur in a unit time interval, so λt is the expected number of events that occur in a time interval of length t and we can apply Eq. (11.42) to deduce that for $k = 0, 1, 2, \ldots$

$$P(k \text{ events occur during a time } t) = e^{-\lambda t} \frac{(\lambda t)^k}{k!} \qquad (11.48)$$

We can use this to find the distribution of times between events. Denote by T the random variable measuring the time between successive events. Then by Eq. (11.48),

$$P(T > t) = P(k = 0 \text{ events occur in } [0, t]) = e^{-\lambda t} \frac{(\lambda t)^0}{0!} = e^{-\lambda t} \qquad (11.49)$$

and so

$$P(T \leq t) = 1 - e^{-\lambda t} \qquad (11.50)$$

Because the time between successive events cannot be negative, we have $P(T \leq t) = P(0 \leq T \leq t)$. Consequently,

$$1 - e^{-\lambda t} = P(0 \leq T \leq t) = \int_0^t f(s) \, ds \qquad (11.51)$$

where $f(s)$ is the probability density function for the distribution of inter-event times T. To find f, differentiate both sides of Eq. (11.51) and apply the fundamental theorem of calculus on the right side.

$$\lambda e^{-\lambda t} = \frac{d}{dt} \left(1 - e^{-\lambda t} \right) = \frac{d}{dt} \int_0^t f(s) \, ds = f(t)$$

We'll check the normalization condition

$$\int_0^\infty f(t) \, dt = \int_0^\infty \lambda e^{-\lambda t} \, dt = -e^{-\lambda t} \Big|_0^\infty = 1$$

Now that we know the probability density function $f(t) = \lambda e^{-\lambda t}$, we can compute the expected value for the interval between events $\mathbb{E}(T) =$

$$\int_0^\infty tf(t)\,dt = \int_0^\infty t\lambda e^{-\lambda t}\,dt = \lim_{S\to\infty}\left(-te^{-\lambda t} - \frac{1}{\lambda}e^{-\lambda t}\right)\Big|_0^S = \frac{1}{\lambda} \quad (11.52)$$

where the second equality follows from integration by parts with $u = \lambda t, dv = e^{-\lambda t}\,dt$.

Poisson-distributed events are memoryless. By this we mean that the distribution of inter-event time intervals T depends only on the length of the time interval investigated, not on the starting time of the interval. We can formulate this by conditional probabilities.

Proposition 11.8.5. The inter-event time intervals T for Poisson-distributed events are memoryless:

$$P(T > s+t \mid T > t) = P(T > s) \quad (11.53)$$

Proof. By the definition of conditional probability, Eq. (11.7),

$$P(T > s+t \mid T > t) = \frac{P(\{T > s+t\} \cap \{T > t\})}{P(T > t)} = \frac{P(T > s+t)}{P(T > t)}$$

where the last equality follows because $T > s+t$ implies $T > t$. With this we can restate Eq. (11.53) as

$$P(T > s+t) = P(T > s)P(T > t)$$

This is easy to show if we use Eq. (11.49):

$$P(T > s+t) = e^{-\lambda(s+t)} = e^{-\lambda s}e^{-\lambda t} = P(T > s)P(T > t)$$

which is what we wanted. □

The memoryless property has interesting implications. Here's one, called the *hitchhiker paradox*. On a road with little traffic, suppose the average time between cars is 10 minutes, and that the arrivals of the cars at a fixed location on the road are Poisson distributed. (This is a reasonable assumption, as long as the inter-event time is not too short.) Here's the paradox: no matter how long it has been since a car passed the point when the hitchhiker arrives, the average time the hitchhiker has to wait before the next car passes that point is 10 minutes.

Poisson-distributed events tend to cluster. In Fig. 11.11 we see 25 (top), 50 (middle), and 100 (bottom) Poisson-distributed events. The clustering is fairly easy to understand from what we know about the distribution of times between events. Start with Eq. (11.50):

Figure 11.11. Poisson events.

$$P(T \leq t) = 1 - e^{-\lambda t} = F(t)$$

Now $F'(t) = \lambda e^{-\lambda t}$ and so from the definition of the derivative, replacing the limit with an approximation for small δt,

$$\frac{P(T \leq t + \delta t) - P(T \leq t)}{\delta t} \approx F'(t) = \lambda e^{-\lambda t}$$

$$P(T \leq t + \delta t) - P(T \leq t) \approx \lambda e^{-\lambda t} \delta t$$

$$P(t < T \leq t + \delta t) \approx \lambda e^{-\lambda t} \delta t$$

where the third line comes from the second because

$$P(T \leq t + \delta t) = P(\{T \leq t\} \cup \{t < T \leq t + \delta t\})$$
$$= P(T \leq t) + P(t < T \leq t + \delta t) \qquad \text{by Eq. (11.5)}$$

Now $F''(t) = -\lambda^2 e^{-\lambda t}$ and so we see $F'(t)$ is monotonically decreasing with t, and because $t \geq 0$, the maximum of $F'(t)$ is $F'(0)$. Then for any fixed δt, the highest probability is for an inter-event time to lie in the interval $0 < T \leq \delta t$. That is, intervals between events are more likely to be small than large, and many small intervals between events mean the events appear to cluster.

Finally, here are brief mentions of a few medical applications of the Poisson distribution.

Example 11.8.1. *DMD and ASD.* In [178] Joyce Wu and her coworkers investigate a relationship between Duchenne muscular dystrophy (DMD) and autism spectrum disorder (ASD). They studied male patients at the Neuromuscular Center of Boston Children's Hospital, where almost all eastern Massachusetts children diagnosed with DMD go. Six of the 158 DMD patients also had ASD. In the general population, the prevalence of ASD is about 1.6 per 1000. The authors of the study asked if, given this prevalence, chance could account for the number of patients with both DMD and ASD? (They did proper statistical hypothesis testing. If you haven't already done so, you should take a statistics course to get a treatment more thorough than we'll give in Sect. 11.11.) How did they answer their question?

We'll assume that ASD patients are Poisson distributed. ASD has about the same low probability for all children, and one child's having ASD has little impact on another child's having it, so the Poisson assumption is reasonable. From Eq. (11.47) and $P_N = 1.6/1000$ we see $\lambda = 1000 \cdot (1.6/1000) = 1.6$. In order to compare populations of the same size, the authors tested the probability that in a population of 1000 DMD patients, at least 6 also would have ASD, numbers consistent with the observation. Denote by X the random variable representing the number of ASD patients. If DMD and ASD are unrelated, then the probability of finding at least 6 ASD cases in a population of 1000 DMD patients should be the same as the probability of finding at least 6 ASD cases in the general population. Now

$$P(X \geq 6) = 1 - P(X = 0) - \cdots - P(X = 5)$$
$$= 1 - \frac{e^{-1.6} \cdot 1.6^0}{0!} - \cdots - \frac{e^{-1.6} \cdot 1.6^5}{5!} \quad \text{by Eq. (11.42)}$$
$$= 1 - 0.994 = 0.006$$

The authors deduced that DMD and ASD likely are associated in some way. If this association can be understood, that may shed some light on the neurobiological basis of ASD. Certainly this is a worthwhile goal. □

Example 11.8.2. *Fiber-fiber contacts.* The Poisson distribution can model fiber-fiber contacts in random fiber networks. Biological examples include connective tissues (cartilage and tendons) and cellular cytoskeleta. A medical reason for studying these networks is that the bacteria *Listeria monocytogenes*, which produces infections that are debilitating and sometimes deadly, spreads by hijacking the fibrous cytoskeleton of the the invaded cell and causing that cell to grow a tail like a pollywog, called a comet tail by many authors. Wiggling the tail moves the infected cell to neighboring cells which then can be infected. See [81, 61] for details.

Pause a moment and appreciate the marvelous inventiveness of coevolution. Mutations of a bacterium enable it to control the actin machinery of the host cell to grow a tail and then repeatedly polymerize and depolymerize filaments in the cross-linked fiber network of the tail, causing it to wiggle and move the cell. And this is just one tiny bacterium. The web of life is far, far, far more complicated than Alexander von Humboldt [179] ever imagined, and he imagined Nature was wonderfully complex.

In this one little example we won't get anywhere near understanding how *Listeria monocytogenes* modifies a cell's fibrous network to grow a pollywog tail. We'll just understand a bit about the network. We'll look at the fibers in a thin slice, essentially 2-dimensional, and estimate the probability of finding a pinhole, that is, a point covered by no fibers. To do this we'll calculate the *coverage*, how many fibers cover a given point in the slice. Among the several ways to quantify this, we'll use

$$\text{coverage} = c = \frac{(\text{network mass/unit area}) \cdot (\text{fiber width})}{\text{fiber mass/unit length}} = \frac{m_n \cdot w}{m_f}$$

$$= \frac{\text{network mass/unit length}}{\text{fiber mass/unit length}} = \text{number of fibers}$$

Looking at a small enough patch of area, we see the pieces of fibers in this area are approximately straight segments, modeled by rectangles rather than line segments to account for their non-zero width. Each is positioned and oriented randomly, all independent of one another (in our simple model). If the population of fibers is low enough, then one fiber covering another has low probability and the distribution of coverings has a Poisson model.

Recall Eq. (11.42), that the Poisson distribution is determined by its parameter λ. But how can we find λ? Easy, once we recall Eq. (11.43). We denote by C the random variable for a measurement of c. Then the average value of the coverage is

$$\langle c \rangle = \left\langle \frac{m_n \cdot w}{m_f} \right\rangle = \frac{\langle m_n \rangle \cdot w}{m_f}$$

because w and m_f are constants. Then

$$P(C = n) = \frac{e^{-\langle c \rangle} \langle c \rangle^n}{n!}$$

So the probability of finding a pinhole is $P(C = 0) = e^{-\langle c \rangle}$. Many more examples, including extensions to 3 dimensions, can be found in [139]. □

Practice Problems

11.8.1. Suppose 4% of the students in a school have the flu.
(a) What is the probability that no student in a class of 25 has the flu?
(b) What is the probability that no student in a class of 35 has the flu?

11.8.2. In a town clinic the number of childbirths per day follows a Poisson distribution with $\lambda = 2.2$. (And yes, there are ways to test from the clinic's daily childbirth records if they follow a Poisson distribution, and if they do, to estimate λ.) Compute the probability that the clinic will see at least 3 childbirths today, given that it has already seen 1.

11.8.3. Suppose we have two Poisson processes, 1 and 2. For both, suppose T denotes the time between successive events.

(a) If $P_1(T \geq t) = 2 \cdot t \cdot P_2(T > t)$, what is the relation between the Poisson parameters λ_1 and λ_2?

(b) If $P_1(T \geq t) = 2^t \cdot P_2(T > t)$, what is the relation between the Poisson parameters λ_1 and λ_2?

Practice Problem Solutions

11.8.1. Using Eq. (11.42) where the random variable X represents the number of students with the flu:

$$P(X = n) = \frac{e^{-\lambda}\lambda^n}{n!} \text{ so } P(X = 0) = \frac{e^{-\lambda}\lambda^0}{0!} = e^{-\lambda}$$

All we need to do is find λ. For this, we'll use Eq. (11.47).

(a) For $N = 25$ and $p_N = 0.04$ (this is the probability of having the flu for the whole population, so holds for every group, assuming there is nothing special about membership in the group), we have $\lambda = N \cdot p_N = 25 \cdot 0.04 = 1$ and so

$$P(X = 0) = e^{-\lambda} = e^{-1} \approx 0.37$$

(b) For $N = 35$ and $p_N = 0.04$, we have $\lambda = N \cdot p_N = 35 \cdot 0.04 = 1.4$ and so

$$P(X = 0) = e^{-\lambda} = e^{-1.4} \approx 0.25$$

Now for both calculations we've used the Poisson distribution. As a check, we'll argue directly. The probability of any student's not having the flu is $1 - 0.04$. Assuming independence of flu cases, by Eq. (11.6) the probability of a group of N students not having the flu is $(1 - 0.04)^N$. So the direct solutions of this problem are

(a) for $N = 25$, $P(X = 0) = (1 - 0.04)^{25} \approx 0.36$

(b) for $N = 35$, $P(X = 0) = (1 - 0.04)^{35} \approx 0.24$

pretty close to the Poisson results.

11.8.2. Take the random variable X to represent the number of childbirths in a given day. We need to compute $P(X \geq 3 | X \geq 1)$. By the definition of conditional probability, Eq. (11.7), we have

$$P(X \geq 3 | X \geq 1) = \frac{P(\{X \geq 3\} \cap \{X \geq 1\})}{P(X \geq 1)} = \frac{P(X \geq 3)}{P(X \geq 1)}$$

because if $X \geq 3$, then automatically $X \geq 1$.

The number of childbirths in a day cannot be negative, so

$$P(X \geq 1) = 1 - P(X < 1) = 1 - P(X = 0)$$
$$P(X \geq 3) = 1 - P(X < 3) = 1 - (P(X = 0) + P(X = 1) + P(X = 2))$$

This last equation follows from Eq. (11.5) because in a given day the events $X = 0$, $X = 1$, and $X = 2$ are mutually exclusive.

The final step is to compute these probabilities by Eq. (11.42). For $k = 0, 1$, and 2 we have $P(X = k) = (e^{-2.2} \cdot 2.2^k)/k!$, so

$$P(X = 0) \approx 0.111, \quad P(X = 1) \approx 0.244, \quad \text{and} \quad P(X = 2) \approx 0.268$$

Combining these bits we find

$$P(X \geq 3 | X \geq 1) \approx \frac{1 - (0.111 + 0.244 + 0.268)}{1 - 0.111} \approx 0.424$$

11.8.3. We'll use Eq. (11.49) to relate the distribution of time between successive events to the Poisson parameter.

(a) Here we have

$$P_1(T \geq t) = 2 \cdot t \cdot P_2(T > t)$$
$$e^{-\lambda_1 t} = 2 \cdot t \cdot e^{-\lambda_2 t}$$
$$\ln(e^{-\lambda_1 t}) = \ln(2 \cdot t \cdot e^{-\lambda_2 t}) = \ln(2 \cdot t) + \ln(e^{-\lambda_2 t})$$
$$-\lambda_1 t = \ln(2 \cdot t) - \lambda_2 t$$
$$\lambda_1 = -\frac{\ln(2 \cdot t)}{t} + \lambda_2$$

not a terribly clean expression because it involves t, the lower bound on the time between events. Maybe (b) will give us a clearer result.

(b) Now the calculation is

$$P_1(T \geq t) = 2^t \cdot P_2(T > t)$$
$$e^{-\lambda_1 t} = 2^t \cdot e^{-\lambda_2 t}$$
$$\ln(e^{-\lambda_1 t}) = \ln(2^t \cdot e^{-\lambda_2 t}) = \ln(2^t) + \ln(e^{-\lambda_2 t})$$
$$-\lambda_1 t = t \cdot \ln(2) - \lambda_2 t$$
$$\lambda_1 = -\ln(2) + \lambda_2$$

A nicer result, don't you think?

Exercises

11.8.1. Suppose 5% of the students in a school have the flu. In a class of size $N = 25$,
(a) find the probability that exactly 1 student has the flu.
(b) Find the probability that $1, 2$, or 3 students have the flu.
(c) Find the probability that at least 5 students have the flu.

11.8.2. In a 25-block neighborhood with 5000 residents evenly distributed among the blocks, suppose 80 people have the flu.
(a) Find the probability that a given block has no cases of the flu.
(b) Find the probability that a given block has exactly one case of the flu.
Do these calculations using the Poisson distribution and also do them directly. (Hint: look at Practice Problem Solution 11.8.1.)

11.8.3. The occurrence of some gene mutations is well fit by a Poisson distribution. For a particular gene, suppose that Poisson distribution has $\lambda = 1.7$. Compute the probability that at least 4 mutations will occur, given we know at least 2 mutations have occurred.

11.8.4. In a rural county the number of reported cases of pneumonia per year follow a Poisson distribution with $\lambda = 5.2$. In a given year, suppose at least 3 cases of pneumonia have been reported. Find the probability that $4, 5$, or 6 cases will be reported that year.

11.8.5. Find λ for all Poisson distributions whose second moments equal their third moments. Remember, λ must be positive.

11.8.6. For Poisson distributions, find the minimum of $\text{sk}(X) + \text{kur}(X)$. Hint: write $\text{kur}(X)$ as a function of $\text{sk}(X)$.

11.8.7. In Example 11.8.1, let X denote the number of ASD cases per 1000 boys with DMD. Assume X is λ-Poisson with $\lambda = 1.6$. Find
(a) $P(X = 0)$, (b) $P(1 \leq X \leq 2)$, (c) $P(X \geq 3)$, and (d) $P(X \leq 5)$.

11.8.8. Suppose the random variable T represents the time between successive Poisson-distributed events. Show that the Poisson parameter λ is determined by

$$\lambda = -\frac{\ln(P(T \geq t))}{t}$$

11.8.9. For events that follow a Poisson distribution with $\lambda = 1.5$,

(a) find the probability of $k = 1, 2, 3$, and 4 events during a time $t = 1$,

(b) What is the effect of doubling the time period for observing k events? That is, compute the ratio

$$P(k \text{ events occur in time } 2t)/P(k \text{ events occur in time } t).$$

11.8.10. Recall the fiber network model of Example 11.8.2. Suppose $\langle c \rangle = 0.7$. Carrying four digits to the right of the decimal, find

(a) $P(C = 0)$,   (b) $P(C = 1)$,   (c) $P(2 \leq C \leq 4)$,   and (d) $P(C > 4)$.

(e) Find $\langle c \rangle$ so that $P(C = 1) = 0.1$.

11.9   THE NORMAL DISTRIBUTION

The normal distribution likely is the most widely used distribution in the natural and social sciences. Some authors credit the discovery to the French mathematician Abraham de Moivre in 1738, but Gauss was the first to understand its importance. (Google Karl Friedrich Gauss if you want to see how many parts of mathematics bear his thumbprint.) Gauss also gave the normal distribution its name to reflect his idea that it occurs often in applications. Some refer to it as the *Gaussian distribution*.

The normal distribution is a continuous probability distribution characterized by two numbers, μ and σ, and with density function

$$f(x) = \frac{1}{\sigma \sqrt{2\pi}} e^{-(x-\mu)^2/(2\sigma^2)} \tag{11.54}$$

In Sect. A.16 we'll work out the normalization factor $1/(\sigma \sqrt{2\pi})$. We'll need double integrals in polar coordinates (Sect. 17.3) to do this calculation.

The numbers μ and σ are the mean (expected value) and standard deviation (square root of the variance) of the normal distribution. The left side of Fig 11.12 shows plots of normal distributions with $\mu = 0$ and $\sigma = 1$ (narrower distribution) and $\sigma = 2$ (wider distribution). The other pictures show the $\mu = 0$, $\sigma = 1$ normal distribution, with (middle) one standard deviation on either side of the mean shaded and (right) two standard deviations on either side of the mean shaded.

The left side of Fig 11.12 illustrates how σ characterizes the spread of the normal density function. The smaller σ, the more tightly the density function is concentrated about the mean.

The normal density function has no antiderivative, but numerical integration gives $\int_{-\sigma}^{\sigma} f(x)\,dx \approx 0.6827$ and $\int_{-2\sigma}^{2\sigma} f(x)\,dx \approx 0.9545$. These results exemplify

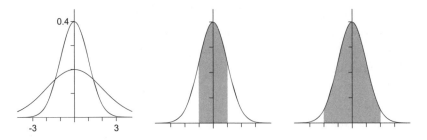

Figure 11.12. Some aspects of some normal distributions.

the rarity of events far from the mean. For instance, the probability of observing a value more than 10σ from the mean is about 10^{-23}. In Example 11.9.1 we'll see how to do this calculation. Some financial data exhibit 10σ events every few years, providing strong evidence that these do not follow the normal distribution.

We'll see that many natural processes are approximately normally distributed, so the normal distribution is important to hypothesis testing, a central avenue of applied statistics. In Sect. 11.11 we'll present a bit more detail. For now here's an illustration: assess the claim that a new anticancer biologic increases the 5-year survival rate of patients. The data are the mean 5-year survival rate of the control group and of the test group (the patients who get the new biologic), together with the standard deviations of both. The standard approach is to adopt the *null hypothesis*, that there is no difference of the means. If the groups' means differ, we use values of normal distributions to scale this difference in terms of standard deviations, and then assess how likely we are to see such a large difference. If this meets or exceeds an agreed upon standard, we reject the null hypothesis and so deduce the population means differ. That is, the biologic significantly alters the 5-year survival rate.

For some years now, computer algebra systems that can do numerical integration have been widely available. But before that, the most common way to find normal distribution probabilities was to look them up in a table. (When I was a student, the same was true for logarithms, trigonometric functions, square roots, and cube roots: use a table or use a slide rule. Google "slide rule" if the term is unfamiliar.) Because the normal distribution is determined by two parameters, mean and standard deviation, did we need a *Webster's Unabridged*-size-book of normal tables? No: if the values x_i follow a normal distribution with mean μ and standard deviation σ, then the

$$z_i = \frac{x_i - \mu}{\sigma} \qquad (11.55)$$

follow the normal distribution with mean 0 and standard deviation 1. Some statistics texts refer to the transformation $x_i \rightarrow z_i$ as "computing the z-score." This should recall the rescaling we did when computing the skewness (Eq. (11.22)) and the kurtosis (Eq. (11.24)). The reason that conversion to z-scores is so useful is this:

Proposition 11.9.1. If the random variable X is normally distributed with mean μ and standard deviation σ, then

$$P(X > a) = P\left(Z > \frac{a - \mu}{\sigma}\right) \tag{11.56}$$

where the random variable $Z = (X - \mu)/\sigma$.

Proof. This is just a change of variables $z = (x - \mu)/\sigma$, so $dx = \sigma\, dz$. Start with Eq. (11.54). Then

$$P(X > a) = \frac{1}{\sigma\sqrt{2\pi}} \int_a^\infty e^{-((x-\mu)/\sigma)^2/2}\, dx = \frac{1}{\sigma\sqrt{2\pi}} \int_{(a-\mu)/\sigma}^\infty e^{-z^2/2}\, \sigma\, dz$$

$$= \frac{1}{\sqrt{2\pi}} \int_{(a-\mu)/\sigma}^\infty e^{-z^2/2}\, dz = P\left(Z > \frac{a - \mu}{\sigma}\right)$$

and we're finished. $\square$

So any probability of any normal distribution can be computed by the normal distribution of z-scores. This is called the *standard normal distribution*.

Now we can compute the likelihood of observing a value more than 10 standard deviations from the mean.

Example 11.9.1. *Probability of a 10σ event.* The probability of a normally distributed random variable being more than 10σ from the mean μ is about 10^{-23}. We'll convert to z-scores and integrate numerically.

$$P(Z > 10) = \frac{1}{\sqrt{2\pi}} \int_{10}^\infty e^{-z^2/2}\, dz \approx \frac{1}{\sqrt{2\pi}} \int_{10}^{1000} e^{-z^2/2}\, dz \approx 7.62 \times 10^{-24}$$

By symmetry (or just do the integral if you don't like the symmetry argument),

$$P(Z < -10) \approx 7.62 \times 10^{-24}$$

and so $P(|Z| > 10) \approx 1.42 \times 10^{-23}$. That is, the probability that a normally distributed random variable is more than 10 standard deviations from the mean is about 10^{-23}. We call this *Avogadro unlikely.* $\square$

Next we'll compute the moments of the normal distribution. First, the expected value.

Proposition 11.9.2. The expected value of the normal density function Eq. (11.54) is μ.

Proof. For this we just evaluate an integral, through a couple of steps.

$$\mathbb{E}(X) = \int_{-\infty}^{\infty} xf(x)\, dx = \frac{1}{\sigma\sqrt{2\pi}} \int_{-\infty}^{\infty} xe^{-(x-\mu)^2/(2\sigma^2)}\, dx$$

Now substitute $w = (x-\mu)/(\sigma\sqrt{2})$, so $x = \sigma\sqrt{2}w + \mu$ and $dx = \sigma\sqrt{2}\, dw$.

$$\mathbb{E}(X) = \frac{1}{\sigma\sqrt{2\pi}} \int_{-\infty}^{\infty} (\sigma\sqrt{2}w + \mu)e^{-w^2}\sigma\sqrt{2}\, dw$$

$$= \frac{\sqrt{2}\sigma}{\sqrt{\pi}} \int_{-\infty}^{\infty} we^{-w^2}\, dw + \frac{\mu}{\sqrt{\pi}} \int_{-\infty}^{\infty} e^{-w^2}\, dw$$

$$= \frac{\sqrt{2}\sigma}{\sqrt{\pi}} \left(-e^{-w^2}\right)\Big|_{-\infty}^{\infty} + \frac{\mu}{\sqrt{\pi}}\sqrt{\pi} = \mu$$

where the result $\int_{-\infty}^{\infty} e^{-w^2}\, dw = \sqrt{\pi}$ is derived in Sect. A.16. □

To make the calculations less messy, we'll compute the second, third, and fourth moments after transforming the variables into z-scores. That is, we'll find the moments of the normal distribution with $\mu = 0$ and $\sigma^2 = 1$. We'll leave the more general case for Exercise 11.9.7.

Proposition 11.9.3. The normal distribution with $\mu = 0$ and $\sigma^2 = 1$ has these moments:

(a) $\mathbb{E}(X) = 0$, (b) $\mathbb{E}(X^2) = 1$, (c) $\mathbb{E}(X^3) = 0$, and (d) $\mathbb{E}(X^4) = 3$.

Proof. With $\mu = 0$ and $\sigma^2 = 1$, Eq. (11.54) becomes $f(x) = \frac{1}{\sqrt{2\pi}}e^{-x^2/2}$.
(a) We have $\mathbb{E}(X) =$

$$\int_{-\infty}^{\infty} xe^{-x^2/2}dx = \frac{1}{\sqrt{2\pi}} \int_{-\infty}^{\infty} xe^{-x^2/2}dx = \frac{1}{\sqrt{2\pi}}\left(-e^{-x^2/2}\right)\Big|_{-\infty}^{\infty} = 0$$

where we evaluated the integral by the substitution $w = -x^2/2$.
(b) For the second moment,

$$\mathbb{E}(X^2) = \frac{1}{\sqrt{2\pi}} \int_{-\infty}^{\infty} x^2 e^{-x^2/2}dx$$

integrate by parts with $u = x$ and $dv = xe^{-x^2/2}dx$

$$= \frac{1}{\sqrt{2\pi}}\left(-xe^{-x^2/2}\right)\Big|_{-\infty}^{\infty} + \frac{1}{\sqrt{2\pi}} \int_{-\infty}^{\infty} e^{-x^2/2}dx = 0 + 1 = 1$$

where the observation that $\int_{-\infty}^{\infty} e^{-x^2}\,dx = \sqrt{2\pi}$ is a special case of the result derived in Sect. A.16.

(c) For the third moment,

$$\mathbb{E}(X^3) = \frac{1}{\sqrt{2\pi}} \int_{-\infty}^{\infty} x^3 e^{-x^2/2}\,dx$$

substitute $w = x^2/2$ and integrate by parts

$$= \frac{1}{\sqrt{2\pi}} \left((-2 - x^2)e^{-x^2/2} \right)\Big|_{-\infty}^{\infty} = 0$$

because as $x \to \infty$ and as $x \to -\infty$, we know that $e^{-x^2/2}$ goes to 0 faster than any power of x goes to ∞. Apply l'Hôpital's rule if you've forgotten why this is true.

(d) For the fourth moment,

$$\mathbb{E}(X^4) = \frac{1}{\sqrt{2\pi}} \int_{-\infty}^{\infty} x^4 e^{-x^2/2}\,dx$$

integrate by parts with $u = x^3$ and $dv = xe^{-x^2/2}\,dx$

$$= \frac{1}{\sqrt{2\pi}} \left(-x^3 e^{-x^2/2}\Big|_{-\infty}^{\infty} + 3\int_{-\infty}^{\infty} x^2 e^{-x^2/2}\,dx \right)$$

integrate by parts again, now with $u = x$ and $dv = xe^{-x^2/2}\,dx$

$$= \frac{1}{\sqrt{2\pi}} \left((-x^3 - 3)e^{-x^2/2}\Big|_{-\infty}^{\infty} + 3\int_{-\infty}^{\infty} e^{-x^2/2}\,dx \right)$$

$$= \frac{1}{\sqrt{2\pi}} \left(0 + 3\sqrt{2\pi} \right) = 3$$

where we've used again the arguments at the ends of the proofs of parts (b) and (c). $\square$

Corollary 11.9.1. For the normal distribution with $\mu = 0$ and $\sigma^2 = 1$, $\mathrm{sk}(X) = 0$ and $\mathrm{kur}(X) = 3$.

Proof. Apply the results of Prop. 11.9.3 to the formulas for skewness, Eq. (11.23), and the formula for kurtosis, Eq. (11.25). $\square$

Now back to an earlier question: Why are so many measured variables normally distributed? The Central Limit theorem (CLT) is the key to our understanding of this phenomenon.

Theorem 11.9.1. Given measurements $X_1, X_2, \ldots, X_n$, all taken independently of one another and from identical distributions having finite mean μ and variance

σ^2, for large n the sum $X_1 + X_2 + \cdots + X_n$ is approximately normally distributed with mean $n\mu$ and variance $n\sigma^2$.

Details of the underlying distribution of the X_i are unimportant, so long as the hypotheses are satisfied. Variables of this sort are called *i.i.d.*, that is, independent, identically-distributed.

The independence hypothesis is essential, but identically distributed can be dropped, provided we modify the conclusion to this: $X_1 + X_2 + \cdots + X_n$ is approximately normally distributed with mean $\mu_1 + \mu_2 + \cdots + \mu_n$ and variance $\sigma_1^2 + \sigma_2^2 + \cdots + \sigma_n^2$, where μ_i and σ_i^2 are the mean and variance of X_i.

As far as we can tell, de Moivre didn't recognize the normal distribution as a separate entity, but rather viewed it as a quick way to approximate the binomial distribution. But we can compute binomial probabilities exactly, so why do we care about a normal approximation? Example 11.9.2 shows one reason.

Example 11.9.2. *Unbalanced coin outcomes.* Suppose a coin is slightly unbalanced, so the probability of heads is $p = 0.48$ and the probability of tails is $q = 1 - p = 0.52$. When this coin is tossed $N = 1000$ times, what is the probability of getting heads between 400 and 455 times?

We could calculate this probability using the binomial distribution

$$\binom{1000}{400} p^{400} q^{600} + \binom{1000}{401} p^{401} q^{599} + \cdots + \binom{1000}{455} p^{455} q^{545} \qquad (11.57)$$

Coding this would be simple, but can you imagine trying to do this calculation before computer algebra systems were common? The normal approximation to the binomial distribution gives a much easier approach.

The sum of the outcomes of the coin tosses has expected value (Eq. (11.29)) and variance (Eq. (11.32))

$$\mu = Np = 480 \quad \sigma^2 = Npq = 249.6$$

The sum of the coin toss outcomes satisfies the hypotheses of the CLT, so we'll compute the z-scores and use the standard normal distribution.

$$z_1 = \frac{399.5 - 480}{\sqrt{249.6}} \approx -5.095 \qquad z_2 = \frac{455.5 - 480}{\sqrt{249.6}} \approx -1.551$$

Then the probability of getting between 400 and 455 heads is

$$\frac{1}{\sqrt{2\pi}} \int_{-5.095}^{-1.551} e^{-x^2/2} dx \approx 0.0604$$

By the way, working out the sum from the expression (11.57) also gives a probability of 0.0604. □

Why 399.5 and 455.5 instead of 400.0 and 455.0? Remember, binomial variables are integer-valued while normal variables are real-valued. On the right we give a simple illustration of the consequence of mixing integer-valued and real-valued variables. To find $P(5 \leq X \leq 7)$ for a binomial random variable X, we want the area of the shaded rectangles on the left. The best match with the area under the curve is for $4.5 \leq X \leq 7.5$.

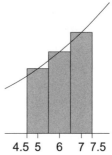

4.5 5   6   7 7.5

Now for an historical example with public health consequences.

Example 11.9.3. *Polio vaccines.* We'll use a real example: the Salk polio vaccine trial. Until the middle of the 20th century, polio was a terrifying disease. It attacked mostly children, and although the incidence was low, about 1 in 2000, the effects could be devastating: paralysis, the rest of one's life in a respirator, or death. Polio was high-profile, because President Franklin Roosevelt was a victim. So in the middle of the 20th century, developing a polio vaccine was a national priority. David Oshinsky's book [113] is excellent. Here we recount the tiniest fraction of this story.

Vaccines have two forms: live virus and killed virus. Live virus vaccines consist of closely related viruses that do not cause the disease. Edward Jenner's 1796 smallpox vaccine, for example, was an injection of live cow pox virus. The word "vaccine" is derived from the Latin word "vacca," which means "cow." Killed virus vaccines consist of the virus itself, killed by a treatment such as formaldehyde, but still able to cause the body to generate antibodies. In 1954 the Salk killed virus vaccine was ready for a test, but what sort of test?

Because polio was such a scary disease, you might expect the test would be to give the vaccine to a large group of kids and see if the number of cases of polio drops. But polio behaves as an epidemic and exhibits wild variability. Before any vaccines were deployed, in 1951 60,000 polio cases were reported, while in 1952 only 35,000 cases were reported. So the test design had to be more subtle. It was a double-blind test of about 400,000 kids. Half were randomly placed in the test group, those who got the Salk vaccine, while the other half got a placebo (an injection of saline). In the test group, 57 children contracted polio; in the control group, 142 got polio. Does this difference support the effectiveness of the vaccine?

Suppose the vaccine had no effect. (This is the null hypothesis.) Then the $57 + 142 = 199$ kids who will get polio are placed randomly into one of two groups, test and control, of about the same size, so the probability of placement into each group is $p = 0.5$, the same for each kid, and each kid's group selection is independent of all the others. These are Bernoulli trials, so by Eqs. (11.29) and (11.32), in each group the expected number of polio cases and the variance of that number are $Np = 199 \cdot 0.5 = 99.5$ and $Np(1-p) = 199 \cdot 0.5 \cdot 0.5 = 49.75$. Obviously, the mean and variance are finite (after all, the population is finite), so we can apply the CLT. Then the z-score (Eq. (11.55)) of the test group polio cases is

$$z = \frac{57 - Np}{\sqrt{Np(1-p)}} = \frac{57 - 199 \cdot 0.5}{\sqrt{199 \cdot 0.5 \cdot 0.5}} \approx -6.025$$

The probability of being more than 6 standard deviations from the mean is about 1 part in 10^9. How do we know this? Because $\int_{-100}^{-6} e^{-x^2/2} dx \approx 10^{-9}$ (we can ignore the $1/\sqrt{2\pi}$ factor), and from the point of view of integrating the normal density function, -100 is close enough to $-\infty$. The conclusion was that the Salk vaccine worked. A large-scale vaccination program began the next year, and polio was almost eliminated. The force of probabilistic reasoning propelled this program.

The story is a bit more complicated. Not long after the Salk vaccine was deployed, the Sabin live virus vaccine was given to all school kids. This vaccine was simpler to administer: a few drops of liquid on a sugar cube. Doctors or nurses weren't required. I recall lining up with my brother and sister in a junior high school in St. Albans, WV, to get our sugar cubes. So many people, kids with parents, in a line. Public service to stop a frightening disease. The atmosphere wasn't festive, like being in line to get into a carnival. Kids didn't run around and yell. People were happy, but quiet. Everyone, and I do mean *everyone*, knew they were doing something good. Curiosity is the very, very best part of us, but the drive to do good is a close second. □

Practice Problems

11.9.1. Suppose $N = 500$ Bernoulli trials give 125 successes. Based on this information, find (a) $P(75 \leq X \leq 100)$ and (b) $P(100 \leq X \leq 110)$.

11.9.2. In 10,000 tosses of a fair coin, how likely is ≤ 4700 heads?

Practice Problem Solutions

11.9.1. With the available information, $N = 500$ Bernoulli trials and 125 successes, our only estimate for the success probability comes from Eq. (11.29), $\mu = Np$, that is, $p = 125/500 = 0.25$. Then by Eq. (11.32), $\sigma^2(X) = Np(1-p) = 93.75$, and so $\sigma = \sqrt{93.75} \approx 9.682$.

(a) To find $P(75 \leq X \leq 100)$ we need the z-scores, z_1 and z_2, that correspond to $x = 74.5$ and to $x = 100.5$ Recall the reason for the additional half-steps when finding the normal bounds corresponding to binomial bounds in Example 11.9.2. These are

$$z_1 = \frac{74.5 - 125}{9.682} \approx -5.216 \qquad z_2 = \frac{100.5 - 125}{9.682} \approx -2.530$$

and the probability $P(75 \leq X \leq 100)$ is

$$P(z_1 \leq Z \leq z_2) = \frac{1}{\sqrt{2\pi}} \int_{-5.216}^{-2.530} e^{-z^2/2} \, dz \approx 0.0057$$

(b) For $P(100 \leq X \leq 110)$ the z-scores of the bounds are

$$z_1 = \frac{99.5 - 125}{9.682} \approx -2.634 \qquad z_2 = \frac{110.5 - 125}{9.682} \approx -1.498$$

and the probability $P(100 \leq X \leq 110)$ is

$$P(z_1 \leq Z \leq z_2) = \frac{1}{\sqrt{2\pi}} \int_{-2.634}^{-1.498} e^{-z^2/2} \, dz \approx 0.0628$$

11.9.2. By the CLT the sum of the outcomes $X_1 + X_2 + \cdots + X_{10000}$ is approximately normally distributed with mean (Eq. (11.29)) $N \cdot p = 10000 \cdot (1/2) = 5000$ and variance (Eq. (11.32)) $N \cdot p \cdot (1-p) = 2500$, so the standard deviation is $\sqrt{2500} = 50$. Consequently, 4700 heads is $300/50 = 6$ standard deviations from the mean. Without any calculation, we know this is a very unlikely event. To see just how unlikely, we compute the probability

$$P(X \leq 4700) = \frac{1}{\sqrt{2\pi}} \int_{-\infty}^{-6} e^{-x^2/2} dx \approx \frac{1}{\sqrt{2\pi}} \int_{-1000}^{-6} e^{-x^2/2} dx \approx 10^{-9}$$

That we will never ever see this is a safe bet.

Exercises

11.9.1. Suppose X is normally distributed with $\mu = 10$ and $\sigma = 5$.
(a) Compute $P(10 \leq X \leq 17.5)$.
(b) Use the symmetry of the normal distribution about its mean to compute $P(-\infty < X < 2.5)$ without evaluating an integral.

11.9.2. Suppose X is normally distributed with $\mu = 0$ and $\sigma^2 = 9$.
(a) Find $P(0 \leq X \leq 9)$.
(b) To within 0.001 find A with $P(9 \leq X \leq A) = 0.5 \cdot P(0 \leq X \leq 9)$.

11.9.3. Suppose X is normally distributed with mean 1. Find the standard deviation for which $P(X > 3) \approx 0.15866$.

11.9.4. Suppose X is normally distributed with $\mu = 2$ and $\sigma = 4$. Find the conditional probability $P(-1 \leq X \leq 3 \mid 2 \leq X \leq 4)$.

11.9.5. As reported in [143], the mean LDL cholesterol level is 4.3 mmol/L with a standard deviation of 0.6. Assuming LDL levels are normally distributed, how many people in the U.S. have LDL levels above 5.5? Assume the U.S. population is 320 million.

11.9.6. Suppose i.i.d. random variables X_i have mean $\mu = 5$ and standard deviation $\sigma = 2$. How large must n be in order for the sum $X_1 + X_2 + \cdots + X_n$ to have a standard deviation of about 10?

11.9.7. Compute the second, third, and fourth moments for the general normal density function Eq. (11.54), $f(x) = 1/(\sigma\sqrt{2\pi})e^{-(x-\mu)^2/(2\sigma^2)}$. Show the variance is σ^2, the skewness is 0, and the kurtosis is 3.

11.9.8. Suppose independent random variables X_i and Y_j have $\mu(X_i) = 2$, $\mu(Y_j) = 3$, $\sigma^2(X_i) = 4$, and $\sigma^3(Y_j) = 2$.
(a) How many combinations of X_i and Y_j have sums with mean 24?
(b) For each of these combinations, what are the variances of the sums?

11.9.9. Suppose independent random variables X_i and Y_j have $\mu(X_i) = 4$, $\mu(Y_j) = k$, $\sigma^2(X_i) = 2$, and $\sigma^3(Y_j) = 3$, for some positive constant k. Add together n copies of X_i and m copies of Y_j. Find n and m in terms of k so that $nX_i + mY_j$ has mean 10 and variance 15.

11.9.10. Suppose X is distributed by the standard normal distribution $P(a \leq X \leq b) = (1/\sqrt{2\pi})\int_a^b e^{-x^2/2}dx$. For each positive integer n, show $P(n \leq X \leq n+2) > P(n+1 \leq X \leq n+3)$. Hint: don't integrate. Look at the graph.

11.10 INFINITE MOMENTS

We've made a big point about the independence hypothesis of the Central Limit theorem, but have mostly ignored the other hypotheses: that the mean and variance of the distributions are finite. But surely this always is true, right? How could it not be? We'll see.

In preparation, we'll think about a game of chance. You toss a fair coin ten times and the house pays you $1 each time H comes up. Now of course the house is not a charity, so you must pay some fee to play the game. The expected value of the payout is

$$\mathbb{E}(X_1) + \cdots + \mathbb{E}(X_{10}) = \sum_{i=1}^{10} (0 \cdot P(X_i = T) + 1 \cdot P(X_i = H))$$

$$= \sum_{i=1}^{10} (0 \cdot (1/2) + 1 \cdot (1/2)) = 5$$

If the house charges anything amount $D > \$5$ to play, over the long term the house will win, though some individuals may win up to $10, giving a profit of $10 - D$. Seems easy enough.

So how about this game? You toss a fair coin until you get H. The house pays

$2	if the first H occurs on toss 1
$4	if the first H occurs on toss 2
$8	if the first H occurs on toss 3
$\cdots$	
2^n	if the first H occurs on toss n
$\cdots$	

As with the previous game, the house must charge at least the expected value of the winnings. That is, the house should charge at least

$$\mathbb{E}(X) = 2 \cdot P(\text{the first H occurs on toss 1})$$

$$+ 4 \cdot P(\text{the first H occurs on toss 2}) + \cdots$$

$$= 2 \cdot \frac{1}{2} + 4 \cdot \frac{1}{4} + 8 \cdot \frac{1}{8} + \cdots$$

This series diverges, so $\mathbb{E}(X) = \infty$. What should the house charge for a buy-in? The expected value calculation says an infinite amount. This is called the St. Petersburg Paradox, proposed by Niklaus Bernoulli in 1713 and with a resolution proposed by Daniel Bernoulli in 1738. The paradox is that most people aren't willing to pay even a modest amount to play the game. Bernoulli's approach was to consider the *utility* of money rather than the amount of money: 1000 means more to a poor person that it does to a wealthy person. (Okay, Bernoulli wrote about ducats instead of dollars, but the point is valid regardless of the units.) Bernoulli used the logarithm of the amount of money to represent utility. But other payoffs make the series of logs diverge, and although this problem is over three centuries old, it still is being discussed. Our point is not to explore this problem any further, but just to illustrate a concrete case of a random variable with infinite moments.

So far as we know, the first-studied probability distribution with infinite moments is the *Cauchy distribution*

$$f(x) = \frac{1}{\pi} \frac{1}{1+x^2}, \quad -\infty < x < \infty \tag{11.58}$$

Despite its name, this distribution was studied first by Poisson.

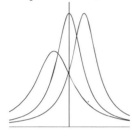

In the same way that the normal distribution is a family of distributions, each member of the family determined by two numbers, μ locating the peak of the distribution and σ^2 determining the width, the Cauchy distribution is a family determined by two parameters, a and b,

Figure 11.13. Three Cauchy distributions.

$$f_{a,b}(x) = \frac{1}{b\pi} \frac{1}{1+\left(\dfrac{x-a}{b}\right)^2} \tag{11.59}$$

In Fig. 11.13 we plot, left to right, $f_{-1,1.5}(x)$, $f_{0,1}(x)$, and $f_{1,1}(x)$. We see that the number a shifts the distribution left or right, while the number b makes it wider or narrower.

Here is the reason we study the Cauchy distribution in this section.

Proposition 11.10.1. All moments of the Cauchy distribution (11.58) are unbounded.

Proof. We'll be thorough with the first moment calculation, then sketch the main steps for the higher moments.

$$\mathbb{E}(X) = \frac{1}{\pi} \int_{-\infty}^{\infty} \frac{x}{1+x^2}\,dx = \frac{1}{2\pi} \ln(1+x^2)\Big|_{-\infty}^{\infty}$$

where the second equality is integration by substitution, $w = x^2 + 1$ and $dw = 2x\,dx$. Now you might think this gives $\mathbb{E}(X) = \infty - \infty = 0$, and some authors do say that the mean of the Cauchy distribution is 0 because the graph of $g(x) = xf(x)$ is antisymmetric: $g(-x) = -g(x)$, so for each $R > 0$,

$$\int_{-R}^{0} \frac{x}{1+x^2}\,dx = -\int_{0}^{R} \frac{x}{1+x^2}\,dx$$

But with improper integrals we must be more careful. If either of

$$\int_{0}^{\infty} \frac{x}{1+x^2}\,dx \quad \text{or} \quad \int_{-\infty}^{0} \frac{x}{1+x^2}\,dx$$

diverges, then the improper integral that defines $\mathbb{E}(X)$ is unbounded.

For each of the higher moments, to evaluate the integral we first divide the fraction (review Practice Problem 4.7.1 (b)) to obtain a polynomial plus either $1/(1+x^2)$ or $x/(1+x^2)$.

$$\mathbb{E}(X^2) = \frac{1}{\pi} \int_{-\infty}^{\infty} \frac{x^2}{1+x^2}\,dx = \frac{1}{\pi} \int_{-\infty}^{\infty} \left(1 - \frac{1}{1+x^2}\right)\,dx$$

$$= \frac{1}{\pi}\left(x - \tan^{-1}(x)\right)\Big|_{-\infty}^{\infty}$$

$$\mathbb{E}(X^3) = \frac{1}{\pi} \int_{-\infty}^{\infty} \frac{x^3}{1+x^2}\,dx = \frac{1}{\pi} \int_{-\infty}^{\infty} \left(x - \frac{x}{1+x^2}\right)\,dx$$

$$= \frac{1}{\pi}\left(\frac{x^2}{2} - \frac{1}{2}\ln\left(1+x^2\right)\right)\Big|_{-\infty}^{\infty}$$

$$\mathbb{E}(X^4) = \frac{1}{\pi} \int_{-\infty}^{\infty} \frac{x^4}{1+x^2}\,dx = \frac{1}{\pi} \int_{-\infty}^{\infty} \left(x^2 - 1 + \frac{1}{1+x^2}\right)\,dx$$

$$= \frac{1}{\pi}\left(\frac{x^3}{3} - x + \tan^{-1}(x)\right)\Big|_{-\infty}^{\infty}$$

All of these integrals diverge, so the moments are unbounded. The same techniques work, though with more tedious division of polynomials, for all (positive integer) moments. □

The *Lévy distribution* is a family, one for each $a > 0$, defined by

$$f_a(x) = \sqrt{\frac{a}{2\pi}} x^{-3/2} e^{-a/2x} \qquad (11.60)$$

for $0 < x < \infty$. In Fig. 11.14 we see plots of $f_a(x)$ for $a = 1, 2, 3, 4$, and 5, the higher a, the lower the maximum of f_a. The constant $\sqrt{a/(2\pi)}$ is the normalization factor. See Exercise 11.10.2 for the $a = 1$ calculation.

The first and second derivatives are

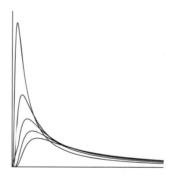

Figure 11.14. Lévy distributions for $a = 1, 2, 3, 4, 5$.

$$f_a'(x) = \sqrt{\frac{a}{2\pi}} \frac{e^{-a/2x}(a - 3x)}{2x^{7/2}}$$

$$f_a''(x) = \sqrt{\frac{a}{2\pi}} \frac{e^{-a/2x}(a^2 - 10ax + 15x^2)}{4x^{11/2}}$$

The maximum value of f_a occurs at $x = a/3$; the inflection points are at $x_\pm = a\left(5 \pm \sqrt{10}\right)/15$. The graph of f_a is concave up for $x < x_-$ and $x > x_+$, and concave down for $x_- < x < x_+$.

Like the Cauchy distribution, all the moments of the Lévy distribution diverge. To simplify the calculations, we'll present the main steps for the $a = 1$ distributions, but the result holds for all $a > 0$.

Proposition 11.10.2. The moments of the Lévy distribution diverge.

Proof. The first moment is

$$\mathbb{E}(X) = \frac{1}{\sqrt{2\pi}} \int_0^\infty x \cdot x^{-3/2} e^{-1/2x} \, dx = \frac{1}{\sqrt{2\pi}} \int_0^\infty x^{-1/2} e^{-1/2x} \, dx$$

Now integrate by parts with $u = e^{-1/2x}$ and $dv = x^{-1/2} \, dx$ to obtain

$$\mathbb{E}(X) = \frac{1}{\sqrt{2\pi}} \left(2x^{1/2} e^{-1/2x} \Big|_0^\infty - \int_0^\infty x^{-3/2} e^{-1/2x} \, dx \right)$$

The integral is the one we do to find the normalization factor in Exercise 11.10.2, that is, the integral is $\sqrt{2\pi}$. Now as $x \to \infty$, $e^{-1/2x} \to e^0 = 1$ and so $2x^{1/2} e^{-1/2x} = \infty$. This means $\mathbb{E}(X) = \infty$.

The other moments can be approached similarly, though the integrals get more complicated. Here's a simpler way. First observe that

$$x > 1 \text{ implies } 2x > 2 \text{ implies } 1/2x < 1/2 \text{ implies } -1/2x > -1/2$$

and so $e^{-1/2x} > e^{-1/2}$. Then for each positive x,

$$x^n \cdot x^{-3/2} e^{-1/2x} > x^n \cdot x^{-3/2} e^{-1/2} = x^{n-3/2} e^{-1/2}$$

and we're ready to show that the moments are unbounded. The second inequality below is an application of the lower bound $e^{-1/2x} > e^{-1/2}$ for $x > 1$ that we just found.

$$\mathbb{E}(X^n) = \frac{1}{\sqrt{2\pi}} \int_0^\infty x^n \cdot x^{-3/2} e^{-1/2x} \, dx > \frac{1}{\sqrt{2\pi}} \int_1^\infty x^n \cdot x^{-3/2} e^{-1/2x} \, dx$$

$$> \frac{1}{\sqrt{2\pi}} \int_1^\infty x^n \cdot x^{-3/2} e^{-1/2} \, dx = \frac{1}{\sqrt{2\pi}} e^{-1/2} \int_1^\infty x^{n-3/2} \, dx$$

Recalling Eq. (5.4), that $\int_1^\infty x^k \, dx$ diverges for $k \geq -1$, we see that $\mathbb{E}(X^n)$ diverges for $n - 3/2 > -1$, that is, for $n > 1/2$. So we've shown the first, second, third, fourth, …moments of the Lévy distribution diverge. □

The most straightforward approach to this proof was to evaluate the integrals, as we did for $\mathbb{E}(X)$. The $e^{-1/2x}$ factor makes these integrals challenging, or at least tedious. Our approach was to find a constant lower bound ($e^{-1/2}$) and show the integral diverges when the lower bound replaces $e^{-1/2x}$. When it can be applied, this is a good trick, and not so subtle that we can't figure it out for ourselves, especially after we've seen this example.

The Lévy distribution is interesting for reasons other than that its moments all diverge, although that has a very interesting consequence we'll mention now. Suppose that to ship an appropriate amount of this season's flu vaccine to a county health department, we need a good estimate of the mean age of county residents, and also the variance of these ages. The usual approach is to sample the population and measure the mean and variance of the sample. If adequate care is taken to obtain a random sample (don't sample only in kindergartens or only in retirement homes), then we expect that the sample parameters should give good estimates of the corresponding population parameters. Moreover, if we take larger samples, the estimates should be better. When it can be applied, the CLT supports these expectations. But if the population follows the Lévy distribution, as sample size increases the means need not settle down, but can wander all over the place, and the variances can grow without bound, interpreted within the confines of finite data sets, of course. For Lévy-distributed populations, we must rethink not just how we estimate population parameters, but what parameters we can estimate.

The tail, or tails, of a distribution are how the distribution behaves for large positive x, or for large $|x|$. The tails of the standard normal distribution decrease as $e^{-x^2/2}$, and they get small very, very rapidly. Remember that in Example 11.9.1 we saw that the probability of observing an event 10σ from the mean is about 10^{-23}.

In contrast, Lévy distributions have a power law tail. For all $a > 0$, we've seen that $\lim_{x \to \infty} e^{-a/2x} = 1$ and so for large x, $f_a(x) \approx x^{-3/2}$. The tail of the Lévy distribution is $P(X > c)$ for large c, and so

$$P(X > c) = \int_c^\infty f_a(x)\, dx \approx \sqrt{\frac{a}{2\pi}} \int_c^\infty x^{-3/2}\, dx$$

$$= \sqrt{\frac{a}{2\pi}} \left(-2x^{-1/2}\right)\Big|_c^\infty = \sqrt{\frac{a}{2\pi}}\, 2c^{-1/2}$$

That is, the tail of the Lévy distribution follows a power law with exponent $-1/2$.

A minor modification of this argument establishes a scaling relation for Lévy distributions. We see that for large c

$$P(X > k \cdot c) \approx \sqrt{\frac{a}{2\pi}}\, 2(k \cdot c)^{-1/2} = k^{-1/2} P(X > c)$$

Scaling occurs in many biological (and physical) systems, so Lévy distributions are good candidates for describing the underlying processes. And the unbounded moments do have some interesting consequences.

For example, in Prop. 11.9.3 we saw that the normal distribution has a kurtosis of 3, and by far the most widely used models of the stock market are based on

Brownian motion, which has differences that follow the normal distribution. Yet between 1970 and 2001, Wim Schoutens [146] reported the Standard and Poor's 500 hds a kurtosis of 43.36, not exactly in support of the Brownian model of stock markets, but consistent with an unbounded kurtosis seen through the window of a finite sample. Mandelbrot and Hudson [92] present an approach more directly based on scaling behavior.

For data sampled from a normal distribution, as the sample size increases, the sample mean and variance converge to the population mean and variance. So central is this belief that the hypothesis usually goes untested, unnoticed even. Because many biological systems exhibit scaling behavior, sample means and variances must be interpreted much more carefully.

Simple discrete data exhibit a power law distribution if the number, $N(n)$, of measurements of size $n = 1, 2, 3 \ldots$, is proportional to $n^{-\alpha}$. That is, $N(n) = k \cdot n^{-\alpha}$ for some positive constant k. Then the mean value is

$$\sum_{n=1}^{\infty} n \cdot (k \cdot n^{-\alpha})$$

and for $r = 1, 2, 3, \ldots$, the rth moment is

$$\sum_{n=1}^{\infty} n^r \cdot (k \cdot n^{-\alpha}) = k \sum_{n=1}^{\infty} \frac{1}{n^{\alpha-r}} \tag{11.61}$$

In Sect. 10.1 we saw that $\sum_{n=1}^{\infty} 1/n^p$ converges for $p > 1$ and diverges for $p \leq 1$. With this, we see

- for $1 < \alpha < 2$, the rth moment diverges for all $r \geq 1$;
- for $2 < \alpha < 3$, the rth moment converges for $r = 1$ and diverges for all $r \geq 2$;
- for $3 < \alpha < 4$, the rth moment converges for $r = 1, 2$ and diverges for all $r \geq 3$;
- and so on.

The continuous distribution version of this construction is explored in Exercise 11.10.9.

In Ch. 6 we saw that power laws often arise in biological systems. More examples can be found in [20].

Practice Problems

11.10.1. Show the tails of the Cauchy distribution $f_{a,b}(x)$ follow a power law with exponent -1.

11.10.2. Recall the Lévy distribution $f_a(x)$ has its maximum at $x = a/3$. Find $P(a/6 \leq X \leq a/2)$ for $a = 1, 2$, and 3.

11.10.3. Suppose the time T to death after exposure to anthrax is Lévy distributed with $a = 1.1$ in units of hours. If a large population is exposed to anthrax, what fraction of those exposed will survive at least two days? Carry two digits to the right of the decimal.

Practice Problem Solutions

11.10.1. For large x, we see $x - a \approx x$ and so $1 + ((x-a)/b)^2 \approx 1 + (x/b)^2 \approx (x/b)^2$. Then

$$f_{a,b}(x) = \frac{1}{b\pi}\frac{1}{1+\left(\dfrac{x-a}{b}\right)^2} \approx \frac{1}{b\pi}\frac{1}{\left(\dfrac{x}{b}\right)^2} = \frac{b}{\pi}\frac{1}{x^2}$$

and consequently, the right tail of the Cauchy distribution is

$$P(X > c) = \int_c^\infty f_{a,b}(x)\,dx \approx \int_c^\infty \frac{b}{\pi}\frac{1}{x^2}\,dx = \frac{b}{\pi}\frac{-1}{x}\bigg|_c^\infty = \frac{b}{\pi c}$$

The left tail is computed in the same way.

11.10.2. These probabilities are

$$P(a/6 \le X \le a/2) = \sqrt{\frac{a}{2\pi}}\int_{a/6}^{a/2} e^{-a/2x}x^{-3/2}\,dx$$

Evaluating these integrals numerically gives 0.142993 for $a = 1, 2$, and 3. This can't be a coincidence, so let's try to do the integral analytically. Substitute $z = \sqrt{a/x}$. Then $x^{-3/2}dx = -2z^{-1/2}dz$, and the limits $x = a/6$ and $x = a/2$ become $z = \sqrt{6}$ and $z = \sqrt{2}$. Then

$$\sqrt{\frac{a}{2\pi}}\int_{a/6}^{a/2} e^{-a/2x}x^{-3/2}\,dx = \sqrt{\frac{2}{\pi}}\int_{\sqrt{2}}^{\sqrt{6}} e^{-z^2/2}\,dz$$

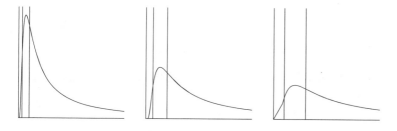

Figure 11.15. The graph of $f_a(x)$ with the lines $x = a/6$ and $x = a/2$, for $a = 1, 2$, and 3. See the solution of Practice Problem 11.10.2.

Because this integral does not depend on a, it is no surprise that the numerical integration gives the same answer for all values of a. In Fig. 11.15 we plot $f_a(x)$ and the lines $x = a/6$ and $x = a/2$. Note these lines are at $a/6 = a/3 - a/6$ and $a/2 = a/3 + a/6$, that is, placed an equal distance $(a/6)$ on either side of the x-coordinate $(a/3)$ of the maximum of the graph. We see that as a increases, the height of the maximum decreases, but the distance between the vertical lines increases, making plausible that the fraction of the total area under the curve that lies between these lines is the same. The proof is the integral we just evaluated by substitution.

11.10.3. Surviving at least two days means the time to death is at least 48 hours. So we must compute $P(T \geq 48)$.

$$P(T \geq 48) = \sqrt{\frac{a}{2\pi}} \int_{48}^{\infty} e^{-a/2t} t^{-3/2} \, dt = \sqrt{\frac{1.1}{2\pi}} \int_{48}^{\infty} e^{-1.1/2t} t^{-3/2} \, dt$$

$$\approx \sqrt{\frac{1.1}{2\pi}} \int_{48}^{1000000} e^{-0.55/t} t^{-3/2} \, dt \approx 0.12$$

That is, after two days about 12% of those exposed still are alive.

Exercises

11.10.1. Explain the factor of $1/\pi$ in Eq. (11.58), the Cauchy distribution.

11.10.2. Explain the $1/\sqrt{2\pi}$ factor for the $a = 1$ Lévy distribution. Hint: substitute $z = x^{-1/2}$ to evaluate $\int_0^{\infty} e^{-1/2x} x^{-3/2} \, dx$, and recall from the normal distribution that $\int_{-\infty}^{\infty} e^{-z^2/2} \, dz = \sqrt{2\pi}$.

11.10.3. For a Cauchy distributed random variable x, $-\infty < x < \infty$, find $P(0 \leq X \leq 1)$, $P(1 \leq X \leq 2)$, and $P(10 \leq X \leq 11)$. Do the last two numerically. Carry four digits to the right of the decimal.

11.10.4. For a Cauchy distributed random variable x, $-\infty < x < \infty$, find the value of T for which $P(0 \leq X \leq T) = 1/3$.

11.10.5. For the Lévy distribution, numerically compute the probabilities of (a) and (b). Carry four digits to the right of the decimal.
(a) $P(0 \leq X \leq 1)$ for $a = 1, 2$, and 3
(b) $P(1 \leq X \leq 2)$ for $a = 1, 2$, and 3
(c) Interpret the results of (a) and (b) by plotting $f_1(x)$, $f_2(x)$, and $f_3(x)$ for $0 \leq x \leq 2$.

(d) Estimate the smallest x beyond which $f_3(x) > f_2(x) > f_1(x)$. Carry four digits to the right of the decimal.

11.10.6. Suppose the viability (measured in hours) of a virus outside its host is Lévy distributed.
(a) Take $a = 1$. Estimate the probability that a virus can survive at least 10 hours outside its host. Carry two digits to the right of the decimal.
(b) Find the value of a for which the virus has a 10% probability of surviving at least 10 hours outside its host. Carry two digits to the right of the decimal.

11.10.7. Suppose the time (in hours) between exposure and the first appearance of symptoms of WNV (West Nile Virus) follows a Lévy distribution with $a = 0.95$.
(a) If you think you might have been exposed to WNV at noon on Monday but have not developed symptoms by noon on Tuesday, estimate the probability that you were not exposed at noon on Monday. Carry two digits to the right of the decimal. (Hint: estimate the probability that you could stay symptom-free for a day if you had been exposed.)
(b) Find the probability that you'll stay symptom-free for 48 hours if you were exposed.
(c) Suppose you stay symptom-free for three weeks. Find the probability that you were not exposed.

11.10.8. The point P lies a distance 1 from the vertical line L. At the point P a direction between $-\pi/2$ and $\pi/2$ is chosen uniformly randomly. Find the distribution of the heights y where the line from P in the chosen direction intersects L. Follow these steps.
(a) Because θ is uniformly distributed between $-\pi/2$ and $\pi/2$, show $P(\theta \le a) = 1/2 + a/\pi$.
(b) Next, show $P(Y \le y) = 1/2 + \tan^{-1}(y)/\pi$.
(c) Finally, show the function

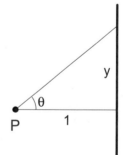

$$f(y) = \frac{d}{dy} P(Y \le y)$$

has the Cauchy distribution.

11.10.9. For constant $\alpha > 1$ and $1 \le x < \infty$,
(a) find the normalization factor k for $f(x) = k \cdot x^{-\alpha}$.

(b) For $n = 1, 2, 3, \ldots$, compute $\mathbb{E}(X^n)$ for the probability distribution f.

(c) If $1 < \alpha < 2$, show all the moments diverge.

(d) If $2 < \alpha < 3$, show the mean exists, but all higher moments diverge.

11.10.10. Find the power law exponent for the right tail of the distribution of Exercise 11.10.9. That is, show $P(X > c) \approx c^\beta$ for some β.

11.11 CLASSICAL HYPOTHESIS TESTS

We've mentioned hypothesis tests in Sects. 11.6 and 11.9. Here we'll say a bit more. Hypothesis tests share their general structure with that of experimental science: experiments can disprove hypotheses, but cannot prove them. The deflection of starlight by the sun's gravity field observed during the May 29, 1919, solar eclipse revealed that Newton's theory of gravity isn't correct. While this observation, and many others including the recent detection of gravity waves, support Einstein's theory of gravity, even all together these observations do not *prove* Einstein was right. Science is a process, roughly dialectical. Hypothesis, experiment, revision, another experiment, another revision, and so on. Right now no evidence suggests this ever will stop, but we just don't know.

Hypothesis tests follow the same outline: design an experiment to test some hypothesis. The possible interpretations of the experimental results are to contradict the hypothesis or not to contradict. With this in mind, some care is useful in the selection of the hypothesis. An example will explain this choice and illustrate the principles of hypothesis tests. We have several concepts to introduce. Easier to follow in the context of a medical example.

Example 11.11.1. *Flu symptom duration.* During last year's flu season, unvaccinated patients exhibited symptoms for a mean duration of $\mu = 5$ days. At the start of this year's flu season, a random sample of $n = 50$ unvaccinated patients exhibited symptoms for a mean duration $\bar{X} = 6$ days with a standard deviation of $s = 2$ days. Do these data suggest that this year's flu patients will exhibit symptoms for a mean $\mu > 5$ days?

First some notation. The random variable X represents the symptom duration of a person infected with the flu. The mean and standard deviation of the whole population are denoted by μ and σ; those of the sample by $\bar{X}$ and s. A consequence of random sampling is that the individuals X_i are selected independently of one another, and all are drawn from the same population, so we can apply the CLT (Theorem 11.9.1) to deduce that the sample mean $X_1 + \cdots + X_n$ is approximately normally distributed with mean $n\mu$ and variance $n\sigma^2$, if n is large enough. In practice, $n \geq 30$ is regarded as large enough to apply the CLT.

We'll take the null hypothesis H_0 to be $\mu = 5$ and the alternate hypothesis H_a to be $\mu \neq 5$. Assuming the null hypothesis is true, how likely are we to observe $\bar{X} = 6$?

Recall the computation of z-scores (Eq. (11.55)): for normally distributed variables, subtracting the mean and dividing by the standard deviation gives normally distributed variables with mean 0 and variance 1. Consequently,

$$\frac{X_1 + \cdots + X_n - n\mu}{\sigma\sqrt{n}} = \frac{(X_1 + \cdots + X_n)/n - \mu}{\sigma/\sqrt{n}} = \frac{\bar{X} - \mu}{\sigma/\sqrt{n}}$$

is normally distributed with mean 0 and variance 1. So we compute

$$Z = \frac{\bar{X} - \mu}{\sigma/\sqrt{n}} \approx \frac{\bar{X} - \mu}{s/\sqrt{n}} = \frac{6 - 5}{2/\sqrt{50}} \approx 3.54$$

where we have replaced the (unknown) population standard deviation σ with the (known) value s for the sample standard deviation, because only rarely do we know σ. Think about it: if our goal is to estimate the population mean, how likely are we to know σ? For large samples, s often is a good surrogate for σ. Alternately, we'll see that in circumstances where we have no clue about σ, we can use the test statistic $(\bar{X} - \mu)/(s/\sqrt{n})$, but it follows the t-distribution, not the normal distribution. □

Now the point is that if the sample mean $\bar{X}$ is far from the proposed population mean μ, we have some reason to doubt the proposed value of μ. But how far is "far enough"? The CLT implies that the sample means $\bar{X}$ are normally distributed around μ, so the values of Z follow the standard normal distribution, $\mu = 0$ and $\sigma = 1$. In Fig. 11.16 we see a plot of the standard normal distribution, more precisely, the part of

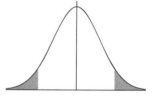

Figure 11.16. Normal distribution and rejection regions.

the plot with $-3 \leq Z \leq 3$. Each shaded region is 5% of the area under the curve. Then we are 90% certain that the mean of a random sample will not lie in the shaded regions. So if the mean of the sample does lie in one of these regions, we are 90% certain that the population mean is not μ, that is, we are 90% certain we can reject the null hypothesis. The shaded area is called the *rejection region*, and the area of the rejection region is denoted α and is called the *level of significance* of the test.

For these data we have $Z = 3.54$. What is the smallest value of α so that Z falls in the rejection region? We could look up values of the normal distribution in a table, but a line of Mathematica code gives the answer. The left-right symmetry of the normal distribution implies $P(Z \le -3.54) = P(Z \ge 3.54)$, so

$$\alpha = 2P(Z \ge 3.54) = 2\frac{1}{2\pi} \int_{3.54}^{\infty} e^{-x^2/2}\, dx \approx \frac{1}{\pi} \int_{3.54}^{100} e^{-x^2/2}\, dx \approx 0.0004$$

That is, assuming the null hypothesis is true, α, the probability of observing a sample mean of at least 6, is tiny. $\square$

Common terminology in hypothesis tests is to say that a *Type I error* is to reject a true null hypothesis, and a *Type II error* is to fail to reject a false null hypothesis. The value of α is the probability of a Type I error.

Hypothesis tests can detect differences in population means and variances, among many other parameters, though for some tests distributions other than the normal must be used. The too brief sketch of this section is the merest glimpse. A proper statistics course will fill these gaps.

Example 11.11.2. *Mean recovery times.* Two drugs, A and B, have been developed to treat a rare ailment. In a clinical trial, $n_A = 35$ patients are treated with drug A, and $n_B = 31$ patients with drug B. Patients in group A have a mean recovery time of $\bar{X}_A = 6$ days with variance $s_A^2 = 2$ days; patients in group B have a mean recovery time of $\bar{X}_B = 5$ days with variance $s_B^2 = 3$ days. With a level of significance $\alpha = 0.01$ do these data support the claim that drug B gives a quicker recovery than does drug A in the population? That is, do the data support $\mu_B < \mu_A$?

Take the null hypothesis H_0 to be $\mu_A - \mu_B \le 0$ and the alternate hypothesis H_a to be $\mu_A - \mu_B > 0$. (Remember, we want to reject H_0 in favor of H_a.) Suppose all the patients are selected independently from the same population so we can apply the CLT.

First, we need to determine the mean and variance of the difference of random variables. Straightforward variations of the proofs of Props. 11.6.1 (c) and 11.6.2 show that $\mathbb{E}(X - Y) = \mathbb{E}(X) - \mathbb{E}(Y)$ and $\sigma^2(X - Y) = \sigma^2(X) + \sigma^2(Y)$. That is, the mean of a difference is the difference of the means, and the variance of a difference is the sum of the variances. Then arguing as in Example 11.11.1, we see that the variance of the difference of these sample means is $s_A^2/n_A + s_B^2/n_B$ so the test statistic is

$$Z = \frac{\mu_A - \mu_B}{\sqrt{s_A^2/n_A + s_B^2/n_B}} = \frac{6-5}{\sqrt{2/35 + 3/31}} \approx 2.55$$

Then

$$\alpha = P(Z \ge 2.55) = \frac{1}{2\pi} \int_{2.55}^{\infty} e^{-x^2/2}\, dx \approx \frac{1}{2\pi} \int_{2.52}^{100} e^{-x^2/2}\, dx \approx 0.0054$$

So the data suggest we reject the null hypothesis with 99.46% confidence. That is, with this confidence the data support the claim that treatment B gives a quicker recovery. $\square$

Three other distributions are important in hypothesis tests: the χ^2-distributions, the t-distributions, and the F-distributions. We'll say a little about each.

The χ^2-distributions have a parameter, d, called the number of *degrees of freedom*. This is the distribution of the sum of the squares of d independent normal random variables, all with mean 0 and variance 1. The probability density function is

$$f_{\chi^2}(x) = \frac{1}{2^{d/2}\Gamma(d/2)} x^{d/2-1} e^{-x/2} \qquad \text{for } x > 0 \qquad (11.62)$$

where for positive integers n, $\Gamma(n) = (n-1)!$. For positive non-integer values x, $\Gamma(x)$ is more complicated to define (Google, if you're interested), but to compute χ^2 probabilities, Mathematica can do whatever we need: for example, Gamma[3.5] returns 3.32335. Computation of the appropriate moments gives

$$\mu = d, \ \sigma^2 = 2d, \ \text{sk} = \sqrt{\frac{8}{d}}, \ \text{kur} = \frac{12 + 3d}{d}$$

For large d the χ^2-distribution approaches a normal distribution (apply the CLT), but the values of skewness and kurtosis mean that this convergence can be slow. See Exercise 11.11.9.

The t-distributions also have a degrees of freedom parameter d. To study the statistics of small samples, these distributions were developed early in the 20th century by William Gosset, an employee of the Guinness brewery in Dublin. Gosset published his work under the pseudonym Student, the source of the name "Student's t-test." The probability density function is

$$f_t(x) = \frac{\Gamma((d+1)/2)}{\sqrt{d\pi}\,\Gamma(d/2)} \left(1 + \frac{x^2}{d}\right)^{-(d+1)/2} \qquad (11.63)$$

Computation of the appropriate moments gives

$$\mu = \begin{cases} 0 & \text{for } d > 1 \\ \text{undefined} & \text{for } d = 1 \end{cases} \qquad \sigma^2 = \begin{cases} \dfrac{d}{d-2} & \text{for } d > 2 \\ \infty & \text{for } d = 2 \\ \text{undefined} & \text{for } d = 1 \end{cases}$$

$$\text{sk} = \begin{cases} 0 & \text{for } d > 3 \\ \text{undefined} & \text{for } d = 1,2,3 \end{cases} \qquad \text{kur} = \begin{cases} \dfrac{3d-12}{d-4} & \text{for } d > 4 \\ \infty & \text{for } d = 3,4 \\ \text{undefined} & \text{for } d = 1,2 \end{cases}$$

The F-distributions are the distribution of the ratio of two independent χ^2 random variables, the numerator variable with d_1 degrees of freedom, the denominator with d_2. The probability density function is

$$f_F(x) = \frac{\Gamma((d_1 + d_2)/2)d_1^{d_1/2}d_2^{d_2/2}x^{(d_1/2)-1}}{\Gamma(d_1/2)\Gamma(d_2/2)(d_2 + d_1 x)^{(d_1+d_2)/2}}, \quad x > 0 \qquad (11.64)$$

From the moments of this density function we find

$$\mu = \frac{d_2}{d_2 - 2} \quad \text{for } d_2 > 2 \qquad \sigma^2 = \frac{2d_2^2(d_1 + d_2 - 2)}{d_1(d_2 - 2)^2(d_2 - 4)} \quad \text{for } d_2 > 4$$

$$\text{sk} = \frac{(2d_1 + d_2 - 2)\sqrt{8(d_2 - 4)}}{(d_2 - 6)\sqrt{d_1(d_1 + d_2 - 2)}} \quad \text{for } d_2 > 6$$

$$\text{kur} = 3 + 12\frac{d_1(5d_2 - 22)(d_1 + d_2 - 2) + (d_2 - 4)(d_2 - 2)^2}{d_1(d_2 - 6)(d_2 - 8)(d_1 + d_2 - 2)} \quad \text{for } d_2 > 8$$

In Fig. 11.17 we see plots of χ^2-, t-, and F-distributions for several degrees of freedom. The χ^2 for $d = 2$, 4, 6, and 10. As d increases, the χ^2-distribution becomes more left-right symmetric about its maximum (its mean) and also more widely spread. The t-distribution is plotted for $d = 2$, 10, 20, and 30, though only the first and last are labeled. Unlike the χ^2-distributions, as d increases, the t-distributions appear to converge. In fact, they converge to the standard normal distribution. For the F-distributions, we have held the numerator degrees of freedom constant at 10 and increased the denominator degrees of freedom. You can explore other combinations if you are interested. Will patterns be revealed?

Why do we need these distributions? The χ^2-distributions are used to test the variance of normally distributed populations. The t-distributions are used to test means of normally distributed populations with unknown variances, and also can be used for small samples. The F-distributions are used to test the ratio of variances of normally distributed populations.

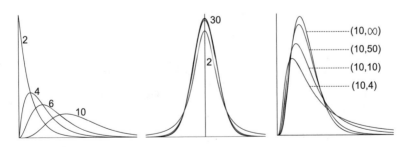

Figure 11.17. Some distributions: χ^2 (left), t (center), and F (right).

In more detail, suppose we guess that the variance of a normally distributed population is σ_0^2, and a random sample of size n has variance s^2. Then $(n-1)s^2/\sigma_0^2$ belongs to a χ^2-distribution with $n-1$ degrees of freedom. The null hypothesis is $H_0 : \sigma^2 = \sigma_0^2$ and the alternate is $H_a : \sigma^2 \neq \sigma_0^2$, $H_a : \sigma^2 < \sigma_0^2$, or $H_a : \sigma^2 > \sigma_0^2$, depending on the problem.

To test the value μ of the mean of a normally distributed population where the population variance is unknown, or when the sample has size $n < 30$, suppose $\bar{X}$ is the sample mean and s^2 is the sample variance. Then $(\bar{X} - \mu)/(s/\sqrt{n})$ belongs to a t-distribution with $n-1$ degrees of freedom.

Suppose independent random samples of sizes n_A and n_B are drawn from normally distributed populations, and the samples have variances s_A^2 and s_B^2. Then to test the null hypothesis $H_0 : \sigma_A^2 = \sigma_B^2$ that the population variances are equal, against the alternative $H_a : \sigma_A^2 \neq \sigma_B^2$, the ratio s_A^2/s_B^2 belongs to an F-distribution with $n_A - 1$ numerator degrees of freedom and $n_B - 1$ denominator degrees of freedom.

These are only a few from a larger collection of hypothesis tests. Not for the first time I'll mention that you should take a proper statistics course. Anyone interested in science, medicine, engineering, economics, law, political science, or history should take a statistics course.

Example 11.11.3. *Recovery time variances.* For a well-studied antibiotic the variance in the recovery times is $\sigma^2 = 1.5$. A new antibiotic, given to a sample of size $n_B = 20$, has a recovery time variance of $s_B^2 = 1.1$. To a level of significance $\alpha = 0.05$, do these data support the notion that recovery times with B have a smaller variance than those with drug A?

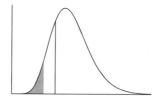

Figure 11.18. Rejection region for Example 11.11.3.

The null hypothesis is $H_0 : \sigma_B^2 \geq 1.5$ and the alternate is $H_a : \sigma_B^2 < 1.5$. Here we compute $Z_0 = (n_B - 1)s_B^2/\sigma^2 = 19 \cdot 1.1/1.5 \approx 13.9$ and test this against the χ^2-distribution with $n_B - 1 = 19$ degrees of freedom. In Fig. 11.18 we see a plot of f_{χ^2}, together with the rejection region for $\alpha = 0.05$. In order to reject H_0, the computed value of Z_0 must lie in the shaded region. But it doesn't: the rejection region is determined by $0 < Z \leq 10.12$, which we find by numerical integration of f_{χ^2} (Eq. (11.62)), and the vertical line is at $Z_0 = 13.9$. The available data are not sufficient to reject the null hypothesis at $\alpha = 0.05$. □

Estimating the variance is important because the smaller the variance, the more tightly sample values cluster about the population mean, and the better the population mean predicts most sample means. I'll mention an example from finance, not medicine. When I was a postdoc, one of my colleagues studied the mean lifetimes of small businesses in Ontario. He found no significant correlation between the mean lifetime and the business size, but he found the variance of the business lifetimes decreases with the size. That's a useful piece of information. He prepared a graph of the variance decrease with size as part of his report to government ministers. On previewing the report before the presentation, a colleague cautioned against showing the slide of the variance graph because, "the ministers don't like to see graphs that go down." So even though the ministers' dislikes were silly, the lesson is clear: keep your audience in mind when you present information. Your audience includes your patients.

Practice Problems

11.11.1. (a) In Example 11.11.3, find the largest value of s_B^2 that allows us to reject the hull hypothesis with $\alpha = 0.05$.
(b) Suppose we want to reject the null hypothesis with $\alpha = 0.01$. What's the largest s_B^2 to reject H_0 with this α?

11.11.2. Two new drugs, A and B, are tested on samples of size $n_A = 20$ and $n_B = 25$. The variances of the observed recovery times are $s_A^2 = 60$ and $s_B^2 = 135$. To $\alpha = 0.05$ do these data suggest that the population variances satisfy $\sigma_A^2 < \sigma_B^2$?

Practice Problem Solutions

11.11.1. (a) In Fig. 11.18 the rejection region is $0 < Z \le 10.12$, so we need to solve

$$10.12 = Z_0 = \frac{(n_B - 1)s_B^2}{\sigma^2} = \frac{19s_B^2}{1.5}$$

for s_B^2. This gives $s_B^2 \approx 0.8$.
(b) First we need to find the Z_0 that gives $\int_0^{Z_0} f_{\chi^2}(x)\,dx = 0.01$. Experiments with numerical integrals give $Z_0 \approx 7.63$. Then $7.63 = 19s_B^2/1.5$ gives $s_B^2 \approx 0.6$. Note that s_B^2 is proportional to Z_0.

11.11.2. We use the F-distribution with $n_A - 1 = 19$ numerator degrees of freedom and $n_B - 1 = 24$ denominator degrees of freedom. To find the Z_0 value that bounds the rejection region (on the left side of the F-distribution), we evaluate $\int_0^{Z_0} f_F(x)\,dx = 0.05$ and obtain $Z_0 \approx 0.47$. Because $s_A^2/s_B^2 \approx 0.44$, the data support rejection of the null hypothesis. That is, the recovery time for patients who take drug B has a higher variance than that for patients who take drug A.

Exercises

11.11.1. In Example 11.11.1 suppose $\bar{X} = 5.5$ and all other numbers are as in that example. How does the value of α change?

11.11.2. In Example 11.11.2 suppose the variances are $\sigma_A^2 = 2^2$ and $\sigma_B^2 = 3^2$. How does the value of α change?

11.11.3. For a sample of size $n = 20$, what is the largest the sample variance s^2 can be to conclude that the population variance $\sigma^2 < 2$ with a level of significance $\alpha = 0.01$?

11.11.4. Suppose a sample of size $n = 20$ is drawn from a normally distributed population whose variance we do not know. If the sample variance is $s^2 = 3$, does a sample mean $\bar{X} = 10$ support the claim that the population mean $\mu > 8$ at the $\alpha = 0.01$ level of significance?

11.11.5. Continuing Exercise 11.11.4, if the sample variance is $s^2 = 4$, does a sample mean $\bar{X} = 10$ support the claim that the population mean $\mu > 8$ at the $\alpha = 0.01$ level of significance?

11.11.6. To what α-level does an $n = 20$ sample with sample variance $s^2 = 5$ suggest the population variance $\sigma^2 < 8$?

11.11.7. (a) For two drug treatments A and B, samples of size $n_A = 10$ and $n_B = 15$ have sample variances $s_A^2 = 5$ and $s_B^2 = 9$. To $\alpha = 0.05$ do these data support $\sigma_A^2 < \sigma_B^2$?
(b) Find the smallest α for which these data do support the rejection of the null hypothesis.

11.11.8. Show that as d, the number of degrees of freedom of a t-distribution, becomes very large, the mean, variance, skewness, and kurtosis approach those of the standard normal distribution.

11.11.9. Show that as d, the number of degrees of freedom of a χ^2-distribution, becomes very large, the skewness and kurtosis approach those of the normal distribution. What happens to the mean and variance?

11.11.10. Find an article in a medical journal that reports hypothesis tests and comment on the experimental design and the interpretation of the results.

11.12 BAYESIAN INFERENCE

In Sect. 11.1 we introduced probability calculations through the relative frequency interpretation. This is called the *frequentist* approach to probability. It is based on such common-sense examples that you may be surprised to learn that there is another approach. In part this surprise comes from how combinatorial probabilities are introduced: repeated coin tosses, repeated dice rolls. The point is that the experiment is repeated many times. (Recall that repetition is a condition to apply the CLT.) But medicine has many situations in which repetition is uncommon. Clinical trials for rare diseases necessitate small samples. Further, the complex interactions of body subsystems, and of individuals in a population, can cast doubt on our ability to control every factor except the one being tested. This can make problematic both the independent and the identically distributed hypotheses of the CLT. Standard hypothesis tests do not apply in some situations of interest.

Bayesian inference can handle small samples. Rather than positing a fixed distribution from which samples are drawn, the Bayesian approach updates the distribution with each set of data. Relative frequencies are replaced with likelihoods. A good introduction is [13].

Here's a sketch of a criticism of the frequentist approach. Suppose 5 tosses of a coin give 5 heads. Then the relative frequency of tails is $0/5 = 0$, not likely to be a good predictor of future behavior. The small sample—five tosses—is one problem. Another is how to use prior information, the story of how the coin's history may influence its motion. Bayesian inference uses prior information and current experiments to estimate probabilities. And the process can be iterated. The new estimates can be incorporated into the prior information, and then combined with the next round of experiments.

Bayesian inference is a way to use experiment or observation, even a very small size data set, to adjust previous assessments of the probability distribution. The probability based on the previous experiments is called the *prior distribution*. The *likelihood* is the probability of obtaining the data of the current experiment, given the prior distribution. The adjusted assessment, based on the prior distribution and the current experiment, is called the *posterior distribution*. We'll see one example with a discrete distribution and one with a continuous distribution.

Example 11.12.1. *Recovery times.* Suppose we want to estimate the proportion p of patients who receive a new therapy for a rare infection that recover in half the time as untreated patients. In a test of 5 patients, 3 saw the recovery time halved (call these "successes"), 2 did not (call these "failures"). Denote these data $(3,2)$. What can we conclude?

Suppose we estimate the proportion p only coarsely: p can be $p_1 = 0$, $p_2 = 0.2$, $p_3 = 0.04$, $p_4 = 0.06$, $p_5 = 0.8$, or $p_6 = 1$. Based on experience with other therapies, we take the prior distribution

$$P(p_1) = 0.1, \ P(p_2) = 0.1, \ P(p_3) = 0.3,$$
$$P(p_4) = 0.3, \ P(p_5) = 0.1, \ P(p_6) = 0.1 \tag{11.65}$$

With this distribution the expected value of p is

$$\mathbb{E}(p) = P(p_1) \cdot p_1 + \cdots + P(p_6) \cdot p_6 = 0.5$$

To find how the data $(3,2)$ alter the prior distribution, we compute the likelihoods $P((3,2)|p_i) = p_i^3 (1 - p_i)^2$. Here we have assumed independence of the five additional samples, so by (11.6) the probability of 3 successes is p_i^3 and that of 2 failures is $(1 - p_i)^2$. For $i = 1, \ldots, 6$ we get

$$P((3,2)|p_i) = 0, \ 0.0051, \ 0.0230, \ 0.0345, \ 0.0205, \ 0$$

The posterior distribution is the adjustment of $P(p_i)$ based on the data $(3,2)$, that is, $P(p_i|(3,2))$. To find this, apply Bayes' theorem (11.10)

$$P(p_i|(3,2)) = \frac{P((3,2)|p_i) \cdot P(p_i)}{P((3,2))} = \frac{\text{likelihood} \cdot \text{prior distribution}}{\text{probability of the data}}$$

We compute the denominator by the law of conditioned probabilities (11.9)

$$P((3,2)) = P((3,2)|p_1) \cdot P(p_1) + \cdots + P((3,2)|p_6) \cdot P(p_6) = 0.0198$$

Then we see that the posterior distribution is

$$P(p_i|(3,2)) = 0.0, \ 0.0258, \ 0.348, \ 0.524, \ 0.104, \ 0.0 \tag{11.66}$$

These are a bit different from the prior distribution (11.65), and when we use these probabilities to compute the expected value, we get $\mathbb{E}(p) = 0.540$, a bit higher than the value we got with the prior distribution, but not much higher. Of course, the data give only weak support of the success of the treatment. We'll continue this example in Exercises 11.12.1, 11.12.2, and 11.12.3. $\square$

For Bayesian inference with continuous variables representing proportions (of successful trials), often a β-distribution is a good model for the prior distribution. For positive constants a and b, the $\beta(a,b)$-distribution has probability density function (pdf)

$$\beta(a,b)(x) = \frac{\Gamma(a+b)}{\Gamma(a)\Gamma(b)}x^{a-1}(1-x)^{b-1} \quad \text{for } 0 \le x \le 1 \qquad (11.67)$$

The Γ factors are for normalization, because

$$\int_0^1 x^{a-1}(1-x)^{b-1}\,dx = \frac{\Gamma(a)\Gamma(b)}{\Gamma(a+b)}$$

(As the presence of the gamma functions suggests, this is not an easy integral. Better just look up the answer.) Some examples of $\beta(a,b)$ are plotted in Fig. 11.19. In all three graphs, the x-range is $0 \le x \le 1$; for the first and third graphs the y-range is $0 \le y \le 3$, for the second $0 \le y \le 2$.

The plots are for these (a,b):

plot	solid graph	wide dash	narrow dash
first	(1,1)	(1,2)	(1,3)
second	(2,1)	(2,2)	(2,3)
third	(3,1)	(3,2)	(3,3)

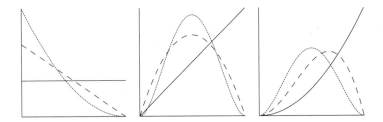

Figure 11.19. Plots of some $\beta(a,b)$.

Not every prior distribution can be fit with a β-distribution. For instance, every β-distribution has at most one local maximum for $0 \le x \le 1$, so distributions with two local peaks cannot be matched.

A random variable X that follows the $\beta(a,b)$-distribution has

$$\mathbb{E}(X) = \frac{a}{a+b} \qquad \sigma^2(X) = \frac{ab}{(a+b)^2(a+b+1)}$$

$$\mathrm{sk}(X) = \frac{2(b-a)\sqrt{a+b+1}}{(a+b)\sqrt{ab}}$$

$$\mathrm{kur}(X) = 3 + 6\frac{(a-b)^2(a+b+1) - ab(a+b+2)}{ab(a+b+2)(a+b+3)}$$

The expected value calculation is known as *Laplace's rule of succession.*

As with any continuous distribution, probabilities are computed by integration. For instance, suppose a random variable X follows the $\beta(4,3)$-distribution. Then integration gives

$$P(0.5 \leq X \leq 0.75) = \int_{0.5}^{0.75} \beta(4,3)(x) \, dx$$

$$= \frac{\Gamma(7)}{\Gamma(4)\Gamma(3)} \int_{0.5}^{0.75} x^3(1-x)^2 \, dx \approx 0.4868$$

Part of the utility of β-distributions in Bayesian inference is the ease of determining the posterior distribution. Here's how: suppose the prior distribution is well fit by $\beta(a,b) = x^{a-1}(1-x)^{b-1}$ and that the new data are s successes and f failures. Assume the prior model has success proportion x, so consequently failure proportion $1-x$, and assume the events of the new data are independent of one another. The likelihood of observing s successes and f failures is $x^s(1-x)^f$. Then by Bayes' theorem, the posterior distribution is the product of the likelihood $x^s(1-x)^f$ and the prior distribution $\beta(a,b) = x^{a-1}(1-x)^{b-1}$, that is,

$$x^s(1-x)^f \cdot x^{a-1}(1-x)^{b-1} = x^{a+s-1}(1-x)^{b+f-1} = \beta(a+s,b+f)$$

This is quite a bit simpler than the calculations of Example 11.12.1. But not so fast, some work still must be done. Specifically, how do we find if the $\beta(a,b)$-distribution is a good match for observed data? Part of this problem is to find the right parameters a and b. That's a bit far afield for us. Google is your friend here.

Example 11.12.2. *Drug success proportions.* Suppose the proportion p of successes in a new drug trial is modeled by the $\beta(5,7)$-distribution. Compute $P(p > 0.75)$. Suppose the next test had $s = 3$ successes and $f = 1$ failure. Find the posterior distribution, and with this new distribution compute $P(p > 0.75)$. How does this additional data change the expected value?

With the $\beta(5,7)$-distribution,

$$P(p > 0.75) = \int_{0.75}^{1} \beta(5,7) \, dx = \frac{\Gamma(12)}{\Gamma(5)\Gamma(7)} \int_{0.75}^{1} x^4(1-x)^6 \, dx \approx 0.00756$$

The additional data convert the distribution to $\beta(5+2,7+1)$ and so

$$P(p > 0.75) = \int_{0.75}^{1} \beta(7,8) \, dx = \frac{\Gamma(15)}{\Gamma(7)\Gamma(8)} \int_{0.75}^{1} x^6(1-x)^7 \, dx \approx 0.01031$$

The expected value changes from $5/(5+7) \approx 0.4167$ to $7/(7+8) \approx 0.4667$. It is little surprise that the expected value increases because in the additional data successes outnumber failures. The small size of the increase may be a surprise. $\square$

Although Bayesian inference has not yet been fully incorporated into medical research, much progress has occurred. See [14, 23, 58] and the epigenetics part of Sect. 15.5.2 for some examples.

Practice Problems

11.12.1. Suppose a new antiviral drug has been developed to treat geriatric pneumonia, and suppose the experimental data are coarse-grained so that in each test group the proportion that recover can take on ten values: $p_1 = 0.0, p_2 = 0.1, \ldots, p_{10} = 0.9$. (We leave out $p_{11} = 1.0$ because we've never observed 100% recovery.) Take the prior distribution to be

$$P(p_i) = 0.05 \text{ for } i = 1, 2, 3, 8, 9, 10, \ P(p_i) = 0.1 \text{ for } i = 4, 5,$$
$$P(p_6) = 0.2, \text{ and } P(p_7) = 0.3$$

(a) Compute $\mathbb{E}(p)$ for the prior distribution.
(b) Suppose in the next sample of 5 who receive this treatment, 4 recover and one does not. Compute the posterior distribution.
(c) Compute $\mathbb{E}(p)$ for the posterior distribution.

11.12.2. Suppose the proportion of successes in a clinical trial is modeled by $\beta(3.5, 5)$.
(a) With this distribution, estimate the expected value $\mathbb{E}(p)$.
(b) Another experiment produces 1 failure and a successes. Find the smallest (integer) a for which the posterior distribution has expected value at least 0.5.

Practice Problem Solutions

11.12.1. (a) This is a simple computation

$$\mathbb{E}(p) = P(p_1) \cdot P(p_1) + \cdots + P(p_{10}) \cdot P(p_{10}) = 0.485$$

(b) For each proportion p_i, the likelihood of 4 recovered and 1 not is $P((4,1)|p_i) = p_i^4(1 - p_i)$. For $i = 1, \ldots, 10$ the likelihood values are

0, 0.00009, 0.00128, 0.00567, 0.01536, 0.03125,

0.05184, 0.07203, 0.08192, 0.06561

The posterior distribution is $P(p_i|(4,1)) = P((4,1)|p_i) \cdot P(p_i)/P((4,1))$ where we write $P(p_i)$ for $P(p = p_i)$. We compute the denominator by

$$P((4,1)) = P((4,1)|p_1) \cdot P(p_1) + \cdots P((4,1)|p_{10}) \cdot P(p_{10})$$
$$= 0 \cdot 0.05 + 0.00009 \cdot 0.05 + \cdots + 0.06561 \cdot 0.05 \approx 0.03495$$

With this we compute the posterior distribution $P(p_i|(4,1))$ for $i = 1,2,\ldots,10$:

0.0, 0.0001288, 0.001831, 0.01622, 0.04395, 0.1788 0.4450,

0.1030, 0.1172, 0.09386

(c) With the posterior distribution, the expected value is

$$\mathbb{E}(p) = P(p_1|(4,1)) \cdot p_1 + \cdots + P(p_{10}|(4,1)) \cdot p_{10} \approx 0.6296$$

This is noticeably higher than the expected value calculated with the prior distribution, but then the additional data favored a higher recovery proportion.

11.12.2. (a) The expected value is $\int_0^1 x \cdot \beta(3.5,5)(x)\, dx \approx 0.412$.

(b) Here we evaluate $\int_0^1 x \cdot \beta(3.5 + a, 5 + 1)(x)\, dx$. For $a = 2$ the integral is about 0.4783; for $a = 3$ the integral is 0.52, so $a = 3$ is the answer. Note we can get these answers from $\mathbb{E}(X) = a/(a+b)$.

Exercises

11.12.1. In the set-up of Example 11.12.1, suppose the prior distribution is $P(p_i) = 1/6$ for $i = 1,\ldots,6$. Suppose a test of 5 patients had 3 successes and 2 failures. Compute the posterior distribution and $\mathbb{E}(p)$. Compare these results with those of Example 11.12.1.

11.12.2. Here we'll continue Example 11.12.1.
(a) Take the posterior distribution of that example as the new prior distribution. Suppose also that the next test of 5 patients also saw 3 successes and 2 failures. Compute the new posterior distribution and $\mathbb{E}(p)$.
(b) Start from the prior distribution of Example 11.12.1 and suppose a test of 10 patients gave 6 successes and 4 failures. Compute the new posterior distribution and $\mathbb{E}(p)$.
(c) Compare the results of (a) and (b).

11.12.3. (a) Once more, start from the prior distribution of Example 11.12.1. Suppose a test of 6 patients saw 6 successes. Compute the posterior distribution and $\mathbb{E}(p)$.

(b) Take the posterior distribution of (a) as the new prior distribution and suppose a new test of 4 patients saw 4 failures. Compute the posterior distribution and $\mathbb{E}(p)$.

(c) Compare this result with that of Exercise 11.12.2 (b).

11.12.4. Suppose we are pessimistic about the effectiveness of a new therapy and make these estimates of the proportion p of treated patients who recover. For $p_1 = 0, p_2 = 0.2, p_3 = 0.4, p_4 = 0.6, p_5 = 0.8$, and $p_6 = 1.0$ we take the probabilities $P(p_1) = 0.1$, $P(p_2) = 0.2$, $P(p_3) = 0.4$, $P(p_4) = 0.2$, $P(p_5) = 0.1$, and $P(p_6) = 0$.

(a) With this prior distribution, compute $\mathbb{E}(p)$.

(b) A round of trials produces 4 successes and 2 failures. Compute the posterior distribution.

(c) Compute $\mathbb{E}(p)$ using the posterior distribution.

(d) Does comparison of the values of $\mathbb{E}(p)$ from (a) and (b) support our initial pessimism?

11.12.5. Suppose we are optimistic about the effectiveness of a new therapy and make these estimates of the proportion p of treated patients who recover. For $p_1 = 0, p_2 = 0.2, p_3 = 0.4, p_4 = 0.6, p_5 = 0.8$, and $p_6 = 1.0$ we take the probabilities $P(p_1) = 0.0$, $P(p_2) = 0.1$, $P(p_3) = 0.2$, $P(p_4) = 0.4$, $P(p_5) = 0.2$, and $P(p_6) = 0.1$.

(a) With this prior distribution, compute $\mathbb{E}(p)$.

(b) A round of trials produces 4 successes and 2 failures. Compute the posterior distribution.

(c) Compute $\mathbb{E}(p)$ using the posterior distribution.

(d) Does comparison of the values of $\mathbb{E}(p)$ from (a) and (b) support our initial optimism?

11.12.6. For $a = b$ show that the $\beta(a, b)$-distribution has $\mathbb{E}(X) = 1/2$ and $\mathrm{sk}(X) = 0$. Further, as $a \to \infty$, $\mathrm{kur}(X) \to 3$. What does this suggest? What happens to $\sigma^2(X)$?

11.12.7. Based on the history of a similar drug, we expect the proportion of successes of a new drug to be modeled by the $\beta(3.5, 6.5)$-distribution.

(a) With this prior distribution, compute $P(0.5 \leq X \leq 0.75)$ and $P(0.75 \leq X)$.

(b) Compute the posterior distribution of a new test that produces 6 successes and 4 failures.

(c) With this posterior distribution, compute $P(0.5 \leq X \leq 0.75)$ and $P(0.75 \leq X)$.

11.12.8. (a) For $\beta(2,2)$ and $\beta(5,5)$ compute the means and variances, which we'll denote $\mu(2,2)$, $\mu(5,5)$, $\sigma^2(2,2)$, and $\sigma^2(5,5)$.

(b) Compute $P(0.5 \leq X \leq 0.75)$ for $\beta(2,2)$ and for the normal distribution with mean $\mu(2,2)$ and variance $\sigma^2(2,2)$.

(c) Compute $P(0.5 \leq X \leq 0.75)$ for $\beta(5,5)$ and for the normal distribution with mean $\mu(5,5)$ and variance $\sigma^2(5,5)$.

(d) Compute the kurtosis for $\beta(2,2)$ and $\beta(5,5)$. Interpret the results of (b) and (c) in light of the kurtosis calculations.

11.12.9. We think the success proportion of a new therapy follows the $\beta(5,7.5)$-distribution. If this model were correct, how likely is an observation of 1 success in a trial of 8 subjects? Does this build confidence in the model?

11.12.10. (a) Plot $\beta(0.1, 0.1)$. Where is the distribution concentrated? (For comparison, the standard normal distribution is concentrated at near the origin.) Find the minimum value of $\beta(0.1, 0.1)$.

But ... this is just a tiny glimpse of an immense field. As we have seen in some of the examples, probability is one of the main tools of statistics. Statistical reasoning is subtle, like mathematical reasoning but also different from it. For example, if someone tosses a coin ten times and gets heads each time, many, maybe most, people will say the next toss is much more likely to be tails because tails "is due." (In the early part of the 20th century, the German philosopher Karl Marbe posited that nature has a memory, so long runs of one result increase the likelihood of the other outcome. This view was not embraced by the statistics community. See page 136 of [37].) A more informed person will say that heads and tails are equally likely. But this is based on the assumption that the coin is fair. In fact, if heads come up ten times in a row, that's a reasonable indication that the coin is not fair, that heads is more likely than tails. To be prepared to understand much of medical literature, and to parse the correctness of drug company claims, you should take a statistics class. Pay careful attention. This is important stuff, useful in all manner of ways, and sometimes really, really surprising. Great fun.

Chapter 12 Why this matters

Part of a physician's training is learning to depersonalize patients. Being familiar and comfortable with your patients can make difficult the selection of a treatment that can hurt them, sometimes quite badly, or worse, the realization that no treatment will help. If you know and like your patient, how could you not be paralyzed with fear when you see a spot on a chest X-ray? For very many reasons, you need not to be emotional; you need to give the most accurate interpretation of the available data. You need to step back from your patient as a person.

I am not a physician. I have no medical training, but as a teacher I've seen a faint hint of this problem. Some colleagues have refused to learn the names of their students because when grading exams they did not want to know which paper went with which student. But giving a poor grade in a course is nothing compared with some of the things physicians must do to their patients. Honestly, I don't know how doctors do this. My respect for them is boundless.

Nevertheless, this chapter is about moving in the other direction. Not to increase the emotional cost of prescribing difficult treatments, but rather to illustrate why it's important to keep trying even when you expect to fail. This chapter is personal. Really personal. There will be no exam questions on its content, so you can skip it if you wish. But I hope you won't. I hope you'll read it and that you'll see a bit about how we should treat one another. If you're a physician, when viewed in the right light, every decision you make is personal to someone. I hope this story helps you find some of that light, and helps you understand that this responsibility need not be paralyzing, but can be liberating.

My brother Steve, pictured in the photograph, is five years younger than I am. We grew up in St. Albans, West Virginia, with our sister Linda and our parents Mary and Walter. Steve and I shared the room our father had built from the attic of the house. Over each bed was a window facing the Kanawha River and the hills on the other side of the valley. On summer nights we read with our windows open, our books accompanied by the sound of trucks downshifting on Route 60 a couple of blocks away, by lonely train whistles from the switching yard across the river. On stormy nights we watched cloud to cloud lightning above the hills, and we talked. At first about concrete things–what we'd done that day, when we would take the bus to Charleston, when Steve and Dad would go fishing. Gradually our conversations became more abstract– what did we want to do with our lives, why does time appear to go at different speeds, just why can't we change the past? Even then, occasionally our recollections were very different, and this generated a few interesting talks. Some weekends the family went for walks in the woods. Linda and her friends played tricks on Steve and me, and we played tricks on them. Kids growing up in the late 50s and early 60s in West Virginia. Not idyllic, but we got to know one another very well, and still spend time together whenever we can.

This story is about Steve's experiences with CLL (chronic lymphocytic leukemia). He was diagnosed in September 2002. As a standard part of his annual exam, Dr. Moore, Steve's primary care physician, ordered a CBC (complete blood count). The WBC (white blood cell count) was 15.6. The normal range is 3.8 to 10.8 (the unit is 1000 cells/μL). Steve was referred to Dr. Arvind Shah, a local oncologist. Dr. Anuradha Thummala, a member of Dr. Shah's group, repeated the CBC. Steve's WBC now was 13.3 and the absolute lymphocyte count was 8.0 (the normal range is 0.85 to 3.9, in units of 1000 cells/μL). Dr. Thummala also did a bone marrow biopsy. (In 2018 my CLL diagnosis was based on flow cytometry–bone marrow biopsies no longer are the standard diagnostic tool for

CLL. Sorry, Steve. You got sick about fifteen years too soon.) The diagnosis was CLL.

Normally, CLL is regarded as an indolent cancer. Counts plateau and the patient can live for years, sometimes many years, without intervention. But Steve's doctor ordered a FISH (fluorescence in situ hybridization) test for genetic abnormalities associated with CLL. The test was read to show a 17p deletion. In this deletion the breakpoint on chromosome 17 always is proximal to the p53 gene. Some evidence suggests a tumor suppressor gene at this location; the deletion inactivates this TSG. For CLL patients this is a bad mutation. It signals that the disease is terminal. With no proven treatments and a mean life expectancy of eight years, this is very bad news.

Steve needed a CLL specialist. Dr. John Byrd at the James Cancer Center at Ohio State University was very highly regarded, so Steve and his wife Kim headed to Columbus. In December of 2002 Dr. Byrd confirmed the CLL diagnosis. Initially Steve saw Dr. Shah or Dr. Thummala every month, then every three months. And every six months, Steve and Kim returned to Columbus for blood work, examinations, sometimes scans. Watch and wait.

In January 2010, while Steve and Kim were at OSU for a scan, Steve collapsed. His WBC had gone up to 600, almost 60 times the maximum of the normal range. He was rushed to the Intensive Care Unit (ICU). His hemoglobin was under 3 (the normal range is 13.2–17.1 g/dL), his kidneys had failed, and Dr. Byrd was at a conference in California. The ICU doctors put him on dialysis and a respirator, and until the Red Cross found a match for transfusions, his doctors drew out Steve's blood, centrifuged it to pull off most of the white cells, rehydrated the blood with plasma, and transfused Steve's own blood back into him. They began treatment with alemtuzumab (Campath), though that is risky for patients with such low hemoglobin. Steve began to thrash around, using oxygen he couldn't spare, so his doctors administered a paralytic.

Despite our fears, Steve survived the night, and the next day, and a few more. Then he was moved from the ICU to the Cancer Center. He had lost the sight in both eyes, but eventually his sight returned. The FISH test was rerun and showed not a 17p mutation, but rather a 13q mutation. This still is bad, but not as bad as the 17p.

In the summer of 2010 Steve began a two-drug clinical trial, and in the summer of 2012 he started on PCI-32765 (ibrutinib) and still is on that treatment. (See the 2019 update at the end of this chapter.) His previously enlarged spleen and lymph nodes have shrunk. Although he bruises easily and bled for a very long time after a recent tooth extraction, and needs IgG infusions (pooled immunoglobulin G from the plasma of a thousand blood donors–yes, medicine has gotten pretty close to magic), he's still here, still active. Steve gardens, works part-time in IT, part-time in construction. He and Kim bowl, care for two yellow Labs, enjoy a lot of time with a lot of friends.

My brother's doctors took extraordinary steps to save him. So far he's gotten eight more years of a complicated, happy, interesting life. Both my parents have died, and although I remember them clearly, I can't touch them, or talk with Mom about the importance of kindness, or joke with Dad about the ancient cowboy movies he watched. No more evenings on the front porch watching storms roll down the valley. No more Mom and Dad and three little kids around the kitchen table for dinner. These evenings live only in our memories now. But Linda still is here and Steve still is here, and that means worlds to me.

Linda's story

Here is my sister Linda's description of Steve's crisis. This is a sketch of how medical situations can be fraught with heartbreak, and yet sometimes, against all expectations, we get a happy ending.

> On the morning of January 28, 2010, in a matter of hours my brother Steve went from preparing to begin a clinical trial for CLL at the James Cancer Center at the OSU Hospital, to vomiting and passing out in the shower of a Columbus hotel and then coding in the OSU parking lot. As my husband David and I rushed to Columbus, the ICU staff worked desperately to stabilize him. Time almost ceased to exist as we waited with Steve's wife Kim for word of his condition. I'll never forget a somber Dr. Joseph Flynn coming in and telling us we needed to call whoever should be called and to prepare ourselves. Then, looking directly at us he forcefully added, "That being said, we're going to throw everything we've got at him, including the kitchen sink. And don't underestimate the power of prayer." With those few words he had given us the priceless gifts we most needed: honesty, his determination, and hope.

> After several calls to family and friends, the most heartbreaking to our 85-year-old father Walter who had lost his precious wife Mary the previous January, I found a quiet hallway, leaned against the wall, and silently cried at the possibility of losing one of my beloved brothers. Not a 53-year-old man, but forever my baby brother. A nurse quickly passed by, stopped, came back, and enfolded me in the most amazing embrace. I'll never forget that hug from a stranger.

> During the following days we became familiar with the terms syncopal episode, blast crisis, tumor lysis syndrome, autoimmune

hemolytic anemia, acute hypoxia respiratory failure. We waited and prayed as medicines like alemtuzumab, cyclophosphamide, prednisone, Rituxan, Campath, paralytics, and phentonol began to have an effect. Our first real glimmer of hope came on Saturday, January 30. Even though Steve still couldn't speak or open his eyes, Dr. Flynn told us he was encouraged. Steve's WBC was down from over 650 to 358, his hemoglobin was up from 3.7 to 5.8 (still low, but much better), and his ventilator had been turned down from 100% to 65% without affecting his blood oxygen level. Since being diagnosed several years ago Steve had researched CLL and possible medicines and treatments, often saying "Knowledge is power." We learned the wisdom of this as we kept track of numbers and medicines. Even before he could respond, we told him over and over that his numbers were going in the right direction, giving him hard data that he could believe in. The doctors and staff at OSU were so patient as we took notes each time something was reported, even spelling words when we asked, so we could report to family and friends.

Slowly Steve continued to improve, and on the evening of February 9, with new treatment plans in place, he was dismissed from OSU. As Steve, Kim and I enjoyed dinner at a nearby restaurant I can't begin to express our thankfulness. Our smiles must have lit up the room. Watching them look at each other I knew they were both rejoicing in having more time together. We spent another night at our hotel so Steve could rest before the trip home. I've never expected to be someone's hero,[1] but being in the car as we pulled up to our father's home, that's exactly how I felt. Being a part of bringing Dad's youngest child back to him from the brink of death was indescribable.

What you say, how you say it, what you do, and your medical knowledge, can make all the difference to your patients and their families.

Why this really matters

To be a physician you must learn a lot of science. Our understanding of medicine is based on biology and chemistry, and on a bit of physics and math. You need

---

1 Linda, you were a brilliant, loving elementary school teacher for many years. Your kids loved you. You were a hero to tons of kids. But at that moment, and many others, you were Dad's hero, too.

to see cells, organs, organisms, maybe populations, depending on the illness or injury. In addition, systems biology teaches us that interactions can be as important as objects. This may be overwhelming. One of my students described the first year of medical school as feeling like having a fire hose stuck in his mouth and the valve opened full.

With all this, why should you want to learn about mathematical models? Certainly, math has given some concrete results: more effective chemotherapy schedules, how models of metabolic rates scale, identification of the gene responsible for cardiac arrhythmia. Sure, but other people figure these things out. Why should every physician want to know something about mathematical models? I'll give two answers, one specific, the other more general.

First, in Sect. 3.4 we mentioned that when Norton and Simon proposed their chemotherapy schedule based on Laird's mathematical model of tumor growth, most oncologists resisted their work. It disagreed with established strategies, and even though the current approaches had some significant problems, most physicians weren't convinced that a mathematical model could productively direct clinical methods. If Norton-Simon trials had begun earlier, and their impressive results had been seen earlier, many people would have survived longer without recurrence. Certainly, not every physician needs to work with mathematical models, but every physician does need to appreciate the power of models. Don't dismiss them out of hand, even if their results directly oppose established thinking. Once upon a time, the established thinking was that quartz is water frozen so hard that it never can thaw. Sometimes established wisdom is right, sometimes it's wrong. Keep an open mind.

The more general answer involves empathy, for the patient and the people close to the patient. But empathy is trickier than you might think. It isn't just trying to understand how the other person feels, because at any interesting level this is impossible. You and the other person have different lives, different interests, different thoughts. The two of you won't respond the same way to Bach's *Goldberg Variations* or to the final pages of Saramago's *Death with Interruptions* or to Bogart and Bergman in *Casablanca*. How can you know what I feel when my CBC comes back with a lymphocyte count of 15? How can I know what you feel when you have to tell me that I didn't get into the clinical trial? If this is what empathy is about, we're out of luck.

In her novel *Man Walks into a Room* [77], Nicole Krauss writes this about empathy, in the voice of one of her characters. "The misery of other people is only an abstraction, …something that can be sympathized with only by drawing from one's own experiences. But as it stands, true empathy remains impossible."

Empathy isn't about knowing how the other person feels. It's about how you would feel if you were in the other person's circumstance. How do you marshall the energy and stamina to do what you can? You do what you'd want your doctors to do if you were the patient. The better you know the patient, the easier this will be.

So far as I know, the best exploration of empathy is Leslie Jamison's *The Empathy Exams* [69]. Near the beginning we find, "Empathy means acknowledging a horizon of context that extends perpetually beyond what you can see" and "Empathy means realizing no trauma has discrete edges," near the middle, "Another person's pain registers as an experience in the perceiver: empathy as forced symmetry, a bodily echo," and near the end, "But I don't believe in a finite economy of empathy; I happen to think that paying attention yields as much as it taxes. You learn to start seeing." In between, Jamison recounts her experiences as a medical actor, at a meeting for Morgellons patients, at a writers' workshop in the middle of the Mexican narco wars, as an assault victim in Nicaragua, as a visitor in a Bolivian silver mine, as a viewer of a reality show on interventions, on a tour of LA led by former gang members, as the support crew when her brother ran the Barkley Marathons, as decoder of saccharine literature and saccharin sweeteners, as interviewer of a prisoner on a mine-ravaged landscape of West Virginia, as student of Frida Kahlo's injuries, as observer of mundane detail amid social upheaval and of the scaling of want, as commenter on the sadness of James Agee, as observer of films about the West Memphis Three. She notes empathy can be a moving target, and from many directions she views the reality of female pain. Often lyrical, these essays opened my eyes to new dimensions of empathy. And I enjoyed the analyses of people whose work I love–Tori Amos, Kate Bush, Joan Didion, Louise Glück, Susan Sontag, Mazzy Star. And the essays are personal. Really personal. This I learned from Jamison: that empathy is not abstract, that "Empathy is easier when it comes to concrete particulars." You need to know the stories.

Dr. Byrd knows this. Early on, he talked with Steve and Kim about their personal lives and interests, about their stories. They talked about their dogs, about fishing (Drs. Byrd and Flynn have taken fishing trips together, with the expected friendly competition), about sports (OSU versus WVU). While one of the goals of these conversations is to put Steve and Kim at ease, they felt that Dr. Byrd's interest in them is genuine. He wants to know his patients and their families. Steve's experience (and mine, too) is that everyone–physicians, nurses, techs, and staff–in cancer hospitals is friendly and helpful. For a disease as frightening as cancer, this sort of care is welcome, necessary. Every physician who deals with mortal illness or injury should know empathy.

Other people, or maybe you wearing a different hat, will decide what's feasible in our world of limited resources. But as far as your energy, your determination, your knowledge, your cleverness, and your math are concerned, you know what to do. Every decision you make is personal to your patient and to your patient's family. If this were your family, what would you do?

2019 update. For a while Steve's lymphocyte and platelet counts had moved in the wrong directions. Eventually mutation would produce a strain of CLL immune to ibrutinib and natural selection would amplify a clone of that strain. Dr. Byrd told Steve that some other treatments are in clinical trials. In late January, Steve developed a persistent headache that a CT scan showed was the result of a subdural hematoma. CLL often is paired with ITP, idiopathic thrombocytopenia. Thrombocytes are platelets, "penia" means "decrease of," and "idiopathic" means "we don't know why." Steve has ITP. While ibrutinib knocks down CLL, ITP also diminishes. When ibrutinib stops being effective against CLL, ITP wakes up and can cause hematomas.

Dr. Byrd stopped Steve's ibrutinib, but because ibrutinib depresses the platelet supply, he needed to wait at least a week to reduce the risk from surgery for pressure relief. Pain was controlled, but not pressure. Steve had a truly miserable week at the James Cancer Center. Then surgery relieved the pressure. Steve and Kim went home to rest.

A "treatment vacation" is common before beginning a new clinical trial, this time for ARQ 531, a drug designed to treat patients whose CLL had developed resistance to ibrutinib. Because Steve's counts began to move in the wrong direction and his lymph nodes and spleen had enlarged, Steve's vacation lasted only four months.

Steve and Kim returned to Columbus for blood work, scans, an ECG. Then the news that surprised me: an independent review board and the trial sponsor, ArQule, would determine who would get places in the trial. A recommendation from Dr. Byrd did not guarantee a place for Steve. We all had a tense week. If Steve didn't get a position in the trial, the backup plan was treatment with venetoclax, but this can have a terrifying side effect, Richter's transformation, that converts CLL into a non-Hodgkin's lymphoma with a poor prognosis. This is a backup plan to avoid, if at all possible.

Steve's platelet level was below 50, the cut-off for participation in the trial. Dr. Byrd recommended a variance of this requirement, and ArQule agreed. Steve began treatment on June 17, 2019.

Initially he responded well, but there were hiccups. A rash, a familiar side effect, caused a temporary reduction in the ARQ 531 dosage. And on an early visit while his vitals were being taken, Steve passed out. The diagnosis was a vasovagal syncope episode, which Steve heard as a "bagelbagel" syncope, and gave him a couple of days in the hospital. Soon the dose was increased back to the approved level. Now, in mid-October of 2019, Steve is doing well. The only persistent side effect is a change in his taste, "salty is saltier and sweet is sweeter."

For us, for the moment, this story is in a happy place. But I'm left with an uneasy feeling. Steve might not have gotten into the trial. Are there places enough for every patient who meets the trial criteria? I don't know. In addition to the James Cancer Center, trials of ARQ 531 are being conducted in at least four other locations. And ArQule has increased the number of patients they will take in each trial. But is this enough? Because Steve got a spot, was someone excluded? I don't know; I don't want to know. If you become a physician, you'll have to deal with these issues. I wish I could provide some guidance, but I can't. I just can't.

Appendix A   Technical notes

Here we'll see some details that are too complicated, or would have taken us too far afield, to include in the main text. Each section of the appendix references the section of the book in which the appendix section's concepts appear.

A.1   INTEGRATING FACTORS

We mentioned the method of integrating factors in Sects. 8.2 and 10.9, and we will use it in Sect. 13.4 of volume 2. With this technique we can solve some differential equations. For example, with integrating factors we can solve equations of the form

$$\frac{dx}{dt} + a(t)x = b(t) \tag{A.1}$$

provided we can evaluate a couple of integrals. Here's how it works.

The left-hand side of Eq. (A.1) looks a bit like the result of applying the product rule for differentiation, except that the term dx/dt needs another factor. So, multiply both sides of Eq. (A.1) by a function $u(t)$, the *integrating factor*. This gives

$$u\frac{dx}{dt} + uax = ub \tag{A.2}$$

Assume the left-hand side of Eq. (A.2) is the result of the product rule

$$\frac{d}{dt}(u \cdot x) = u\frac{dx}{dt} + \frac{du}{dt}x \qquad (A.3)$$

For the right-hand side of Eq. (A.3) to match the left-hand side of Eq. (A.1), we must have $du/dt = ua$. This we can solve by separation of variables (Sect. 4.6)

$$\frac{du}{u} = a\,dt$$

(The inclusion of a constant of integration for this equation doesn't change the final solution for $x(t)$.) This gives $\ln(u) = \int a\,dt$ and so the integrating factor is

$$u = e^{\int a\,dt} \qquad (A.4)$$

Now replace the left-hand side of Eq. (A.2) with the left-hand side of Eq. (A.3) and obtain

$$\frac{d}{dt}(u \cdot x) = ub$$

Integrate both sides

$$u \cdot x = \int ub\,dt + C$$

and so

$$x(t) = \frac{1}{u}\int ub\,dt + \frac{C}{u} \qquad (A.5)$$

We can solve for x, *if* we can find the function u, and *if* we can evaluate the integral on the right-hand side of Eq. (A.5).

The pair of equations (A.4) and (A.5) constitute the method of integrating factors. We'll illustrate this with two examples, one from Sect. 8.2 and one from Sect. 10.9.

Example A.1.1. *A simple non-autonomous system.* For the system $x' = -x + y$, $y' = -y$ of Example 8.2.5, we know $y' = -y$ gives $y(t) = y(0)e^{-t}$ and the x' equation becomes $x' = -x + y(0)e^{-t}$. In the form of Eq. (A.1), we have $a = +1$ and $b = y(0)e^{-t}$. Apply Eq. (A.4),

$$u(t) = e^{\int 1\,dt} = e^t$$

and then Eq. (A.5)

$$x(t) = \frac{1}{u}\int ub\,dt + \frac{C}{u} = \frac{1}{e^t}\int e^t y(0)e^{-t}dt + \frac{C}{e^t}$$

$$= y(0)e^{-t}\int dt + Ce^{-t} = y(0)e^{-t}t + Ce^{-t}$$

Substituting $t = 0$ gives $C = x(0)$ and so $x(t) = x(0)e^{-t} + ty(0)e^{-t}$. $\square$

Example A.1.2. *A simple non-autonomous equation.* In Practice Problem 10.9.2 we used power series to solve the differential equation $x' = x + t^3$ subject to $x(0) = 1$. Now we'll use an integrating factor. In the format of Eq. (A.1) we have $a = -1$ and $b = t^3$. Then Eq. (A.4) gives

$$u(t) = e^{\int -1 \, dt} = e^{-t}$$

and Eq. (A.5) gives

$$x(t) = \frac{1}{u} \int ub \, dt + \frac{C}{u} = \frac{1}{e^{-t}} \int e^{-t} t^3 \, dt + \frac{C}{e^{-t}}$$

$$= e^t \left(e^{-t}(-6 - 6t - 3t^2 - t^3) \right) + Ce^t = -6 - 6t - 3t^2 - t^3 + Ce^t$$

where the penultimate equality comes from integration by parts, three times. Substituting $t = 0$ we find $C = x(0) + 6 = 7$, so the solution of the differential equation is

$$x(t) = -6 - 6t - 3t^2 - t^3 + 7e^t$$

That is, we recover the result of Practice Problem 10.9.2. □

So …why did we bother with learning to solve non-autonomous differential equations by power series when integrating factors appear to work? For one thing, who would have thought to multiply both sides of the differential equation by a function $u(t)$? The argument presented in my sophomore differential equations class–that the left-hand side of the differential equation looks a bit like the result of applying the product rule of differentiation–seemed inadequate. That the trick worked was apparent, but it was just too much magic.

The second reason is that power series are an interesting and useful tool, and solving differential equations is a good way to get some practice with series. A complicated technique is easier to learn in the context of some application.

Still, we didn't get good result with every attempt to use power series. The series solution we found in Example 10.9.2 we did not recognize as the series of a familiar function. Maybe an integrating factor will give a better result. Let's see.

Example A.1.3. *A difficult non-autonomous system.* Let's try an integrating factor to solve $x' = tx + t^2$, the equation of Example 10.9.2. In the form of Eq. (A.1) we have $a = -t$ and $b = t^2$. By Eq. (A.4) we find

$$u = e^{\int a \, dt} = e^{\int -t \, dt} = e^{-t^2/2}$$

and then by Eq. (A.4),

$$x = \frac{1}{u} \int ub \, dt + \frac{C}{u} = \frac{1}{e^{-t^2/2}} \int e^{-t^2/2} t^2 \, dt + \frac{C}{e^{-t^2/2}}$$

How close can we come to evaluating this integral? First, substitute $w = t^2/2$. This gives $e^{-t^2/2}t^2\,dt = e^{-w}\sqrt{2w}\,dw$. Next, integrate by parts with $u = \sqrt{2w}$ and $dv = e^{-w}\,dw$. This gives

$$\int e^{-t^2/2}t^2\,dt = \int e^{-w}\sqrt{2w}\,dw = -\sqrt{2w}e^{-w} - \int -e^{-w}\frac{1}{\sqrt{2w}}\,dw$$

$$= -te^{-t^2/2} + \int e^{-t^2/2}\,dt$$

That's as far as we can go. The last integral is related to the normal density function, Eq. (11.54). $\square$

So we see that the caveats about being able to evaluate the two integrals (A.4) and (A.5) do indeed need to be taken seriously.

As a final example, we'll solve the virion differential equation for HIV treated with a reverse transcriptase inhibitor, discussed in Sect. 13.4 of volume 2.

Example A.1.4. *A virion integrating factor.* In Sect. 13.4 of volume 2 we will express the virion population equation as

$$\frac{dz}{dt} + \delta_z z = \kappa y(0)e^{-\delta_y t}$$

So in the form of Eq. (A.1), we have $a(t) = \delta_z$ and $b(t) = \kappa y(0)e^{-\delta_y t}$. Then the integrating factor is

$$u = e^{\int a\,dt} = e^{\int \delta_z\,dt} = e^{\delta_z t}$$

and the solution is

$$z = \frac{1}{u}\int ub\,dt + \frac{C}{u} = \frac{1}{e^{\delta_z t}}\int e^{\delta_z t}\kappa y(0)e^{-\delta_y t}\,dt + \frac{C}{e^{\delta_z t}}$$

$$= e^{-\delta_z t}\kappa y(0)\int e^{\delta_z t - \delta_y t}\,dt + Ce^{-\delta_z t} = \frac{\kappa y(0)}{\delta_z - \delta_y}e^{-\delta_z t}e^{\delta_z t - \delta_y t} + Ce^{-\delta_z t}$$

$$= \frac{\kappa y(0)}{\delta_z - \delta_y}e^{-\delta_y t} + Ce^{-\delta_z t}$$

Set $t = 0$ to evaluate C in terms of $z(0)$, and then simplify to obtain

$$z(t) = \frac{\kappa y(0)}{\delta_z - \delta_y}\left(e^{-\delta_y t} - e^{-\delta_z t}\right) + z(0)e^{-\delta_z t}$$

as claimed in Sect. 13.4. $\square$

A.2 EXISTENCE AND UNIQUENESS

Here we'll show that if f is continuous, the differential equation $x' = f(t,x)$ with $x(t_0) = x_0$ has a unique solution on some rectangle $t_0 \le t \le t_0 + \delta_1$ and

$x_0 - \delta_2 \leq x \leq x_0 + \delta_2$. That is, close enough to the initial point, the differential equation has one, and only one, solution. This result was mentioned in Sects. 3.1 and 10.9, and will be used in Sect. 13.8 of volume 2. This result is part of a large collection of of existence and uniqueness theorems that in one way or another all have the same proof. We'll look at the general proof because it organizes the central concepts in ways that might appear unmotivated for the particular case of differential equations.

Before the proof, we need four notions.

First, a *Cauchy sequence* (the same Augustin Cauchy of the Cauchy distribution of Sect. 11.10) is a sequence $x_1, x_2, x_3, \ldots$ whose elements eventually get as close together as we like. That is, for every $\epsilon > 0$ (this is the "as close together as we like" part), there is a number N (this is the "eventually" part) so that if $m, n \geq N$, then $|x_n - x_m| < \epsilon$.

For example, the harmonic sequence (*not* to be confused with the harmonic series, though they are related: the series is the sum of the sequence terms) $x_n = 1/n$ is Cauchy because for any $\epsilon > 0$ we can take N to be the smallest integer $N > 1/\epsilon$. Then for $m, n \geq N$, say $m > n$

$$|x_n - x_m| = \left| \frac{1}{n} - \frac{1}{m} \right| = \frac{|m-n|}{nm} < \frac{m}{nm} = \frac{1}{n} \leq \frac{1}{N} < \epsilon$$

Second, the *sup norm* of a function f defined on an interval $[a, b]$ is

$$\|f\|_\infty = \sup\{|f(t)| : a \leq t \leq b\}$$

For example, take $[a, b] = [0, 1]$. Then $\|1 + x^2\|_\infty = 2$ and $\|x - x^2\|_\infty = 1/4$. (Sketch the graphs, or evaluate f at its critical points and at the endpoints of the interval.)

Third, a space is called *complete* if every Cauchy sequence of elements of the space converges to an element of the space.

For example, $(0, 2)$ is not complete because the Cauchy sequence $x_n = 1/n$ converges to 0, not an element of $(0, 2)$. The space $[0, 2]$ is complete.

Hidden in this definition is that the notion of a Cauchy sequence depends on how we measure the distance between elements of the sequence. For sequences of numbers we'll just use the absolute value of the difference of the numbers. For points (a, b) and (c, d) in the plane we'll use the Euclidean distance $\sqrt{(a-c)^2 + (b-d)^2}$. For sequences of functions, we'll use the sup norm of the difference of the functions.

Fourth, a function T is a *contraction* with contraction factor $C < 1$ if $|T(x) - T(y)| \leq C|x - y|$, that is, if T reduces the distances between points by a factor of at least C. Here the distance is the absolute value if x and y are numbers, and the Euclidean distance if x and y are points in the plane.

For example, in the space of real numbers and distance measured by the absolute value, $T(x) = x/2$ is a contraction because $|T(x) - T(y)| = |x/2 - y/2| = (1/2)|x - y|$.

The main result we'll use is that the space of continuous functions on an interval $[a, b]$ is complete when distances are measured by the sup norm. We won't prove this, but we'll prove a result called the *Contraction Mapping Principle* (CMP) and apply this to the space of continuous functions.

Theorem A.2.1. On a complete space a contraction T has a unique fixed point.

Proof. Start with any point x_0 and form the sequence $x_1 = T(x_0)$, $x_2 = T(x_1)$, $x_3 = T(x_2), \ldots$. Then

$$|x_{n+1} - x_n| = |T(x_n) - T(x_{n-1})| \le C|x_n - x_{n-1}|$$
$$= C|T(x_{n-1}) - T(x_{n-2})| \le C^2|x_{n-1} - x_{n-2}| \le \cdots \le C^n|x_1 - x_0|$$

Recall the *triangle inequality*: $|a + b| \le |a| + |b|$. For $n > m$, applications of this inequality give

$$|x_n - x_m| \le |x_n - x_{n-1}| + |x_{n-1} - x_{n-2}| + \cdots + |x_{m+1} - x_m|$$
$$\le C^{n-1}|x_1 - x_0| + C^{n-2}|x_1 - x_0| + \cdots + C^m|x_1 - x_0|$$
$$= C^m(1 + C + \cdots + C^{n-m-1})|x_1 - x_0|$$
$$= C^m \frac{1 - C^{n-m}}{1 - C}|x_1 - x_0| < \frac{C^m}{1 - C}|x_1 - x_0|$$

The last equality follows from the sum of a finite geometric series $1 + a + \cdots + a^n = (1 - a^{n+1})/(1 - a)$. (To see this, multiply both sides by $1 - a$ and expand the left side.)

For any $\epsilon > 0$, any x_0 (recall $x_1 = T(x_0)$), and large enough m we see $(C^m/(1 - C))|x_1 - x_0| < \epsilon$ because $C < 1$. Then for large enough N, if $n, m > N$ we have $|x_n - x_m| < \epsilon$. That is, the sequence $x_0, x_1, x_2 \ldots$ is Cauchy and consequently converges to a point x_*. We must show that x_* is a fixed point of T and that it is unique.

First, to see that x_* is a fixed point, note

$$|x_* - T(x_*)| = |x_* - x_n + x_n - T(x_*)| \le |x_* - x_n| + |x_n - T(x_*)|$$
$$= |x_* - x_n| + |T(x_{n-1}) - T(x_*)| \le |x_* - x_n| + C|x_{n-1} - x_*|$$

where the first inequality is an application of the triangle inequality and the last an application of the definition of contraction. Because $x_n \to x_*$, both terms of the last inequality go to 0 as $n \to \infty$ and consequently $|x_* - T(x_*)| = 0$, that is, $T(x_*) = x_*$.

Finally, suppose T has two fixed points x_* and y_*. Then

$$|x_* - y_*| = |T(x_*) - T(y_*)| \le C|x_* - y_*|$$

Because $C < 1$, this is impossible unless $x_* = y_*$. $\square$

Now we'll apply the CMP to show that

$$x' = f(t, x), \qquad x(t_0) = x_0 \tag{A.6}$$

has a unique solution near the point (t_0, x_0).

Suppose $\partial f / \partial x$ is continuous. Then for every rectangle

$$R = \{(t, x) : t_0 \le t \le t_0 + \delta_1, x_0 - \delta_2 \le x \le x_0 + \delta_2\}$$

there is a constant K for which

$$\left| \frac{\partial f}{\partial x}(t, x) \right| \le K \quad \text{for } (t, x) \in R \tag{A.7}$$

Now integrate both sides of $x' = f(t, x)$ from t_0 to $t \le t_0 + \delta_1$

$$\int_{t_0}^t x' \, ds = \int_{t_0}^t f(s, x) \, ds$$

$$x(t) - x(t_0) = \int_{t_0}^t f(s, x) \, ds \quad \text{so} \quad x(t) = x_0 + \int_{t_0}^t f(s, x) \, ds$$

If we define a function

$$T(x) = x_0 + \int_{t_0}^t f(s, x) \, ds \tag{A.8}$$

then a solution of Eq. (A.6) is a fixed point of the function T of Eq. (A.8). We'll show this by Theorem A.2.1, once we show that T is a contraction.

For any x_1 and x_2 in the interval $[x_0 - \delta_1, x_0 + \delta_1]$, by the mean value theorem there is an x_3 between x_1 and x_2 for which

$$\frac{f(t, x_2) - f(t, x_1)}{x_2 - x_1} = \frac{\partial f}{\partial x}(t, x_3)$$

Then for $t_0 < t < t_0 + \delta_1$,

$$|f(t, x_1) - f(t, x_2)| = \left| \frac{\partial f}{\partial x}(t, x_3) \right| |x_1(t) - x_2(t)| \le K|x_1(t) - x_2(t)| \tag{A.9}$$

where the inequality follows from Eq. (A.7). With this we'll show T is a contraction in the sup norm. For all $t \in [t_0, t_0 + \delta_1)$,

$$|T(x_1)(t) - T(x_2)(t)| = \left| \int_{t_0}^t f(s, x_1(s)) \, ds - \int_{t_0}^t f(s, x_2(s)) \, ds \right|$$

$$= \left| \int_{t_0}^t \left(f(s, x_1(s)) - f(s, x_2(s)) \right) ds \right| \le \int_{t_0}^t |f(s, x_1(s)) - f(s, x_2(s))| \, ds$$

$$\leq \int_{t_0}^{t} K |x_1(s) - x_2(s)| \, ds \leq \int_{t_0}^{t} K \|x_1 - x_2\|_\infty \, ds$$

$$= K \|x_1 - x_2\|_\infty \int_{t_0}^{t} ds = K \|x_1 - x_2\|_\infty (t - t_0)$$

Here the first inequality follows because $|\int f(s) \, ds| \leq \int |f(s)| \, ds$ (if f changes sign, the area under $|f|$ doesn't suffer the subtractions that arise in computing the area under f), the second from Eq. (A.9), and the third from the fact that for all s in $[t_0, t_0 + \delta_1]$, $|x_1(s) - x_2(s)| \leq \|x_1 - x_2\|_\infty$.

We've shown that for all t in $[t_0, t_0 + \delta_1]$,

$$\big| T(x_1)(t) - T(x_2)(t) \big| \leq K \|x_1 - x_2\|_\infty (t - t_0)$$

$$\leq K \|x_1 - x_2\|_\infty (t_0 + \delta_1 - t_0) = K \|x_1 - x_2\|_\infty \delta_1$$

Consequently,

$$\| T(x_1) - T(x_2) \|_\infty \leq K \delta_1 \|x_1 - x_2\|_\infty$$

Take δ_1 small enough that $K\delta_1 < 1$ and we see T is a contraction. (Wait: didn't we use δ_1 to bracket the possible values for x_1 and x_2? If the condition $K\delta_1 < 1$ makes δ_1 even smaller, that's no problem because it further restricts the possible choices for x_1 and x_2. If $K\delta_1 < 1$ doesn't decrease the size of δ_1, just keep δ_1 at its original value.) Then by the CMP, T has a unique fixed point and this fixed point is the unique solution of the differential equation (A.6). This finishes our sketch of the proof of the existence and uniqueness theorem $\square$.

The sequence generated in the application of the CMP to the existence and uniqueness theorem gives another way to generate the power series solutions of Sect. 10.9. A simple example illustrates this.

Example A.2.1. *Series solution by CMP.* For the differential equation $x' = x$ with $x(0) = 1$, begin with $x_0(t) = 1$ and write the sequence of functions $x_{n+1}(t) = 1 + \int_0^t x_n(s) \, ds$.

$$x_0(t) = 1$$

$$x_1(t) = 1 + \int_0^t 1 \, ds = 1 + t$$

$$x_2(t) = 1 + \int_0^t (1 + s) \, ds = 1 + t + \frac{t^2}{2}$$

$$x_3(t) = 1 + \int_0^t \left(1 + s + \frac{s^2}{2} \right) ds = 1 + t + \frac{t^2}{2} + \frac{t^3}{3!}$$

$$\cdots$$

We see that the sequence of functions generated by the CMP converges to the solution $x(t) = e^t$. $\square$

Isn't this simpler than the series method presented in Sect. 10.9? Not so much. This approach just replaces the term-matching to find the series coefficients. To recognize the function represented by the series is just as much a challenge here as it was in Sect. 10.9. Don't be misled by the simple example presented here.

This application of the CMP is satisfying because it both gives the direction to prove the existence and uniqueness theorem and provides a way to generate successive approximations of the solution. But it does something else, something you won't see now. Many other existence and uniqueness results are derived with the CMP. For example, the iterated function systems we will discuss briefly in Sect. 19.3 of volume 2 were used first to generate fractals. That an iterated function system always generates a unique fractal (if all the functions are contractions) is another application of the CMP, where instead of continuous functions on a closed interval, the space is pictures (technically, compact sets) in the plane, and instead of being measured by the sup norm, the distance between pictures is measured by a function called the Hausdorff metric. Also, a proof of the Perron-Frobenius theorem is based on the CMP, and way more besides. That so many different parts of math come from applications of one idea in different contexts is one of the very best reasons to study math. We uncover hidden similarities in the structure of ideas.

A.3 THE THRESHOLD THEOREM

Here we derive Kermack and McKendrick's epidemic threshold theorem mentioned in Example 3.2.5 of Sect. 3.2. We'll use the constant-population SIR model (Eq. (3.9))

$$\frac{dS}{dt} = -\epsilon SI \quad \frac{dI}{dt} = \epsilon SI - \gamma I \quad \frac{dR}{dt} = \gamma I$$

We can view the constant-population condition as a result of the short time duration of the epidemic.

Recall that ϵ and γ are positive constants, and that the populations S, I, and R are non-negative. Then the dS/dt and dR/dt equations show that S decreases monotonically and R increases monotonically. A bit more work is needed to figure out the behavior of I.

Note that the dS/dt and dI/dt equations do not involve R, so we can use the chain rule to compute

$$\frac{dI}{dS} = \frac{dI/dt}{dS/dt} = \frac{\epsilon SI - \gamma I}{-\epsilon SI} = -1 + \frac{\gamma}{\epsilon S} \tag{A.10}$$

For $S < \gamma/\epsilon = \tau$ we see that I increases. Because S decreases, I increases to its maximum, at $S = \tau$, then decreases. We call τ the *threshold* of the infection because if $S_0 < \tau$ the epidemic doesn't take off: the number of infecteds decreases from I_0 to 0.

To find more detail, we need an expression of I as a function of S. Integrate both sides of Eq. (A.10) from an initial point (S_0, I_0) to a later point (S, I) to obtain

$$\int \frac{dI}{dS} dS = -S + \tau \ln(S) \quad \text{so } I(S) = I_0 + S_0 - S + \tau \ln\left(\frac{S}{S_0}\right) \tag{A.11}$$

From this we see that as $S \to 0$, $I \to -\infty$ and consequently for some $S = S_* > 0$, $I(S_*) = 0$. That is, the epidemic dies out before every susceptible is infected. (This is an alternate derivation of the $Z_\infty < 1$ result of Example 3.2.5.) Moreover, the epidemic dies out because of a lack of infecteds. This may be a surprise, but it is a straightforward consequence of a simple model.

With Eq. (A.11) we can prove the threshold theorem of Kermack and McKendrick.

Theorem A.3.1. Suppose $S_0 = \tau + \delta$ and $\delta/\tau \ll 1$. (That is, S_0 is larger that the threshold τ, but only a little larger.) If I_0 is small, the number of individuals who become infected is about 2δ.

Proof. Recall that as $t \to \infty$, $S \to S_*$ and $I \to 0$. Consequently, as $t \to \infty$, Eq. (A.11) gives

$$0 = I_0 + S_0 - S_* + \tau \ln\left(\frac{S_*}{S_0}\right)$$

Because I_0 is small, this equation becomes

$$0 \approx S_0 - S_* + \tau \ln\left(\frac{S_*}{S_0}\right) = S_0 - S_* + \tau \ln\left(1 - \frac{S_0 - S_*}{S_0}\right)$$

Now $S_0 - S_*$ is the number of susceptibles who become infected. We can't solve this equation for $S_0 - S_*$, so we'll use the quadratic approximation of the Taylor series $\ln(1 - x) = -x - x^2/2 - x^3/3 - \cdots$. This is sensible if $S_0 - S_*$ is small compared to S_0. Take $x = (S_0 - S_*)/S_0$ in the quadratic approximation. Then

$$0 = (S_0 - S_*) - \tau\left(\frac{S_0 - S_*}{S_0}\right) - \frac{\tau}{2}\left(\frac{S_0 - S_*}{S_0}\right)^2$$

$$= (S_0 - S_*)\left(1 - \frac{\tau}{S_0} - \frac{\tau(S_0 - S_*)}{2S_0^2}\right)$$

Because $S_* < \tau$ and $S_0 > \tau$, we see $S_0 - S_* \neq 0$ and so the equation reduces to

$$0 = 1 - \frac{\tau}{S_0} - \frac{\tau(S_0 - S_*)}{2S_0^2}$$

Solve for $(S_0 - S_*)$ and simplify, with the substitution

$$S_0 - S_* = 2S_0\left(\frac{S_0}{\tau} - 1\right) = 2(\tau + \delta)\frac{\delta}{\tau} = 2\tau\left(1 + \frac{\delta}{\tau}\right)\frac{\delta}{\tau} \approx 2\delta$$

where the approximation follows because $\delta/\tau \ll 1$. $\square$

This topic is explored carefully in Sect. 4.12 of [19]. A further point is that from a public health point of view, direct measurement of dI/dt is problematic because only infecteds who seek medical help generate data. Usually data on dR/dt are more readily available.

The SIR model can make predictions of $R(t)$. First, by the chain rule $dS/dR = (dS/dt)/(dR/dt) = -S/\tau$, so $S(R) = S_0 e^{-R/\tau}$. Next, recall $S + I + R = N$ is constant. Then the R' equation becomes

$$R' = \gamma I = \gamma(N - R - S) = \gamma(N - R - S_0 e^{-R/\tau})$$

Use the quadratic approximation from the Taylor series for $e^{-R/\tau}$. Then separation of variables gives a differential equation of the form $dR/(-AR^2 - BR + C) = dt$, but the coefficients A, B, and C are complicated. Details are on pages 462–464 of [19]. For a plague epidemic on the island of Bombay from December 17, 1905, to July 21, 1906, between 80% and 90% if the infecteds died (remember, for the purposes of this model of this epidemic, the R population represents "removed" infecteds, those who recover and those who die), so public health data were available. For this epidemic, Kermack and McKendrick's model is $R' = 890\text{sech}^2(0.2t - 3.4)$. The agreement of prediction and observed data was very good.

If a population density is only slightly above the epidemic threshold, an epidemic will reduce the population only slightly below the threshold. The population drops from $\tau + \delta$ to $\tau - \delta$. But if the population increases considerably above this threshold, an epidemic can have devastating effects. Every death is heartbreak for someone, and we must do what we can to reduce the number of broken hearts.

A.4   VOLTERRA'S TRICK

Recall the Lotka-Volterra predator-prey equations of Sect. 7.3. At the non-zero fixed point $(c/d, a/b)$ the eigenvalues of the derivative matrix are $\pm i\sqrt{ac}$, so the Hartman-Grobman theorem gives no information about the stability of this fixed point. In Example 9.5.2 we found a Liapunov function that shows this fixed point is surrounded by closed trajectories, but this argument was fairly involved. In this section, we'll learn Volterra's trick [158] to determine the stability of the non-zero fixed point. This is a beautiful, clever trick, useful to see because it shows the power of thinking in unexpected directions. We promised this trick in Sect. 7.3.

BASIC LOTKA-VOLTERRA VECTOR FIELD DIRECTIONS

Recall the Lotka-Volterra equations (7.5) and (7.4):

$$x' = -ax + bxy = x(-a + by) \qquad y' = cy - dxy = y(c - dx)$$

In Sect. 7.5 we saw the x-nullcline is the lines $x = 0$ and $y = a/b$, and the y-nullcline is the lines $y = 0$ and $x = c/d$. The fixed points are $(0,0)$ and $(c/d, a/b)$. The phase plane for this system is the first quadrant, because x and y denote populations, never negative. The nullclines divide the phase plane into four regions, on each of which x' and y' have constant signs.

$$x < c/d: \quad y' = y(c - dx) > 0 \qquad\qquad x > c/d: \quad y' = y(c - dx) < 0$$
$$y < a/b: \quad x' = x(-a + by) < 0 \qquad\qquad y > a/b: \quad x' = x(-a + by) > 0$$

Combining this information in a graph, in Fig. A.1 we show the general directions of the vector field. For example, in the region $x > c/d$, $y > a/b$, in the left graph of the figure we have $x' > 0$ and $y' < 0$, so in the right graph we see the vector field points to the right ($x' > 0$) and down ($y' < 0$).

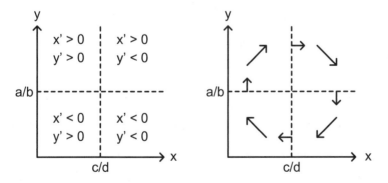

Figure A.1. Determining the directions of the Lotka-Volterra vector field.

From this we deduce the solution curves can take one of two possible forms: periodic solutions around the fixed point $(c/d, a/b)$, or spirals into the fixed point. These have considerably different biological implications. In the first, both predator and prey populations are periodic, but are out of phase, the increase or decrease of one leading that of the other. In the second, both populations converge to the values at the fixed point. That the vector field is vertical on $y = a/b$ and horizontal on $x = c/d$ makes the first seem more likely, because for the spiral, the solution curves would have to move closer to the fixed point as they travel between nullclines, but slow their approach to 0 while crossing each nullcline, then continue moving closer to the fixed point. This fairly complicated behavior seems unlikely for these simple equations.

HOW VOLTERRA'S TRICK WORKS

"Seems unlikely" hardly can be taken for a proof. Volterra gave a clever argument that the solutions are periodic. Here are the main points.

$$\frac{dy}{dx} = \frac{dy/dt}{dx/dt} \quad \text{by the chain rule } dy/dt = dy/dx \cdot dx/dt$$

$$= \frac{y(c - dx)}{x(-a + by)} \quad \text{using Eqs. (7.5) and (7.4) for } dy/dt \text{ and } dx/dt$$

$$= \left(\frac{c - dx}{x}\right) \cdot \left(\frac{y}{-a + by}\right) \quad \text{separating variables} \qquad \text{(A.12)}$$

This gives

$$\frac{-a + by}{y} dy = \frac{c - dx}{x} dx \quad \text{that is,} \quad \left(-\frac{a}{y} + b\right) dy = \left(\frac{c}{x} - d\right) dx$$

Integrate both sides to obtain

$$-a\ln(y) + by = c\ln(x) - dx + k$$

Now exponentiate both sides and write $K = e^k$. This gives

$$y^{-a} e^{by} = Kx^c e^{-dx} \qquad \text{(A.13)}$$

Straightforward so far, not much of a trick. But now we run into a problem. We can't solve this equation for either x or y. Here's Volterra's trick: define two new variables, w and z, by

$$w = y^{-a} e^{by} \quad \text{and} \quad z = Kx^c e^{-dx} \qquad \text{(A.14)}$$

Plotting z as a function of y and w as a function of x we see graphs as illustrated in Fig. A.2. Here we have taken $a = 1$, $b = 1$, $c = 1$, $d = 1$, and $K = 10$. What can we learn from these? We don't need all parts of these graphs. From the definitions (A.14) of w and z, the solution (A.13) can be written as

$$z = w \tag{A.15}$$

So we are interested in which points of the graphs of Fig. A.2 correspond to $z = w$.

Being populations, x and y take only non-negative values. By Eq. (A.14), w and z take only non-negative values. Then each of the four graphs x vs y, w vs y, z vs x, and w vs z occupies only one quadrant, and so all four can be plotted in a single graph. See the left

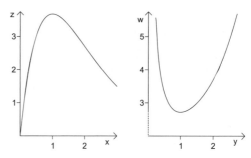

Figure A.2. Graphs of z and of w.

side of Fig. A.3. Note the orientation of the axes, each shared by two quadrants. In each quadrant, the two variables of that quadrant, indicated by the axes bounding that quadrant, take on only non-negative values. The wy quadrant contains the graph $w = y^{-a}e^{by}$ (Eq. (A.14)), the zx quadrant contains the graph $z = Kx^c e^{-dx}$ (Eq. (A.14)), and the zw quadrant contains the graph of $z = w$ (Eq. (A.15)). Our task is to find what goes in the xy quadrant.

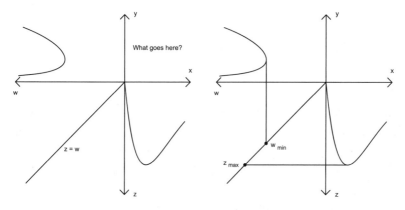

Figure A.3. Left: the quadrants of Volterra's graph. Right: determining the range of $z = w$.

On the right side of Fig. A.3 we show the range of values on the $z = w$ line given by the intersection of $[w_{min}, \infty)$ from the w vs y graph, and $[0, z_{max}]$ from the z vs x graph.

In Fig. A.4 we use the information in these graphs to begin plotting points on a solution curve to the Lotka-Volterra equations. On the left graph of this figure, we begin with the point (w_{min}, w_{min}) and draw a vertical line to the minimum point on the w vs y graph. This determines the value of y corresponding to w_{min}. Extend a horizontal line with this y-value into the xy quadrant. Next, from the point (w_{min}, w_{min}) draw a horizontal line into the zx quadrant. Notice this line intersects the z vs x graph at two points. Draw vertical lines from these points into the xy quadrant. The intersection of each of these vertical lines with the horizontal line drawn into this quadrant determines two points on the solution curve.

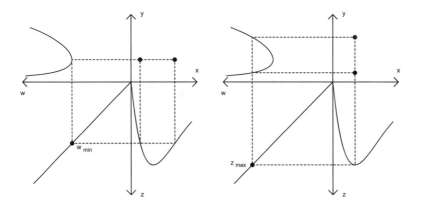

Figure A.4. Finding four points on a solution curve.

In the right graph of Fig. A.4 we construct two more points on the solution curve, this time starting from the point (z_{max}, z_{max}). Then starting from any point lying between (w_{min}, w_{min}) and (z_{max}, z_{max}), we generate four points on the solution curve. See the left side of Fig. A.5. Repeating this process for many points between (w_{min}, w_{min}) and (z_{max}, z_{max}), by the same process we generate many points lying on the solution curve. See the right side of Fig. A.5.

How do we generate solution curves passing through different points? The curve $w = y^{-a}e^{by}$ depends on only a and b, parameters of the model, representing the mortality rate of the predator and the positive impact predator-prey interactions have on the predator population. This curve does not depend on the initial values of the populations. The curve $z = Kx^c e^{-dx}$ depends on model

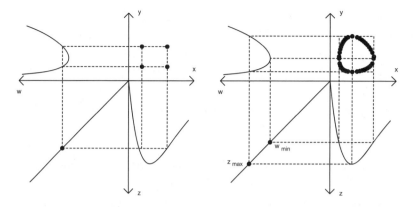

Figure A.5. Left: finding four more points on a solution curve. Right: many points on a solution curve.

parameters c and d, but also on K, the constant of integration. In Volterra's approach, different solution curves are the result of different curves $z = Kx^c e^{-dx}$, varying with changing K.

This is very clever. Volterra added 2 more dimensions, pushing the problem up to $\mathbb{R}^4$. Then he took a manageable subset of that problem, and showed the solutions of the predator-prey equations are closed trajectories. The main reason for including this section is that Volterra's trick is beautiful. And it gives you an example not only of "thinking outside the box," but of thinking outside the space containing the box.

A.5 EULER'S FORMULA

We used Euler's formula, $e^{it} = \cos(t) + i\sin(t)$, first in Sect. 8.5. This formula has many proofs. For example, apply the power series of e^x to $x = it$. The real terms constitute the power series for $\cos(t)$, the imaginary terms are i times the power series for $\sin(t)$. But here's a simple, elegant proof. Take

$$f(t) = e^{-it}(\cos(t) + i\sin(t))$$

Then differentiate.

$$f'(t) = -ie^{-it}(\cos(t) + i\sin(t)) + e^{-it}(-\sin(t) + i\cos(t)) = 0$$

So f is constant. Because $f(0) = 1$, we have $f(t) = 1$ for all t. This proves Euler's formula.

Sadly, I've forgotten where I first saw this argument. Maybe my undergraduate complex analysis course, but I don't know.

But I do have a story about Euler's formula. On a drive back to Connecticut from a summer visit with my father in West Virginia, we stopped at a rest area on I-84 east of Scranton. As we walked in, two parents and their teenage daughter walked out. On the daughter's t-shirt was written, "Talk nerdy to me." That was interesting. A bit later in the day, we stopped at the rest area in Danbury. As we walked in to the rest area, the same trio walked out. The daughter smiled and waved. I am balding, have an unkempt beard, walk with a cane, and at that time wore an eye patch. (Now I've just painted over one lens of my glasses.) That she recognized me is not such a surprise. As we passed, I said to the daughter, "If you're serious about what's on your shirt, how's this? $e^{i\pi} = -1$." The daughter grinned and said, "Euler's formula. Cool!" As they continued to their car, the daughter began to explain Euler's formula to her parents. I suppose I'm lucky the parents didn't call the police.

A.6 PROOF OF THE SECOND-DERIVATIVE TEST

Here we'll sketch a proof of the second-derivative test for functions of two variables. Suppose $(\partial f/\partial x)(x_0,y_0) = (\partial f/\partial y)(x_0,y_0) = 0$, and

$$Q(x_0,y_0) = \left(\frac{\partial^2 f}{\partial x^2}(x_0,y_0)\right) \cdot \left(\frac{\partial^2 f}{\partial y^2}(x_0,y_0)\right) - \left(\frac{\partial^2 f}{\partial x\partial y}(x_0,y_0)\right)^2 > 0$$

Then (x_0,y_0) is a local minimum if $(\partial^2 f/\partial x^2)(x_0,y_0) > 0$, and is a local maximum if $(\partial^2 f/\partial x^2)(x_0,y_0) < 0$. If $Q(x_0,y_0) < 0$, the point (x_0,y_0) is a saddle point. If $Q(x_0,y_0) = 0$, the test fails.

We used this test in Sect. 9.4 to prove Prop. 9.4.1, a criterion to determine if a quadratic Liapunov function $ax^2 + bxy + cy^2$ is positive definite, negative definite, or neither. To understand the general architecture of the two-variable second-derivative test, recall the second-derivative test from single-variable calculus. Suppose $f'(x_0) = 0$. If $f''(x_0) > 0$ then x_0 is a local minimum, if $f''(x_0) < 0$ then x_0 is a local maximum, and if $f''(x_0) = 0$ then the test fails. The reason is that $f''(x_0)$ detects the concavity of the graph at x_0: $f''(x_0) > 0$ means the graph is concave up, and $f''(x_0) < 0$ means the graph is concave down. A point where the graph has a horizontal tangent and is concave up is a local minimum because the graph opens upward at that point; concave down is a local maximum because the graph opens downward. This simple idea will guide us in the second-derivative test for functions of two variables.

Our proof of the two-variable second-derivative test is based on this single-variable result:

Lemma A.6.1. For the function $g(x) = Ax^2 + 2Bx + C$,
(1) if $AC - B^2 > 0$ and $A > 0$, then $g(x) > 0$ for all x,
(2) if $AC - B^2 > 0$ and $A < 0$, then $g(x) < 0$ for all x, and
(3) if $AC - B^2 < 0$, then $g(x) > 0$ for some x and $g(x) < 0$ for some x.

Proof. We see $g(x) = 0$ at $x_{\pm} = (-B \pm \sqrt{B^2 - AC})/A$. In cases (1) and (2), the expression in the square root is negative so the graph of g does not cross the x-axis. Consequently, $g(x)$ is positive for all x or is negative for all x. Now the condition $AC - B^2 > 0$ means AC must be positive, so A and C have the same sign. Then in case (1), $A > 0$ means $C > 0$ and so $g(0) = C > 0$. Because the graph of g doesn't cross the x-axis, we see $g(x) > 0$ for all x. An analogous argument establishes case (2). For case (3), $AC - B^2 < 0$ means that $g(x) = 0$ at two points, x_+ and x_-. Because g is quadratic, its graph is a parabola that crosses the x-axis at x_+ and x_-. Consequently, $g(x)$ switches sign at x_+ and x_-. $\square$

To simplify the arithmetic, translate the point (x_0, y_0) to the origin by the change of variables $u = x - x_0$ and $v = y - y_0$. By the two-variable Taylor theorem (Google, if you're curious), after this change of variables we can write

$$f(u, v) = z_0 + Au^2 + 2Buv + Cv^2 + \text{higher-order terms}$$

where $z_0 = f(x_0, y_0)$, and $A = \partial^2 f / \partial x^2$, $B = \partial^2 f / \partial x \partial y$, and $C = \partial^2 f / \partial y^2$, all three evaluated at $(x, y) = (x_0, y_0)$. The higher-order terms (abbreviated HOTs by one of my differential equations teachers at RPI) involve u^3, uv^2, $u^2 v^3$, and the like, so very near $(u, v) = (0, 0)$, the shape of the graph of f is dominated by the quadratic terms.

Why doesn't $f(u, v)$ have linear terms? Translation of the critical point to the origin removes the linear terms. An example will illustrate this. The function $h(x, y) = -5x - 7y + 2x^2 + xy + 3y^2$ has one critical point, at $(x, y) = (1, 1)$. Then the change of variables is $u = x - 1, v = y - 1$. That is, $x = u + 1, y = v + 1$ and so

$$h(x, y) = h(u + 1, v + 1)$$
$$= -5(u + 1) - 7(v + 1) + 2(u + 1)^2 + (u + 1)(v + 1) + 3(v + 1)^2$$
$$= -6 + 2u^2 + uv + 3v^2$$

Back to the second-derivative test. Write

$$h(u, v) = f(u, v) - z_0 = Au^2 + 2Buv + Cv^2$$

and note $h(0, 0) = 0$. At $(u, v) \neq (0, 0)$ we have $u \neq 0$, $v \neq 0$, or both. Suppose $v \neq 0$. (Use an analogous argument if $u \neq 0$.) Then we can write $h(u, v)$ as

$$h(u, v) = v^2 (A(u/v)^2 + 2B(u/v) + C)$$

The sign of $h(u,v)$ is determined by the sign of $A(u/v)^2 + 2B(u/v) + C$. By Lemma A.6.1 with $x = u/v$,

(1) if $AC - B^2 > 0$ and $A > 0$, then $h(u,v) > 0$, and so $f(x_0, y_0)$ is a local minimum,

(2) if $AC - B^2 > 0$ and $A < 0$, then $h(u,v) < 0$, and so $f(x_0, y_0)$ is a local maximum, and

(3) if $AC - B^2 < 0$, then some $h(u,v) < 0$ and some $h(u,v) > 0$, so $f(x_0, y_0)$ is a saddle point.

This finishes our sketch of the proof of the second-derivative test for functions of two variables.

The addition of a second variable produced another case, the saddle point. Near a local minimum the function increases in every direction. Near a local maximum the function decreases in every direction. Near a saddle point the function increases in some directions and decreases in others. Appropriately interpreted, still more variables add nothing new, just more directions in which the function can increase or decrease. How to visualize these is a different matter.

A.7 SOME LINEAR ALGEBRA

We've introduced concepts from linear algebra, as needed, mostly in Sects. 8.3, 8.4, and 8.5. Here we'll organize and extend a few notions that we've used, and we'll present a tiny bit of the underlying theory.

LINEAR INDEPENDENCE, SPAN, BASIS

Vectors $\vec{v}_1, \vec{v}_2, \ldots, \vec{v}_n$ are *linearly dependent* if there are constants $c_1, c_2, \ldots, c_n$, not all 0, that satisfy the equation

$$c_1\vec{v}_1 + c_2\vec{v}_2 + \cdots + c_n\vec{v}_n = 0 \qquad (A.16)$$

Here's why this relation is called linear dependence. We know not all of the c_i are 0, so let's suppose $c_1 \neq 0$. Then we can solve Eq. (A.16) for $\vec{v}_1$,

$$\vec{v}_1 = -\frac{c_2}{c_1}\vec{v}_2 - \cdots - \frac{c_n}{c_1}\vec{v}_n$$

That is, $\vec{v}_1$ lies in the space determined by $\vec{v}_2, \ldots, \vec{v}_n$. For example, the vectors $\vec{v}_1 = \langle 1/2, 1/3 \rangle$, $\vec{v}_2 = \langle 1, 1 \rangle$, and $\vec{v}_3 = \langle -1, 1 \rangle$ are linearly dependent because

$$\langle 1/2, 1/3 \rangle = \frac{5}{12}\langle 1, 1 \rangle - \frac{1}{12}\langle -1, 1 \rangle$$

If the vectors $\vec{v}_1, \vec{v}_2, \ldots, \vec{v}_n$ are not linearly dependent, they are *linearly independent*: $c_1 = c_2 = \cdots = c_n = 0$ is the only solution of Eq. (A.16).

The *span* of a collection of vectors $\vec{v}_1, \vec{v}_2, \ldots, \vec{v}_n$ is the set of all linear combinations of these vectors,

$$\text{span}(\vec{v}_1, \vec{v}_2, \ldots, \vec{v}_n) = \{c_1\vec{v}_1 + c_2\vec{v}_2 + \cdots + c_n\vec{v}_n : \text{for all real } c_1, c_2, \ldots, c_n\}$$

For example, $\text{span}(\langle 1,1 \rangle, \langle 2,3 \rangle) = \mathbb{R}^2$ because for any a and b,

$$\langle a, b \rangle = (3a - 2b)\langle 1,1 \rangle + (-a + b)\langle 2,3 \rangle$$

A *spanning set* of vectors in $\mathbb{R}^n$ is a set of vectors whose span is all of $\mathbb{R}^n$.

If we remove vectors from a linearly independent set, the resulting smaller set of vectors still is linearly independent. Adding vectors to a linearly independent set may destroy independence.

If we add vectors to a spanning set, the resulting larger set still is spanning. Removing vectors from a spanning set may destroy its spanning.

A central concept in linear algebra is a basis: a set of vectors is a *basis* if it is linearly independent and spanning. If a spanning set is not a basis, we can remove some vectors (we need to be careful about which vectors we remove so we don't destroy the spanning of the remaining vectors) until the remaining vectors are a basis. If a linearly independent set is not a basis, we can add some vectors (we need to be careful about which vectors we add so we don't destroy the linear independence of the larger set of vectors) until the larger set is a basis.

Linear algebra works on spaces besides $\mathbb{R}^n$. A *vector space* is a collection of vectors $\vec{v}$ and scalars s, along with two operations, vector addition and scalar multiplication, that work pretty much the way familiar vectors and real numbers work. Here are the rules. For all vectors $\vec{u}$, $\vec{v}$, and $\vec{w}$, and for all scalars s and t,

$$\vec{u} + \vec{v} = \vec{v} + \vec{u} \quad (\vec{u} + \vec{v}) + \vec{w} = \vec{u} + (\vec{v} + \vec{w}) \quad \vec{v} + \vec{0} = \vec{v}$$
$$\vec{v} + -\vec{v} = \vec{0} \quad s(\vec{u} + \vec{v}) = s\vec{u} + s\vec{v} \quad (s+t)\vec{v} = s\vec{v} + t\vec{v}$$
$$(st)\vec{v} = s(t\vec{v}) \quad 1\vec{v} = \vec{v}$$

where $\vec{0}$ is the unique zero vector, $-\vec{v}$ is the additive inverse of $\vec{v}$, and 1 is the unit scalar.

Example A.7.1. Here are three examples of vector spaces that appear different from $\mathbb{R}^n$.

(a) *Matrix space.* The 2×2 matrices form a vector space with these operations

$$\begin{bmatrix} a & b \\ c & d \end{bmatrix} + \begin{bmatrix} e & f \\ g & h \end{bmatrix} = \begin{bmatrix} a+e & b+f \\ c+g & d+h \end{bmatrix} \qquad s\begin{bmatrix} a & b \\ c & d \end{bmatrix} = \begin{bmatrix} sa & sb \\ sc & sd \end{bmatrix}$$

(b) *Polynomial space.* The polynomials of degree n form a vector space with these operations

$$(a_0 + a_1x + \cdots + a_nx^n) + (b_0 + b_1x + \cdots + b_nx^n)$$
$$= (a_0 + b_0) + (a_1 + b_1)x + \cdots + (a_n + b_n)x^n$$
$$s(a_0 + a_1x + \cdots + a_nx^n) = sa_0 + sa_1x + \cdots + sa_nx^n$$

(c) *Function space.* The real-valued functions defined on $[0,1]$ form a vector space with these operations

$$(f+g)(x) = f(x) + g(x) \qquad (sf)(x) = sf(x)$$

We can use functions defined on any interval $[a,b]$, not just $[0,1]$. $\square$

A subset W of a vector space V that is itself a vector space is called a *subspace* of V. To check that W is a subspace, all we need to do is show that the sum of every pair of elements of W is an element of W, and that all scalar multiples of all elements if W are elements of W.

Example A.7.2. Subspaces of the vector spaces of Example A.7.1.
(a) The upper triangular matrices form a subspace of the 2×2 matrices because

$$\begin{bmatrix} a & b \\ 0 & c \end{bmatrix} + \begin{bmatrix} d & e \\ 0 & f \end{bmatrix} = \begin{bmatrix} a+d & b+e \\ 0 & c+f \end{bmatrix} \quad \text{and} \quad s\begin{bmatrix} a & b \\ 0 & c \end{bmatrix} = \begin{bmatrix} sa & sb \\ 0 & sc \end{bmatrix}$$

(b) The even polynomials (polynomials with all exponents even) form a subspace because the sum of polynomials with even exponents is another polynomial with even exponent, and because scaler multiplication doesn't change polynomial exponents.
(c) The continuous functions $f : [0,1] \to \mathbb{R}$ with $f(0) = 0$ form a subspace because

$$(f+g)(0) = f(0) + g(0) = 0 \quad \text{and} \quad (sf)(0) = s(f(0)) = 0$$

Similar arguments show that for any point $x \in [0,1]$ the set of continuous functions f with $f(x) = 0$ form a subspace. On the other hand, the set of continuous functions f with $f(0) = 1$ do not form a subspace. $\square$

The number of elements in a basis of a vector space is the *dimension* of the vector space. Although examples make it plausible that all bases of a given vector space have the same number of elements (and so the definition of dimension makes sense), some proof is required. For example, see Theorem 6.2.3 of [154], my favorite linear algebra book. If you're interested, try to figure this out on your own. Suppose two bases have different numbers of elements. How can this cause trouble?

Here's a useful application of the notion of dimension. If a vector space has dimension n, then

- A collection of n linearly independent vectors is a basis.
- A spanning collection of n vectors is a basis.

So in an n-dimensional space, to show that a set of n vectors is a basis, it suffices to check either that the vectors are linearly independent, or that the set spans the vector space. Without using the the notion of dimension, we must check both.

The vector space of Example A.7.1 (a) has dimension 4 because it has basis

$$\begin{bmatrix} 1 & 0 \\ 0 & 0 \end{bmatrix}, \quad \begin{bmatrix} 0 & 1 \\ 0 & 0 \end{bmatrix}, \quad \begin{bmatrix} 0 & 0 \\ 1 & 0 \end{bmatrix}, \quad \text{and} \quad \begin{bmatrix} 0 & 0 \\ 0 & 1 \end{bmatrix}$$

Here's why these matrices span: for any a, b, c, and d,

$$\begin{bmatrix} a & b \\ c & d \end{bmatrix} = a\begin{bmatrix} 1 & 0 \\ 0 & 0 \end{bmatrix} + b\begin{bmatrix} 0 & 1 \\ 0 & 0 \end{bmatrix} + c\begin{bmatrix} 0 & 0 \\ 1 & 0 \end{bmatrix} + d\begin{bmatrix} 0 & 0 \\ 0 & 1 \end{bmatrix}$$

And here's why these matrices are linearly independent

$$a\begin{bmatrix} 1 & 0 \\ 0 & 0 \end{bmatrix} + b\begin{bmatrix} 0 & 1 \\ 0 & 0 \end{bmatrix} + c\begin{bmatrix} 0 & 0 \\ 1 & 0 \end{bmatrix} + d\begin{bmatrix} 0 & 0 \\ 0 & 1 \end{bmatrix} = \begin{bmatrix} 0 & 0 \\ 0 & 0 \end{bmatrix}$$

implies that $a = b = c = d = 0$.

The vector space of Example A.7.1 (b) has dimension $n+1$ because it has basis

$$1, \quad x, \quad x^2, \quad \ldots, \quad x^n$$

Certainly these span this vector space because any polynomial of degree n is a linear combination of $1, x, \ldots, x^n$. To see these are independent, observe that any equation $c_0 + c_1 x + \cdots + c_n x^n = 0$ holds for at most n distinct values of x (this is the fundamental theorem of algebra), and so does not hold for all x in $[0,1]$.

Extending the argument for (b), we see that the vector space of Example A.7.1 (c) is infinite-dimensional because it has a basis $1, x, x^2, \ldots$ with infinitely many elements.

EIGENVECTORS, REVISITED

Here we'll gather some observations about eigenvalues, eigenvectors, and linear independence.

Theorem A.7.1. Suppose the $n \times n$ matrix M has eigenvectors $\vec{e}_1, \ldots, \vec{e}_n$ that form a basis for $\mathbb{R}^n$. Let W denote the matrix whose columns are these eigenvectors, $W = [\vec{e}_1 \cdots \vec{e}_n]$. Then

$$W^{-1}MW = \text{diag}(\lambda_1,\ldots,\lambda_n) \tag{A.17}$$

where $\text{diag}(\lambda_1,\ldots,\lambda_n)$ is the matrix with the eigenvalues along the diagonal and all other matrix entries are 0.

Proof. We'll prove Eq. (A.17) for $n = 2$. The more general result uses the same ideas but just takes more room. First note

$$MW = \begin{bmatrix} a & b \\ c & d \end{bmatrix} \begin{bmatrix} e_{11} & e_{12} \\ e_{21} & e_{22} \end{bmatrix} = \begin{bmatrix} ae_{11} + be_{21} & ae_{12} + be_{22} \\ ce_{11} + de_{21} & ce_{12} + de_{22} \end{bmatrix}$$

Next,

$$M\vec{e}_1 = \begin{bmatrix} a & b \\ c & d \end{bmatrix} \begin{bmatrix} e_{11} \\ e_{21} \end{bmatrix} = \begin{bmatrix} ae_{11} + be_{21} \\ ce_{11} + de_{21} \end{bmatrix}$$

$$M\vec{e}_2 = \begin{bmatrix} a & b \\ c & d \end{bmatrix} \begin{bmatrix} e_{12} \\ e_{22} \end{bmatrix} = \begin{bmatrix} ae_{12} + be_{22} \\ ce_{12} + de_{22} \end{bmatrix}$$

That is, the first column of MW is $M\vec{e}_1$, the second column is $M\vec{e}_2$.

Now we know that

$$M\vec{e}_1 = \lambda_1\vec{e}_1 = \begin{bmatrix} \lambda_1 e_{11} \\ \lambda_1 e_{21} \end{bmatrix} \quad \text{and} \quad M\vec{e}_2 = \lambda_2\vec{e}_2 = \begin{bmatrix} \lambda_2 e_{12} \\ \lambda_2 e_{22} \end{bmatrix}$$

and consequently

$$MW = \begin{bmatrix} \lambda_1 e_{11} & \lambda_2 e_{12} \\ \lambda_1 e_{21} & \lambda_2 e_{22} \end{bmatrix} = \begin{bmatrix} e_{11} & e_{12} \\ e_{21} & e_{22} \end{bmatrix} \begin{bmatrix} \lambda_1 & 0 \\ 0 & \lambda_2 \end{bmatrix} = W\text{diag}(\lambda_1,\lambda_2)$$

Multiplying both sides of $MW = W\text{diag}(\lambda_1,\lambda_2)$ on the left by W^{-1} gives Eq. (A.17). $\square$

When Eq. (A.17) holds we say that the matrix M is *diagonalizable*. Iteration of diagonalizable matrices is particularly simple. First, multiply Eq. (A.17) on the left by W and on the right by W^{-1} to obtain $M = W\text{diag}(\lambda_1,\ldots,\lambda_n)W^{-1}$. For the moment, write $D = \text{diag}(\lambda_1,\ldots,\lambda_n)$. Then for all positive integers k,

$$M^k = (WDW^{-1})^k = (WDW^{-1})(WDW^{-1})\cdots(WDW^{-1})$$

$$= WD(W^{-1}W)D(W^{-1}W)\cdots(W^{-1}W)DW^{-1} = WD^kW^{-1}$$

This is simpler because

$$D^k = (\text{diag}(\lambda_1,\ldots,\lambda_n))^k = \text{diag}(\lambda_1^k,\ldots,\lambda_n^k)$$

Then if M is diagonalizable, for any n-dimensional vector $\vec{v}$, the long-term behavior $M^k\vec{v}$ is easy to compute:

$$M^k\vec{v} = WD^kW^{-1}\vec{v} = W\text{diag}(\lambda_1^k,\ldots,\lambda_n^k)W^{-1}\vec{v}$$

How can we tell if the eigenvectors $\vec{e}_1, \vec{e}_2, \ldots, \vec{e}_n$ of M form a basis for $\mathbb{R}^n$? Because $\dim(\mathbb{R}^n) = n$, the eigenvectors form a basis if and only if they are linearly independent. That is, if and only if the only solution of

$$c_1\vec{e}_1 + c_2\vec{e}_2 + \cdots + c_n\vec{e}_n = \vec{0} \tag{A.18}$$

is $c_1 = c_2 = \cdots = c_n = 0$. Again denote by W the matrix whose columns are these eigenvectors. Then Eq. (A.18) is equivalent to

$$W \begin{bmatrix} c_1 \\ \cdots \\ c_n \end{bmatrix} = \vec{0} \tag{A.19}$$

because their left-hand sides are equal. Write out the $n = 2$ case if this isn't clear. We know that $\det(W) \neq 0$ implies that W is invertible, so we can multiply both sides of Eq. (A.19) on the left by W^{-1} to obtain $c_1 = c_2 = \cdots = c_n = 0$. Note that we haven't used the hypothesis that the $\vec{e}_i$ are eigenvectors, so we can apply it to any square matrix M. That is, we have shown

Proposition A.7.1. The columns of an $n \times n$ matrix M are linearly independent if $\det(M) \neq 0$.

This condition is easy enough to check. Here's one that's even easier to check.

Proposition A.7.2. If an $n \times n$ matrix M has n distinct eigenvalues (no eigenvalue is equal to any of the other $n-1$ eigenvalues), then the n eigenvectors are linearly independent.

Proof. The $n = 2$ case demonstrates all the points of the general proof. So suppose a 2×2 matrix M has eigenvalues λ_1 and λ_2, with eigenvectors $\vec{e}_1$ and $\vec{e}_2$. If $\lambda_1 \neq \lambda_2$, then $\vec{e}_1$ and $\vec{e}_2$ are linearly independent.

Suppose $c_1\vec{e}_1 + c_2\vec{e}_2 = \vec{0}$. Linear dependence of the eigenvectors requires that at least one of c_1 and c_2 must be non-zero. Say $c_1 \neq 0$. Then $\vec{e}_1 = -(c_2/c_1)\vec{e}_2 = \alpha\vec{e}_2$ and so

$$M\vec{e}_1 = M(\alpha\vec{e}_2) = \alpha M\vec{e}_2$$
$$\lambda_1\vec{e}_1 = \alpha\lambda_2\vec{e}_2 \tag{A.20}$$

Then multiply both sides of $\vec{e}_1 = \alpha\vec{e}_2$ by λ_1 and we have

$$\lambda_1\vec{e}_1 = \alpha\lambda_1\vec{e}_2 \tag{A.21}$$

Subtract Eq. (A.21) from Eq. (A.20) to find

$$\vec{0} = \alpha\lambda_2\vec{e}_2 - \alpha\lambda_1\vec{e}_2 = \alpha(\lambda_2 - \lambda_1)\vec{e}_2$$

This is impossible, because $\alpha \neq 0$, $\lambda_2 \neq \lambda_1$, and no eigenvector is $\vec{0}$. Consequently, the eigenvectors are linearly independent. □

Two more concepts are related to the independence of eigenvectors. Suppose λ_0 is an eigenvalue of an $n \times n$ matrix M. The *algebraic multiplicity* $A(\lambda_0)$ of λ_0 is the number of times λ_0 occurs as a root of $\det(M - \lambda I) = 0$, that is, the exponent of the factor $(\lambda - \lambda_0)$ in the decomposition

$$\det(M - \lambda I) = (\lambda - \lambda_0)^m P(\lambda)$$

where $(\lambda - \lambda_0)$ is not a factor of $P(\lambda)$. The *geometric multiplicity* $G(\lambda_0)$ of λ_0 is the dimension of the span of the eigenvectors of λ_0, called the *eigenspace* $E(\lambda_0)$ of λ_0. A simple example will illustrate these concepts.

Example A.7.3. *Algebraic and geometric multiplicities.* The matrix

$$M = \begin{bmatrix} 1 & 2 & 0 \\ 0 & 1 & 0 \\ 0 & 3 & 1 \end{bmatrix} \text{ has } \det(M - \lambda I) = (1 - \lambda)^3$$

So the matrix M has only one eigenvalue, $\lambda = 1$. Because $\lambda - 1$ occurs three times as a factor of $\det(M - \lambda I)$, the algebraic multiplicity of $\lambda = 1$ is 3.

For the geometric multiplicity, use the eigenvector equation $M\vec{v} = \lambda\vec{v}$,

$$\begin{bmatrix} 1 & 2 & 0 \\ 0 & 1 & 0 \\ 0 & 3 & 1 \end{bmatrix} \begin{bmatrix} x \\ y \\ z \end{bmatrix} = \begin{bmatrix} x \\ y \\ z \end{bmatrix} \quad \text{that is,} \quad \begin{aligned} x + 2y &= x \\ y &= y \\ 3y + z &= z \end{aligned}$$

These give one condition, $y = 0$, so the eigenvectors of $\lambda = 1$ span the xz-plane and the geometric multiplicity of $\lambda = 1$ is 2. □

This example illustrates a relation between the algebraic and geometric multiplicities.

Proposition A.7.3. For every eigenvalue λ of an $n \times n$ matrix M, the geometric multiplicity $G(\lambda) \leq A(\lambda)$, the algebraic multiplicity.

Proof. Suppose the geometric multiplicity $G(\lambda) = m$ and that the eigenvectors $\vec{e}_1, \ldots, \vec{e}_m$ are a basis for the eigenspace of λ. Take $\vec{v}_{m+1} \ldots, \vec{v}_n$ to be any vectors that make $\{\vec{e}_1, \ldots, \vec{e}_m, \vec{v}_{m+1}, \ldots, \vec{v}_n\}$ a basis for $\mathbb{R}^n$. Then the matrix $A = [\vec{e}_1, \ldots, \vec{e}_m, \vec{v}_{m+1}, \ldots, \vec{v}_n]$ is invertible and the first m columns of $A^{-1}MA$ are λ times the first m columns of the $n \times n$ identity matrix I. To see this, let the rows of A^{-1} be the row vectors $\vec{c}_1,\ldots,\vec{c}_n$. Then $A^{-1}A=I$ gives $\vec{c}_1\vec{e}_1=1$, $\vec{c}_1\vec{e}_2=0$, and so on. Then

$$A^{-1}MA = A^{-1}(MA) = A^{-1}[M\vec{e}_1 \cdots M\vec{e}_m\ M\vec{v}_{m+1} \cdots M\vec{v}_n]$$
$$= A^{-1}[\lambda\vec{e}_1 \cdots \lambda\vec{e}_m\ M\vec{v}_{m+1} \cdots M\vec{v}_n]$$

so the first m columns of $A^{-1}MA$ are λ times the first m columns of I, and consequently λ is an eigenvalue of $A^{-1}MA$ with algebraic multiplicity at least m.

The final step is to show that $A^{-1}MA$ and M have the same eigenvalues. For this we use $\det(PQ) = \det(P)\det(Q)$ and so $\det(A^{-1}) = 1/\det(A)$. The eigenvalues of $A^{-1}MA$ are the solutions of $\det(A^{-1}MA - \lambda I) = 0$ and we calculate

$$\det(A^{-1}MA - \lambda I) = \det(A^{-1}MA - A^{-1}\lambda IA) = \det(A^{-1}(M - \lambda I)A)$$
$$= \det(A^{-1})\det(M - \lambda I)\det(A) = \det(M - \lambda I)$$

That is, the geometric multiplicity of λ is $\leq$ the algebraic multiplicity of λ. $\square$

REPEATED EIGENVALUES

Repeated eigenvalues were encountered in Sect. 8.5. There the question is whether or not the differential equation has enough solutions of the form $\vec{v}e^{\lambda t}$, where $\vec{v}$ is an eigenvector for the eigenvalue λ. In the language of the previous section, this reduces to the question of whether for each eigenvalue the geometric multiplicity equals the algebraic multiplicity. If the multiplicities are equal, these solutions $\vec{v}e^{\lambda t}$ suffice; if not, we'll see that we need to use solutions that include $t\vec{v}e^{\lambda t}$.

First, we'll show how to use the Eq. (8.8) formulation, $(M - \lambda I)\vec{v} = \vec{0}$, of the eigenvector equation to find the eigenspace $E(\lambda)$ of an eigenvector λ. For this we'll use another notion from linear algebra. For vector spaces V and W a function $T : V \to W$ is a *linear transformation* if for all v_1 and v_2 in V and for all scalars s,

$$T(v_1 + v_2) = T(v_1) + T(v_2) \quad \text{and} \quad T(sv_1) = sT(v_1)$$

For example, multiplication by a 2×2 matrix is a linear transformation on $\mathbb{R}^2$.

The *kernel* of a linear transformation $T : V \to W$ is defined by

$$\ker(T) = \{\vec{v} \in V : T(\vec{v}) = \vec{0}\}$$

That $\ker(T)$ is a subspace of V follows easily from the linearity of T. For example, for $T(x,y) = (x - y, y - x)$ we see $\ker(T) = \{(x,y) : x = y\}$.

By Eq. (8.8) we see that the eigenspace $E(\lambda)$ is a kernel

$$E(\lambda) = \ker(M - \lambda I)$$

For example, with the matrix M of Example A.7.3 we find the eigenspace of $\lambda = 1$ is

$$\ker(M - 1I) = \ker \begin{bmatrix} 0 & 2 & 0 \\ 0 & 0 & 0 \\ 0 & 3 & 0 \end{bmatrix} \quad \text{and} \quad \begin{bmatrix} 0 & 2 & 0 \\ 0 & 0 & 0 \\ 0 & 3 & 0 \end{bmatrix} \begin{bmatrix} x \\ y \\ z \end{bmatrix} = \begin{bmatrix} 0 \\ 0 \\ 0 \end{bmatrix}$$

gives $y = 0$. We knew this already, but the kernel calculation is a bit quicker.

To end this section, let's find the general solution of the $\vec{v}' = M\vec{v}$, where M is the matrix of Example A.7.3. From the calculation of eigenvalues and eigenvectors of M we know two solutions are

$$\vec{v}_1(t) = \begin{bmatrix} x_1(t) \\ y_1(t) \\ z_1(t) \end{bmatrix} = \begin{bmatrix} 1 \\ 0 \\ 0 \end{bmatrix} e^t \quad \text{and} \quad \vec{v}_2(t) = \begin{bmatrix} x_2(t) \\ y_2(t) \\ z_2(t) \end{bmatrix} = \begin{bmatrix} 0 \\ 0 \\ 1 \end{bmatrix} e^t$$

To find the third solution, use case 5 of Sect. 8.5 as a model. So we take

$$\vec{v}_3(t) = \begin{bmatrix} a_1 \\ a_2 \\ a_3 \end{bmatrix} te^t + \begin{bmatrix} b_1 \\ b_2 \\ b_3 \end{bmatrix} e^t = \begin{bmatrix} b_1 + ta_1 \\ b_2 + ta_2 \\ b_3 + ta_3 \end{bmatrix} e^t$$

Then equating the coefficients of like powers of t in each of the three vector coordinates in the equation $\vec{v}'_3 = M\vec{v}_3$ gives six equations:

constant terms	$b_1 + a_1 = b_1 + 2b_2$	$b_2 + a_2 = b_2$	$b_3 + a_3 = 3b_2 + b_3$
coefficients of t	$a_1 = a_1 + 2a_2$	$a_2 = a_2$	$a_3 = 3a_2 + a_3$

These give $a_1 = 2b_2$, $a_2 = 0$, and $a_3 = 3b_2$. Take $b_1 = b_2 = b_3 = 1$. (Remember that our goal is to construct three solutions. Linear combinations of these three give the general solution.) This gives the third solution

$$\vec{v}_3(t) = \begin{bmatrix} x_3(t) \\ y_3(t) \\ z_3(t) \end{bmatrix} = \begin{bmatrix} 1 + 2t \\ 1 \\ 1 + 3t \end{bmatrix} e^t$$

These solutions, $\vec{v}_1$, $\vec{v}_2$, and $\vec{v}_3$, are independent and in fact are a basis for all solutions of $\vec{v}' = M\vec{v}$.

We'll present another approach in Sect. A.12.2 of volume 2.

Sects. A.8, A.9, and A.10 are much more abstract and mathematically demanding than are the other sections of the appendices. Sects. A.17 and A.21

also are more abstract, but A.21 uses line integrals, and A.17 is applied in our study of Markov chains, a topic of the second volume. Those two sections of Appendix A are in that book. Understanding them isn't essential to follow the arguments of other sections. I've included these sketches of proofs because the ideas involved, the interlocking of logic and geometry, are so very pretty. Too often we do not emphasize the aesthetics of our field. When they fit perfectly in our heads, when we see exactly how they work and why, geometric constructs reveal a remarkable beauty. I've included these sections to give a hint of why I fell in love with geometry, why I've stayed smitten by it for the last half century.

A.8   A PROOF OF LIAPUNOV'S THEOREM

In this section we sketch the main ideas of a proof of the first two parts of Liapunov's theorem of Sect. 9.4. The use of this theorem to determine the stability of fixed points is called Liapunov's direct method or Liapunov's second method. There is an indirect method, also called the first method. We mention this to clarify any references to other sources.

Suppose V is positive definite on a disc D with center $(0,0)$ and $V(0,0) = 0$, and V has continuous partial derivatives.

- If V' is negative definite, the origin is asymptotically stable.
- If V' is negative semidefinite, the origin is stable.

We'll suppose that the origin is the only critical point of V in the disc D with radius r and center $(0,0)$. For any value V_i that V takes for points in D, the level set of V for this value is

$$\{(x,y) : V(x,y) = V_i \text{ and } (x,y) \in D\}$$

Topologically, the connected components of the level set are closed curves (topologically equivalent to circles) and possibly arcs with endpoints on the boundary of D. Working through a few examples will make this plausible, but the proof is subtle. One approach is to use a result called the implicit function theorem. Rather than take this route, which would involve quite a long detour, here we'll be content with our intuition. By continuity of V, if V_1 is small enough, a component of the level set is a closed curve C_1 that contains the origin. Here "contains" means that the origin is in the bounded component of $\mathbb{R}^2 - C_1$.

Similarly, each of a sequence of values $V_1 > V_2 > V_3 > \ldots$ with $V_i \to 0$ determines a closed curve C_i that contains the origin. Observe that if $V_i \neq V_j$ then C_i and C_j are disjoint. If $p \in C_i \cap C_j$, then both $V(p) = V_i$ and $V(p) = V_j$

would be true. Impossible because $V_i \neq V_j$. Then $C_1, C_2, C_3, \ldots$ form a nested family of closed curves, sketched in Fig. A.6.

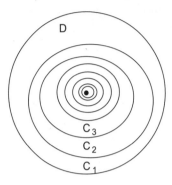

Figure A.6. Nested C_i.

Next we need to establish a relation between the tangents of the curves C_i and the gradient of V. Take $\vec{\sigma}(t) = \langle x_\sigma(t), y_\sigma(t) \rangle$ to be a parameterization of (perhaps part of) C_i. Then for all t, $V(\vec{\sigma}(t)) = V_i$, a constant, and so

$$0 = \frac{d}{dt}V(\vec{\sigma}(t)) = \frac{\partial V}{\partial x}x'_\sigma(t)$$
$$+ \frac{\partial V}{\partial y}y'_\sigma(t) = \nabla(V) \cdot \vec{\sigma}'(t)$$

(The second equality is an application of the chain rule.) That is, the gradient $\nabla(V)$ is perpendicular to the tangents of C_i, so we can call $\nabla(V)$ the normal vector $\vec{N}$ field for the family of level curves of V.

Now suppose $\vec{\rho}(t)$ is a trajectory of the differential equation. Because V' is negative definite,

$$0 > \frac{d}{dt}V(\vec{\rho}(t)) = \frac{\partial V}{\partial x}x'_\rho(t) + \frac{\partial V}{\partial y}y'_\rho(t) = \nabla(V) \cdot \vec{\rho}'(t) = \|\nabla(V)\| \|\vec{\rho}'\| \cos(\theta)$$

The last equality is the familiar dot product formula. We'll review the derivation in Eq. (17.5) of volume 2. Then θ, the angle between $\nabla(V)$ and $\vec{\rho}'$, satisfies $\pi/2 < \theta < 3\pi/2$. Because ∇V is perpendicular to the tangent vectors of C_i, this condition on θ means that the trajectory $\vec{\rho}$ crosses C_i from the outside to the inside and points inward.

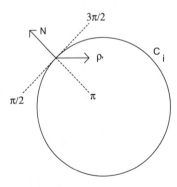

Figure A.7. Direction of $\vec{\rho}$.

If V' is negative semidefinite the only difference in the argument is that now $\pi/2 \leq \theta \leq 3\pi/2$. This means that once a trajectory reaches C_i it can never escape, so the fixed point at the origin is stable. See Fig. A.7. However, the possibility $\theta = \pi/2$ or $\theta = 3\pi/2$ means that the trajectory can be tangent to C_i and so can orbit round and round C_i, never approaching the origin. If V is positive definite and V' negative semidefinite, the origin is stable but need not be asymptotically stable.

Back to the case where V' is negative definite. To show $\vec{\rho}(t) \to (0,0)$, for contradiction suppose $\lim_{t\to\infty} V(\vec{\rho}(T)) = \epsilon > 0$. We know that along every trajectory V decreases (this is the negative definiteness of V') and that V is bounded below by 0, so the limit exists, provided that the trajectory $\vec{\rho}(t)$ continues for $t \to \infty$. This isn't difficult, but we'll skip it here. See page 206 of [18].

Continuity of V implies there is a $\delta > 0$ with $V(x,y) < \epsilon$ whenever $\|(x,y)\| < \delta$. Consequently, $V(\vec{\rho}(t)) \geq \epsilon$ implies $\|\vec{\rho}(t)\| \geq \delta$. The annulus $S = \{(x,y) : \delta \leq \|(x,y)\| \leq r\}$, where r is the radius of the disc D, is closed and bounded, so the continuous function V' has $\mu = \max\{V'(x,y) : (x,y) \in S\} < 0$. That is,

$$\frac{d}{dt} V(\vec{\rho}(t)) = V'(\vec{\rho}(t)) \leq \mu < 0$$

Integrating

$$\int_0^T \frac{d}{dt} V(\vec{\rho}(t))\, dt = \int_0^T V'(\vec{\rho}(t))\, dt \leq \int_0^T \mu\, dt = \mu T$$

and we see

$$V(\vec{\rho}(T)) - V(\vec{\rho}(0)) \leq \mu T \quad \text{that is, } V(\vec{\rho}(T)) \leq V(\vec{\rho}(0)) + \mu T$$

Because $V(\vec{\rho}(0))$ is a positive constant and μ is a negative constant, for large enough T this inequality gives $(\vec{\rho}(T)) < 0$, impossible because V is positive definite.

Consequently, the hypothesis $\lim_{t\to\infty} V(\vec{\rho}(T)) = \epsilon > 0$ is wrong and so $\lim_{t\to\infty} V(\vec{\rho}(T)) = 0$. Because V is positive definite, this means that $\lim_{t\to\infty} \vec{\rho}(t) = (0,0)$. This is true for all trajectories that approach $(0,0)$ closely enough, so $(0,0)$ is asymptotically stable.

A.9 A PROOF OF THE HARTMAN-GROBMAN THEOREM

This theorem was proved by D. M. Grobman and independently by Philip Hartman [60]. The theorem is more extensive than the portion we presented in Sect. 9.2. The full statement is this.

Theorem A.9.1. Suppose $\vec{v}\,' = \vec{F}(\vec{v})$ has a fixed point at the origin O and every eigenvalue of the derivative matrix $D\vec{F}|O$ has non-zero real part. Then near the origin the trajectories of $\vec{v}\,' = \vec{F}(\vec{v})$ can be mapped continuously to the trajectories of $\vec{w}\,' = [D\vec{F}|O]\vec{w}$, and vice versa.

Technically, this map is called a homeomorphism, a continuous map that is 1-1 and onto, and has a continuous inverse map. Not only do both systems have the same type of fixed point (saddle point, unstable spiral, unstable node, asymptotically stable spiral, or asymptotically stable node), but their trajectories can be matched one for one.

The proof is complicated and subtle. See Sect. 2.8 of [124]. Here we have a more modest goal. Assume that for the linear system $\vec{w}\,' = [D\vec{F}|O]\vec{w}$ the origin is an asymptotically stable fixed point and construct a positive definite Liapunov function V with V' negative definite. Then if $\vec{v}\,' = \vec{F}(\vec{v})$ is not too badly nonlinear, the same Liapunov function has V' negative definite for the nonlinear differential equation.

What do we mean by the phrase "not too badly nonlinear"? The system

$$x' = ax + by + G(x,y) \qquad y' = cx + dy + H(x,y) \tag{A.22}$$

is called *quasilinear* if

$$\lim_{r \to 0} \frac{G(x,y)}{r} = 0 \quad \text{and} \quad \lim_{r \to 0} \frac{H(x,y)}{r} = 0 \tag{A.23}$$

where as usual $r = \sqrt{x^2 + y^2}$.

Example A.9.1. *A quasilinear system.* The system

$$x' = x + y + x^2 \qquad y' = -x + 2y$$

is quasilinear because by converting x^2 to polar coordinates we find

$$\lim_{r \to 0} \frac{x^2}{r} = \lim_{r \to 0} \frac{r^2 \cos^2(\theta)}{r} = \lim_{r \to 0} r \cos^2(\theta) = 0$$

To see the last equality, note that $0 \le r \cos^2(\theta) \le r$. □

Example A.9.2. *Another quasilinear system.* The system

$$x' = x + y + \sin(x^2 + y^2) \qquad y' = -x + 2y$$

is quasilinear because converting to polar coordinates

$$\lim_{r \to 0} \frac{\sin(x^2 + y^2)}{r} = \lim_{r \to 0} \frac{\sin(r^2)}{r} = 0$$

by l'Hôpital's rule, or by the power series for $\sin(x)$ with $x = r^2$. □

Example A.9.3. *A system that isn't quasilinear.* The system

$$x' = x + y + \sqrt{x^2 + y^2} \qquad y' = -x + 2y$$

is not quasilinear because by converting to polar coordinates we find

$$\lim_{r \to 0} \frac{\sqrt{x^2 + y^2}}{r} = \lim_{r \to 0} \frac{r}{r} = 1$$

But it would be quasilinear if the exponent of $x^2 + y^2$ were $> 1/2$. $\Box$

Suppose $(0,0)$ is asymptotically stable for the associated linear system

$$\begin{bmatrix} x' \\ y' \end{bmatrix} = \begin{bmatrix} a & b \\ c & d \end{bmatrix} \cdot \begin{bmatrix} x \\ y \end{bmatrix} = M \begin{bmatrix} x \\ y \end{bmatrix} \tag{A.24}$$

Recalling Eq. (8.19), we see $\text{tr}(M) < 0$ and $\det(M) > 0$. By constructing an appropriate Liapunov function for the linear system, we show that $(0,0)$ is asymptotically stable for the quasilinear system (A.22) using the same Liapunov function.

The derivative of the system (A.22) is

$$D\vec{F}(0,0) = \begin{bmatrix} a + \left.\dfrac{\partial G}{\partial x}\right|_{(0,0)} & b + \left.\dfrac{\partial G}{\partial y}\right|_{(0,0)} \\[2ex] c + \left.\dfrac{\partial H}{\partial x}\right|_{(0,0)} & d + \left.\dfrac{\partial H}{\partial y}\right|_{(0,0)} \end{bmatrix}$$

To show $D\vec{F}(0,0) = M$, we must show that each of these partial derivatives is 0.

Now $(0,0)$ belongs to the domain of both G and H, so the quasilinearity condition Eq. (A.23) implies $G(0,0) = H(0,0) = 0$. Then

$$\left.\frac{\partial G}{\partial x}\right|_{(0,0)} = \lim_{h \to 0} \frac{G(0+h,0) - G(0,0)}{h} = \lim_{h \to 0} \frac{G(h,0)}{h}$$

This limit is 0, being a special case, the path along the positive x-axis, of the limit in Eq. (A.23).

We construct a quadratic Liapunov function

$$V(x,y) = Ax^2 + Bxy + Cy^2$$

for the associated linear system (A.24). By Prop. 9.4.1, V is positive definite if and only if $A > 0$ and $4AC - B^2 > 0$. We must find values for A, B, and C in terms of the system parameters a, b, c, and d, making V positive definite and

$$V' = \frac{\partial V}{\partial x}(ax + by) + \frac{\partial V}{\partial y}(cx + dy) \tag{A.25}$$

negative definite, so $(0,0)$ is asymptotically stable for the linear differential equation (A.24). Many Liapunov functions satisfy these conditions; we seek one for which also

$$V' = \frac{\partial V}{\partial x}(ax + by + G(x,y)) + \frac{\partial V}{\partial y}(cx + dy + H(x,y))$$

is negative definite, so $(0,0)$ is asymptotically stable for the quasilinear system (A.22). This is a lot to manage. How shall we proceed?

For the linear system

$$V' = \frac{\partial V}{\partial x} \cdot (ax + by) + \frac{\partial V}{\partial y} \cdot (cx + dy)$$

$$= (2Ax + By) \cdot (ax + by) + (Bx + 2Cy) \cdot (cx + dy)$$

$$= (2Aa + Bc)x^2 + (2Ab + Ba + Bd + 2Cc)xy + (Bb + 2Cd)y^2 \qquad \text{(A.26)}$$

One way to guarantee that V' is negative definite is to take A, B, and C so that

$$2Aa + Bc = Bb + 2Cd = -1 \quad \text{and} \quad 2Ab + Ba + Bd + 2Cc = 0$$

That is, for the linear system (A.24),

$$V'(x,y) = -x^2 - y^2 \qquad \text{(A.27)}$$

Can we do this? Solving these equations for A, B, and C we find

$$A = \frac{c^2 + d^2 + ad - bc}{2(a+d)(bc-ad)} = \frac{c^2 + d^2 + \det(M)}{2\text{tr}(M)(-\det(M))}$$

$$B = \frac{ac + bd}{\text{tr}(M)\det(M)} \qquad C = \frac{a^2 + b^2 + \det(M)}{2\text{tr}(M)(-\det(M))}$$

Recalling that $\text{tr}(M) < 0$ and $\det(M) > 0$, we see that $A > 0$ and

$$4AC - B^2 = \frac{a^2 + b^2 + c^2 + d^2 + 2\det(M)}{(\text{tr}(M))^2 \det(M)} > 0$$

and so V is positive definite by Prop. 9.4.1.

In order to use the same Liapunov function for the quasilinear system, we must show that V' is negative definite. With this system,

$$V' = \frac{\partial V}{\partial x} \cdot (ax + by + G(x,y)) + \frac{\partial V}{\partial y} \cdot (cx + dy + H(x,y))$$

$$= -(x^2 + y^2) + \frac{\partial V}{\partial x} G + \frac{\partial V}{\partial y} H \quad \text{using (A.25) and (A.27)}$$

$$= -(x^2 + y^2) + (2Ax + By)G + (Bx + 2Cy)H \qquad \text{(A.28)}$$

So we must show that for (x,y) close enough to the origin, the second and third terms of (A.28) are smaller than $(x^2 + y^2)$.

Converting to polar coordinates,

$$V' = -r^2 + (2Ar\cos(\theta) + Br\sin(\theta))G + (Br\cos(\theta) + 2Cr\sin(\theta))H$$

$$\leq -r^2 + |2Ar\cos(\theta) + Br\sin(\theta)||G| + |Br\cos(\theta) + 2Cr\sin(\theta)||H|$$

$$\leq -r^2 + r(2|A| + |B|)|G| + r(|B| + 2|C|)|H|$$

$$\leq -r^2 + r2M|G| + r2M|H|$$

where we have taken $M = \max\{2|A|, |B|, 2|C|\}$.

Recall that $\lim_{r \to 0} G/r = 0$ means that for every $\epsilon > 0$, there is a $\delta_G > 0$ with the property that for all r, $0 < r < \delta_G$,

$$|G/r| < \epsilon \quad \text{or equivalently} \quad |G| < \epsilon r$$

Similarly, for every $\epsilon > 0$, there is a $\delta_H > 0$ with $|H| < \epsilon r$ whenever $0 < r < \delta_H$.

Take $\epsilon < 1/5M$, and $\delta = \min\{\delta_G, \delta_H\}$. Then for all $r < \delta$,

$$V' < -r^2 + r2M(r/5M) + r2M(r/5M) = -r^2/5$$

That is, for the quasilinear system, V' is negative definite on the disc of radius δ. So this part of the Hartman-Grobman theorem for quasilinear functions is a consequence of Liapunov's theorem.

A.10   A PROOF OF THE POINCARÉ-BENDIXSON THEOREM

The Poincaré-Bendixson theorem, presented in Sect. 9.7, was proved in the late 19th century by Henri Poincaré [125] and Ivar Bendixson [11]. The theorem places significant restrictions on the geometry of solution curves for differential equations in the plane. Some students respond to a statement of the theorem with, "What else could happen? Why is there anything to prove?" So you let them talk for a while, convince themselves that the result is clear, that Poincaré and Bendixson were slacking. Then ask what happens in 3 dimensions. More or less the same thing. Okay, saddle points are a bit more complicated: there can be two stable directions and one unstable, or two unstable and one stable, but that's about the only complication. Then show a picture of the Lorenz attractor. If you use Mathematica, then you can freely rotate the attractor, emphasize just how complicated it is. This is not a surface. So ...intuition is useful to guide what you think is true, but it is *not* a substitute for a proof. Not in 3 dimensions, and not in 2, either. Let's sketch how to prove it. More detail is provided in Chapter 11 of [63], Sect. 5.4 of [66], and Sect. 3.7 of [124], for example.

In order to sketch the proof of the Poincaré-Bendixson theorem we need four topological concepts, not so difficult because we can restrict our attention to sets in the plane. Some of these notions we have mentioned earlier; we collect them here.

Recall that a set $A \subset \mathbb{R}^2$ is open if for every point $(x_0, y_0) \in A$, there is some $r > 0$ for which $D_r(x_0, y_0) = \{(x,y) : d((x,y), (x_0, y_0)) < r\} \subset A$, where

$$d((x,y), (x_0, y_0)) = \sqrt{(x - x_0)^2 + (y - y_0)^2}$$

is the familiar Euclidean distance between the points (x,y) and (x_0, y_0).

A set $A \subset \mathbb{R}^2$ is *closed* if its complement $\mathbb{R}^2 - A$ is open.

A set $A \subset \mathbb{R}^2$ is *bounded* if for some $(x_0, y_0) \in A$ there is a large enough r for which $A \subset D_r(x_0, y_0)$. That is, A does not run off to infinity in any direction.

A point (x_0, y_0) is a *limit point* of a set A if for every $\epsilon > 0$ there is some point $(x, y) \in A$ for which $0 < d((x, y), (x_0, y_0)) < \epsilon$.

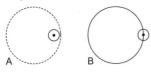

Figure A.8. The set A is open; the set B is not.

Whenever we give a definition, we should provide an example that satisfies the definition, so we know we are not talking about nothing, and also an example that does not satisfy the definition, so we know we are not talking about everything, because not much of interest can be said about everything.

The disc $A = D_1(0,0)$ with radius 1 and center $(0,0)$ is open. To see this, take any point $(x_0, y_0) \in A$ and let $r_0 = d((x_0, y_0), (0,0))$. Then $(1 - r_0)/2$ is half the distance between (x_0, y_0) and the boundary of A, so $D_{(1-r_0)/2}(x_0, y_0) \subset A$. The set $B = \{(x, y) : d((x, y), (0,0)) \leq 1\}$ is not open, because $(x_0, y_0) = (1, 0) \in B$ and yet for every $r > 0$, the disc $D_r(1,0)$ does not lie in B. See Fig. A.8. The dotted curve on the left side of the figure indicates that the circle $x^2 + y^2 = 1$ is not part of A; the solid curve on the right side that the circle does belong to B. In both, the dot is the point (x_0, y_0).

The set $A' = \mathbb{R}^2 - A = \{(x, y) : x^2 + y^2 \geq 1\}$ is closed because its complement A is open. The set $B' = \mathbb{R}^2 - B$ is not closed because its complement B is not open.

The set A is bounded: $A \subset D_2(0,0)$. On the other hand, the set $C = \{(x, y) : 0 \leq x, 0 \leq y\}$ is not bounded: for any point $(x_0, y_0) \in C$ and for any $r > 0$, the point $(2rx_0, 2ry_0) \notin D_r(x_0, y_0)$ and yet $(2rx_0, 2ry_0) \in C$.

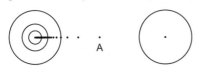

Figure A.9. The point $(0,0)$ is a limit point of A; the point $(1,0)$ is not.

Now consider the set $A = \{(1/n, 0) : n = 1, 2, 3, \ldots\}$. The point $(0,0)$ is a limit point of A, because every disc centered on $(0,0)$ contains points of A having a positive distance from $(0,0)$. The point $(1,0)$, right-most in the figure, is not a limit point of A, because for every $r < 1/2$, the disc $D_r(1,0)$ contains no point of A having a positive distance from $(1,0)$. See Fig. A.9. Note that every disc centered at a limit point of A must contain infinitely many distinct points of A.

A useful characterization of closed sets in terms of their limit points is

Proposition A.10.1. The set A is closed if and only if every limit point of A belongs to A.

This is Corollary 6.7 of [97].

Now we are ready to sketch the proof of this form of the Poincaré-Bendixson theorem, Thm. 9.7.1. Suppose R is a closed and bounded region in the plane, and R contains no fixed points of $x' = f(x,y)$, $y' = g(x,y)$ where f and g have continuous partial derivatives throughout R. Suppose a solution curve $\vec{\gamma}(t)$ remains in R for all $t \geq t_0$. Then either $\vec{\gamma}(t)$ is a closed trajectory in R, or $\vec{\gamma}(t)$ converges to a limit cycle in R.

Proof. First, suppose the trajectory $\vec{\gamma}(t)$ is closed, and $\vec{\gamma}(t_0)$, represented by a dot in Fig. A.10, lies in R. If part of the trajectory lies outside of R, then for some $t' > t_0$, $\vec{\gamma}(t')$ lies outside of R, contradicting the hypothesis that $\vec{\gamma}(t)$ remains in R for all $t \geq t_0$. So this closed trajectory must lie entirely inside R. That is, the right side of Fig. A.10 is possible; the left side is not.

What can we do if $\vec{\gamma}(t)$ is not a closed trajectory? First, define a sequence of points $(x_i,y_i) = \vec{\gamma}(t_0 + i)$ for $i = 0,1,2,\ldots$. Because $\vec{\gamma}(t)$ is not a closed trajectory, all the points (x_i,y_i) are distinct. We show that R contains a limit point of the set $A = \{(x_i,y_i) :$

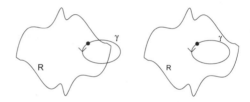

Figure A.10. Closed trajectories not contained in R (left) and contained in R (right).

$i = 1,2,3,\ldots\}$. Because R is bounded, we can divide it into a finite collection of pieces. See the right side of Fig. A.11. The infinite set A of points is spread among the finite number of pieces of R, so at least one of these pieces must contain infinitely many points of A. (This is just an extension of the *pigeonhole principle*: if $n > m$ items are divided into m groups, then at least one group must contain more than one item. But really, the extension is pretty obvious.) Take a piece of R containing infinitely many points of A and divide this piece into a finite number of still smaller pieces. At least one of these smaller pieces must contain infinitely many points of A. Continue, selecting a sequence of ever smaller subpieces B_j, with the size of the subpieces going to 0. Now from the subpiece B_j select a point (x_{i_j}, y_{i_j}). This sequence of points converges to a point (x_*, y_*) in the plane. This is a fairly subtle property, called the *completeness* of the real numbers, roughly, that the real numbers have no holes. This point (x_*, y_*)

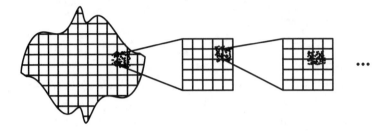

Figure A.11. Showing the sequence (x_i, y_i) has a limit point.

is a limit point of the sequence (x_{i_j}, y_{i_j}), and so a limit point of A. Because R is closed and $A \subset R$, it follows from Prop A.10.1 that $(x_*, y_*) \in R$.

Let $\vec{\rho}(t)$ be the solution curve satisfying $\vec{\rho}(0) = (x_*, y_*)$. That there is such a trajectory, and only one, is a consequence of the existence and uniqueness theorem, sketched in Sect. A.2.

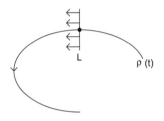

Figure A.12. The solution curve passing through the limit point (x_*, y_*), indicated by the dot, and the perpendicular line L.

Denote by L the straight line perpendicular to $\vec{\rho}'(0)$ and passing through the limit point (x_*, y_*). Continuity of f and g guarantees that if L is short enough, the vector field along L will be approximately parallel to $\vec{\rho}'(0)$. See Fig. A.12.

For every $r > 0$, some of the points (x_i, y_i) on $\vec{\gamma}(t)$ lie in $D_r(x_*, y_*)$, the trajectory $\vec{\gamma}(t)$ must cross L infinitely many times, at points arbitrarily close to (x_*, y_*). Then $\vec{\rho}(t)$ must cross L again, for if not, by continuity no trajectory sufficiently close to $\vec{\rho}(t)$ can cross L infinitely often, and $\vec{\gamma}(t)$ gets arbitrarily close to $\vec{\rho}(t)$.

Let $(\hat{x}, \hat{y})$ be the first point after (x_*, y_*) where $\vec{\rho}(t)$ crosses L. Our goal is to show that $\vec{\rho}(t)$ is a periodic trajectory, certainly true if $(\hat{x}, \hat{y}) = (x_*, y_*)$. So let's suppose $(\hat{x}, \hat{y}) \neq (x_*, y_*)$ and see what trouble this causes.

The point $(\hat{x}, \hat{y})$ must lie on one side of (x_*, y_*) on L. The left image of Fig. A.13 illustrates one choice. Here the dot indicates (x_*, y_*) and the circle $(\hat{x}, \hat{y})$.

Now $\vec{\gamma}(t)$ must cross L on one side or the other of (x_*, y_*). On the right side of Fig. A.13 we show $\vec{\gamma}(t)$ crossing L between (x_*, y_*) and $(\hat{x}, \hat{y})$. The other possibilities are handled similarly. (There are some subtle points, but this case illustrates the main ideas.) Also in this figure we have thickened the part of $\vec{\rho}(t)$ from (x_*, y_*) to $(\hat{x}, \hat{y})$, plus the part of L from $(\hat{x}, \hat{y})$ to (x_*, y_*). These thickened curves divide the plane into two parts, an inside I and an outside O. Once $\vec{\gamma}(t)$

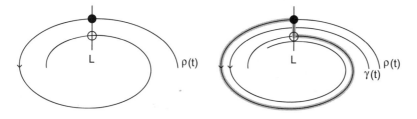

Figure A.13. Left: finding the point $(\hat{x},\hat{y})$. Right: the problem that results from $(\hat{x},\hat{y}) \neq (x_*,y_*)$

enters I, it cannot leave. It cannot cross the part of L between (x_*,y_*) and $(\hat{x},\hat{y})$ because the vector field along L points into I. It cannot cross the part of $\vec{\rho}(t)$ between (x_*,y_*) and $(\hat{x},\hat{y})$, because trajectories can intersect only at fixed points, and R contains no fixed points. Consequently, once $\vec{\gamma}(t)$ enters I, it cannot leave.

Why is this a problem? Recall $\vec{\gamma}(t)$ must cross L infinitely often, at points arbitrarily close to (x_*,y_*). The trajectory $\vec{\gamma}(t)$ enters I at some finite time t_0, so for all $t < t_0$, $\vec{\gamma}(t)$ can cross L only finitely many times. Once $\vec{\gamma}(t)$ enters I, it can never approach (x_*,y_*) more closely than $(\hat{x},\hat{y})$, so cannot cross L at points arbitrarily close to (x_*,y_*). This contradicts our earlier observation.

Where did we go wrong? Every step after the assumption $(\hat{x},\hat{y}) \neq (x_*,y_*)$ is correct, so that assumption must be wrong. Consequently, $(\hat{x},\hat{y}) = (x_*,y_*)$, $\vec{\rho}(t)$ is a closed trajectory, a limit cycle of $\vec{\gamma}(t)$. $\square$

An essential step in this proof was the construction of the thickened curve on the right side of Fig. A.13. The completely plausible result that this curve divides the plane into an inside and an outside is a consequence of the surprisingly difficult to prove Jordan Curve theorem, that every simple (non self-intersecting) closed curve in the plane divides the plane into two disjoint regions.

Certainly, the Jordan curve theorem cannot be extended to 3 dimensions. Although there are some intricate examples, square-filling Peano-Hilbert curves cobbled together to form the faces of a cube, almost all curves in space do not divide space into two regions. However, the failure of one method of proof does not imply the result is false. Perhaps other tools can be applied to prove a version of the Poincaré-Bendixson theorem in 3 dimensions. Alas, this proof never will be found. In Ch. 13 of volume 2 we will see that the ordered world of fixed points and limit cycles does not exhaust every possibility for differential equations having three variables. In some sense, the reason is typically curves cannot divide the plane into two disjoint open subsets. Some surfaces can. In general, to divide a n-dimensional space into two disjoint open pieces, an $(n-1)$-dimensional space is needed.

A.11 THE LAW OF LARGE NUMBERS

The Law of Large Numbers, mentioned in Sect. 11.1, is one of the foundations of modern probability theory. Good treatments are presented in [27, 37, 129, 156]. Here we'll sketch just a few of the main points.

To begin, the LLN has two versions, the weak LLN and the strong LLN. We'll state both, describe how they differ, and sketch the proof of the weak LLN.

Weak LLN A sequence $X_1, \ldots, X_n, \ldots$ of i.i.d. random variables with $\mu = \mathbb{E}(X_n)$ for each n satisfies the weak LLN if for every $\epsilon > 0$,

$$\lim_{n \to \infty} P\left(\left| \frac{X_1 + \cdots + X_n}{n} - \mu \right| < \epsilon \right) = 1 \tag{A.29}$$

This result was proved by Aleksandr Khintchine [74].

Strong LLN A sequence $X_1, \ldots, X_n, \ldots$ of i.i.d. random variables with $\mu = \mathbb{E}(X_n)$ for each n satisfies the strong LLN if for every $\epsilon > 0$ and every $\delta > 0$, there is an N with

$$P\left(\left| \frac{X_1 + \cdots + X_n}{n} - \mu \right| < \epsilon \right) \geq 1 - \delta \text{ for } n = N, N+1, \ldots, N+r$$

for all $r > 0$. This result was proved by Émile Borel [17].

How are these different? William Feller [37] points out that the weak LLN says that $|(X_1 + \cdots + X_n)/n - \mu|$ is likely to be small for large n, but it does not say that $|(X_1 + \cdots + X_n)/n - \mu|$ will *remain* small for all large n. There are sequences of i.i.d. random variables that satisfy the weak LLN and yet $|(X_1 + \cdots + X_n)/n - \mu|$ continues to fluctuate, perhaps without bound, though certainly at infrequent moments, as n increases. On the other hand, the strong LLN says that $|(X_1 + \cdots + X_n)/n - \mu|$ almost surely remains small for all sufficiently large n.

Now we'll derive a form of the weak LLN. For a random variable X and for any $\delta > 0$, define another random variable Y_δ by

$$Y_\delta = \begin{cases} 1 & \text{if } |X| \geq \delta \\ 0 & \text{if } |X| < \delta \end{cases}$$

Then

$$\mathbb{E}(Y_\delta) = 1 \cdot P(|X| \geq \delta) + 0 \cdot P(|X| < \delta) = P(|X| \geq \delta)$$

and $X^2 \geq X^2 \cdot Y_\delta \geq \delta^2 Y_\delta$. Consequently,

$$\mathbb{E}(X^2) \geq \mathbb{E}(\delta^2 Y_\delta) = \delta^2 \mathbb{E}(Y_\delta) = \delta^2 P(|X| \geq \delta)$$

where the penultimate equality is an application of Prop. 11.6.1 (b). Dividing both sides by δ^2 we obtain

$$P(|X| \geq \delta) \leq \frac{1}{\delta^2} \mathbb{E}(X^2) \tag{A.30}$$

This is called *Chebyshev's inequality*, the main tool we need to derive the weak LLN. The only hypothesis is that the second moment $\mathbb{E}(X^2)$ exists and is finite.

If Y is a random variable with mean μ and variance σ^2, apply Eq. (A.30) to the random variable $X = (Y - \mu)/\sigma$:

$$P\left(\left|\frac{Y-\mu}{\sigma}\right| \geq \delta\right) \leq \frac{1}{\delta^2}\mathbb{E}\left(\left(\frac{Y-\mu}{\sigma}\right)^2\right) = \frac{1}{\delta^2}\frac{1}{\sigma^2}\mathbb{E}((Y-\mu)^2) = \frac{1}{\delta^2}$$

where we've used the definition $\mathbb{E}(Y - \mu^2) = \sigma^2$ from Sect. 11.6. From this we see another form of Chebyshev's inequality,

$$P(|Y - \mu| \geq \delta\sigma) \leq \frac{1}{\delta^2} \tag{A.31}$$

This inequality was derived by the Russian mathematician Pafnuty Chebyshev. Chebyshev's students at St. Petersburg University included Andrey Markov, familiar to us for his work on Markov chains, and Aleksandr Liapunov, known for Liapunov functions and Liapunov exponents. Chebyshev is rightly considered to be one of the founders of Russian mathematics.

Now we'll use Chebyshev's inequality to derive the weak Law of Large Numbers. Suppose $X_1, X_2, \ldots$ is a sequence of i.i.d. random variables, and for each n, $n = 1, 2, 3, \ldots$, take $Z_n = (X_1 + \cdots + X_n)/n$. Then by Prop. 11.6.1 (c),

$$\mathbb{E}(X_1 + \cdots + X_n) = \mathbb{E}(X_1) + \cdots + \mathbb{E}(X_n) = n\mu \quad \text{so } \mathbb{E}(Z_n) = \mu$$

Denote $\mathbb{E}(Z_n)$ by μ_{Z_n}. Because the X_i are independent, by Prop. 11.6.2

$$\sigma^2(X_1 + \cdots + X_n) = \sigma^2(X_1) + \cdots + \sigma^2(X_n) = n\sigma^2$$

From Eq. (11.17) and Prop. 11.6.1 (b) again, for any constant A

$$\sigma^2(AX) = \mathbb{E}((AX)^2) - (\mathbb{E}(AX))^2 = A^2\mathbb{E}(X^2) - (A\mathbb{E}(X))^2 = A^2\sigma^2(X)$$

Consequently,

$$\sigma^2(Z_n) = \sigma^2\left(\frac{X_1 + \cdots + X_n}{n}\right) = \frac{n\sigma^2}{n^2} = \frac{\sigma^2}{n}$$

Denote the standard deviation $\sqrt{\sigma^2(Z_n)} = \sigma^2/\sqrt{n}$ by σZ_n.

Now apply Chebyshev's inequality (A.31) to Z_n.

$$P(|Z_n - \mu_{Z_n}| \geq \delta\sigma Z_n) = P\left(|Z_n - \mu_{Z_n}| \geq \delta\frac{\sigma}{\sqrt{n}}\right) \leq \frac{1}{\delta^2}$$

This holds for every positive δ, so take $\delta = \epsilon\sqrt{n}/\sigma$. Then the application of Chebyshev's inequality becomes

$$P(|Z_n - \mu_{Z_n}| \geq \epsilon) \leq \frac{\sigma^2}{n\epsilon^2}$$

Because σ and δ are fixed, as $n \to \infty$ this gives $P(|Z_n - \mu_{Z_n}| \geq \epsilon) \to 0$, and so $P(|Z_n - \mu_{Z_n}| < \epsilon) \to 1$. Because $Z_n = (X_1 + \cdots + X_n)/n$ and $\mu_{Z_n} = \mu$, this last expression is the weak LLN (A.29).

Some thought about the CLT suggests it is related to at least one of the LLNs. I'll leave this for you.

Appendix B   Some Mathematica code

Here we'll give examples of Mathematica code to generate some of the figures in the text, integrate numerically and symbolically, solve differential equations numerically, and find eigenvalues and eigenvectors. You can adapt the examples to perform similar calculations and simulations.

Now I admit that I am not a skilled coder. Certainly these results can be obtained in ways more subtle and elegant. My goal was to try to find the simplest code, the most easily understood, not the smartest. If the clunkiness of my algorithms bothers you, I'm sorry. Only rarely do autodidacts achieve a mastery that can be called graceful. I am not one of those rare cases.

When I ran some of this code, the differential equations solvers in particular, on two different versions of Mathematica, while the basic curve forms agreed, they differed in detail. Probably this is a result of differences in the underlying numerical algorithms. Copies of this code that can be copied and pasted into Mathematica are at https://gauss.math.yale.edu/~frame/BiomathMma.html.

B.1   CYCLES AND THEIR STABILITY

In this section we'll study concepts presented in Sect. 2.3. We'll show code to detect a 2-cycle for a function $f(x)$ that depends on a parameter r and to test the stability of this 2-cycle. With obvious adaptations we can detect and test the stability of fixed points and n-cycles for $n > 2$. A bit more complicated code automates the search, but we prefer a simpler approach that includes experimentation by the user. For definiteness, we'll use the function $f(x) = r\sin(\pi x)$. Substitute another function, and its derivative, as you wish.

```
r = 0.75;
f[x_]:=r*Sin[Pi*x]
df[x_]:=r*Pi*Cos[Pi*x]
z = FindRoot[x == f[f[x]],{x, 0.5}][[1,2]]
f[z]
Abs[N[df[f[z]]*df[z]]]
```

Note the double equal signs, ==, in the FindRoot command. Arguments of functions are enclosed in square brackets []. In the definition of the function, the _ in x_ indicates that x is the variable, while := indicates the definition of a function. Names of familiar functions are capitalized. That is, f[x_]: = Sin[x] is Mathematica for f(x) = sin(x). The derivative of f is denoted df[x_].

In the FindRoot command, the {x, 0.5} carries the information that the variable is x, and that the initial guess is x = 0.5. FindRoot uses Newton's method, so must be provided with an initial guess. Most anything close to the root will work. The [[1,2]] is included so z is assigned the value determined by FindRoot. Here's why.

In Mathematica lists are indicated by curly brackets, for example {1, 5, 17} is the list whose members are 1, 5, and 17. A double bracketed number after a list selects the list element whose position in the list is that number. So {1, 5, 17}[[2]] returns 5. If (some of) the elements of the list are themselves lists, the double bracketed numbers can be iterated: {a, {b,c,d}}[[2,3]] returns d. Then

FindRoot[x == f[f[x]],{x, 0.5}]	returns {x → 0.540177}
FindRoot[x == f[f[x]],{x, 0.5}][[1]]	returns x → 0.540177
FindRoot[x == f[f[x]],{x, 0.5}][[1,1]]	returns x (not what we want)
FindRoot[x == f[f[x]],{x, 0.5}][[1,2]]	returns 0.540177

After a point z of the 2-cycle is found, the line f[z] prints the other point of the cycle. Then Abs[N[df[f[z]]*df[z]]] prints the absolute value of the product df[f[z]]*df[z], the derivative of f^2 evaluated on the 2-cycle. The N[is included to give the numerical value. Without this, Mathematica may return a very long and unenlightening algebraic expression.

B.2 CHAOS

In Sect. 2.4 we do two computations, numerical approximation of the Liapunov exponent of Eq. (2.10) and from Practice Problem 2.4.1 we count the number of iterates of f defined by Eq. (2.16) that land in the interval $[0, 1/3]$ and in the interval $(1/3, 1]$.

First, we compute numerical approximations for the Liapunov exponent. We could use any differentiable function, but for the example we'll use a logistic map.

```
r=3.5;
f[x_]:=r*x*(1-x)
df[x_]:=r - 2*r*x
x=Random[Real,{,}];
num=1000; (* THE NUMBER OF ITERATES *)
derlst={};
ptlst={PointSize[.01]};
Do[{x=f[x],AppendTo[derlst,Log[Abs[df[x]]]],
AppendTo[ptlst,Point[{n,Apply[Plus,derlst]/n}]]},{n,1,num}]
Show[Graphics[ptlst]]
```

Comments are bracketed by $(*, *)$. Here derlst is the list

$$\{\ln |f'(x_1)|, \ln |f'(x_2)|, \dots\}$$

where x_0 is a random number in $[0, 1]$, $x_1 = f(x_0)$, $x_2 = f(x_1), \dots$. When n terms have been generated in derlst, Apply[Plus,derlst]/n sums the terms of derlst and divides the sum by n. The point with this value as y-coordinate and n as x-coordinate is added to ptlst and these points are plotted. If as n increases, this list of points converges to a horizontal line, the y-coordinate of that line is the Liapunov exponent.

Here is code to generate iterates of f and count the number $ct1$ and $ct2$ of times the iterates land in the interval $[0, 1/3]$ and $(1/3, 1]$.

```
f[x_]:= If[x > 1/3,(3/2)*x - 1/2,3*x]
x = Random[Real{0,1}]
num=100000; (* THE NUMBER OF ITERATES *)
ct1=0; ct2=0;
If[x > 1/3, ct2=ct2+1,ct1=ct1+1]
```

```
Do[{x=f[x],If[x > 1/3, ct2=ct2+1,ct1=ct1+1]},{i,1,num}]
ct1
ct2
```

The If statement If[a,b,c] works this way. If condition a is true, then action b is taken; if condition a is false, then action c is taken. In this case, if the current iterate x is greater than 1/3, then ct2 is incremented by 1; if $x \leq 1/3$, then ct1 is incremented by 1.

B.3   A SIMPLE CARDIAC MODEL

To plot a graph of $f(V)$ and $f(f(V))$ for

$$f(V) = \begin{cases} aV + u & \text{for } V < V_c/a \\ aV & \text{for } V \geq V_c/a \end{cases}$$

from Practice Problem 2.6.2 of Sect. 2.6, this code works. We've inserted sample parameter values for concreteness.

```
a = 0.5; u = 0.5; vc = 0.25;
f[x_]:=If[x < vc/a, a*x + u, a*x]
Plot[{x, f[x]},{x,0,1},AspectRatio— >Automatic]
Plot[{x, f[f[x]]},{x,0,1},AspectRatio— >Automatic]
```

Though Mathematica draws a vertical line segment connecting the branches of the graph at the jumps. Eliminating these segments required a little bit of trickery.

B.4   THE NORTON-SIMON MODEL

Here is Mathematica code to plot a solution curve of the differential equation $y'(t) = f(t)$. In the applications of Sect. 3.4, we'll plot solutions of several forms of the Norton-Simon model equation. For differential equations of the form $y' = f(t)$ Mathematica can generate solution curves with the NDSolve command. Note the use of double equal signs, as in FindRoot.

```
sol = NDSolve[{y'==f[t], y[0]==initial value},y,{t,0,tmax}]
```

The y after the first pair of curly brackets defines the variable for which NDSolve solves. To plot the graph of the solution curve use this command.

```
Plot[Evaluate[y[t]/.sol],{t,0,tmax}]
```

The sol in the Evaluate command refers to the previously run code, so NDSolve must be evaluated before Evaluate. They can be in the same program, so long as the NDSolve line occurs before the Evaluate line.

In Practice Problem 3.4.2 we use a function $L(t)$ defined in pieces by Eq. (3.30). Mathematica code to define this function is

```
L[t_]:=If[t < 6,0,1]
```

In Practice Problem 3.4.3 we plot the solution curve $N(t)$ of the Norton-Simon model differential equation. Here's Mathematica code to plot the solution curve and the line $y = 20$.

```
a=1; k=2; ninf=100;
L[t_]:=If[t<5,0,1]
sol = NDSolve[{y'[t]==a*y[t]*Log[ninf/y[t]] - k*L[t]*y[t],y[0]==0.01},
    y,{t,0,15}]
Plot[{Evaluate[y[t]/.sol],20},{t,0,15},PlotRange->All]
```

In Exercise 3.4.9 we use the chemotherapy schedule $L(t)$ given by Eq. (3.31). In Mathematica we can code this with nested If functions:

```
L[t_]:=If[t<5,0,If[t<10,1,0]]
```

B.5 LOG-LOG PLOTS

To find the best-fitting line for the log-log plot through the points of Practice Problem 6.2.1, first make a list of points $(\log(i), \log(x(i)))$. In Mathematica, Log[x] means ln(x). The base-10 logarithm is Log[10.,x]. The decimal after 10 is to give a decimal approximation of the base-10 log. Writing Log[10,3] returns Log[3]/Log[10]; writing Log[10.,3] returns 0.477121.

In the practice problem we are given the y-values of points on the graph. In Mathematica, we enter these as

```
ylst = {1.1, 1.61, 2.01, 2.36, 2.67, 2.95, 3.21, 3.45, 3.68, 3.90, 4.11, 4.31,
4.51, 4.70, 4.88, 5.05, 5.23, 5.39, 5.56, 5.71};
```

The corresponding x-values are 1 through 20. The list of points for the log-log plot is made by this command

```
data = {};
Do[AppendTo[data, {Log[10., i], Log[10., xlst[[i]]]}], {i, 1, Length[xlst]}]
```

Then the best-fitting line through these points is found by the command

```
LinearModelFit[data, x, x]
```

Mathematica returns

```
FittedModel[0.0414057 + 0.549974x]
```

We'll keep three digits, the accuracy of the presented data. Then the data suggest that x and y are related by $y = 0.550x + 0.041$.

B.6 VECTOR FIELDS AND TRAJECTORIES

Here we'll learn the Mathematica code to plot vector fields in the plane for Sect. 7.1. In the next section we'll learn to plot solution curves.

To plot a vector field $\langle f(x,y), g(x,y)\rangle$ we can use the VectorPlot function of Mathematica.

For example, to plot the field $\langle x = y^2, xy \rangle$ with this code

```
f[x_,y_]:= x + y^2
g[x_,y_]:= x*y
VectorPlot[{f[x,y], g[x,y]},{x,-2,2},{y,-2,2}]
```

Two vector fields can be plotted together by placing both fields in brackets:

```
{{a[x,y],b[x,y]},{c[x,y],d[x,y]}}
```

To tell which arrow goes with which field, add the command

VectorStyle –>{Red,Blue}

which plots the first vector field arrows red and the second vector field arrows blue.

The size of the vector arrowheads can be controlled by the command VectorScale. The arguments of this command are Small, Medium, and Large. For example, VectorScale->Small gives the smallest arrowheads.

B.7 DIFFERENTIAL EQUATIONS SOFTWARE

Here we'll learn Mathematica commands to plot nullclines and trajectories of differential equations in the plane for Chapters 7, 8, and 9. Throughout the text we'll use modifications of this code to investigate properties of planar differential equations solutions.

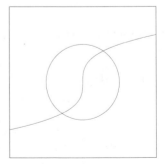

To plot the nullclines of the differential equation

$$x' = f(x,y) \quad y' = g(x,y)$$

we use the ContourPlot function to plot the curve or curves determined by $f(x,y) = 0$, the x-nullcline, and by $g(x,y) = 0$, the y-nullcline. For example, we can find the nullclines of $x' = x - y^3$, $y' = x^2 + y^2 - 1$ with this code:

```
f[x_,y_]:= x - y∧3
g[x_,y_]:= x∧2 + y∧2 - 1
ContourPlot[{f[x,y]==0,g[x,y]==0},{x,-2,2},{y,-2,2}]
```

With Mathematica we can solve differential equations

$$x' = f(x,y), \ y' = g(x,y)$$

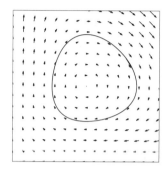

numerically with the NDSolve command. But plotting the results takes an additional command. We might plot both $x(t)$ and $y(t)$ as functions of time t with the Plot[{x[t],y[t]}] command, or we might plot the trajectory $(x(t),y(t))$ with the ParametricPlot[{x[t],y[t]}] command.

Sometimes we'll want to superimpose the trajectory on the vector field {f[x,y], g[x,y]}. We'll illustrate this for a predator-prey system.

```
sol = NDSolve[{x'[t]==-x[t]+x[t]*y[t], y'[t]==y[t]-x[t]*y[t], x[0]==0.65,
y[0]==0.65}, {x,y},{t,0,10}]
g1 = ParametricPlot[Evaluate[{x[t], y[t]}/.sol],{t,0,10}]
```

With these two commands we can plot a trajectory. To superimpose the trajectory on the vector field, first we plot the vector field

```
g2 = VectorPlot[{-x + x*y, y - x*y}, {x,0,2}, {y,0,2}]
```

and finally combine the two plots g1 and g2,

```
Show[{g1, g2}, PlotRange −>{{0,2}, {0,2}}]
```

B.8 PREDATOR-PREY MODELS

To plot the curves $x(t)$ and $y(t)$ for the system $x' = -ax+bxy, \ y' = cy - dxy$, that is, Eqs. (7.4) and (7.5) of Sect. 7.3, first generate the solutions as in Sect. B.7. In order to generate any solution curves, we must assign values to the system parameters a, b, c, and d, to the initial values xinit and yinit of x and y, and to tmax, the upper time limit of the solution. Remember, NDSolve just generates the solution. Other commands must be used to plot trajectories.

```
a = 1; b = 1; c = 1; d = 1; xinit = 0.65; yinit = 0.65; tmax = 10;
sol = NDSolve[{x'[t]==-a*x[t] + b*x[t]*y[t], y'[t]==c*y[t] - d*x[t]*y[t],
x[0]==xinit, y[0]==yinit}, {x,y},{t,0,tmax}]
```

Now plot the $x(t)$ and $y(t)$ curves and name the plots so they can be shown together.

Plot[{Evaluate[x[t]/.sol],Evaluate[y[t]/.sol]},
  {t,0,tmax}]

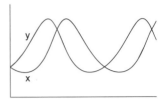

As we expect, the predator population first drops as the prey population rises. With more prey, the predator population rises until it begins to decrease the prey population. And so on.

Another way to view this behavior is to plot the trajectory, the curve $(x(t), y(t))$ for $0 \le t \le$ tmax. This we can do with the ParametricPlot command, using the solution, sol, already generated. The Mathematica command is this.

ParametricPlot[Evaluate[{x[t],y[t]}/.sol],
  {t,0,tmax}]

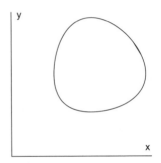

Now Mathematica has some odd ideas about the placement of axes. As plotted the axes do not necessarily cross at the origin. For example, Mathematica's default is to plot the x-axis at $y = 0.6$ and the y-axis at $x = 0.6$. Don't let this confuse you.

B.9 THE FITZHUGH-NAGUMO EQUATIONS

To plot the population curves $x(t)$ and $y(t)$, and the trajectory $(x(t), y(t))$, of the Fitzhugh-Nagumo equations (7.15) of Sect. 7.6,

$$x' = c(y + x - x^3/3 + z) \qquad y' = (-x + a - by)/c$$

define z[t_]:=0 unless we want to impose a current. For example, a single pulse

$$z(t) = \begin{cases} 0 & \text{for } t < 10, \\ 1 & \text{for } 10 \le t \le 20, \text{ and} \\ 0 & \text{for } t > 20 \end{cases}$$

can be defined by nested If statements

z[t_]:=If[t < 10,0,If[t < 20,1,0]]

Then to solve the Fitzhugh-Nagumo equations with initial values $x(0) = 0.6$ and $y(0) = 0.4$, and with parameters $a = 0.8$, $b = 0.35$, and $c = 1$, for $0 \le t \le 10$, use this command:

a = 0.8; b = 0.35; c = 1; tmax = 30;
sol = NDSolve[{x′[t]==c*(y[t] + x[t] - x[t]∧3/3 + z[t]),
    y′[t]==(-x[t] + a - b*y[t])/c, x[0]==0.6, y[0]==0.6},
    {x,y}, {t,0,tmax}, MaxSteps— >1000]

The MaxSteps command is included to guarantee that the solution curves are adequately smooth.

To plot the population curves $x(t)$ and $y(t)$ in a single graph, and to plot the trajectory, use the commands

Plot[{Evaluate[x[t]/.sol],Evaluate[y[t]/.sol]},{t,0,tmax},PlotRange->All]
ParametricPlot[Evaluate[{x[t],y[t]}/.sol],{t,0,tmax},PlotRange->All]

These will reproduce the graphs of Fig. 7.26 of Sect. 7.6.

B.10 EIGENVALUES AND EIGENVECTORS

In Sect. 8.4 we learned to find eigenvalues and eigenvectors of a 2×2 matrix by hand. To do this with Mathematica, first we must learn how to encode a matrix. In Mathematica a matrix is a list of rows, each of which is a list of entries. For example,

$$m = \begin{bmatrix} 1 & 1 \\ 4 & 1 \end{bmatrix} \quad \text{is written} \quad m = \{\{1,1\},\{4,1\}\};$$

The commands to find the eigenvalues and eigenvectors of m are

Eigenvalues[m]   Eigenvectors[m]

These return {3,-1} and {{1,2},{-1,2}}. That is, the eigenvalues are $\lambda = 3$ and $\lambda = -1$, an eigenvector of $\lambda = 3$ is a column vector with first entry 1 and second entry 2, and an eigenvector of $\lambda = -1$ is a column vector with first entry -1 and second entry 2. Remember that every (non-zero) multiple of an eigenvector is an eigenvector for the same eigenvalue. Multiplying this eigenvector by -1 gives the eigenvector we found in Sect. 8.4.

Note that the first eigenvector corresponds to the first eigenvalue and the second eigenvector corresponds to the second eigenvalue.

Appendix C Some useful integrals and hints

Double angle formulas $\sin^2(x) = \dfrac{1-\cos(2x)}{2}$ $\quad\cos^2(x) = \dfrac{1+\cos(2x)}{2}$

1. $\displaystyle\int \sin^2(x)\,dx = \frac{1}{2}x - \frac{1}{4}\sin(2x) + C$

2. $\displaystyle\int \cos^2(x)\,dx = \frac{1}{2}x + \frac{1}{4}\sin(2x) + C$

3. $\displaystyle\int \tan^2(x)\,dx = \tan(x) - x + C$

4. $\displaystyle\int \cot^2(x)\,dx = -\cot(x) - x + C$

5. $\displaystyle\int \sec^2(x)\,dx = \tan(x) + C$

6. $\displaystyle\int \csc^2(x)\,dx = -\csc(x) + C$

7. $\displaystyle\int \sin^n(x)\,dx = -\frac{1}{n}\sin^{n-1}(x)\cos(x) + \frac{n-1}{n}\int \sin^{n-2}(x)\,dx$

8. $\displaystyle\int \cos^n(x)\,dx = \frac{1}{n}\cos^{n-1}(x)\sin(x) + \frac{n-1}{n}\int \cos^{n-2}(x)\,dx$

9. $\int \tan^n(x)\, dx = \dfrac{1}{n-1} \tan^{n-1}(x) - \int \tan^{n-2}(x)\, dx$

10. $\int \cot^n(x)\, dx = -\dfrac{1}{n-1} \cot^{n-1}(x) - \int \cot^{n-2}(x)\, dx$

11. $\int \sec^n(x)\, dx = \dfrac{1}{n-1} \sec^{n-2}(x)\tan(x) + \dfrac{n-2}{n-1} \int \sec^{n-2}(x)\, dx$

12. $\int \csc^n(x)\, dx = -\dfrac{1}{n-1} \csc^{n-2}(x)\cot(x) + \dfrac{n-2}{n-1} \int \csc^{n-2}(x)\, dx$

13. $\int \sin^m(x)\cos^{2k+1}(x)\, dx = \int \sin^m(x)(1-\sin^2(x))^k \cos(x)\, dx$ and substitute $u = \sin(x)$

14. $\int \sin^{2k+1}(x)\cos^m(x)\, dx = \int (1-\cos^2(x))^k \sin(x)\cos^m(x)\, dx$ and substitute $u = \cos(x)$

15. $\int \sin^{2m}(x)\cos^{2n}(x)\, dx$ use both double angle formulas

For integrals involving $\sqrt{a^2 - x^2}$ substitute $x = a\sin(\theta)$.
For integrals involving $\sqrt{a^2 + x^2}$ substitute $x = a\tan(\theta)$.
For integrals involving $\sqrt{x^2 - a^2}$ substitute $x = a\sec(\theta)$.

16. $\int \sqrt{a^2 + x^2}\, dx = \dfrac{x}{2}\sqrt{a^2 + x^2} + \dfrac{a^2}{2} \ln\left|x + \sqrt{a^2 + x^2}\right| + C$

17. $\int x^2\sqrt{a^2 + x^2}\, dx = \dfrac{x}{8}(a^2 + 2x^2)\sqrt{a^2 + x^2} - \dfrac{a^4}{8} \ln\left|x + \sqrt{a^2 + x^2}\right| + C$

18. $\int \dfrac{\sqrt{a^2 + x^2}}{x}\, dx = \sqrt{a^2 + x^2} - a\ln\left|\dfrac{a + \sqrt{a^2 + x^2}}{x}\right| + C$

19. $\int \dfrac{\sqrt{a^2 + x^2}}{x^2}\, dx = -\dfrac{\sqrt{a^2 + x^2}}{x} + \ln\left|x + \sqrt{a^2 + x^2}\right| + C$

20. $\int \dfrac{dx}{\sqrt{a^2 + x^2}} = \ln\left|x + \sqrt{a^2 + x^2}\right| + C$

21. $\int \dfrac{x^2\, dx}{\sqrt{a^2 + x^2}} = \dfrac{x}{2}\sqrt{a^2 + x^2} - \dfrac{a^2}{2} \ln\left|x + \sqrt{a^2 + x^2}\right| + C$

22. $\int \dfrac{dx}{x\sqrt{a^2 + x^2}} = -\dfrac{1}{a} \ln\left|\dfrac{\sqrt{a^2 + x^2} + a}{x}\right| + C$

23. $\int \dfrac{dx}{x^2\sqrt{a^2 + x^2}} = -\dfrac{\sqrt{a^2 + x^2}}{a^2 x} + C$

24. $\int \sqrt{a^2 - x^2}\, dx = \dfrac{x}{2}\sqrt{a^2 - x^2} + \dfrac{a^2}{2} \sin^{-1}\left(\dfrac{x}{a}\right) + C$

25. $\int x^2\sqrt{a^2 - x^2}\, dx = \dfrac{x}{8}(2x^2 - a^2)\sqrt{a^2 - x^2} + \dfrac{a^4}{8} \sin^{-1}\left(\dfrac{x}{a}\right) + C$

26. $\displaystyle\int \frac{\sqrt{a^2 - x^2}}{x}\, dx = \sqrt{a^2 - x^2} - a\ln\left|\frac{a + \sqrt{a^2 - x^2}}{x}\right| + C$

27. $\displaystyle\int \frac{\sqrt{a^2 - x^2}}{x^2}\, dx = -\frac{\sqrt{a^2 - x^2}}{x} - \sin^{-1}\left(\frac{x}{a}\right) + C$

28. $\displaystyle\int \frac{dx}{\sqrt{a^2 - x^2}} = \sin^{-1}\left(\frac{x}{a}\right) + C$

29. $\displaystyle\int \frac{x^2\, dx}{\sqrt{a^2 - x^2}} = -\frac{x}{2}\sqrt{a^2 - x^2} + \frac{a^2}{2}\sin^{-1}\left(\frac{x}{a}\right) + C$

30. $\displaystyle\int \frac{dx}{x\sqrt{a^2 - x^2}} = -\frac{1}{a}\ln\left|\frac{\sqrt{a^2 - x^2} + a}{x}\right| + C$

31. $\displaystyle\int \frac{dx}{x^2\sqrt{a^2 - x^2}} = -\frac{\sqrt{a^2 - x^2}}{a^2 x} + C$

32. $\displaystyle\int \sqrt{x^2 - a^2}\, dx = \frac{x}{2}\sqrt{x^2 - a^2} - \frac{a^2}{2}\ln\left|x + \sqrt{x^2 - a^2}\right| + C$

33. $\displaystyle\int x^2\sqrt{x^2 - a^2}\, dx = \frac{x}{8}(2x^2 - a^2)\sqrt{x^2 - a^2} - \frac{a^4}{8}\ln\left|x + \sqrt{x^2 - a^2}\right| + C$

34. $\displaystyle\int \frac{\sqrt{x^2 - a^2}}{x}\, dx = \sqrt{x^2 - a^2} - a\cos^{-1}\left(\frac{a}{|x|}\right) + C$

35. $\displaystyle\int \frac{\sqrt{x^2 - a^2}}{x^2}\, dx = -\frac{\sqrt{x^2 - a^2}}{x} + \ln\left|x + \sqrt{x^2 - a^2}\right| + C$

36. $\displaystyle\int \frac{dx}{\sqrt{x^2 - a^2}} = \ln\left|x + \sqrt{x^2 - a^2}\right| + C$

37. $\displaystyle\int \frac{x^2\, dx}{\sqrt{x^2 - a^2}} = \frac{x}{2}\sqrt{x^2 - a^2} + \frac{a^2}{2}\ln\left|x + \sqrt{x^2 - a^2}\right| + C$

38. $\displaystyle\int \frac{dx}{x\sqrt{x^2 - a^2}} = \frac{1}{a}\sec^{-1}\left(\frac{x}{a}\right) + C$

39. $\displaystyle\int \frac{dx}{x^2\sqrt{x^2 - a^2}} = \frac{\sqrt{x^2 - a^2}}{a^2 x} + C$

40. $\displaystyle\int \frac{Ax + B}{(ax + b)(cx + d)}\, dx = \frac{E}{a}\ln|ax + b| + \frac{F}{c}\ln|cx + d| + C$, where E and F
are the solutions of $Ax + B = E(cx + d) + F(ax + b)$.

41. $\displaystyle\int \frac{1}{x^2 + 1}\, dx = \tan^{-1}(x) + C$

42. $\displaystyle\int \sec(x)\, dx = \ln|\sec(x) + \tan(x)| + C$

43. $\displaystyle\int \tan(x)\, dx = \ln|\sec(x)| + C$

44. $\displaystyle\int \csc(x)\, dx = \ln|\csc(x) - \cot(x)| + C$

45. $\displaystyle\int \cot(x)\, dx = \ln|\sin(x)| + C$

46. $\displaystyle\int \sin^{-1}(x)\, dx = x\sin^{-1}(x) + \sqrt{1-x^2} + C$

47. $\displaystyle\int \cos^{-1}(x)\, dx = x\cos^{-1}(x) - \sqrt{1-x^2} + C$

48. $\displaystyle\int \tan^{-1}(x)\, dx = x\tan^{-1}(x) - \frac{1}{2}\ln(1+x^2) + C$

49. $\displaystyle\int \ln(x)\, dx = -x + x\ln(x) + C$

Some differentiation formulas

1. $\dfrac{d}{dx}\sin^{-1}(x) = \dfrac{1}{\sqrt{1-x^2}}$

2. $\dfrac{d}{dx}\cos^{-1}(x) = -\dfrac{1}{\sqrt{1-x^2}}$

3. $\dfrac{d}{dx}\tan^{-1}(x) = \dfrac{1}{1+x^2}$

4. $\dfrac{d}{dx}\csc^{-1}(x) = -\dfrac{1}{x\sqrt{x^2-1}}$

5. $\dfrac{d}{dx}\sec^{-1}(x) = \dfrac{1}{x\sqrt{x^2-1}}$

6. $\dfrac{d}{dx}\cot^{-1}(x) = -\dfrac{1}{1+x^2}$

Figure credits

Fig. 6.4 is a NASA photograph, with our lines added.

Fig. 6.8 is a photograph I took of a dog lung cast provided by Dr. Robert Henry, DMV.

Fig. 6.10, the skull suture photograph, was taken by Donna Laine.

Figs. 6.17 and 6.18 are courtesy of Dr. Ewald Weibel.

I generated all the other figures in both volumes.

References

[1] P. Achermann, R. Hartmann, A. Gunzinger, W. Guggenbühl, A. Borbély, "Estimation of the correlation dimension of all-night sleep EEG data with a personal supercomputer," pgs. 283–290 of [103].

[2] K. Aihara, G. Matsumoto, "Chaotic oscillations and bifurcations in the squid giant axon," pgs. 257–269 of [65].

[3] E. Allman, J. Rhodes, *Mathematical Models in Biology: An Introduction*, Cambridge Univ. Pr., Cambridge, 2004.

[4] U. Alon, *An Introduction to Systems Biology: Design Principles of Biological Circuits*, Chapman & Hall, Boca Raton, 2007.

[5] R. Anderson, R. May, *Infectious Diseases of Humans: Dynamics and Control*, Oxford Univ. Pr., Oxford, 1991.

[6] O. Arah, "The role of causal reasoning in understanding Simpson's paradox, Lord's paradox, and the suppression effect: covariate selection in the analysis of observational studies," *Emerging Themes in Epidemiology* 5 (2008), doi:10.1186/1742-7622-5-5.

[7] A. Babloyantz, A. Destexhe, "Strange attractors in the human cortex," pgs. 48–56 of [132].

[8] R. Baillie, A. Cecen, C. Erkal, "Normal heartbeat series are non-chaotic, nonlinear, and multifractal: new evidence from semiparametric and perametric tests," *Chaos* 19 (2009), 028503.

[9] J. Banavar, A. Maritan, A. Rinaldo, "Size and form in efficient transportation networks," *Nature* 399 (1999), 130–132.

[10] J. Bassingthwaighte, L. Liebovitch, B. West, *Fractal Physiology*, Oxford Univ. Pr., Oxford, 1994.

[11] I. Bendixson, "Sur les courbes définies par des équations différentielles," *Acta Math.* **24** (1901), 1–88.

[12] J. Berkson, "Limitations of the application of fourfold table analysis of hospital data," *Biometrics Bulletin* **2** (1946), 47–53.

[13] D. Berry, *Statistics: A Bayesian Approach*, Wadsworth, Belmont, 1996.

[14] D. Berry, "Bayesian clinical trials," *Nature Reviews. Drug Discovery* **5** (2006), 27–36.

[15] G. Birkhoff, "Quelques théorèms sur le mouvement des systèmes dynamiques," *Bull. Soc. Math. France* **40** (1912), 305–323.

[16] C. Blyth, "On Simpson's paradox and the sure-thing principle," *J. Amer. Statistical Assoc.* **67** (1972), 364–366.

[17] E. Borel, "Les probabilités dénombrables et leurs applications arithmetique," *Rend. Circ. Met. Palermo* **27** (1909), 247–271.

[18] F. Brauer, J. Nohel, *The Qualitative Theory of Differential Equations: An Introduction*, Dover, New York, 1969.

[19] M. Braun, *Differential Equations and Their Applications*, 4th ed., Springer, New York, 1993.

[20] J. Brown, G. West, *Scaling in Biology*, Oxford Univ. Pr., Oxford, 2000.

[21] K. Bush, "π" track 2 of "A Sea of Honey," from *Ariel*, Columbia Records, 2005.

[22] C. Caldwell, J. Rosson, J. Surowiak, T. Hearn, "Use of the fractal dimension to characterize the structure of cancellous bone in radiographs of the proximal femur," pgs. 300–306 of [103].

[23] G. Campbell, "Bayesian methods in clinical trials with applications to medical devices," *Commun. for Stat. Appl. and Meth.* **24** (2017), 561–581.

[24] L. Cartwright, J. Littlewood, "On non-linear differential equations of the second order I: The equation $y'' + k(1 - y^2) + y = b\lambda k \cos(\lambda t + a)$, k large," *J. London Math. Soc.* **s1-20** (1942), 180–189.

[25] C. Charig, D. Webb, S. Payne, J. Wickham, "Comparison of treatment of renal calculi by open surgery, percutaneous nephrolithotomy, and extracorporeal shockwave lithography," *Brit. Med. J.* **292** (1986), 879–882.

[26] L. Chen, M. Wang, "The relative position and number of limit cycles of quadratic differential systems," *Acta Math. Sinica* **22** (1979), 751–758.

[27] K. Chung, *Elementary Probability Theory with Stochastic Processes*, Springer, New York, 1974.

[28] M.-O. Coppens, "Nature inspired chemical engineering–Learning from the fractal geometry of nature in sustainable chemical engineering," pgs. 507–531 in vol. 2 of *Fractal Geometry and Applications: A Jubilee of Benoit Mandelbrot*, M. Lapidus and M. van Frankenhuijsen, eds., Amer. Math. Soc., Providence, 2004.

[29] M. Courtemanche, L. Glass, J. Bélair, D. Scagliotti, D. Gordon, "A circle map in a human heart," *Phys. D* **40** (1989), 299–310.

[30] M. Crichton, *Jurassic Park*, Knopf, New York, 1990.

[31] R. Devaney, *An Introduction to Chaotic Dynamical Systems*, 2nd ed., Addison-Wesley, Redwood City, 1989.

[32] H. Dulac, "Sur les cycles limites," *Bull. Soc. Math. France* **51** (1923), 45–188.

[33] J. Écalle, "Finitude des cycles limites et accéléro-sommation de l'application de ratour," pgs. 74–159 of [42].

[34] S. Ellner, J. Guckenheimer, *Dynamic Models in Biology*, Princeton Univ. Pr., Princeton, 2006.

[35] B. Enquist, G. West, J. Brown, "Extensions and evaluations of a general quantitative theory of forest structure and dynamics," *Proc. Nat. Acad. Sci. USA* **106** (2009), 7047–7051.

[36] C. Evertsz, C. Zahlten, H.-O. Peitgen, I. Zuna, G. van Kaick, "Distribution of local-connected fractal dimension and the degree of liver fattiness from ultrasound," pgs. 307–314 of [103].

[37] W. Feller, *An Introduction to Probability Theory and Its Applications, Vol. 1*, 2nd ed., Wiley, New York, 1957.

[38] R. Feynman, *The Pleasure of Finding Things Out*, Basic Books, New York, 1999.

[39] R. Fitzhugh, "Thresholds and plateaus in the Hodgkin-Huxley nerve equation," *J. Gen. Physiol.* **43** (1960), 867–896.

[40] R. Fitzhugh, "Impulses and physiological states in theoretical models of nerve membrane," *Biophysical J.* **1** (1961), 445–466.

[41] M. Frame, A. Urry, *Fractal Worlds: Grown, Built, and Imagined*, Yale Univ. Pr., New Haven, 2016.

[42] J.-P. Francoise, *Bifurcations of Planar Vector Fields*, Springer, New York, 1990.

[43] W. Freeman, "The physiology of perception," *Sci. Amer.* **264** (February 1991), 78–85.

[44] M. Gardner, *Time Travel and Other Mathematical Bewilderments*, Freeman, New York, 1988.

[45] L.-A. Gershwin, *Jellyfish: A Natural History*, Univ. Chicago Press, Chicago, 2016.

[46] L. Glass, M. Guevara, A. Shrier, R. Perez, "Bifurcation and chaos in a periodically stimulated cardiac oscillator," *Physica D* **7** (1983), 89–101.

[47] L. Glass, A. Shrier, J. Bélair, "Chaotic cardiac rhythms," pgs. 237–256 of [65].

[48] L. Glass, "Cardiac arrhythmias and circle maps–A classical problem," *Chaos* **1** (1991), 13–19.

[49] L. Glass, "Introduction to controversial topics in nonlinear science: Is the normal heart rate chaotic?" *Chaos* **19** (2009), 028508.

[50] L. Glass, M. Mackey, *From Clocks to Chaos: The Rhythms of Life*, Princeton Univ. Pr., Princeton, 1988.

[51] J. Gleick, *Chaos. Making a New Science*, Viking, New York, 1987.

[52] A. Goldberger, D. Rigney, B. West, "Chaos and fractals in human physiology," *Sci. Am.* Feb. 1990, 42–49.

[53] B. Gompertz, "On the nature of the function expressive of the law of human mortality, and on a new mode of determining the value of life contingencies," *Philos. Trans. Roy. Soc. London* **115** (1825), 513–585.

[54] L. Greenemeier, "Virtual ventricle: computer predicts dangers of arrhythmia drugs better than animal testing," *Sci. Am.* Sept. 2011.
https://www.scientificamerican.com/article/computer-heart-simulation-arrhythmia

[55] S. Greenland, J. Pearl, J. Robins, "Causal diagrams for epidemiologic research," *Epidemiology*, **1** (1999), 37–48.

[56] A. Grosberg, S. Nechaev, E. Shakhnovich, "The role of topological constraints in the kinetics of collapse of macromolecules," *J. Phys. France* **49** (1988), 2095–2100.

[57] A. Grosberg, Y. Rabin, S. Havlin, A. Neer, "Crumpled globule model of the three-dimensional structure of DNA," *Europhys. Lett.*, **23** (1993), 373–378.

[58] S. Gupta, "Use of Bayesian statistics in drug development: Advantages and challenges," *Int. J. Appl. Basic Med. Res.* **2** (2012), 3–6.

[59] J. Hadamard, "Les surfaces à courbures opposées et leur lignes geodesics," *J. de Math.* **4** (1898), 27–73.

[60] P. Hartman, *Ordinary Differential Equations*, John Wiley, New York, 1964.

[61] H. Hatami-Marbini, C. Picu, "Modeling the mechanics of semi-flexible biopolymer networks: non-affine deformations in the presence of long-range correlations," chapter 4 of [82].

[62] W. Hess, "Das Prinzip des kleinsten Kraftverbrauches im Dienste hämodynamischer Forschung," *Arch. Anat. Physiol.*, Physiologische Abteilung, 1914.

[63] M. Hirsch, S. Smale, *Differential Equations, Dynamical Systems, and Linear Algebra*, Academic Press, New York, 1974.

[64] J. Hofbauer, K. Sigmund, *The Theory of Evolution and Dynamical Systems*, Cambridge Univ. Pr., Cambridge, 1998.

[65] A. Holden, *Chaos*, Princeton Univ. Pr., Princeton, 1986.

[66] W. Hurewicz, *Lectures on Ordinary Differential Equations*, Wiley, New York, 1958.

[67] Y. Ilyashenko, "Finiteness theorems for limit cycles," *Uspekhi Mat. Nauk.* **45** (1991), 143–200.

[68] M. Jakobson, "Absolutely continuous invariant measures for one-parameter families of one-dimensional maps," *Commun. Math. Phys.* **81** (1981), 39–88.

[69] L. Jamison, *The Empathy Exams*, Graywolf Press, Minneapolis, 2014.

[70] S. Julious, M. Mullee, "Confounding and Simpson's paradox," BMJ 1994; 309: 1480.

[71] H. Kaplan, "A cartoon-assisted proof of Sarkovskii's theorem," *Amer. J. Phys.* **55** (1987) 1023-1032.

[72] D. Kaplan, L. Glass, *Understanding Nonlinear Dynamics*, Springer-Verlag, New York, 1995.

[73] W. Kermack, A. McKendrick, "A contribution to the mathematical theory of epidemics," *Proc. Roy. Soc. London. A* **115** (1927), 700–721.

[74] A. Khintchine, "Sur la loi des grands nombres," *Compt. Rendus l'Acad. Sci.* **189** (1929), 477–479.

[75] T. Kirkwood, "Why can't we live forever?" *Sci. Am.* **303** (Sept. 2010), 42–49.

[76] M. Kleiber, "Body size and metabolism," *Hilgardia* **6** (1932), 315–353.

[77] N. Krauss, *Man Walks into a Room*, Random House, New York, 2002.

[78] A. Laird, "Dynamics of tumor growth," *British J. Cancer* **18** (1964), 490–502.

[79] A. Laird, "Dynamics of tumor growth: comparison of growth rates and extrapolation of growth curve to one cell," *British J. Cancer* **19** (1965), 278–291.

[80] A. Laird, "Dynamics of tumor growth: comparison of growth and cell population dynamics," *Math. Biosci.* **185** (2003), 153–167.

[81] A. Lambrechts, K. Gevaert, P. Cossart, J. Vandekerckhove, M. Van Troys, "Listeria comet tails: the actin-based motility machinery at work," *Trends Cell Biol.* **18** (2008), 220–227.

[82] S. Li, B. Sun, *Advances in Soft Matter Mechanics*, Springer, New York, 2012.

[83] E. Lieberman-Aiden, N. van Berkum, L. Williams, M. Imakaev, T. Ragoczy, A. Telling, I. Amit, B. Lajoie, P. Sabo, M. Dorschner, R. Sandstrom, B. Bernstein, M. Bender, M. Groudine, A. Gnirke, J. Stamatoyannopoulos, L. Mirny, E. Lander, J. Dekker, "Comprehensive mapping of long-range interactions reveals folding principles of the human genome," *Science* **326** (2009), 289–293.

[84] D. Lind, B. Marcus, *Introduction to Symbolic Dynamics and Coding*, Cambridge Univ. Pr., Cambridge, 1995.

[85] D. Lindley, M. Novick, "The role of exchangeability in inference," *Ann. Stats.* **9** (1981), 45–58.

[86] J. Logan, J. Allen, "Nonlinear dynamics and chaos in insect populations," *Ann. Rev. Entomology* **37** (1992), 455–477.

[87] E. Lorenz, "Deterministic non-periodic flows," *J. Atmos. Sci.* **20** (1963), 130–141.

[88] G. Losa, T. Nonnenmacher, E. Weibel, eds., *Fractals in Biology and Medicine, Vol. II*, Birkhäuser, Basel, 1998.

[89] D. Mackenzie, "New clues to why size equals destiny," *Science* **284** (1999), 1607–1608.

[90] B. Mandelbrot, *The Fractal Geometry of Nature*, Freeman, New York, 1983.

[91] B. Mandelbrot, "How long is the coast of Britain? Statistical self-similarity and fractional dimension," *Science* **156** (1967), 636–638.

[92] B. Mandelbrot, R. Hudson, *The (Mis)Behavior of Markets. A Fractal View of Risk, Ruin, and Reward*, Basic Books, 2004.

[93] B. Mauroy, M. Filoche, E. Weibel, B. Sapoval, "An optimal bronchial tree may be dangerous," *Nature* **427** (2004), 633–636.

[94] R. May, "Simple mathematical models with very complicated dynamics," *Nature* **261** (1976), 459–467.

[95] R. May, G. Oster, "Bifurcation and dynamic complexity in simple ecological models," *Amer. Naturalist* **110** (1976), 573–599.

[96] J. Moreno, Z. Zhu, P.-C. Yang, J. Brakston, M.-T. Jeng, C. Kang, L. Wang, J. Bayer, D. Christini, N. Trayanova, C. Ripplinger, R. Kass, C. Clancy, "A computational model to predict the effects of class I anti-arrhythmic drugs on ventricular rhythms," *Sci. Transl. Med.* **3**, 98ra83 (2011).

[97] J. Munkres, *Topology: A First Course*, Prentice-Hall, Englewood Cliffs, 1975.

[98] C. Murray, "The physiological principle of minimum work. I. The vascular system and the cost of blood," *Proc. Nat. Acad. Sci. USA* **12** (1926), 207–214.

[99] J. Nagumo, S. Arimoto, S. Yoshizawa, "An active pulse transmission line simulating nerve axon," *Proc. I.R.E.* **50** (1962), 2061–2070.

[100] S. Nayyeri, "Analyzing electrocardiography (ECG) signal using fractal method," *Int. J. Current Eng. and Tech.* **7** (2017), 498–505.

[101] N. Neger, M. Frame, "Clarifying compositions with cobwebs," *College Math. J.* **34** (2003), 196–204.

[102] N. Neger, M. Frame, "Visualizing the domain and range of a composition of functions," *Math. Teacher* **98** (2005), 306–311.

[103] T. Nonnenmacher, G. Losa, E. Weibel, eds., *Fractals in Biology and Medicine*, Birkhäuser, Basel, 1994.

[104] L. Norton, R. Simon, H. Brereton, A. Bogden, "Predicting the course of Gompertzian growth," *Nature* **264** (1976), 542–545.

[105] L. Norton, R. Simon, "Growth curve of an experimental solid tumor following radiotherapy," *J. National Cancer Institute* **58** (1977), 1735–1741.

[106] L. Norton, R. Simon, "Tumor size, sensitivity to therapy, and design of treatment schedules," *Cancer Treatment Reports* **61** (1977), 1307–1316.

[107] L. Norton, R. Simon, "The Norton-Simon hypothesis revisited," *Cancer Treatment Reports* **70** (1986), 163–168.

[108] L. Norton, "A Gompertzian model of human breast cancer growth," *Cancer Research* **48** (1988), 7067–7071.

[109] L. Norton, R. Simon, "The Norton-Simon hypothesis revisited: designing more effective and less toxic chemotherapeutic regimens," *Nature Clinical Practice Oncology* **3** (2006), 406–407.

[110] M. Nowak, *Evolutionary Dynamics. Exploring the Equations of Life*, Harvard Univ. Pr., Cambridge, 2006.

[111] M. Nowak, R. May, *Virus Dynamics: Mathematical Principles of Immunology and Virology*, Oxford Univ. Pr., Oxford, 2000.

[112] P. Nyren, B. Pettersson, M. Uhlen, "Solid phase DNA minisequencing by an enzymatic luminometric inorganic pyrophosphate detection assay," *Analytical Biochem.* **208** (1993), 171–175.

[113] D. Oshinsky, *Polio: An American Story*, Oxford Univ. Pr., Oxford, 2006.

[114] D. Paumgartner, G. Losa, E. Weibel, "Resolution effect on the stereological estimation of surface and volume and its interpretation in terms of fractal dimension," *J. Micros.* **121** (1981), 51–63.

[115] D. Peak, M. Frame, *Chaos Under Control: The Art and Science of Complexity*, Freeman, New York, 1994.

[116] J. Pearl, *Causality: Models, Reasoning, and Inference*, 2nd ed., Cambridge Univ. Pr., Cambridge, 2009.

[117] J. Pearl, "Understanding Simpson's paradox," UCLA Cog. Sys. Lab. Tech. Report (2013), R-414.

[118] J. Pearl, "Comment: understanding Simpson's paradox," *Amer. Statistician* **68** (2014), 8–13.

[119] J. Pearl, "Simpson's paradox: an anatomy," UCLA Cog. Sys. Lab. Tech. Report (1999), R-264.

[120] J. Pearl, D. Mackenzie, *The Book of Why: The New Science of Cause and Effect*, Basic Books, New York, 2018.

[121] H.-O. Peitgen, H. Jürgens, D. Saupe, *Chaos and Fractals: New Frontiers in Science*, 2nd ed., Springer-Verlag, 2004.

[122] A. Perelson, "Modelling viral and immune system dynamics," *Nature Reviews Immunology* **2** (2002), 28–36.

[123] A. Perelson, P. Essunger, Y. Cao, M. Vesanen, A. Hurley, K. Saksela, M. Markowitz, D. Ho, "Decay characteristics of HIV-1-infected compartments during combination therapy," *Nature* **387** (1997), 188–191.

[124] L. Perko, *Differential Equations and Dynamical Systems*, Springer, New York, 1991.

[125] H. Poincaré, "Mémoire sur les courbes définies par une equation différentielle," *J. Mathématiques*, **7** (1881), 375–422.

[126] H. Poincaré, *New Methods in Celestial Mechanics*, ed. D. Goroff, Amer. Inst. Physics, 1993.

[127] H. Poincaré, *Science and Method*, transl. F. Maitland, Dover Publ., New York, 1952. Original French edition, 1914.

[128] S.-D. Poisson, *Recherches sur la Probabilité des Jugements en Matiè Criminelle et en Matière Civile*, Bachelier, Paris, 1837.

[129] D. Pollard, *A User's Guide to Measure Theoretic Probability*, Cambridge Univ. Pr., Cambridge, 2002.

[130] G. Pólya, *Mathematical Methods in Science*, 2nd ed., Math. Assoc. Amer., 1977.

[131] T. Pynchon, *Gravity's Rainbow*, Bantam Books, New York, 1974.

[132] L. Rensing, ed., *Temporal Disorder in Human Oscillatory Systems*, Springer, Berlin, 1987.

[133] L. Richardson, "The problem of contiguity: An appendix to statistics of deadly quarrels," *General Systems Yearbook* **6** (1961), 139–187.

[134] R. Ross, "An application of the theory of probabilities to the study of a priori pathometry. Part I," *Proc. Rol. Soc. London. A* **92** (1916), 204–226.

[135] R. Ross, H. Hudson,"An application of the theory of probabilities to the study of a priori pathometry. Part II," *Proc. Royl. Soc. London. A* **93** (1917), 212–225.

[136] R. Ross, H. Hudson, "An application of the theory of probabilities to the study of a priori pathometry. Part III," *Proc. Roy. Soc. London. A* **93** (1917), 225–240.

[137] M. Rubner, "Über den Einfluss der Körpergrösse auf Stoff-und Kraftwechsel," *Zeitschrift für Biologie* **19** (1883), 535–562.

[138] B. Saltzman, "Finite amplitude free convection as an initial value problem–I," *J. Atmos. Sci.* **19** (1962), 329–341.

[139] W. Sampson, *Modelling Stochastic Fibrous Materials with Mathematica*, Springer, London, 2009.

[140] B. Sapoval, T. Gobron, A. Margolina, "Vibrations of fractal drums," *Physical Review Letters* **67** (1991), 2974- 2977.

[141] R. Sassi, M. Signorini, S. Cerutti, "Multifractality and heart rate variability," *Chaos* **19** (2009), 028507.

[142] W. Schaffer, M. Kot, "Differential systems in ecology and epidemiology," pgs. 158–178 of [65].

[143] G. Schectman, M. Patsches, E. Sasse, "Variability in cholesterol measurements: comparison of calculated and direct LDL cholesterol determinations," *Clinical Chemistry* **42** (1996), 732–737.

[144] C. Schmidt, "The Gompertzian view: Norton honored for role in establishing cancer treatment approach," *J. Nat. Cancer Inst.* **96** (2004), 1492–1493.

[145] C. Scholz, B. Mandelbrot, *Fractals in Geophysics*, Birkhäuser, Basel, 1989.

[146] W. Schoutens, *Lévy Processes in Finance: Pricing Financial Derivatives*, Wiley, New York, 2003.

[147] E. Sel'kov, "Self-oscillations in glycolysis. I A simple kinematic model," *European J. Biochem.* **4** (1968), 79–86.

[148] M. Sernetz, J. Wubbeke, P. Wlczek, "Three-dimensional image analysis and fractal characterization of kidney arterial vessels," *Physica A* **191** (1992), 13–16.

[149] S. Shi, "A concrete example of the existence of four limit cycles for plane quadratic systems," *Sci. Sinica* **23** (1980), 153–158.

[150] E. Simpson. "The interpretation of interaction in contingency tables," *J. Roy. Statistical Soc. B* **13** (1051), 238–241.

[151] *The Simpsons*, "Time and punishment," from the *Little Treehouse of Horror V* episode. Oct. 30, 1994.

[152] C. Skarda, W. Freeman, "How brains make chaos in order to make sense of the world," *Behavioral and Brain Sci.* **10** (1987), 161–195.

[153] H. Skipper, F. Schabel Jr., W. Wilcox, "Experimental evaluation of potential anticancer agents. XIII. On the criteria and kinetics associated with 'curability' of experimental leukemia," *Cancer Chemotherapy Reports* **35** (1964), 1–111.

[154] L. Smith, *Linear Algebra* 3rd ed., Springer, New York, 1998.

[155] S. Strogatz, *Nonlinear Dynamics and Chaos with Applications to Physics, Biology, Chemistry, and Engineering*, 2nd ed., Westview Press, Philadelphia, 2015.

[156] H. Tijms, *Understanding Probability: Chance Rules in Everyday Life*, 2nd ed., Cambridge Univ. Pr., Cambridge, 2007.

[157] D. Turcotte, "Fractals in geology and geophysics," pgs 171–196 in [145].

[158] V. Volterra, "Variations and fluctuations of the number of individuals in animal species living together," *I C E S J. of Marine Science* **3** (1928), 3–51.

[159] L. von Bortkiewicz, *Das Gesetz der Kleinen Zahlen*, Teubner, Leipzig, 1898.

[160] R. Voss, "Evolution of long-range fractal correlations and $1/f$ noise in DNA base sequences," *Phys. Rev. Lett.* **68** (1992), 3805–3808.

[161] R. Voss, "Long-range fractal correlations in DNA introns and exons," *Fractals* **2** (1994), 1–6.

[162] R. Voss, J. Clarke, "$1/f$ noise in music and speech," *Nature* **258** (1975), 317–318.

[163] R. Wallace, "A fractal model of HIV transmission on complex sociogeographic networks: towards analysis of large data sets," *Environment and Planning A* **25** (1993), 137–148.

[164] O. Warburg, "On the origin of cancer cells," *Science* **123** (1956), 309–314.

[165] P. Watters, F. Martin, "A method for estimating long-range power law correlations from the electroencephalogram," *Biol. Psych.* **66** (2004), 79–89.

[166] E. Weibel, "Fractal geometry: a design principle for living organisms," *Am. J. Physiol. (Lung Cell. Mol. Physiol. 5)* **261** (1991), L361-L369.

[167] E. Weibel, *Stereological Methods 1: Practical Methods for Biological Morphometry*, Academic Press, London, 1981.

[168] E. Weibel, D. Gomez, "A principle for counting tissue structures on random sections," *J. Appl. Physiol.* **17** (1962), 343–348.

[169] N. Wessel, M. Riedl, J. Kurths, "Is the normal heart rate 'chaotic' due to respiration?" *Chaos* **19** (2009), 028508.

[170] B. West, *Fractal Physiology and Chaos in Medicine*, World Scientific, Singapore, 1990.

[171] G. West, J. Brown, B. Enquist, "A general model for the origin of allometric scaling laws in biology," *Science* **276** (1997), 122–126.

[172] G. West, J. Brown, B. Enquist, "The fourth dimension of life: fractal geometry and allometric scaling of organisms," *Science* **284** (1999), 1677–1679.

[173] G. West, B. Enquist, J. Brown, "A general quantitative theory of forest structure and dynamics," *Proc. Nat. Acad. Sci. USA* **106** (2009), 7040–7045.

[174] E. Wigner, "The unreasonable effectiveness of mathematics in the natural sciences," *Commun. Pure Appl. Math.* **13** (1960), 1–14.

[175] R. Wilding, M. Ferguson, N. Parr, G. Mckellar, B. Adams, "Changes in bone strength during repair predicted by fractal analysis of radiographs," pgs. 335–344 of [88].

[176] A. Winfree, *The Geometry of Biological Time*, Springer-Verlag, Berlin, 1990.

[177] A. Winfree, *When Time Breaks Down: The Three-Dimensional Dynamics of Electrochemical Waves and Cardian Arrhythmias*, Princeton, Univ. Pr., Princeton, 1987.

[178] J. Wu, K. Kuban, E. Allred, F. Shapiro, B. Darras, "Association of Duchenne muscular dystrophy with autism spectrum disorder," *J. Child Neurology* **20** (2005), 290–295.

[179] A. Wulf, *The Invention of Nature: Alexander Von Humboldt's New World*, Vintage, New York, 2016.

[180] J. Yerushalmy, J. Harkness, J. Cupe, B. Kennedy, "The role of dual reading in mass radiography," *Am. Rev. Tuber.* **61** (1950), 443–464.

[181] P. Yu, M. Han, "Twelve limit cycles in a cubic order planar system with $\mathbb{Z}_2$ symmetry," *Commun. Pure Appl. Anal.* **3** (2004), 515–525.

[182] J. Zhang, A. Holden, O. Monfredi, M. Boyett, H. Zhang, "Stochastic vagal modulation of cardiac pacemaking may lead to erroneous identification of cardiac 'chaos' " *Chaos* **19** (2009), 028509.

Index